Hetero Diels–Alder
Methodology
in Organic Synthesis

This is Volume 47 of
ORGANIC CHEMISTRY
A series of monographs
Editor: HARRY H. WASSERMAN

A complete list of the books in this series can be obtained from the publisher on request.

Hetero Diels–Alder Methodology in Organic Synthesis

Dale L. Boger

Department of Chemistry
Purdue University
West Lafayette, Indiana

Steven M. Weinreb

Department of Chemistry
The Pennsylvania State University
University Park, Pennsylvania

ACADEMIC PRESS, INC.
Harcourt Brace Jovanovich, Publishers

San Diego   New York   Berkeley   Boston
London   Sydney   Tokyo   Toronto

QD
281
.R5
B64
1987

ACADEMIC PRESS, INC.
1250 Sixth Avenue, San Diego, California 92101

United Kingdom Edition published by
ACADEMIC PRESS INC. (LONDON) LTD.
24–28 Oval Road, London NW1 7DX

Library of Congress Cataloging in Publication Data

Boger, Dale L.
    Hetero Diels-Alder methodology in organic synthesis.

    (Organic chemistry;   )
    Includes index.
    1. Diels-Alder reaction.   2. Ring formation
(Chemistry)   3. Heterocyclic compounds.   I. Weinreb,
Steven M.   II. Title.   III. Series: Organic chemistry
(New York, N.Y.) ;   .
QD281.R5B64   1987   547'.590459   86-32140
ISBN 0—12—110860—0 (alk. paper)

PRINTED IN THE UNITED STATES OF AMERICA

87 88 89 90   9 8 7 6 5 4 3 2 1

Contents

Preface

In the course of our individual research programs, we have had occasion to utilize hetero Diels–Alder reactions as pivotal steps in natural product total syntheses. It became clear to us during the early stages of our work that this type of cycloaddition, although a potentially powerful synthetic tool, had found relatively little general use. Moreover, it was also evident that reviews on various aspects of this chemistry were often out-of-date or nonexistent. In response to this situation, we have written several journal reviews on different facets of hetero Diels–Alder methodology during the past few years. With an apparent increase in the appreciation of the value of [4 + 2] cycloadditions in heterocyclic synthesis, we felt it was appropriate to write this monograph.

The emphasis in this work is on the scope and preparative synthetic utility of the hetero Diels–Alder reaction. No attempt has been made to carefully define or delineate the important mechanistic questions, many of which are as yet unanswered, of the various [4 + 2] cycloaddition reactions other than to try to provide a rationale for the facility with which the cycloadditions proceed and to provide a basis for the stereo- and regiochemical observations. We have purposely excluded reactions of singlet oxygen as a dienophile, since extensive surveys are available elsewhere. Many miscellaneous heterodienophiles and heterodienes have not been covered if in our opinion they are not of general synthetic value. A comprehensive treatment of all recorded hetero [4 + 2] cycloadditions is beyond the scope of this monograph. However, we do hope to provide a broad survey of this reaction type as it exists today in order to furnish a foundation for continued development.

We express our special thanks to Ms. Beth Swisher for painstakingly preparing the original manuscript on top of her daily hectic responsibilities. We also thank our graduate students and postdoctoral colleagues for their suggestions, encouragement, and continued stimulating interest in this topic. We are grateful to Stacy Remiszewski for help in preparing Chapter 1 and to Dan Yohannes, Mona Patel, Robert Mathvink, Rob Coleman, Thomas Hayes, Rosanna Villani, Scott Bell, Michael Melnick, Gary Lee, Thomas Lessen, and Rick Joyce for their proofreading efforts.

Dale L. Boger
Steven M. Weinreb

Hetero Diels–Alder
Methodology
in Organic Synthesis

N-Sulfinyl Compounds and Sulfur Diimides

INTRODUCTION

The [4 + 2] cycloaddition of N-sulfinylaniline with a conjugated diene was first described by Wichterle and Rocek in 1953.[1] Since then, there have been a number of Diels–Alder cycloadditions reported for a variety of N-sulfinyl compounds [Eq. (1)].[2]

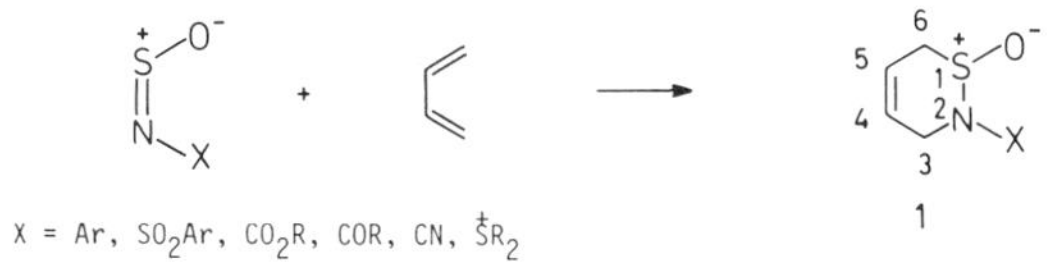

$$\overset{+}{\underset{N}{\overset{\displaystyle \parallel}{S}}}\diagdown_{X} \quad + \quad \diagup\diagdown \quad \longrightarrow \quad \underset{3}{\overset{6}{\underset{\displaystyle 4}{\bigodot}}} \overset{O^-}{\underset{N}{\overset{+}{S}}}\diagdown_X \tag{1}$$

X = Ar, SO₂Ar, CO₂R, COR, CN, ⁺SR₂

These [4 + 2] cycloadditions are unique in that they generally proceed rapidly, and often exothermically, at relatively low temperatures when

1

there is an electron-withdrawing group X on nitrogen. When X is an electron-donating substituent, such cycloadditions usually do not occur. Highly electron-deficient *N*-sulfinyl compounds react most rapidly with 1,3-dienes, and *N*-sulfinylanilines are the least reactive type of dienophile.[2] The product of a *N*-sulfinyl Diels–Alder cycloaddition is a 2-substituted 3,6-dihydro-1,2-thiazine 1-oxide (**1**).

Similarly, sulfur diimides are also reactive dienophiles provided they bear at least one electron-withdrawing substituent on nitrogen [Eq. (2)].[2]

$$\begin{array}{ccc} \underset{\underset{\displaystyle N\diagdown_{X'}}{\overset{\displaystyle \|}{S}}}{\overset{+\diagup\overset{-}{N}X}{}} & + & \text{\large (} \end{array} \longrightarrow \qquad \underset{2}{\overset{6}{\underset{4}{\overset{5}{\bigcirc}}}\overset{+\diagup\overset{-}{N}X}{\underset{3}{\underset{N\diagdown_{X'}}{\overset{1}{S}}}}} \qquad (2)$$

The adducts formed in these cases are the analogous dihydrothiazine imines **2**. Symmetrical and unsymmetrical sulfur diimides are available, and both can act as heterodienophiles. In adducts **1** and **2** the sulfur center is chiral, and the implications of this fact will be discussed in some of the following sections.

Kresze has written several excellent reviews on *N*-sulfinyl and sulfur diimide cycloadditions.[2a-c] This chapter will concentrate most heavily on the more recent developments in the area, and previous summaries should be consulted for additional information.

1. PREPARATION AND STRUCTURE OF DIENOPHILES

A multitude of *N*-sulfinyl compounds and sulfur diimides bearing a variety of substituents on the nitrogen atom(s) have been prepared.[2c] The *N*-sulfinyl compounds are commonly generated by treatment of the parent aniline, amine, sulfonamide, etc., with thionyl chloride and pyridine [Eq. (3)].

$$\text{RNH}_2 \qquad \overset{\text{SOCl}_2}{\underset{\text{C}_5\text{H}_5\text{N}}{\xrightarrow{\hspace{1cm}}}} \qquad \text{RNSO} \qquad (3)$$

The resulting *N*-sulfinyl derivatives can often be distilled and/or crystallized but are water sensitive and thus are frequently prepared *in situ* for subsequent cycloaddition reactions. Symmetrically substituted sulfur diimides can usually be prepared by the reaction of SCl_2 or S_2Cl_2 with the parent NH_2 compound in the presence of a base such as pyridine or triethylamine [Eq. (4)].[2c]

$$RNH_2 \xrightarrow[\substack{S_2Cl_2 \\ R'_3N}]{SCl_2 \text{ or}} RNSNR \tag{4}$$

N,N-Dichlorosulfonamides or N,N-dichlorocarbamates, when treated with elemental sulfur, also afford diimides [Eq. (5)].

$$RNCl_2 \xrightarrow{S_8} RNSNR \tag{5}$$

Unsymmetrical sulfur diimides are commonly prepared by two methods. Treatment of a N,N'-bis(arylsulfonyl)sulfur diimide with an equivalent of an amine leads to the displacement of one sulfonyl group [Eq. (6)].[3]

$$(ArSO_2N)_2S + RNH_2 \longrightarrow ArSO_2NSNR + ArSO_2NH_2 \tag{6}$$

Alternatively, an amine and an N-substituted dichlorosulfimide will react to yield an unsymmetrical sulfur diimide [Eq. (7)].[4]

$$RN=SCl + R'NH_2 \longrightarrow RNSNR' \tag{7}$$

Recently, cationic N-sulfinyl amines and sulfur diimides have been reported. These species are readily prepared by alkylation of a N-sulfinyl compound or a sulfur diimide with a trialkyloxonium tetrafluoroborate [Eq. (8)].[5]

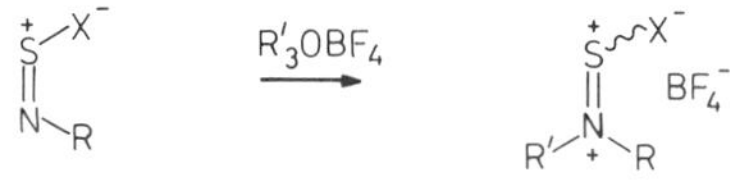

$$X = O, NR$$

Sulfur diimides and N-sulfinyl compounds have nonlinear structures and are configurationally stable at room temperature. Two geometric isomers are possible for N-sulfinyl compounds [i.e., (Z)-**3** and (E)-**4**]. Sulfur

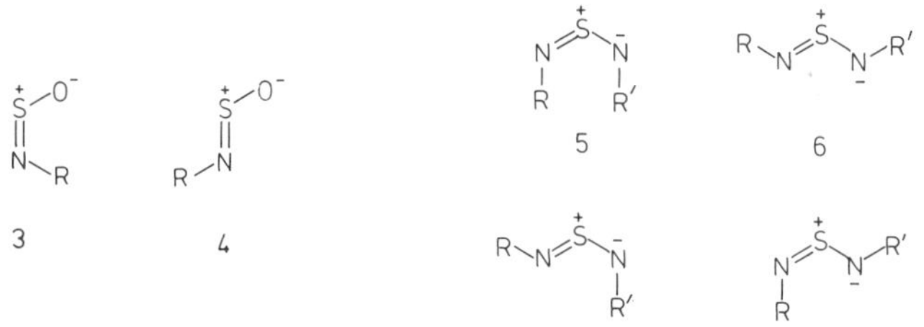

diimides can theoretically exist as four geometric isomers [i.e., (Z,Z)-**5**, (E,E)-**6**, and two E,Z forms **7** and **8**].

It has been found through X-ray, neutron diffraction, and electron scattering analyses that *N*-sulfinylamines, -anilines, -hydrazines, and -sulfonamides have the (*Z*)-**3** configuration in the solid state.[2c] The structure of *N*-sulfinyl compounds in solution is known with less certainty. It has been reported that a series of aryl-substituted *N*-sulfinylanilines exist solely as the *Z* geometric isomers in solution based on analysis of ^{1}H-NMR spectra and dipole moments.[6] However, microwave spectroscopy and ^{13}C-NMR spectroscopy indicate that an *E/Z* equilibrium exists for some *N*-sulfinylamines in solution.[7,8]

In the case of sulfur diimides, X-ray and electron scattering studies have shown that the *E,Z* geometry **7** or **8** is favored in the solid state.[2c,9] ^{1}H- and ^{13}C-NMR spectroscopy at low temperatures in solution indicated two isomeric forms are present: an *E,Z* isomer **7** or **8** and a symmetrical isomer, presumed to be the *E,E* isomer **6**. From coalescence of the ^{1}H-NMR signals at higher temperatures, it has been concluded that rapid interconversion of the (*E,Z*)-**7** or -**8** and (*E,E*)-**6** forms occurs.

Thus, the most stable configuration of *N*-sulfinyl compounds is apparently (*Z*)-**3** and for sulfur diimides, (*E,Z*)-**7** or -**8**. Interactions within a specific molecule can alter the relative stability of the isomeric forms so that in solution configurational equilibria may exist.

2. REGIOCHEMICAL, STEREOCHEMICAL, AND MECHANISTIC ASPECTS

Kresze and Wagner examined the regioselectivity of the cycloaddition of *N*-sulfinyl-*p*-toluenesulfonamide (**9**) with several unsymmetrical dienes [Eq. (9)].[10]

$$R = Ph, Cl, CH_3$$ (9)

The cycloaddition of **9** with some 2-substituted dienes was found to yield only the 5-substituted dihydrothiazine oxides. Cycloadditions of **9** with 1-substituted 1,3-dienes [Eq. (10)]

(10)

$$R = Ph, \underline{p}NO_2Ph, \underline{p}CH_3OPh, CH_3$$

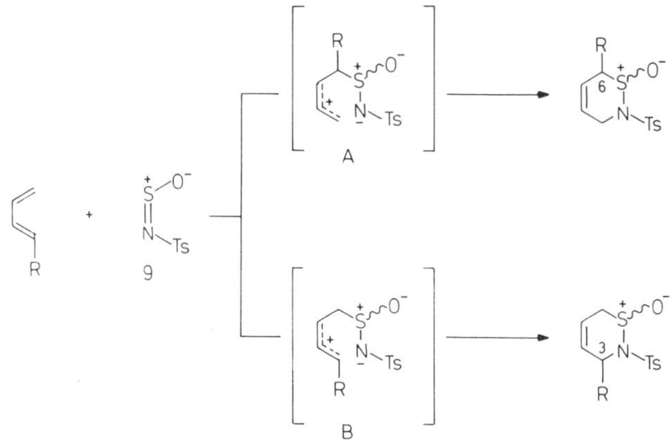

Scheme 1-I

are often dependent on the reaction temperature. At low temperatures, 3-substituted dihydrothiazine oxides are usually formed, but at higher temperatures the 6-substituted heterocycles are produced. For example, the dienes shown in Eq. (10) gave the 3-substituted adducts as the kinetic products. However, these products could be isomerized via a retro-Diels–Alder process to the less sterically crowded thermodynamic isomers by heating in benzene. In the cases of R equal to tBu and CO_2Me, the 6-substituted isomer was produced even at low reaction temperatures.

Kresze and Wagner offered a mechanistic model for the [4 + 2] cycloadditions of N-sulfinyl dienophiles to rationalize the kinetically formed regioisomeric products.[10] They proposed a concerted mechanism for the reaction via a transition state which has dipolar character (Scheme 1-I). For 1-substituted dienes, "transition states" **A** and **B** can be considered. If R is an electron-donating group which stabilizes the cationic center, a 3-substituted product will result. If R is electron-withdrawing (e.g., CO_2Me), the 6-substituted isomer will be the kinetic product of the cycloaddition. A similar argument can be made for 2-substituted and more complex dienes.

Mock and Nugent investigated the mechanism of the [4 + 2] cycloaddition of N-sulfinyl-p-toluenesulfonamide (9) and the isomeric 2,4-hexadienes **10, 11,** and **12** in detail (Scheme 1-II).[11] These workers determined the relative stereochemistry of the resulting adducts and proposed a stepwise dipolar mechanism based primarily on the difference in sulfur stereochemistry between adducts produced from dienes **11** and **12**.

Dihydrothiazine oxide **15** was proposed to result from initial addition of the electrophilic sulfur atom of the N-sulfinyl compound to (E,Z)-diene **11** to form a dipolar intermediate [Eq. (11)].

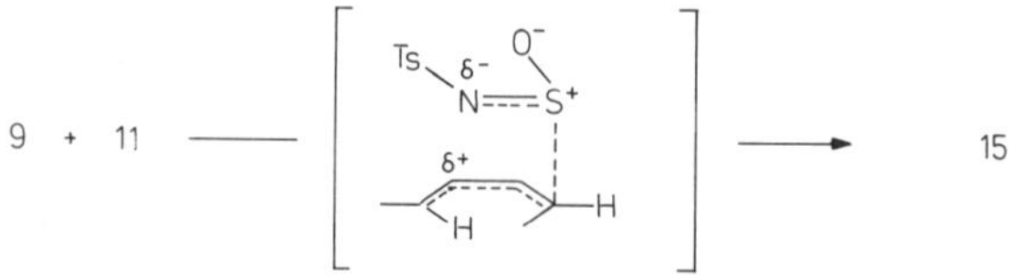

$$9 + 11 \longrightarrow \left[\begin{array}{c} \text{Ts} \diagdown \overset{\delta^-}{\text{N}} \underset{}{===} \overset{O^-}{\underset{|}{\overset{|}{\text{S}^+}}} \\ \overset{\delta^+}{\diagup} \diagdown \overset{}{\text{H}} \text{H} \end{array} \right] \longrightarrow \quad 15 \qquad (11)$$

Addition of **9** to the *Z* olefinic bond of the diene would be favored since a transoid ("sickle") allylic carbonium ion would be formed rather than the less stable cisoid ("U-form") allylic moiety. Closure would afford dihydrothiazine oxide **15**.

The rationalization for the formation of adduct **16** was somewhat more complex. Isomerization of (*Z,Z*)-hexadiene to (*E,Z*)-hexadiene could be discounted since the cycloaddition of the (*E,Z*)-diene and **9** led to the formation of sulfur epimer **15** *exclusively*. Assuming the reactive conformation of the (*Z,Z*)-diene is cisoid, only a nonplanar, helical skew arrangement can be achieved [Eq. (12)].

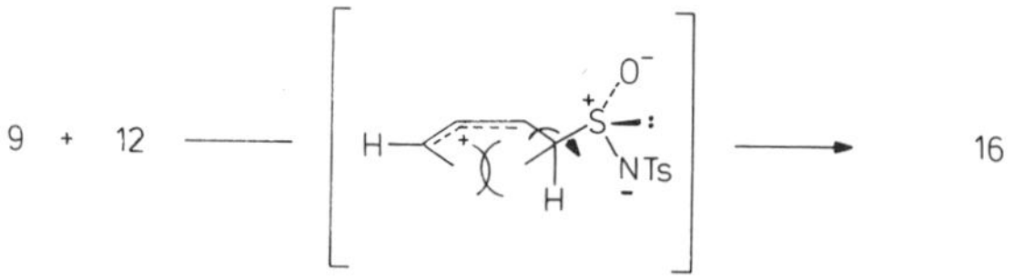

$$9 + 12 \longrightarrow \left[\begin{array}{c} \text{H} \diagdown \diagup \overset{O^-}{\underset{\text{NTs}}{\overset{|}{\underset{|}{\text{S}}}}} \\ \overset{}{\text{H}} \end{array} \right] \longrightarrow \quad 16 \qquad (12)$$

Attack of the electrophile at one end of the helical diene would result in a zwitterion in which direct closure to a 3,6-dihydrothiazine oxide is blocked by a severe methyl–methyl group steric interaction. Rotation of the C-1—C-2 bond in the opposite direction requires only that the methine hydrogen pass the C-4 methyl group. This rotation also would

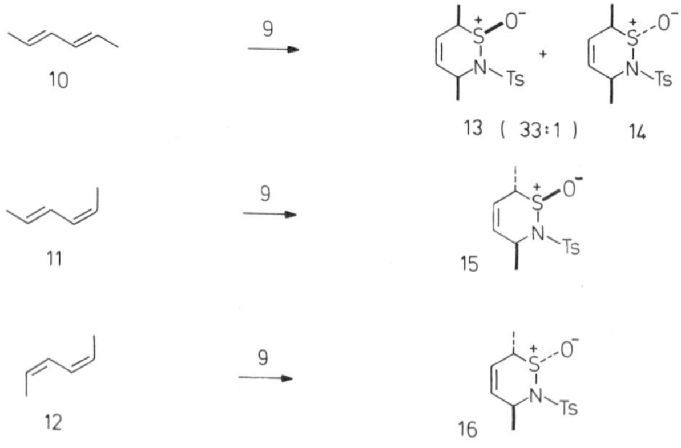

Scheme 1-II

bring the nitrogen atom to a position that would afford the observed sulfur epimer **16** after cyclization. For consistency, Mock and Nugent proposed that the two-step dipolar mechanism must also apply to the cycloadditions that produce **13** and **14**, since the experimental conditions for all of these reactions were similar.

Hanson and Stockburn recently examined the mechanism of the cycloaddition of *N*-sulfinylethyl carbamate and 1,1′-bicyclohexenyl which yields dihydrothiazine oxide **17** exclusively [Eq. (13)].[12]

$$\tag{13}$$

The relative stereochemistry of adduct **17** was determined by single crystal X-ray diffraction.

The cycloaddition reaction rate revealed activation parameters which are characteristic of a pericyclic reaction. The entropy of activation is large ($\Delta S^{\ddagger} = -176.95$ K^{-1} mol^{-1}), and the enthalpy of activation is small ($\Delta H^{\ddagger} = 30.3$ kJ mol^{-1}). These data imply a highly ordered, early transition state with concerted bond making and breaking. Solvent polarity had only a small effect on the rate of the reaction, indicating little separation of charge in the transition state, consistent with a concerted mechanism.

Based on these results, Hanson and Stockburn proposed that other dienophiles of electrophilicity comparable to that of *N*-sulfinylethyl carbamate (i.e., *N*-sulfinylsulfonamides) cycloadd via a similar pericyclic mechanism. They suggested that the observations of Mock and Nugent[11] could be best explained by a concerted mechanism for the cycloaddition of *N*-sulfinylsulfanamides **9** to dienes **10** and **11** and a nonconcerted, two-step mechanism for the cycloaddition of **9** to the highly hindered (*Z,Z*)-diene **12**. These workers have proposed that less electrophilic dienophiles, such as *N*-sulfinylanilines, also undergo cycloaddition via a concerted process. Such concerted [4 + 2] cycloadditions for *N*-sulfinyl compounds are in accord with orbital symmetry considerations.[13]

As can be seen from some of the examples cited above, a single or predominant sulfur stereoisomer often results from the cycloaddition process, but sometimes mixtures of sulfur epimers are produced. In many of these cycloadducts the configuration at sulfur has not been determined, and thus not enough data are currently available to clarify what factors control the configuration established at sulfur. These cycloadditions do show the usual Diels–Alder syn stereoselectivity with respect to the 1,3-diene component.

It has also not been established what importance secondary orbital effects have in these reactions. For example, the cycloaddition of 1,3-cyclohexadiene and *N*-sulfinylbenzenesulfonamide afforded a 1 : 1 mixture of sulfur exo/endo epimers [Eq. (14)].[11]

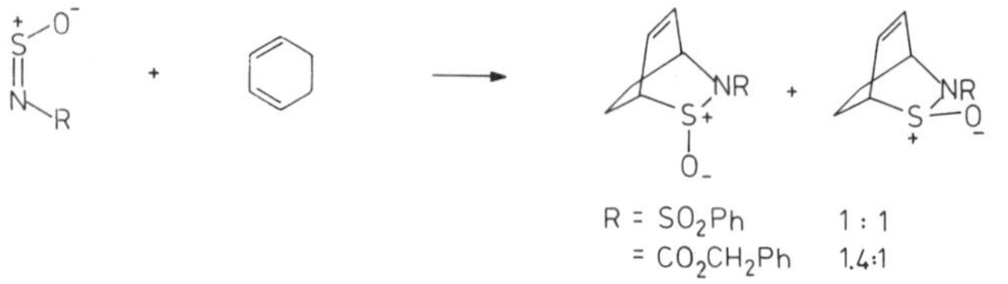

$$R = SO_2Ph \qquad 1:1$$
$$= CO_2CH_2Ph \qquad 1.4:1$$
(14)

Weinreb *et al.* have found a similar lack of stereoselectivity in the cycloaddition of cyclohexadiene and *N*-sulfinylbenzyl carbamate.[14] The mechanistic situation is further complicated in these cases by the fact that one cannot determine whether the (*E*)- or (*Z*)-sulfinyl dienophile is the reacting species.

Levchenko and Balon[15] have examined the regiochemistry of cycloaddition of several symmetrical bis(aryl)sulfonyl sulfur diimides with (*E*)-piperylene and isoprene which afforded the C-3 and C-5 substituted adducts, respectively (Scheme 1-III). These results were confirmed by Wucherpfennig and Kresze, who also found that chloroprene gave the 5-substituted product.[16]

Mock and Nugent investigated the reaction of *N,N'*-bis(*p*-toluene-sulfonyl)sulfur diimide (**18**) and the isomeric 2,4-hexadienes **10, 11,** and **12** in an attempt to elucidate the mechanism of sulfur diimide cycloadditions (Scheme 1-IV).[11] The cycloaddition of **18** and **10** afforded two adducts epimeric at sulfur in a 43 : 1 ratio, while only one adduct of indeterminate stereochemistry was obtained from the cycloaddition of **18** and **11**. No reaction was observed with (*Z,Z*)-diene **12**. Unfortunately, not enough information resulted from these experiments to draw any mechanistic conclusions with regard to the cycloaddition.

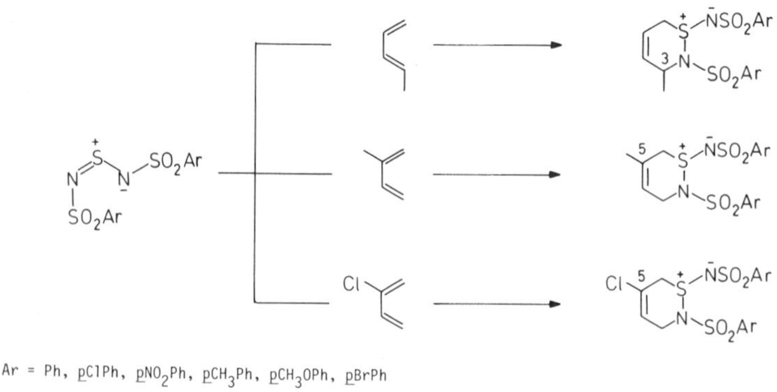

Ar = Ph, *p*ClPh, *p*NO$_2$Ph, *p*CH$_3$Ph, *p*CH$_3$OPh, *p*BrPh

Scheme 1-III

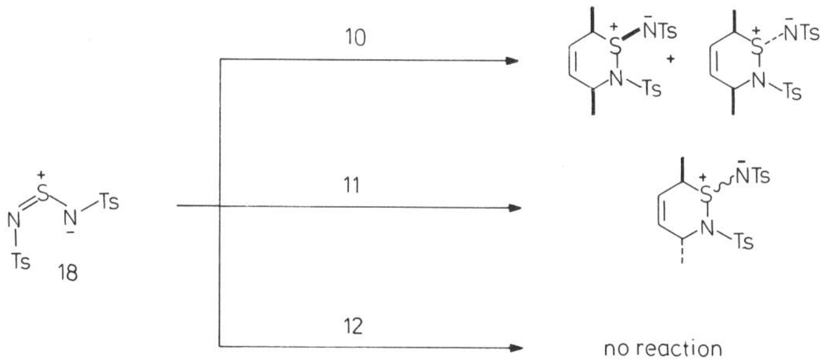

Scheme 1-IV

As mentioned above, only those unsymmetrical sulfur diimides that have at least one nitrogen atom substituted by a strongly electron-withdrawing group undergo [4 + 2] cycloadditions. In principle, a cycloaddition could occur with either of the two sulfur–nitrogen double bonds, which would lead to regioisomeric adducts being produced (Scheme 1-V). Surprisingly, only the nitrogen–sulfur double bond which does *not* bear the most electron-withdrawing (X) group is involved in the cycloaddition (*vide infra*).

3. *N*-SULFINYL DIENOPHILE CYCLOADDITIONS

As stated in the Introduction, *N*-sulfinyl compounds bearing electron-withdrawing substituents react as heterodienophiles. Arylsulfinyl derivatives usually require heating for a reaction to occur, whereas other types of *N*-sulfinyl dienophiles will often cycloadd near room temperature or below. In fact, these cycloadditions are sometimes dangerously exothermic when run in the absence of a solvent, and usually an inert solvent such as benzene, toluene, or cyclohexane is used.[2a]

A number of representative examples of *N*-sulfinyl dienophile cycloadditions can be found in Table 1-I. As noted above, some adducts undergo

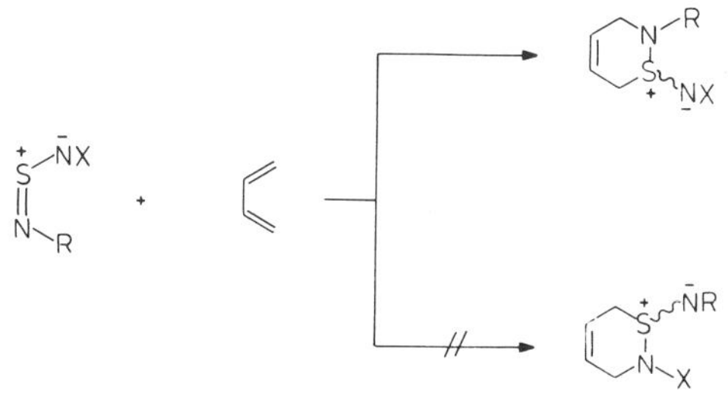

Scheme 1-V

TABLE 1-I
Cycloadditions of *N*-Sulfinyl Compounds

Entry	Dienophile	Diene	Conditions	Product	Yield	Ref.
1	PhNSO	(2,3-dimethylbutadiene)	C_6H_{12}, 80°C, 7 hr	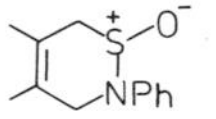 (ring S$^+$–O$^-$, NPh)	72%	1, 17
2	*p*NO$_2$PhNSO	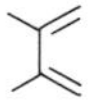 (2,3-dimethylbutadiene)	Neat, Δ, 10 hr	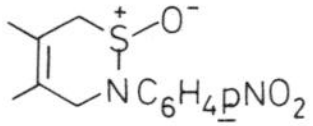 (ring S$^+$–O$^-$, NC$_6$H$_4$*p*NO$_2$)	40%	1
3	MeSO$_2$NSO	 (butadiene)	C_6H_6, Δ, 0.5–1 hr	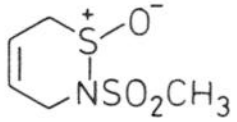 (ring S$^+$–O$^-$, NSO$_2$CH$_3$)	90%	2a, 17
4	TsNSO	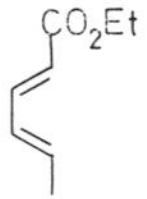 (CO$_2$Et diene)	C_6H_6, RT, 1–2 days	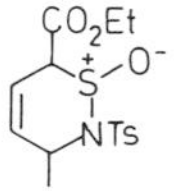 (ring CO$_2$Et, S$^+$–O$^-$, NTs)	53%	2a, 17
5	TsNSO	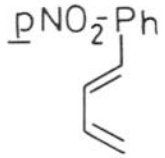 (*p*NO$_2$Ph diene)	C_6H_6, 5°C, 12 hr	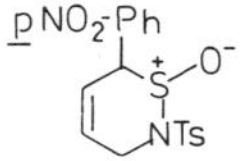 (ring *p*NO$_2$Ph, S$^+$–O$^-$, NTs)	63%	2a, 17
6	*p*NO$_2$PhSO$_2$NSO	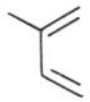 (isoprene)	C_6H_6, RT	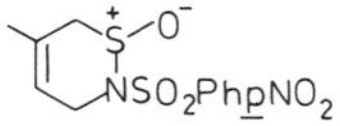 (ring S$^+$–O$^-$, NSO$_2$Ph*p*NO$_2$)	80–85%	18
7	Me$_2$NSO$_2$NSO	(isoprene)	C_6H_6, RT	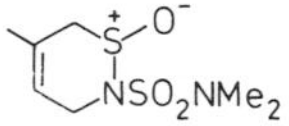 (ring S$^+$–O$^-$, NSO$_2$NMe$_2$)	90%	19
8	PhCONSO	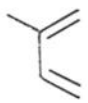 (isoprene)	RT	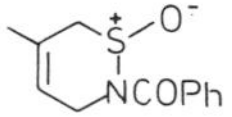 (ring S$^+$–O$^-$, NCOPh)	44%	13b, 20, 24b

10

	Reagent	Diene	Conditions	Product	Yield	Ref.
9	MeO–N=C(Ph)–NSO	(diene)	—	$\overset{O^-}{\underset{+}{S}}$–N=CPh, NOCH$_3$	—	17
10	MeO$_2$CNSO	(diene)	C$_6$H$_6$, 20°C	$\overset{O^-}{\underset{+}{S}}$–NCO$_2CH_3$	70–90%	20, 21, 22
11	(EtO)$_2$PNSO	(diene)	C$_6$H$_6$, 80°C	$\overset{O^-}{\underset{+}{S}}$–NPO(OEt)$_2$	85%	23, 29
12	N≡CNSO	(diene)	Et$_2$O, 20°C	$\overset{O^-}{\underset{+}{S}}$–NCN	90%	24
13	Me$_2$$\overset{+}{N}$SO BF$_4^-$	(diene)	CH$_3$CN	$\overset{O^-}{\underset{+}{S}}$–$\overset{+}{N}Me_2$ BF$_4^-$	—	25
14	Me$_2$N$\overset{+}{N}$SO BF$_4^-$	(diene)	CH$_3$CN, RT, 7–14 days	$\overset{O^-}{\underset{+}{S}}$–N$\overset{+}{N}Me_3$ BF$_4^-$	—	26
15	Me$_2$$\overset{+}{S}$NSO BF$_4^-$	(diene)	CH$_3$CN, RT	$\overset{O^-}{\underset{+}{S}}$–N$\overset{+}{S}Me_2$ BF$_4^-$	95%	26
16	Me$_2$$\overset{+}{N}$SO SbCl$_6^-$	(diene)	CH$_3$CN, −40°C, 1 hr	$\overset{O^-}{\underset{+}{S}}$–$\overset{+}{N}Me_2$ SbCl$_6^-$	45%	27
17	(imidazolyl)–$\overset{+}{N}$NSO	(diene)	CH$_3$CN, RT, 14 days	$\overset{O^-}{\underset{+}{S}}$–N(imidazolyl)	57%	28

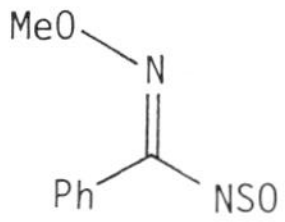

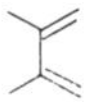

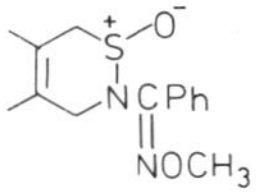

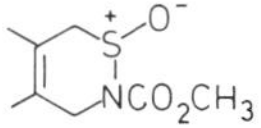

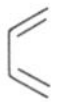

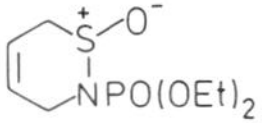

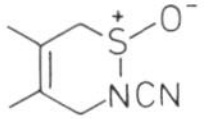

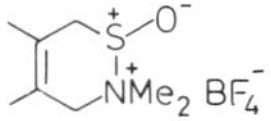

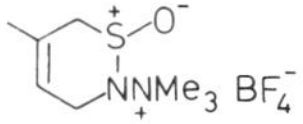

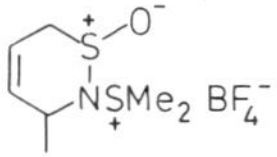

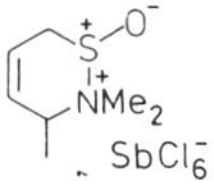

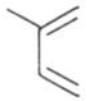

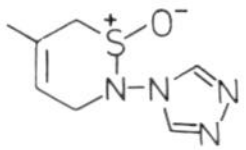

(*continued*)

11

TABLE 1-I (Continued)

Entry	Dienophile	Diene	Conditions	Product	Yield	Ref.
18	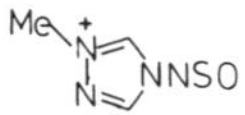 Me–N⁺=N–NSO (imidazolium)	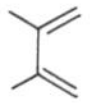 (2,3-dimethylbutadiene)	CH₃CN, 20°C, 10 days	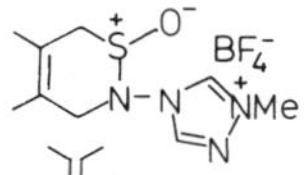 (product structure, BF₄⁻ salt)	82%	28
19	PhNSO	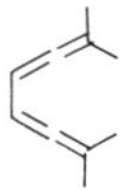 (2,5-dimethyl-2,4-hexadiene)	C₆H₁₂, RT, 17 hr	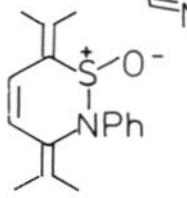 (product structure)	64%	30
20	pMePhNSO	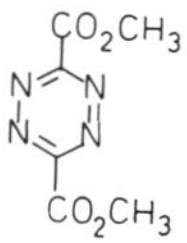 (triazine with CO₂CH₃ groups)	Neat, 80°C, 2 days	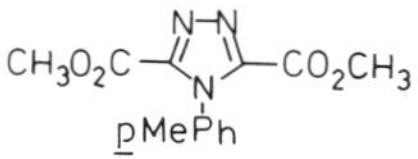 CH₃O₂C—(pyrazole)—CO₂CH₃, pMePh	29%	32

retro-Diels–Alder processes at relatively low temperatures. This is particularly true of the adducts of *N*-sulfinyl-*p*-toluenesulfonamide with ethyl sorbate (entry 4) and cyclopentadiene.[2b] *N*-sulfinyl compounds also undergo other types of pericyclic reactions which can occasionally compete with [4 + 2] cycloadditions. For example, both electron-deficient *N*-sulfinyl compounds and sulfur diimides undergo facile ene reactions.[9] Moreover, *N*-sulfinylanilines occasionally act as 4π components in cycloadditions with alkenes.[31] An example exists of an inverse electron demand *N*-sulfinyl dienophile cycloaddition (entry 20), although in this case the initial Diels–Alder adduct was not isolated.[32] Examples of cycloadditions between diazadienes and sulfinyl dienophiles have also been reported.[33]

Whitesell and co-workers recently reported the first enantioselective *N*-sulfinyl-8-phenylmenthol carbamate (**19** and (*E,E*)-2,4-hexadiene (Scheme 1-VI).[34] When the cycloaddition was done at room temperature in the absence of a Lewis acid, a mixture of all four possible diastereomeric products was formed. In the presence of stannic chloride at 0°C, however, the cycloaddition of *N*-sulfinyl-8-phenylmenthol carbamate and (*E,E*)-hexadiene afforded *only* adduct **20** (42% yield, in greater than 97% diastereomeric excess), whose structure was proved by X-ray crystallography.

The formation of adduct **20** can be rationalized if one considers the (*Z*)-sulfinyl carbamate reacting via parallel conformer **19A** (Scheme 1-VI). If this dienophile conformer is attacked by the diene from the more exposed face in an endo transition state, adduct **20** will result. Another possibility is for conformer **19B** of the *E* dienophile to react via an exo transition state to give **20.** Although the second possibility would seem less likely than the first, it has not been established that secondary orbital effects are important in cycloaddition reactions of *N*-sulfinyl carbamates and thus the exo case should not be summarily dismissed at this time.

Weinreb and co-workers have also utilized *N*-sulfinyl-8-phenylmenthol

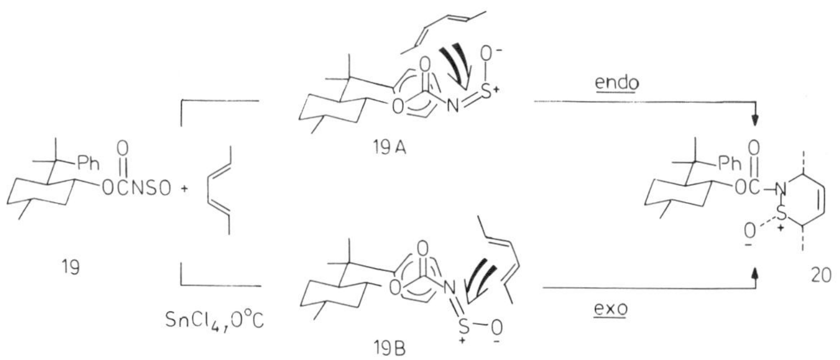

Scheme 1-VI

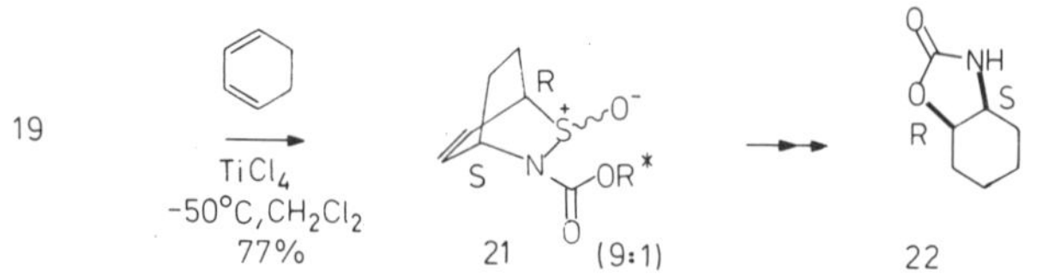

Scheme 1-VII

carbamate (**19**) in an enantioselective Diels–Alder reaction using 1,3-cyclohexadiene.[35] The uncatalyzed cycloaddition of the components afforded a complex mixture of diastereomeric adducts. However, when dienophile **19** was first treated with TiCl₄ at −50°C, followed by addition of cyclohexadiene, adducts **21**, which are epimeric at sulfur, were produced in 77% yield. These dihydrothiazine oxides were shown to have the 3*S*,6*R* configuration by conversion to the known cyclic carbamate **22** (Scheme 1-VII). The formation of **21** can be rationalized as shown in Scheme 1-VI for (*E,E*)-2,4-hexadiene.

The Weinreb group has recently examined the reaction of chiral *N*-sulfinyl dienophile **23**, prepared from (+)-camphor, with 1,3-cyclohexadiene (Scheme 1-VIII).[35] Whereas the uncatalyzed cycloaddition afforded a mixture of diastereomeric adducts, the reaction promoted by TiCl₄ gave a *single* adduct **24** having the 3*S*,6*R* configuration. Stereochemistry at sulfur in this compound could not be determined. As in the phenylmenthol series, one can reasonably consider two reacting dienophile conformations **23A** and **23B** (Scheme 1-VIII). If conformer **23A** is attacked by the diene in an endo manner from the most exposed face, the observed adduct **24** will be formed. Similarly, if conformer **23B** reacts with cyclohexadiene via an exo transition state, **24** will result.

The first examples of intramolecular *N*-sulfinyl dienophile cycloadditions have recently been reported by Weinreb *et al.*[36] Treatment of (*E,E*)-

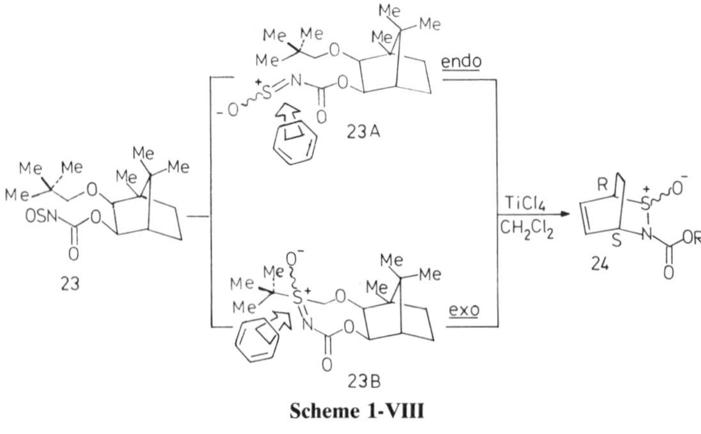

Scheme 1-VIII

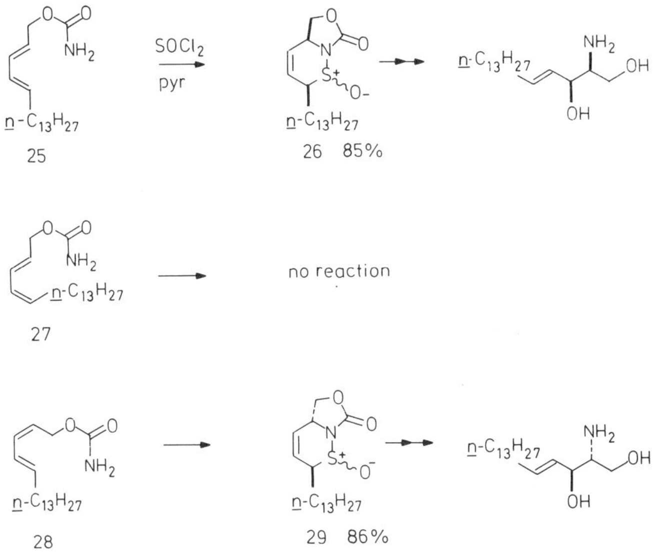

Scheme 1-IX

diene carbamate **25** (Scheme 1-IX) with thionyl chloride in pyridine afforded bicyclic dihydrothiazine oxide **26,** presumably via a transient N-sulfinyl compound. This adduct was transformed to *threo*-sphingosine in a few steps. The (E,Z)-diene carbamate **27** reacted with thionyl chloride to produce a N-sulfinyl carbamate, but the intermediate did not cyclize. It was suggested that this lack of reactivity was due to an inability of the 1,3-diene to achieve a s-cis conformation. Diene **28,** on the other hand, reacted with thionyl chloride to afford dihydrothiazine oxide **29** in excellent yield. This adduct was converted stereospecifically to *erythro*-sphingosine. Both adducts **26** and **29** were single stereoisomers at sulfur, but configuration was not determined.

This type of intramolecular N-sulfinyl carbamate cycloaddition methodology has also been used in synthesis of some simple amino sugars.[37,38] For example, a synthesis of epidesosamine has been developed as shown in Eq. (15).[37]

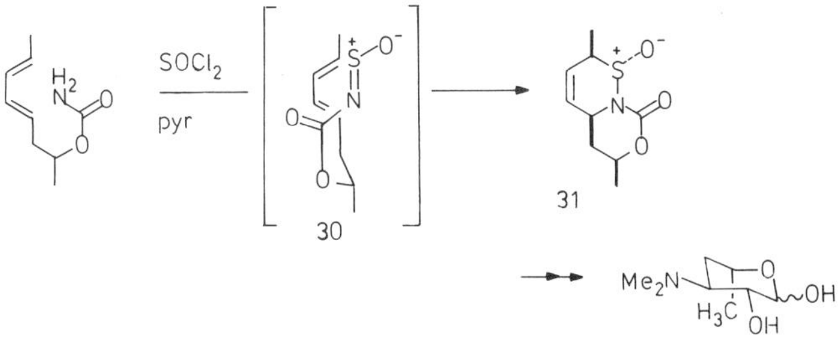

(15)

The key step here involved cycloaddition of the *N*-sulfinyl carbamate **30** to give exclusively adduct **31**, whose structure was proved by X-ray crystallography. It was suggested that the cyclization occurred via the conformer shown in **30** to minimize nonbonded interactions. It seems reasonable that the reaction proceeded through an intermediate (*E*)-*N*-sulfinyl carbamate to provide the sulfur stereochemistry shown in structure **31.** This cycloadduct was subsequently converted to the amino sugar by a short series of transformations.

In a similar study, *N*-sulfinyl carbamate **32** [Eq. (16)]

$$(16)$$

was found to cyclize via the conformation shown to produce dihydrothiazine oxides **33** as a mixture of sulfur epimers.[38] The adduct was then converted to the arabino sugar **34.**

4. SULFUR DIIMIDE CYCLOADDITIONS

Various types of electron-deficient sulfur diimides react as heterodienophiles. In general, these cycloadditions occur under conditions similar to those used for *N*-sulfinyl compounds. However, fewer types of sulfur diimides have been utilized in this process relative to *N*-sulfinyl compounds. Some examples of symmetrical sulfur diimide Diels–Alder reactions are listed in Table 1-II. It should again be noted that the orientational selectivity in these cycloadditions is the same as that shown by *N*-sulfinyl systems (cf. Table 1-I). Several examples of cycloadditions with unsymmetrical sulfur diimides are shown in Table 1-III. In all cases, these reactions were totally regioselective, and as noted above, reactions occurred at the *least* electron-deficient nitrogen–sulfur bond. Frontier molecular orbital (FMO) theory has been used to rationalize the regioselectivity of addition of the cationic sulfur diimide shown in entry 12.[47]

TABLE 1-II

Cycloadditions of Symmetrical Sulfur Diimides (RNSNR)

R	Diene	Conditions	Product	Yield	Ref.
Ts		C_6H_6, RT	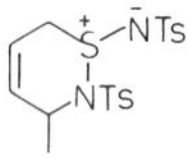	85–90%	15
Ts	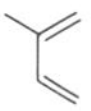	C_6H_6, RT	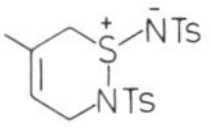	85–90%	15
pClPhSO$_2$		C_6H_6, RT	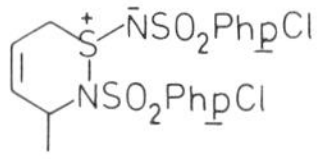	85–90%	15
pClPhSO$_2$	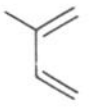	C_6H_6, RT	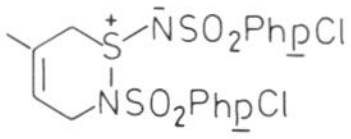	85–90%	15
pFPhSO$_2$	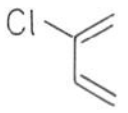	C_6H_6, RT	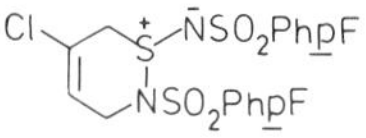	—	16
CO$_2$Et	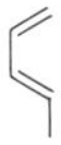	C_6H_6, RT	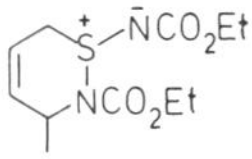	55%	39
CO$_2$Et	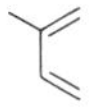	C_6H_6, RT	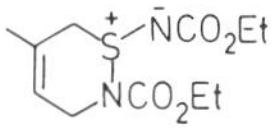	49%	39
PhCO		C_6H_6, RT	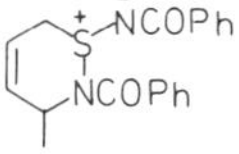	62%	40
PhCO	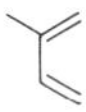	C_6H_6, RT	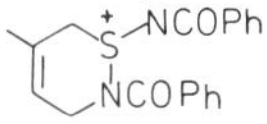	94%	40
pNO$_2$PhCO	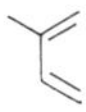	C_6H_6, RT	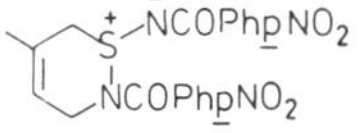	61%	41

5. CONFORMATIONS OF DIHYDROTHIAZINE OXIDES AND IMINES

During the past few years several groups have probed the conformations of [4 + 2] cycloadducts of N-sulfinyl compounds and sulfur diimides. X-Ray crystallography has been particularly useful in this re-

TABLE 1-III

Cycloadditions of Unsymmetrical Sulfur Diimides

Entry	Sulfur Diimide	Diene	Conditions	Product	Yield	Ref.
1	PhSO$_2$NSNPh	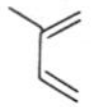	C$_6$H$_6$, Δ	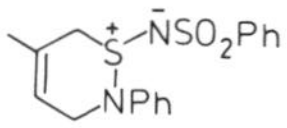 (S$^+$–$\bar{\text{N}}$SO$_2$Ph / NPh)	57%	42
2	PhSO$_2$NSNpNO$_2$Ph	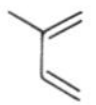	C$_6$H$_6$, Δ	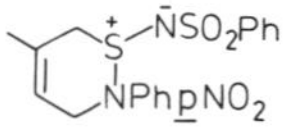 (S$^+$–$\bar{\text{N}}$SO$_2$Ph / NPh$\underline{p}$NO$_2$)	70%	42
3	PhSO$_2$NSNmNO$_2$Ph		C$_6$H$_6$, Δ	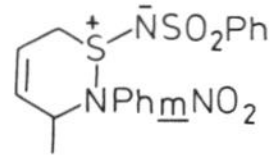 (S$^+$–$\bar{\text{N}}$SO$_2$Ph / NPh$\underline{m}$NO$_2$)	69%	42
4	PhSO$_2$NSNpNO$_2$Ph	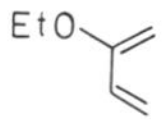 (EtO)	C$_6$H$_6$, Δ	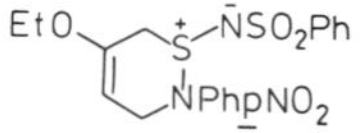 (EtO, S$^+$–$\bar{\text{N}}$SO$_2$Ph / NPh$\underline{p}$NO$_2$)	92%	42
5	pNO$_2$PhSO$_2$NSNC(Me)$_2$CN	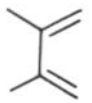	C$_6$H$_6$, 85–90°C	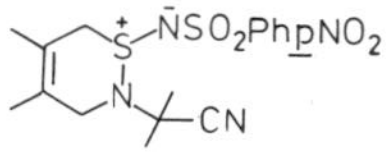 (S$^+$–$\bar{\text{N}}$SO$_2$Ph$\underline{p}$NO$_2$ / N–C(Me)$_2$CN)	74%	43

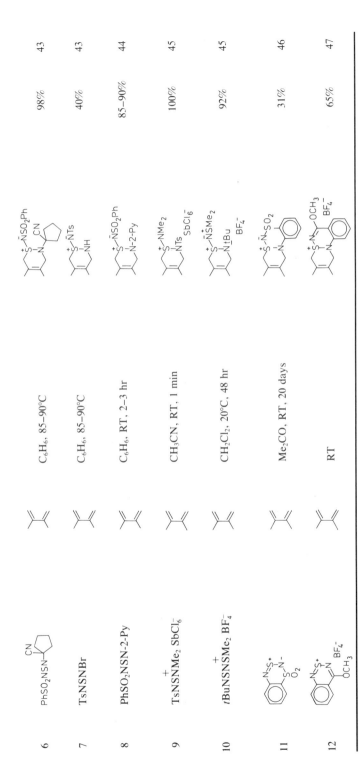

#	Reagent	Diene	Conditions	Product	Yield	Ref.
6	PhSO$_2$NSN (CN, cyclopentyl)	(2,3-dimethylbutadiene)	C$_6$H$_6$, 85–90°C		98%	43
7	TsNSNBr	(2,3-dimethylbutadiene)	C$_6$H$_6$, 85–90°C		40%	43
8	PhSO$_2$NSN-2-Py	(2,3-dimethylbutadiene)	C$_6$H$_6$, RT, 2–3 hr		85–90%	44
9	TsNSNMe$_2$ SbCl$_6^-$	(2,3-dimethylbutadiene)	CH$_3$CN, RT, 1 min		100%	45
10	tBuNSNSMe$_2$ BF$_4^-$	(2,3-dimethylbutadiene)	CH$_2$Cl$_2$, 20°C, 48 hr		92%	45
11	(benzo cyclic sulfur diimide structure)	(2,3-dimethylbutadiene)	Me$_2$CO, RT, 20 days		31%	46
12	(benzo cyclic structure, BF$_4^-$, OCH$_3$)	(2,3-dimethylbutadiene)	RT		65%	47

19

gard. For example, adduct **35,** which was the major adduct of cycloaddition of (*E,E*)-2,4-hexadiene and *N*-sulfinylbenzyl carbamate, was found to

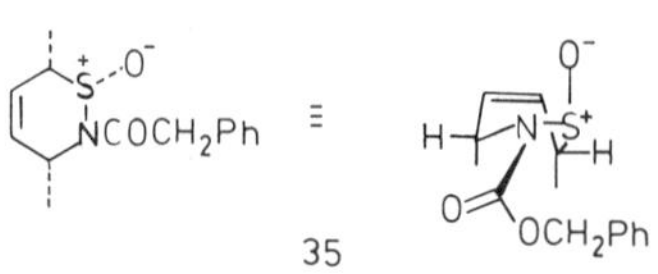

have the boat-like conformation shown with the oxygen on sulfur in a quasi-axial position.[36] Adduct **17** [Eq. (13)] has a half-chair conformation, once again with a quasi-axial sulfur–oxygen bond.[48] Similarly, the adduct **31** [Eq. (15)] was also found by X-ray crystallography to have a half-chair conformation and a quasi-axial sulfur–oxygen bond.[37] It has been suggested that this propensity for the oxygen in these systems to be quasi-axial is due to an anomeric effect.[36,49]

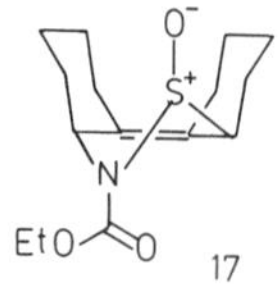

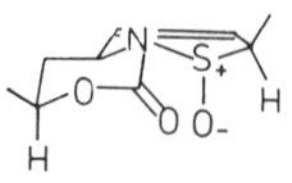

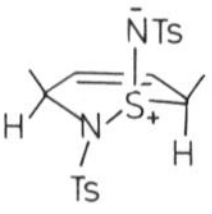

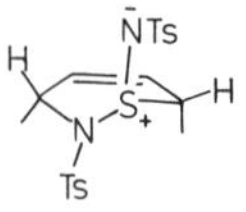

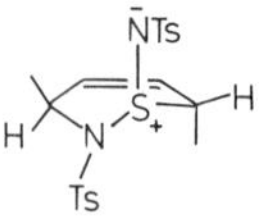

X-Ray crystallography has also established that dihydrothiazine imines have similar conformations. For example, compounds **36** and **38** were found to have half-chair conformations (cf. Scheme 1-IV), once again with the sulfur–heteroatom bond quasi-axial.[49] Adduct **37** was correlated with **36** by lanthanide-induced shift ¹H-NMR experiments, and the data have been interpreted as supporting this same type of conformation.[49]

6. REACTIONS OF CYCLOADDUCTS

There are several well-known transformations of 3,6-dihydrothiazine oxides and 3,6-dihydrothiazine imines.[2] In most of these reactions, either the cyclic sulfinamide moiety and/or the carbon–carbon double bond is involved. For example, a 3,6-dihydrothiazine oxide can be hydrolyzed under either acidic or basic conditions to afford a homoallylic amine derivative [Eq. (17)].

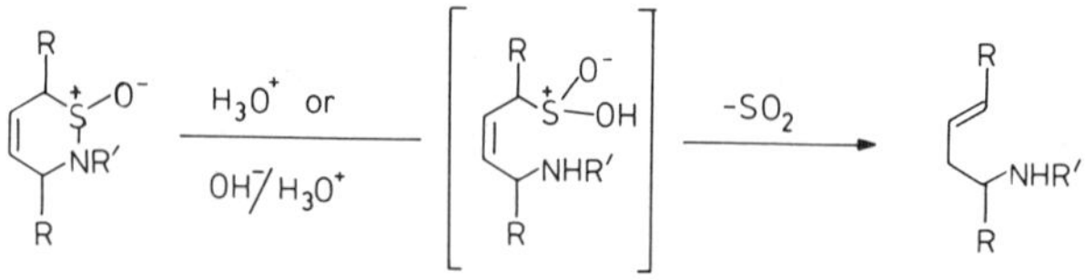

$$-SO_2 \qquad (17)$$

This transformation probably proceeds through an allylic sulfinic acid which loses SO_2 via a retro-ene process. The mechanism of this reaction was first investigated by Mock and Nugent[50] and was examined further by Weinreb et al.[51,52]

The Weinreb group investigated the hydrolysis of dihydrothiazine oxide **39**, prepared from (E,E)-tetramethylbutadiene, which was demonstrated to afford only sulfonamide **40** having an E double bond and the erythro configuration [Eq. (18)].

$$\text{39} \quad \xrightarrow[\text{2) 5\% HCl}]{\text{1) 5\% NaOH}} \quad \text{40} \qquad (18)$$

Similarly, hydrolysis of **41,** derived from (E,Z)-tetramethylbutadiene, yielded only sulfonamide **42** with an E double bond and the threo configuration [Eq. (19)].

$$\text{41} \quad \xrightarrow[\text{2) 5\% HCl}]{\text{1) 5\% NaOH}} \quad \text{42} \qquad (19)$$

These results are best rationalized by a concerted retro-ene reaction of intermediate sulfinic acid **43** [Eq. (20)]

$$\text{39} \quad \xrightarrow[\text{5\% HCl}]{\text{5\% NaOH}} \quad [\text{43}] \quad \longrightarrow \quad \text{40} \qquad (20)$$

as suggested initially by Mock and Nugent.[50] The substituent on C-1 of intermediate **43** assumes a pseudo-equatorial position to avoid $A^{1,3}$ strain with the substituents on C-4. This "anchor effect" controls to which diastereotopic face of the double bond the proton is transferred, and also establishes the E geometry of the newly formed double bond. Retro-ene reaction of adduct **41** proceeds through a similar conformation epimeric at C-4 to afford isomer **42**. This methodology is useful in synthesis of stereochemically complex homoallylic amine derivatives.

It has been known for over 30 years that N-aryldihydrothiazine oxides, when treated with alkali under vigorous conditions, afford N-arylpyrroles [Eq. (21)].[53]

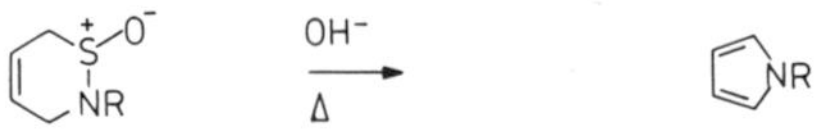

$$\text{(structure)} \quad \xrightarrow[\Delta]{\text{OH}^-} \quad \text{(pyrrole)} \quad \quad (21)$$

R = Ar, CO$_2$Et, PO(OEt)$_2$

This transformation has been extended to a number of other *N*-substituted adducts of *N*-sulfinyl dienophiles,[2,29,46,54] although yields of product are quite variable depending on the nature of the substituents. Russian workers have found that *N*-acyldihydrothiazine oxides can be converted to dihydro-1,3-oxazines in good yield (Scheme 1-X), perhaps via the homoallylic amine derivative [cf. Eq. (17)].[55,56]

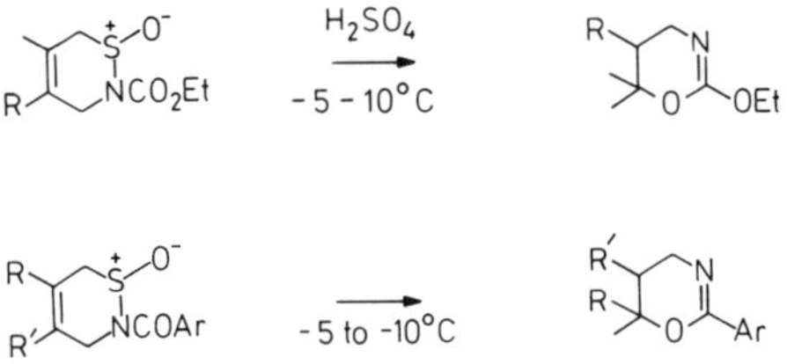

R, R' = H, CH$_3$

Scheme 1-X

The cycloadducts of sulfur diimides exhibit hydrolytic behavior very similar to that of dihydrothiazine oxides (Scheme 1-XI).[2] Alkaline hydrolysis of a dihydrothiazine imine affords an intermediate sulfinamide which, on treatment with aqueous acid, yields a homoallylic amine, presumably via a retro-ene process. A homoallylic amine is formed directly from a cycloadduct on acidic hydrolysis.[2]

Adducts of unsymmetrical sulfur diimides, in which the nitrogen atom in the heterocyclic ring bears an aryl substituent, undergo acidic hydrolysis to produce *N*-aryl-substituted homoallylic amines (Scheme 1-XII). Alkaline hydrolysis of these adducts leads to *N*-arylpyrroles, similar to the reaction shown in Eq. (21).[40a,42,44]

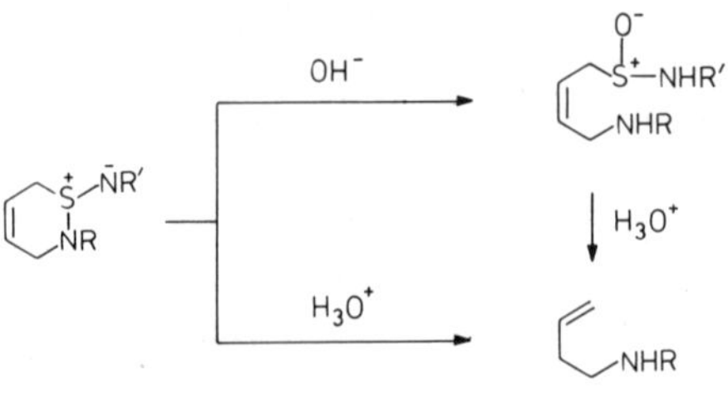

Scheme 1-XI

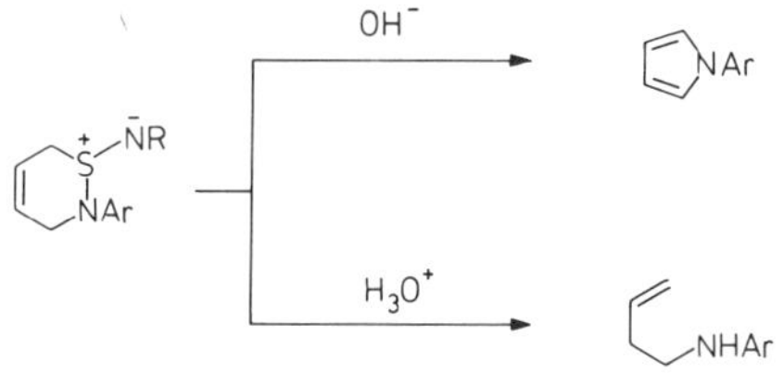

Scheme 1-XII

The sulfur atom of both 3,6-dihydrothiazine oxides and 3,6-dihydrothiazine imines is readily attacked by nucleophiles other than water or hydroxide ion. Wucherpfennig has reported the cleavage of dihydrothiazine oxides and dihydrothiazine imines with methoxide ion to afford sulfinate esters or imidosulfinate esters [Eq. (22)].[54]

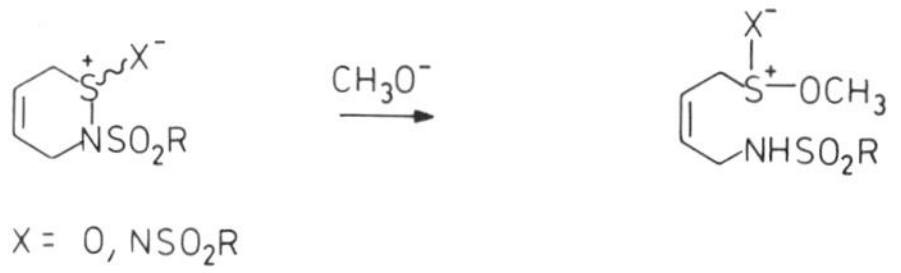

$$X = O, NSO_2R$$ (22)

Similarly, treatment of a dihydrothiazine oxide with thiophenoxide ion and thiophenol yielded an intermediate thiosulfinate ester, which underwent further reaction to produce a disulfide [Eq. (23)].

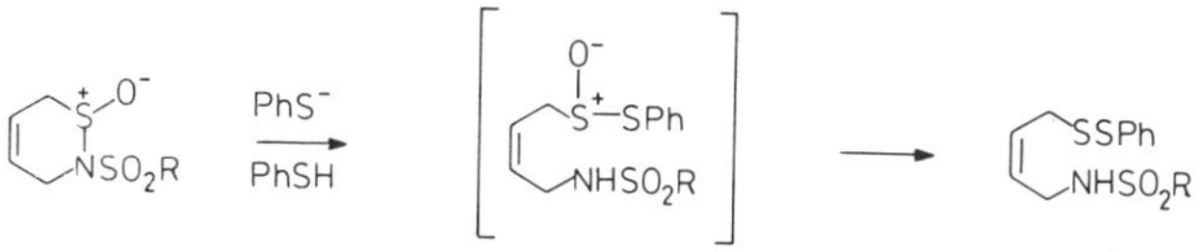

(23)

It has been reported that the adduct of N-sulfinylaniline and 2,3-dimethylbutadiene is stable to ethylmagnesium bromide.[2a] However, Weinreb and co-workers[36–38,52] have recently found that a number of N-acyldihydrothiazine oxides are susceptible to ring opening by a variety of reactive carbon nucleophiles. For example, when dihydrothiazine oxide **44** was treated with phenylmagnesium bromide, followed by trimethyl phosphite, (E)-threo-hydroxycarbamate **45** was produced exclusively [Eq. (24)].[36]

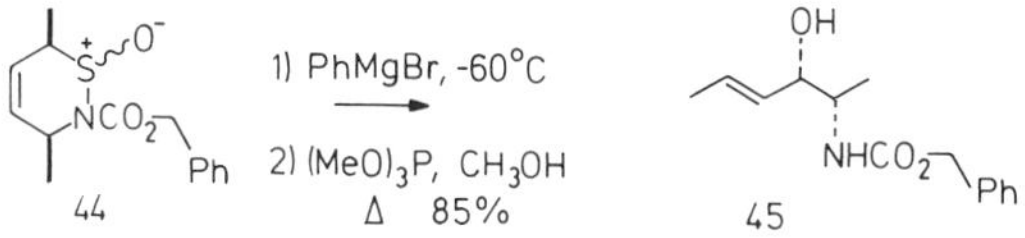

(24)

Similarly, adduct **46** afforded *only* (E)-*erythro*-hydroxycarbamate **47** [Eq. (25)].

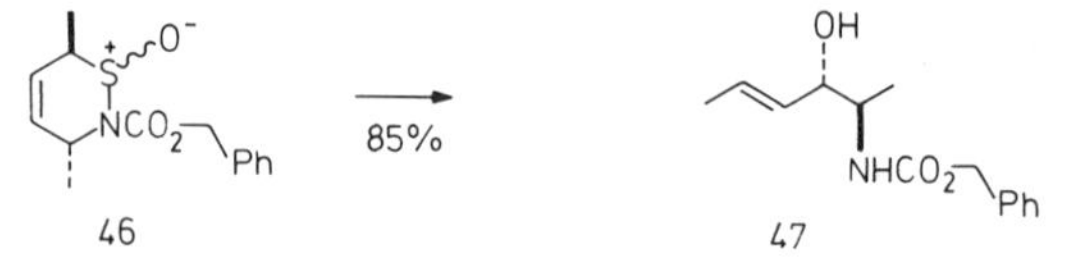

$$(25)$$

Weinreb *et al.* offered the mechanism for the transformation of **44** to **45** shown in Scheme 1-XIII.[36] When **44** is treated with phenylmagnesium bromide, an allylic sulfoxide **48** is produced. A [2,3]-sigmatropic rearrangement of this sulfoxide to a sulfenate ester **49** proceeds stereoselectively via an envelope-like transition state with the C-1 methyl substituent quasi-equatorial to minimize nonbonded interactions. ^{1}H-NMR studies demonstrated that in the absence of phosphite, the (*Z*)-allylic sulfoxide **48** rapidly rearranged to (*E*)-allylic sulfoxide **51** (4 hr at 40°C), while the desulfurization process was slower (12 hr at 60°C). This transformation presumably occurs by an initial rearrangement of **48** to sulfenate ester **49**. This intermediate can equilibrate to conformer **50**, which rearranges to the more stable (*E*)-sulfoxide **51**. When pure sulfoxide **51** was treated with trimethyl phosphite in refluxing methanol *only* (*E*)-*threo*-hydroxycarbamate **45** was produced. A similar sequence of steps would give rise to (*E*)-*erythro*-hydroxycarbamate **47** from adduct **46** via intermediates epimeric to **44** at C-4. It should be noted that sulfur stereochemistry in the dihydrothiazine oxide had no effect on this cleavage and rearrangement sequence. Weinreb and Garigipati have utilized the methodology outlined above in the syntheses of *threo-* and *erythro-* sphingosines as shown in Scheme 1-IX.[36]

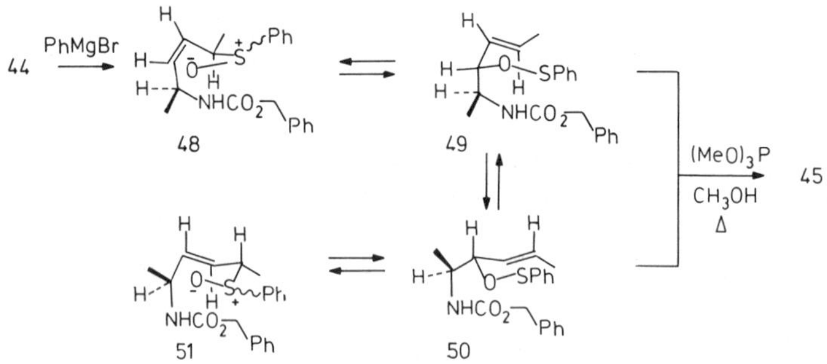

Scheme 1-XIII

Another nucleophile which was used by the Weinreb group to cleave the sulfur–nitrogen bond in dihydrothiazine oxide **44** was vinylmagnesium bromide, directly affording sulfine **52** [Eq. (26)].[14]

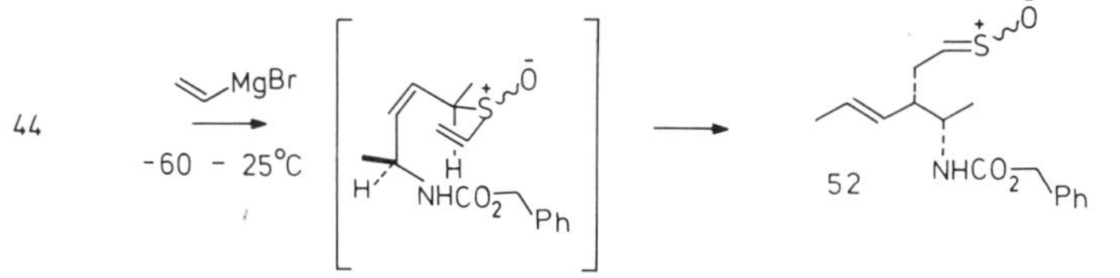

$$(26)$$

This transformation was presumed to occur via an intermediate vinyl sulfoxide, which underwent a [3,3]-sigmatropic rearrangement via a chair-like transition state with the methyl group quasi-equatorial to afford **52**. Since the rearrangement occurred under very mild thermal conditions, it was proposed that this may be a type of alkoxy accelerated pericyclic reaction.[57]

Dihydrothiazine oxide **44** was also stereoselectively converted to sulfide **53**, as outlined in Scheme 1-XIV.[56] Treatment of **44** with trimethylsilylmethylmagnesium chloride afforded a sulfoxide which was converted to a sulfonium salt. On treatment with potassium *tert*-butoxide, this intermediate underwent a [2,3]-sigmatropic rearrangement via an envelope-like transition rate (cf. Scheme 1-XIII) to yield compound **53** in good overall yield.

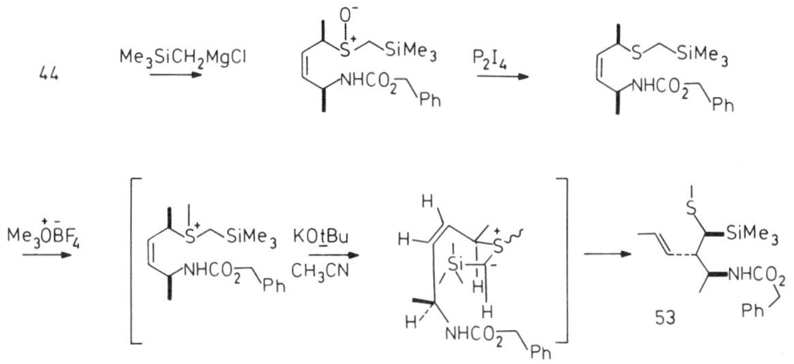

Scheme 1-XIV

The Weinreb group has recently developed stereoselective methodology for synthesis of unsaturated vicinal diamine derivatives from dihydrothiazine imines.[49] For example, when cycloadduct **54** prepared from (*E,E*)-2,4-hexadiene was treated with phenylmagnesium bromide followed by trimethyl phosphite, *E-threo* vicinal diamine **57** was formed cleanly in good yield [Eq. (27)].

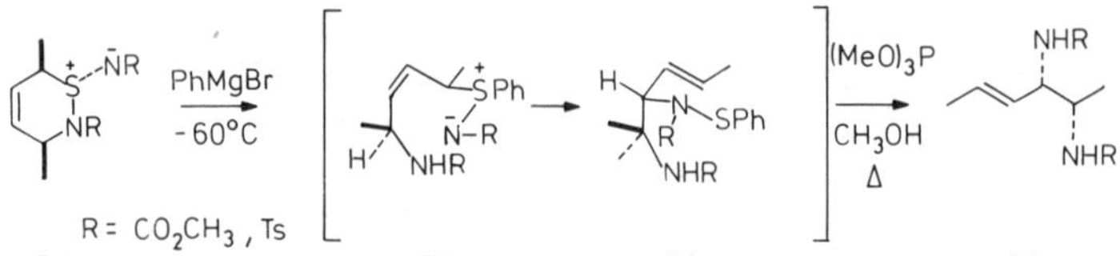

$$R = CO_2CH_3 , Ts$$

54   55   56   57   (27)

This transformation presumably proceeds with initial formation of allylic sulfilimine **55**, which rearranges via an envelope-like transition state to sulfenamide **56** (cf. Scheme 1-XIII). Interestingly, NMR analysis showed that this [2,3]-sigmatropic rearrangement lies totally on the side of this sulfenamide, unlike the allylic sulfoxide–sulfenate ester system, which lies predominantly to the side of the sulfoxide. Similarly, C-3 epimeric dihydrothiazine imine **58** can be converted by an identical pathway to E-erythro vicinal diamine **59** in an efficient and totally stereoselective manner [Eq. (28)].

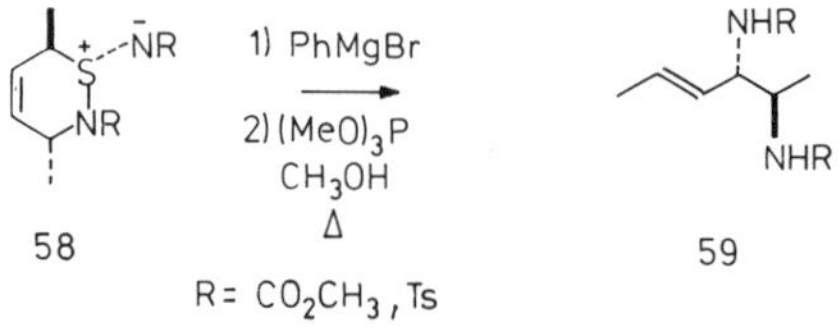

58   59   (28)

$$R = CO_2CH_3 , Ts$$

Surprisingly, the sulfur epimer **60** prepared from (E,E)-2,4-hexadiene was found to be unreactive toward phenylmagnesium bromide. Compound **60**-reacted with methyllithium to give **57**, but the yield was poor. This result was rationalized by considering the conformations of adducts **54** versus **60**. Adduct **54** exists as half-chair conformer **54A** (*vide supra*),

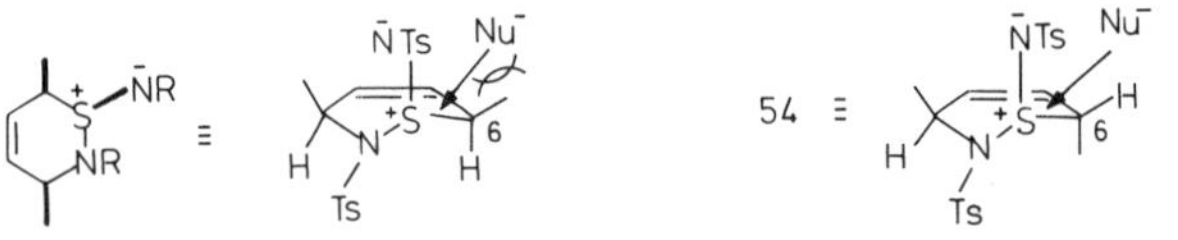

60   60A   54   54A

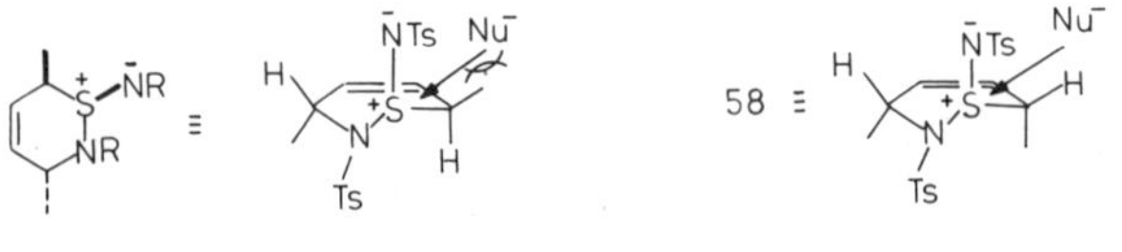

61   61A   58   58A

and thiazine imine **60**, epimeric to **54** at sulfur, exists as half-chair **60A**. Assuming that nucleophilic ring cleavage occurs through a S_N2-like process, the entering carbon nucleophile would experience a steric interaction with the C-6 methyl group in **60A**. In epimer **54**, however, the nucleophile attacking sulfur passes by a sterically less demanding hydrogen atom.

A similar situation exists with sulfur epimer **61**, prepared from (*E,Z*)-2,4-hexadiene, which is unreactive toward phenylmagnesium bromide and gives *E*-erythro vicinal diamine **59** in only poor yield with methyllithium. Inspection of the conformations of the adducts **61A** and **58A** indicates that the path of the attacking nucleophile in the former case is more sterically crowded than in the latter.

Since the Diels–Alder cycloadditions used to prepare various dihydrothiazine imines generally produced mixtures of sulfur epimers, it was desirable to find a procedure to also convert the "unreactive" adducts to the vicinal diamines. It was discovered that adducts **60** rearranged thermally by a novel [2,3]-sigmatropic process to afford thiadiazolidines **62** stereoselectively (Scheme 1-XV). Reduction of **62** with sodium borohydride yielded the *E*-threo vicinal diamine derivative **57**. Similarly, "unreactive" adducts **61** cleanly gave thiadiazolidines **63** on heating, which could be converted to *E*-erythro vicinal diamines **59**. This methodology thus allows both sulfur epimers from (*E,E*)-2,4-hexadiene to be converted to the *E*-threo product, and the epimers from (*E,Z*)-2,4-hexadiene to be used to prepare the *E*-erythro series of compounds.

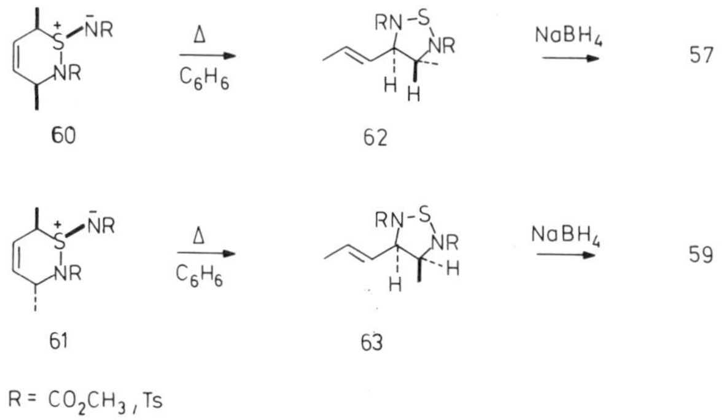

$R = CO_2CH_3$, Ts

Scheme 1-XV

Dihydrothiazine oxides can be oxidized at the carbon–carbon double bond and/or at sulfur. Oxidation of these adducts is frequently accomplished with peracid to afford sultams [Eq. (29)].[2]

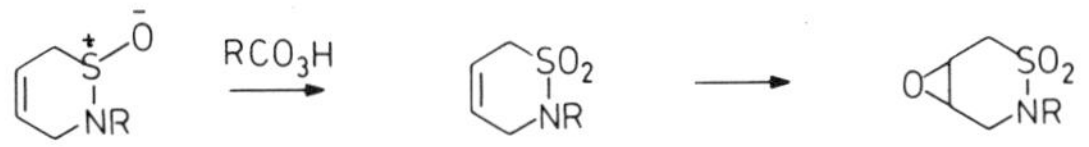

$$\tag{29}$$

The sultams can be oxidized further to epoxysultams on exposure to excess peracid. However, it has been reported that the adduct of 2,3-dimethylbutadiene and N-sulfinylaniline could first be epoxidized by reaction with perbenzoic acid to afford an epoxythiazine oxide [Eq. (30)].

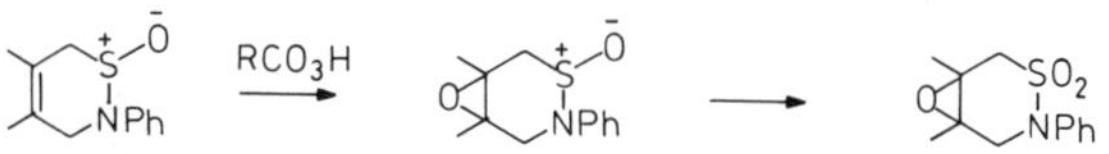

$$
\text{(30)}
$$

Treatment of this epoxide with excess perbenzoic acid yielded an epoxysultam.

Weinreb and co-workers have found some interesting directing effects of the sulfur–oxygen bond in some oxidations of dihydrothiazine oxides.[52] Adduct **64** on oxidation with pertrifluoroacetic acid stereoselectively afforded β-epoxysultam **67** [Eq. (31)].

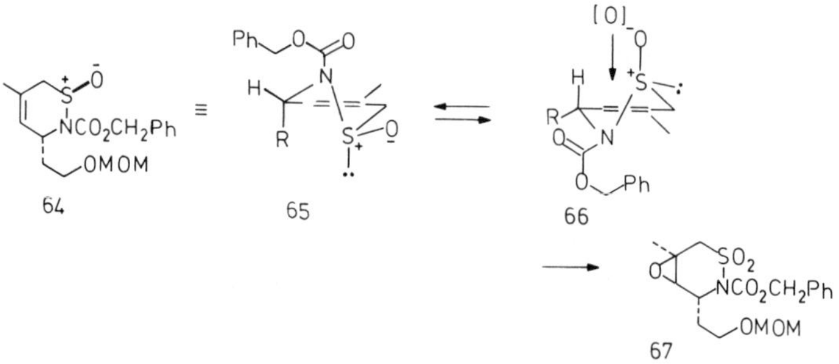

$$
\text{(31)}
$$

On the other hand, epimeric adduct **68** yielded only α-epoxides **71** and **72** [Eq. (32)].

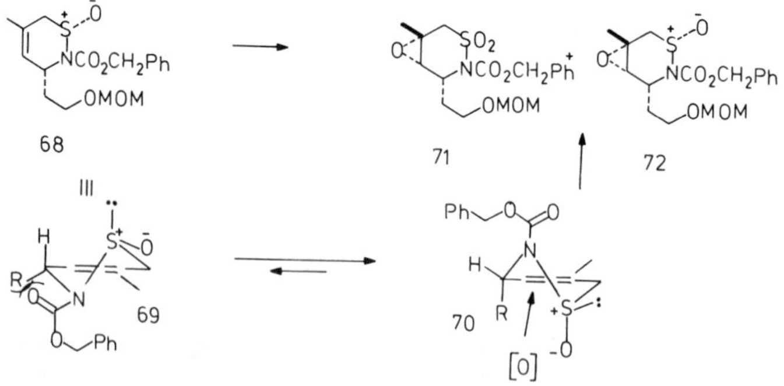

$$
\text{(32)}
$$

It was suggested that epoxidation of **64** occurs via conformer **66** having the sulfur–oxygen bond quasi-axial. As discussed above, this conformer is apparently more stable than **65** which has a quasi-equatorial sulfur–oxygen bond. A syn directing effect of the oxygen in **66** would lead to the observed β-epoxide **67**. Similarly, one would expect adduct **68** to exist as conformer **70** rather than **69**. Syn epoxidation of **70** would produce α-epoxides **71** and **72**. Epoxidation of sultam **73** was also investigated and was found to give exclusively β-epoxysultam **67** [Eq. (33)].

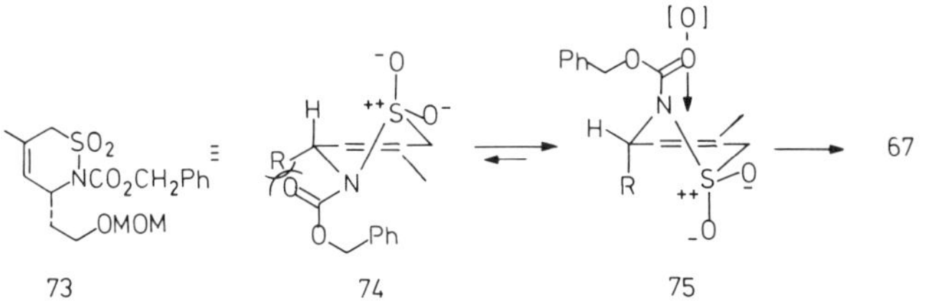

$$\text{(33)}$$

It seems reasonable that in this series conformer **75** is preferred over **74** having minimized $A^{1,3}$ strain. A sterically controlled epoxidation of **75** would afford the observed β product.

N-Aryl-3,6-dihydrothiazine 1-oxides can be reduced to the corresponding 3,6-dihydrothiazines by treatment with lithium aluminum hydride in ether [Eq. (34)].[2a]

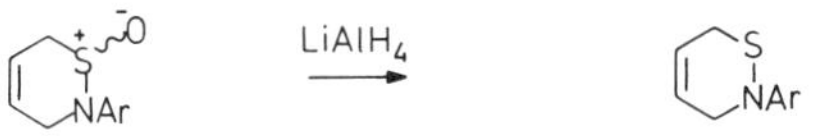

$$\text{(34)}$$

However, dihydrothiazine oxides that have a strong electron-withdrawing group on nitrogen reportedly do not undergo this reduction.

7. N-SULFONYL COMPOUNDS

A few examples have been described of Diels–Alder-like reactions with imides formally derived from sulfur trioxide.[58] *N*-Sulfonylimides are quite reactive in a [2 + 2] fashion with alkenes, but their ability to generally act as dienophiles is yet to be established. Sulfonylimines do react with highly oxygenated dienes to give adducts after hydrolysis [Eq. (35)].

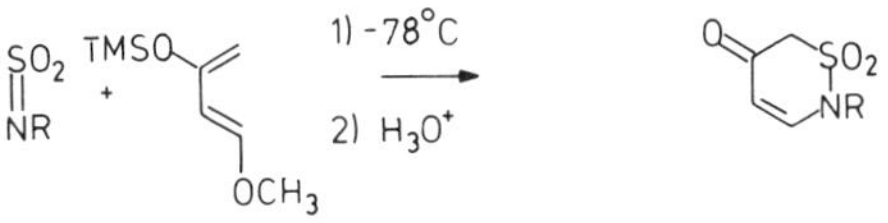

$$\text{(35)}$$

However, product formation via a true pericyclic mechanism may not be occurring in these cases.

8. SULFUR DIOXIDE AND RELATED COMPOUNDS

Cycloaddition reactions of 1,3-dienes with sulfur dioxide, unlike its
mono and bis imino derivatives, usually produce five-membered ring 2,5-
dihydrothiophene dioxides rather than Diels–Alder adducts [Eq. (36)].

$$\underset{25°C}{\overset{120°C}{\rightleftharpoons}} \qquad + \quad SO_2 \quad \underset{< 0°C}{\overset{fast}{\rightleftharpoons}} \qquad \qquad (36)$$

Durst has shown that a Diels–Alder adduct is in fact the kinetic product of
reactions of sulfur dioxide with conjugated dienes, but is thermally unstable and readily undergoes a retro-Diels–Alder reaction to butadiene and
sulfur dioxide.[59] A thermodynamically more favorable addition then en-
sues, producing a stable five-membered ring adduct. Thus, the apparent
inability of sulfur dioxide to form Diels–Alder adducts is not one of reactivity, but of product instability.

Hogeveen and Heldeweg found a rare example of a Diels–Alder reaction with sulfur dioxide where the [4 + 2] adduct is reasonably stable.[60]
Addition of sulfur dioxide to diene 76 at low temperature yielded [4 + 2]
adduct 77 as the kinetic product. At temperatures above 20°C products
resulting from the dihydrothiophene dioxide 78 were observed (Scheme 1-
XVI).

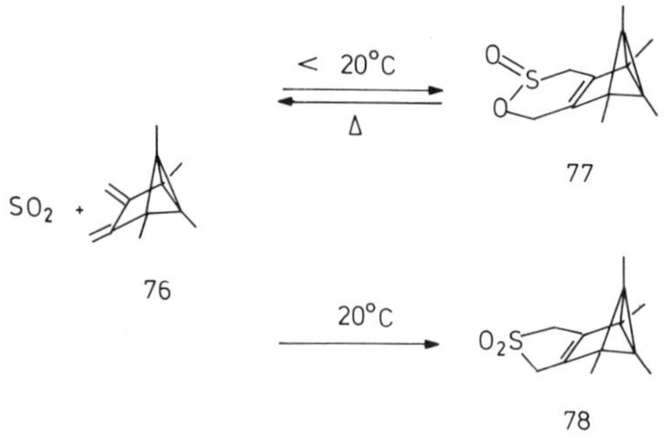

Scheme 1-XVI

In a recent experiment, Durst has shown that six-membered ring ox-
athiins are, in fact, the primary kinetic products of sulfur dioxide–diene
reactions.[61] An o-quinodimethane, which was generated from a diazo
compound [Eq. (37)].

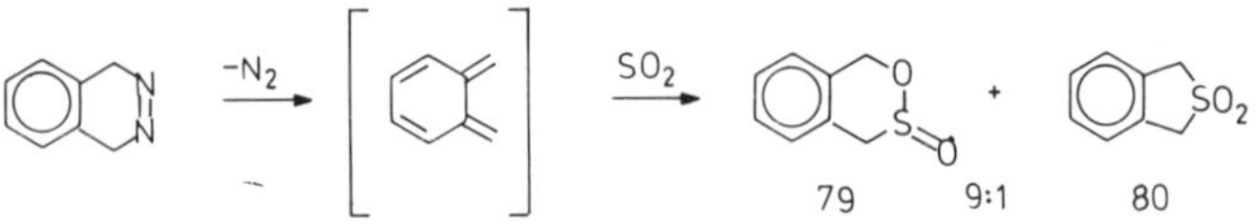

was found to react with sulfur dioxide to afford a 9 : 1 mixture of **79** and **80,** respectively. This ratio reflects the fact that formation of **79** has a $\Delta\Delta G\ddagger$ value about 12 kcal/mol lower than that for formation of **80.** In some related work, Dodson *et al.* have generated disulfur monoxide (S_2O) by pyrolysis of a thiirane oxide and found that this species can be trapped by conjugated dienes to give Diels–Alder products [Eq. (38)].[62]

$$\begin{array}{c} O \\ \parallel \\ S \\ \triangle \end{array} \xrightarrow{\Delta} \left[S_2O \right] \xrightarrow{\begin{array}{c} R \diagdown \\ R \diagup \end{array}} \begin{array}{c} R \diagdown \quad S \\ R \diagup \quad S \diagdown O \end{array} \qquad (38)$$

Soviet workers recently discovered that under high pressures (3–5 kbar) isoprene reacts with phenylsulfinyl chloride to give (*Z*)-chlorosulfoxide **82** [Eq. (39)].[63]

$$\diagup \!\!\!\!\diagdown \;\; + \;\; \begin{array}{c} Cl \diagdown_S\diagup^{Ph} \\ \parallel \\ O \end{array} \xrightarrow{CHCl_3} \left[\begin{array}{c} Cl \\ \mid \\ S-Ph \\ \parallel \\ O \end{array} \atop 81 \right] \xrightarrow{75\%} \begin{array}{c} \diagdown SOPh \\ Cl \end{array} \qquad (39)$$

It was suggested that this product arises via an intermediate Diels–Alder adduct **81.**

9. SELENIUM DIOXIDE

In a reinvestigation of earlier work,[64] Mock and McCausland established that the products of cycloaddition of dienes and selenium dioxide have six-membered ring seleninic ester structures [Eq. (40)].[65]

$$\begin{array}{c} R \diagdown \\ R \diagup \end{array} \;\; + \;\; SeO_2 \;\; \longrightarrow \;\; \begin{array}{c} R \diagdown \quad Se \diagdown\!\!=O \\ R \diagup \quad \diagdown O \end{array} \qquad (40)$$

R = alkyl, aryl

However, little additional information is available on this interesting Diels–Alder reaction.

REFERENCES

1. Wichterle, O., and Rocek, J. (1953). *Chem. Listy* **47,** 1768.
2. Previous reviews: a. Kresze, G. (1967). *In* "1,4-Cycloaddition Reactions, the Diels–Alder Reaction in Heterocyclic Syntheses" (J. Hamer, ed.), p. 453. Academic Press,

New York. b. Kresze, G., and Wucherpfennig, W. (1967). *Angew, Chem. Int. Ed. Engl.* **6,** 149. c. Bussas, R., Kresze, G., Munsterer, H., and Schwobel, A. (1983). *Sulf. Rep.* **2,** 215. d. Weinreb, S. M., and Staib, R. R. (1982). *Tetrahedron* **38,** 3087.

3. Levchenko, E. S., and Kirsanov, A. V. (1965). *J. Org. Chem. USSR (Engl. Transl.)* **1,** 290.

4. Levchenko, E. S., and Dorokhova, E. M. (1972). *J. Org. Chem. USSR (Engl. Transl.)* **8,** 2573.

5. Kresze, G., and Rossert, M. (1978). *Angew. Chem. Int. Ed. Engl.* **17,** 63.

6. van Woerden, H. F., and Bijl-Vieger, S. H. (1974). *Recl. Trav. Chim. Pays-Bas* **93,** 85.

7. Beagley, B., Chantrell, S. J., Kirby, R. G., and Schmidling, D. G. (1975). *J. Mol. Struct.* **25,** 319.

8. a. Kresze, G., Berger, M., Claus, P. K., and Rieder, W. (1976). *Org. Magn. Reson.* **8,** 170. b. Yavari, I., Staral, J. S., and Roberts, J. D. (1979). *Org. Magn. Reson.* **12,** 340.

9. a. Munsterer, H., Kresze, G., Lamm, V., and Gieren, A. (1983). *J. Org. Chem.* **48,** 2833. b. Sharpless, K. B., and Singer, S. P. (1976). *J. Org. Chem.* **41,** 2504.

10. Kresze, G., and Wagner, U. (1972). *Liebigs Ann. Chem.* **762,** 93; Kresze, G., and Wagner, U. (1972). *Liebigs Ann. Chem.* **762,** 106.

11. Mock, W. L., and Nugent, R. M. (1975). *J. Am. Chem. Soc.* **97,** 6521.

12. Hanson, P., and Stockburn, W. A. (1985). *J. Chem. Soc., Perkin Trans.* **2,** 589.

13. a. Grunwell, J. R., and Danison, W. C. (1973). *Int. J. Sulf. Chem.* **8,** 379. b. Carpanelli, C., and Gaiani, G. (1982). *Gazz. Chim. Ital.* **112,** 187, 191.

14. Garigipati, R. S., Cordova, R., Parvez, M., and Weinreb, S. M. (1986). *Tetrahedron* **42,** 2979.

15. Levchenko, E. S., and Balon, Y. G. (1965). *J. Org. Chem. USSR (Engl. Transl.)* **1,** 146.

16. Wucherpfennig, W., and Kresze, G. (1966). *Tetrahedron Lett.,* 1671.

17. Kresze, G., Maschke, A., Albrecht, R., Bederke, K., Patzchke, H. P., Smalla, H., and Trede, A. (1962). *Angew, Chem. Int. Ed. Engl.* **1,** 89.

18. Levchenko, E. S., Balon, Y. G., and Kisilenko, A. V. (1965). *J. Org. Chem. USSR (Engl. Transl.)* **1,** 151.

19. Balon, Y. G., Levchenko, E. S., and Kirsanov, A. V. (1967). *J. Org. Chem. USSR (Engl. Transl.)* **3,** 2009.

20. Wucherpfennig, W. (1967). *Tetrahedron Lett.,* 3235; Wald, L., and Wucherpfennig, W. (1971). *Liebigs Ann. Chem.* **746,** 28.

21. Horhold, H. H. (1967). *Angew. Chem. Int. Ed. Engl.* **6,** 357.

22. Levchenko, E. S., Balon, Y. G., and Kirsanov, A. V. (1967). *J. Org. Chem. USSR (Engl. Transl.)* **3,** 1838.

23. Wieczorkowski, J. (1963). *Chem. Ind. (London),* 825; Balon, Y. G., and Levchenko, E. S. (1967). *J. Org. Chem. USSR (Engl. Transl.)* **3,** 2181.

24. a. Scherer, O. J., and Schmitt, R. (1967). *Angew. Chem. Int. Ed. Engl.* **6,** 701. b. Scherer, O. J., and Schmitt, R. (1968). *Chem. Ber.* **101,** 3302.

25. Kresze, G., and Rossert, M. (1978). *Angew. Chem. Int. Ed. Engl.* **17,** 63.

26. Rossert, M., Kraus, W., and Kresze, G. (1978). *Tetrahedron Lett.,* 4669.

27. Kresze, G., and Rossert, M. (1981). *Liebigs Ann. Chem.,* 58; Niedlein, R., and Walser, P. (1982). *Chem. Ber.* **115,** 2428.

28. Malach, H. P., Bussas, R., and Kresze, G. (1982). *Liebigs Ann. Chem.,* 1384; see also Butler, R. N., O'Donoghue, D. A., and Halloran, G. A. (1986). *J. Chem. Soc., Chem. Commun.,* 800; Butler, R. N., and O'Halloran, G. A. (1986). *Chem. Ind.,* 750.

29. Ichimura, K., Ichikawa, S., and Imamura, K. (1976). *Bull. Soc. Chem. Jpn.* **49,** 1157.

30. Boan, C., and Skattebol, L. (1978). *J. Chem. Soc., Perkin Trans. 1,* 1568.

31. Hanson, P., and Stone, T. W. (1984). *J. Chem. Soc., Perkin Trans. 1,* 2429, and references cited therein.

32. Seitz, G., and Kampchen, T. (1977). *Arch. Pharm.* **310**, 269.
33. Sommer, S. (1977). *Synthesis*, 305.
34. Whitesell, J. K., James, D., and Carpenter, J. F. (1985). *J. Chem. Soc., Chem. Commun.*, 1449.
35. Remiszewski, S. W., Yang, J., and Weinreb, S. M. (1986). *Tetrahedron Lett.* **27**, 1853.
36. Garigipati, R. S., and Weinreb, S. M. (1983). *J. Am. Chem. Soc.* **105**, 4499; Garigipati, R. S., Freyer, A. J., Whittle, R. R., and Weinreb, S. M. (1984). *J. Am. Chem. Soc.* **106**, 7861.
37. Remiszewski, S. W., Whittle, R. R., and Weinreb, S. M. (1984). *J. Org. Chem.* **49**, 3243.
38. Remiszewski, S. W., Stouch, T. R., and Weinreb, S. M. (1985). *Tetrahedron* **41**, 1173.
39. Levchenko, E. S., Balon, Y. G., and Kirsanov, A. V. (1967). *J. Org. Chem. USSR (Engl. Transl.)* **3**, 2014.
40. Dorokhova, E. M., Levchenko, E. S., and Lavrenyuk, T. Y. (1974). *J. Org. Chem. USSR (Engl. Transl.)* **10**, 1877.
41. Borovikova, G. S., Levchenko, E. S., and Dorokhova, E. M. (1979). *J. Org. Chem. USSR (Engl. Transl.)* **15**, 424.
42. Levchenko, E. S., and Balon, Y. G. (1965). *J. Org. Chem. USSR (Engl. Transl.)* **1**, 295.
43. Markovskii, L. N., Fedyuk, G. S., and Balon, Y. G. (1974). *J. Org. Chem. USSR (Engl. Transl.)* **10**, 1444.
44. Levchenko, E. S., and Slyusarenko, E. I. (1975). *J. Org. Chem. USSR (Engl. Transl.)* **11**, 860.
45. Schwobel, A., and Kresze, G. (1985). *Liebigs Ann. Chem.*, 453.
46. Grill, H., and Kresze, G. (1971). *Liebigs Ann. Chem.* **749**, 171.
47. Kosbahn, W., and Schafer, H. (1977). *Angew, Chem. Int. Ed. Engl.* **16**, 780.
48. Dauter, Z., Hanson, P., Reynolds, C. D., and Stockburn, W. A. (1984). *Acta Crystallogr.* **C40**, 521.
49. Natsugari, H., Whittle, R. R., and Weinreb, S. M. (1984). *J. Am. Chem. Soc.* **106**, 7867. Natsugari, H., Turos, E., Weinreb, S. M., and Cvetovich, R. J. (1987). *Heterocycles* **25**, 19.
50. Mock, W. L., and Nugent, R. M. (1978). *J. Org. Chem.* **43**, 3433.
51. Garigipati, R. S., Morton, J. A., and Weinreb, S. M. (1983). *Tetrahedron Lett.* **24**, 987.
52. Weinreb, S. M., Garigipati, R. S., and Gainor, J. A. (1984). *Heterocycles* **21**, 309.
53. Rocek, J. (1953). *Chem. Listy* **47**, 1781; Wichterle, O., and Rocek, J. (1954). *Coll. Czech. Chem. Commun.* **19**, 282.
54. Wucherpfennig, W. (1971). *Liebigs Ann. Chem.* **746**, 16.
55. Slyusarenko, E. I., Pelkis, N. P., and Levchenko, E. S. (1978). *J. Org. Chem. USSR (Engl. Transl.)* **14**, 1016.
56. Levchenko, E. S., and Dorskhova, E. M. (1976). *J. Org. Chem. USSR (Engl. Transl.)* **12**, 432.
57. See also Block, E., and Ahmad, S. (1985). *J. Am. Chem. Soc.* **107**, 6731.
58. Kloek, J. A., and Leschinsky, K. L. (1979). *J. Org. Chem.* **44**, 305.
59. Jung, F., Molin, M., Van Den Elzen, R., and Durst, T. (1974). *J. Am. Chem. Soc.* **96**, 935.
60. Heldeweg, R. F., and Hogeveen, H. (1976). *J. Am. Chem. Soc.* **98**, 2341.
61. Durst, T., and Tetreault-Ryan, L. (1978). *Tetrahedron Lett.*, 2353.
62. Dodson, R. M., Srinivasan, V., Sharma, K. S., and Sauers, R. F. (1972). *J. Org. Chem.* **37**, 2367.
63. Moiseenkov, A. M., Veselovsky, V. V., Makarova, Z. G., Zhulin, V. M., and Smit, W. A. (1984). *Tetrahedron Lett.* **25**, 5929.
64. Backer, H. J., and Strating, J. (1934). *Recl. Trav. Chim. Pays-Bas* **53**, 1113.
65. Mock, W. L., and McCausland, J. H. (1968). *Tetrahedron Lett.*, 391.

Chapter **2**

Imino Dienophiles

INTRODUCTION

The initial example of an imino compound acting as a dienophile was briefly mentioned by Alder in 1943.[1] Reaction of amino diester **1** with various "aliphatic dienes" did not produce the carbocyclic adducts which

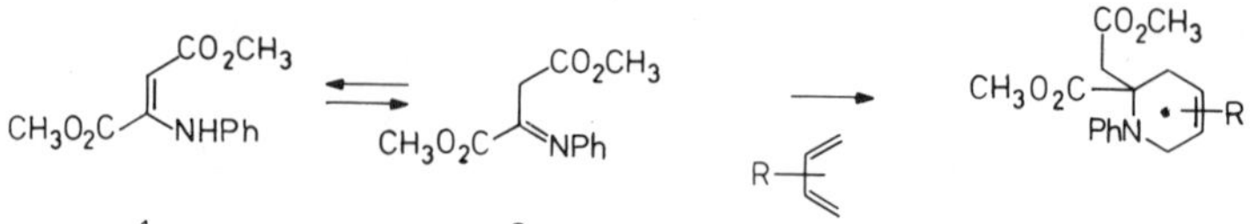

were presumably expected, but instead gave the tetrahydropyridines **3** via imino tautomer **2**. This initial observation, made over 40 years ago, stimulated relatively little research until recently.[2] Even now, systematic studies of the ability of different types of imines to act as dienophiles are scarce and mechanistic studies in this area are virtually nonexistent. In general, electron-deficient imines are the most reactive dienophiles, particularly those of the N-sulfonyl, N-acyl, and iminium salt types. Other kinds of imino compounds have been used successfully as dienophiles on occasion but do not appear to be as generally reliable reaction partners as th aforementioned types. It should also be noted that some imino Diels–Alder reactions are catalyzed by acids, although again systematic studies and definitive data are lacking.

1. MECHANISTIC, REGIOCHEMICAL, AND STEREOCHEMICAL CONSIDERATIONS

As alluded to above, no detailed studies of the mechanism of the imino Diels–Alder reaction have been reported to date. However, it has been suggested that [4 + 2] cycloadditions of most C=N, N=N, C=O and N=O dienophiles are HOMO$_{\text{diene}}$–LUMO$_{\text{dienophile}}$ controlled.[3] Whether, in fact, these reactions are concerted pericyclic processes has yet to be firmly established. In any case, a simple dipolar mechanistic model (Scheme 2-I) can often be used to qualitatively predict the regiochemistry of imino dienophile cyclizations with unsymmetrical dienes, which usually proceed with high selectivity.[2] This model may, in fact, be indicative

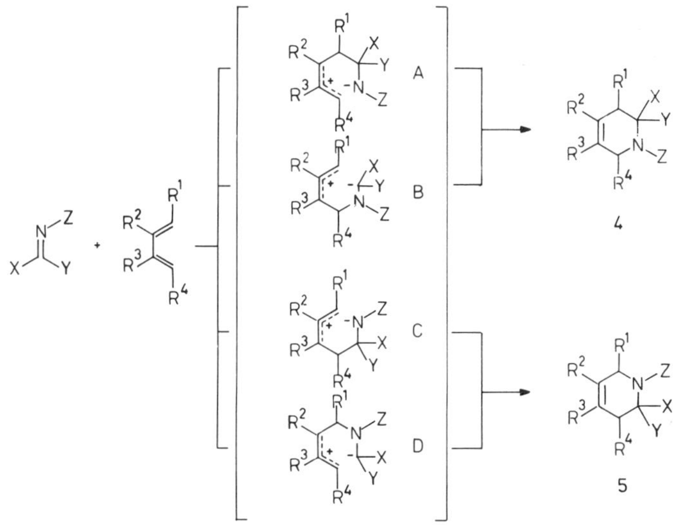

Scheme 2-I

of dipolar character in a concerted pericyclic transition state. For example, in thermal cycloadditions of neutral imines one can consider four possible dipolar forms **A–D**. Forms **A** and **B** lead to regioisomeric tetrahydrophyridine **4**, whereas **C** and **D** give isomer **5**. The major product of the reaction will derive from the most stable of these four possibilities. In general, unless *both* X and Y on the imino dienophile are good carbanion-stabilizing functions, only forms **A** and **C** need be considered. Numerous examples of cycloadditions with unsymmetrical dienes are included in the individual discussions of various types of imino dienophiles.

Imino Diels–Alder reactions also show excellent stereoselectivity, and although extensive studies are lacking some generalizations can be made. The cycloadditions show the usual Diels–Alder syn stereoselectivity with respect to the diene component, which would seem to indicate a concerted, if not synchronous, reaction mechanism. As pointed out by Krow,[4] lone pair inversion of nitrogen in both reactant imine and product tetrahydropyridine intorduces stereochemical ambiguities not present in "all carbon" Diels–Alder reactions. For example, with acyclic imines, one cannot be certain whether the reacting dienophile had the E or the Z configuration and therefore application of the Alder rule of endo addition becomes difficult, if not impossible. Imino dienophiles in which configuration is fixed by virtue of ring restraints usually give stereochemically predictable adducts resulting from endo addition (*vide infra*).

2. N-SULFONYLIMINES

Albrecht and Kresze reported the first examples of [4 + 2] cycloadditions of N-sulfonylimines.[5] Trihalomethylimines **6** and **7**, prepared from chloral or fluoral and p-toluenesulfonamide, reacted with 2,3-dimethylbutadiene to give cycloadducts in high yields [Eq. (1)].

$$
\underset{\substack{6 \quad X = Cl \\ 7 \quad = F}}{\overset{NTs}{\underset{H}{\bigvee}}_{CX_3}} \quad + \quad \bigvee \quad \xrightarrow[\ C_6H_6\ ]{\Delta} \quad \underset{CX_3}{\overset{NTs}{\bigvee}} \quad (1)
$$

These imines also reacted cleanly with cyclic dienes such as cyclopentadiene and 1,3-cyclohexadiene, but in these cases product stereochemistry was not elucidated.

A reinvestigation of these latter reactions by Krow *et al.*[6] established that kinetic addition of trichloromethylimine **6** to cyclopentadiene yields a 78 : 22 ratio of endo adduct **8** to *exo*-**10**. Curiously, with trifluoromethyl-imine 7 the exo isomer 11 is kinetically favored over endo adduct 9 by 57:43. However, with 7 and 1,3-cyclohexadiene a 56:44 ratio of *endo*-12 to *exo*-13 was produced. These workers offered a rationale for these results base on the assumption that an (*E*)-imine is the reactive species and that steric interactions between the trihalomethyl and/or tosyl groups with the 1,4 substituents on the diene are important in determining product stereochemistry.

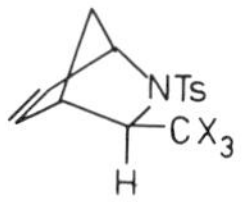

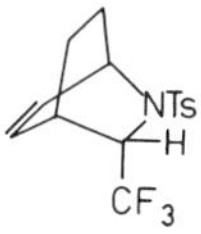

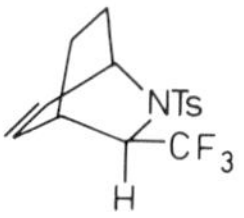

Several examples of cycloadditions of trichloromethylimine 6 with unsymmetrical dienes have been reported,[7-9] and some examples are listed in Table 2-I. These reactions all produce only single regioisomeric products which can be rationalized based on the model shown in Scheme 2-I. In the cases of the dienes in entry 3, the stereochemical outcome has been

TABLE 2-I
Thermal Cycloadditions of Trichloromethyl-*N*-sulfonylimine 6 with Unsymmetrical Dienes

Entry	Diene	Conditions	Product(s)	Yield	Ref.
1			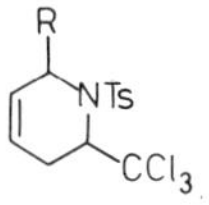		
	R = CH$_3$	C$_6$H$_6$, reflux		72%	7
	R = Ph	C$_6$H$_6$, reflux		70%	7
2	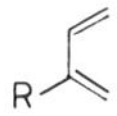		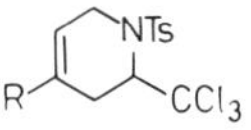		
	R = CH$_3$	C$_6$H$_6$, reflux		82%	7
	R = Ph	C$_6$H$_6$, reflux		86%	7
	R = OCH$_3$	—		"Good"	8
3	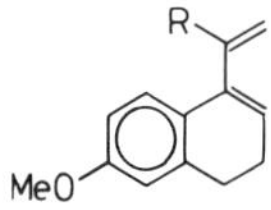		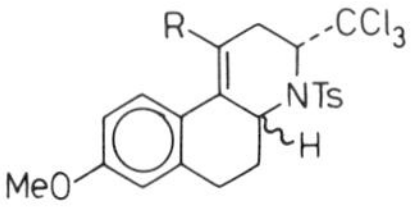		
	R = H	C$_6$H$_6$, 20°C		72% (100% cis)	9
	R = OCH$_3$	C$_6$H$_6$, 20°C		80% (1.5:1 cis:trans)	9

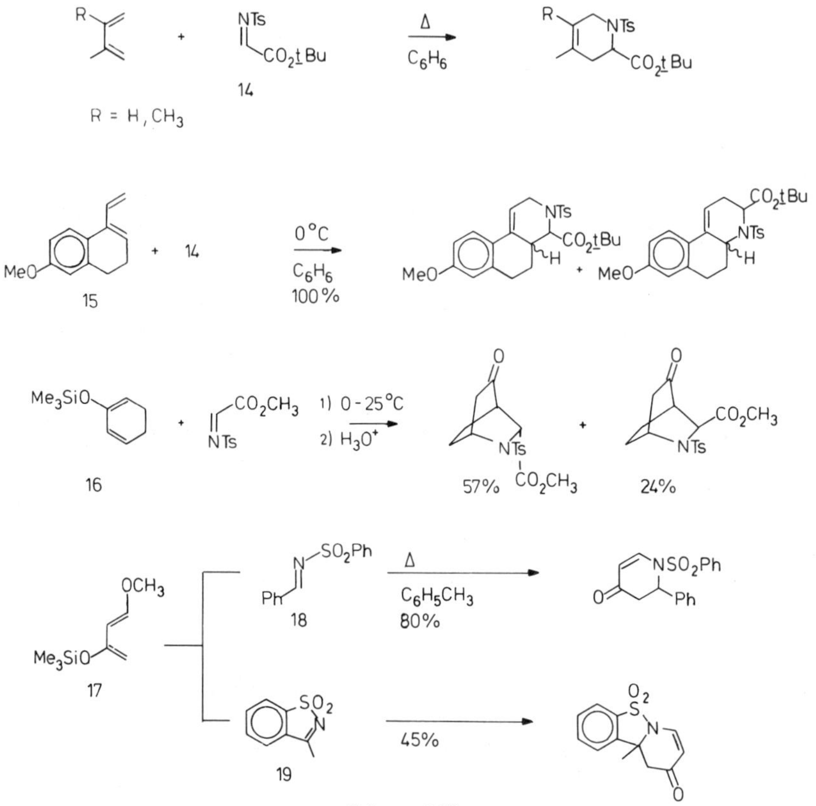

Scheme 2-II

rationalized[9] using arguments similar to those offered by Krow et al.[6] for formation of adducts **8–13**.

Carboxyl-substituted **N**-sulfonylimines have also been used in [4 + 2] cycloadditions, and some examples are shown in Scheme 2-II.[10–12] Imine **14** adds regioselectively to isoprene to give only one adduct, but diene **15** surprisingly afforder a 3 : 1 mixture of adducts. The major exo adduct produced from siloxydiene **16** was recently used in total synthesis of the piperidine alkaloids isoprosopinine B and deoxyprosopinine.[12] The highly oxygenated diene **17** was found to react with both acyclic N-sulfonylimine **18** and cyclic imine **19** to afford cycloadducts.[13]

3. N-ACYLIMINES

Acyclic Systems

N-Acylimines and N-acyliminium species are the most thoroughly studied and widely used class of imino dienophiles.[2] The first reports of Diels–

Adler reaction of this type were by Merten and Muller,[14a,b] who found that biscarbamates react with 1,3-dienes in the presence of a Lewis acid to give *N*-acyltetrahydropyridines [Eq. (2)].[14-18]

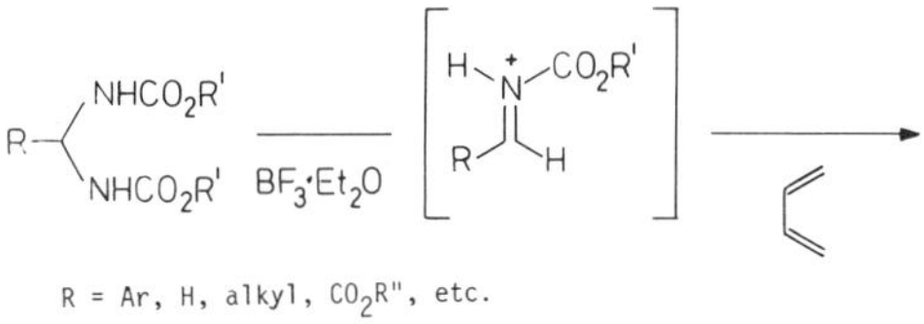

$$R = Ar, H, alkyl, CO_2R'', etc.$$

(2)

A transient *N*-acyliminium intermediate is presumably involved in this process. The structures of these species have been investigated by ^{1}H NMR, and protonated *N*-acylaldimines appear to have the *E* configuration.[19]

In addition to generating *N*-acylimines from biscarbamates, other methods can be used. For example, Lewis acid-catalyzed elimination of methanol from α-alkoxycarbamates [Eq. (3)]

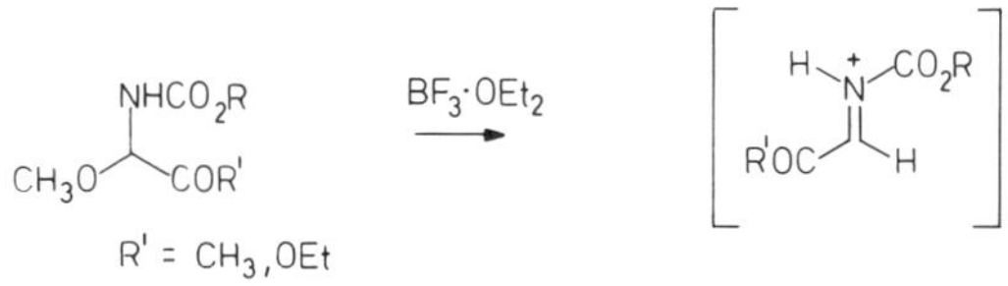

$$R' = CH_3, OEt$$

(3)

can be used to produce *N*-acyliminium intermediates for [4 + 2] cycloadditions.[19-21] One general route to neutral *N*-acylimines (Scheme 2-III) is based on aza-Wittig chemistry.[22,23] Both diacyl and triacyl compounds could be generated *in situ* and used in subsequent Diels–Alder reactions.

Table 2-II lists a number of representative imino Diels–Alder cycloadditions using acyclic *N*-acylimines and *N*-acyliminium dienophiles. In general, cycloadditions using *C,N*-diacylimino systems (entries 1–10) and unsymmetrical dienes show good regioselectivity. The adducts from these reactions have the regiochemistry one would predict based on Scheme 2-I

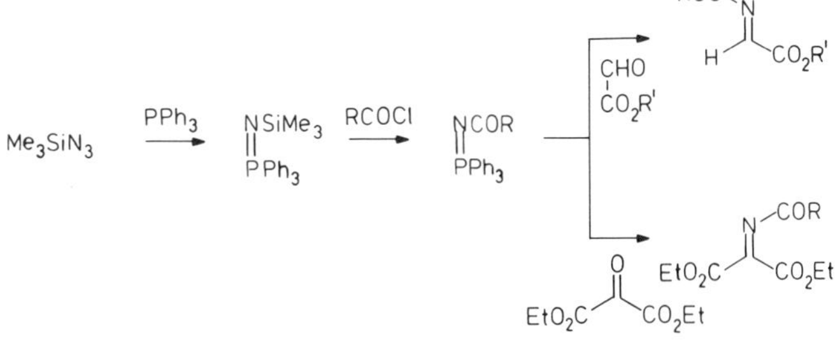

Scheme 2-III

TABLE 2-II

Cycloadditions of Acyclic *N*-Acylimines with 1,3-Dienes

Entry	Imine or precursor	Diene	Conditions	Product(s)	Yield	Ref.
1	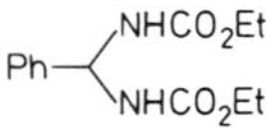 Ph⟨NHCO$_2$Et / NHCO$_2$Et	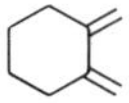 (methylenecyclohexane diene)	BF$_3$, C$_6$H$_6$, Δ	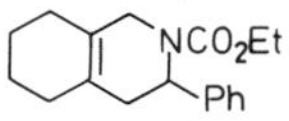 (NCO$_2$Et, Ph bicyclic)	80%	14a
2	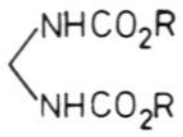 ⟨NHCO$_2$R / NHCO$_2$R R = Et	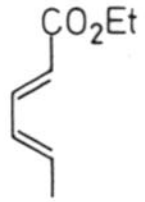 CO$_2$Et diene	BF$_3$, C$_6$H$_6$, Δ	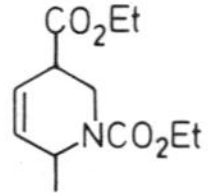 CO$_2$Et, NCO$_2$Et ring	Not specified	14a
3	R = CH$_2$Ph	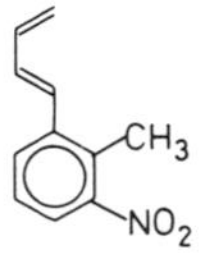 CH$_3$, NO$_2$ aryl diene	BF$_3$, C$_6$H$_6$, Δ	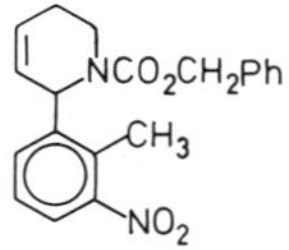 NCO$_2$CH$_2$Ph, CH$_3$, NO$_2$	52%	14c
4	R = Et	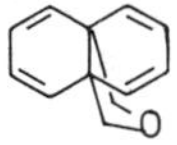 (tricyclic diene with O)	BF$_3$, C$_6$H$_6$, Δ, 20 hr	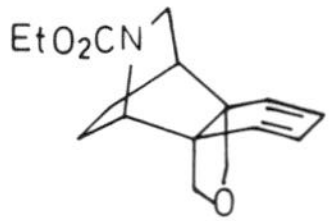 EtO$_2$CN bicyclic	69%	14d
5	R = Et	 (cyclohexadiene)	BF$_3$, C$_6$H$_6$, Δ	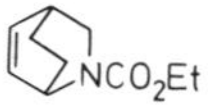 NCO$_2$Et bicyclic	27%	15
6	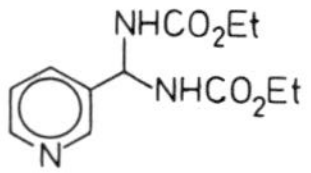 NHCO$_2$Et / NHCO$_2$Et pyridyl	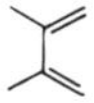 (2,3-dimethylbutadiene)	BF$_3$, glyme, Δ	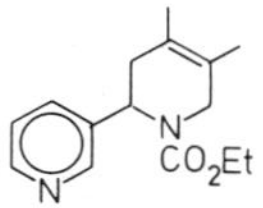 N CO$_2$Et ring, pyridyl	55%	16
7	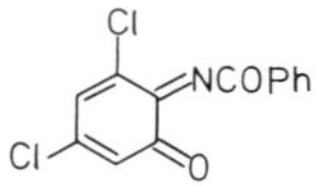 Cl, Cl NCOPh quinone	 (isoprene-type diene)	CH$_2$Cl$_2$, RT	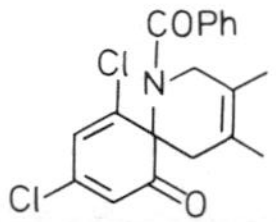 COPh N, Cl, Cl	92%	17

40

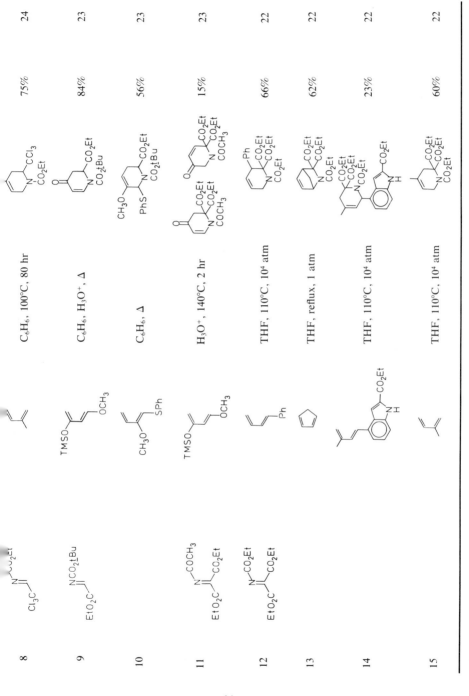

and the dipolar forms **A** and **C**. The triacylimino dienophile type shown in entries 11–15 also shows good regioselectivity in many of the cases reported to date. However, predicting the regiochemistry of addition with this dienophile is not straightforward, since it appears that products can formally arise from all four dipolar forms **A–D** (Scheme 2-I). What is also interesting is that this dienophile is quite reactive with highly electron-rich dienes, but in other cases requires high pressure to effect cycloaddition.

Cycloadditions with acyclic *N*-acylimines show reasonably good kinetic stereochemical control. In an important series of papers, Krow *et al.*[4,25] have thoroughly investigated the stereochemistry of the Diels–Alder reaction of 1,3-cyclohexadiene with a large number of C-substituted *N*-acyliminium dienophiles (**20**) [Eq. (4)].

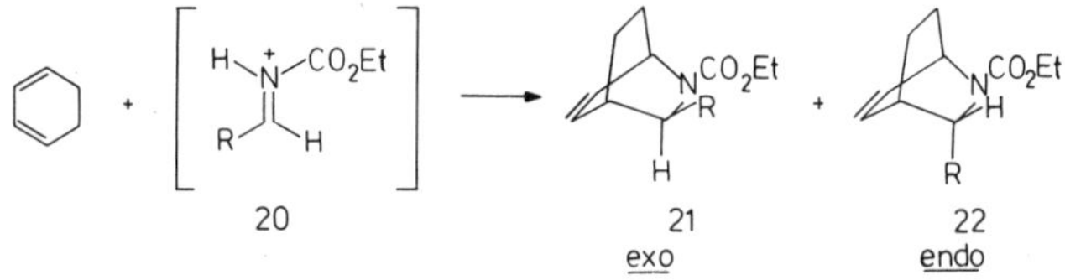

$$\text{(4)}$$

These imines were generated *in situ* from the precursors shown in Eqs. (2) and (3) using BF_3 as catalyst, and the ratios of exo and endo adducts were determined for a variety of substituents (R). The results of this work are summarized in Table 2-III. In general, exo adducts **21** are usually favored

TABLE 2-III

Stereochemistry of the Boron
Trifluoride-Catalyzed Cycloaddition of
Mono-C-Substituted *N*-Acylimines with
1,3-Cyclohexadiene

R in Imine **20**	Exo adduct **21**	Ref.
CO_2Et	70%	21
CO_2CH_3	80%	20, 21
Ph	80%	4
pNO_2Ph	80%	4
pCH_3Ph	85%	4
$COCH_3$	67%	4, 18
CH_3	82%	25, 26
$(CH_2)_3Cl$	81%	25, 27
Et	81%	25
*i*Bu	79%	25
*i*Pr	36%	25
cyclohexyl	54%	25
CCl_3	38%	24, 28

by about 4 : 1, except in those cases where R is α-branched or is highly substituted. These results can be rationalized by assuming that an (*E*)-*N*-acyliminium compound[19] is involved in the cycloaddition and that the reaction is concerted.[4,25] If this is the case, then the acyl group on the imine nitrogen is endo in the Diels–Alder transition state in preference to almost all other substituents on the carbon terminus. Whether this preference derives from secondary orbital interactions[21] or from some other effect has not been established.[29] When the group R on the carbon of the imine is bulky, it appears that steric factors become important in controlling stereochemistry. Interaction of R with the ethano bridge of the diene is destabilizing in a transition state with the *N*-acyl group endo, and thus a reversal of stereochemistry occurs. This same type of rationale has been used to explain the stereochemical results obtained with adducts **8–13**.[6,9]

Cyclic Systems

Cyclic *N*-acylimines have been used effectively as dienophiles in imino Diels–Alder reactions. The most common type of cyclic imino dienophile is dehydrohydantoin [Eq. (5)],

$$
\underset{\substack{\text{HN} \quad \text{NR} \\ \text{O}}}{\overset{\text{CH}_3\text{O} \quad \text{O}}{\diagup}} \xrightarrow[\substack{\text{acid} \\ -\text{CH}_3\text{OH}}]{\Delta \text{ or}} \left[\underset{\substack{\text{N} \quad \text{NR} \\ \text{O}}}{\overset{\text{O}}{\diagup}} \right] \xleftarrow[\substack{2)\ \text{R}_3'\text{N}}]{1)\ \underline{t}\text{BuOCl}} \underset{\substack{\text{HN} \quad \text{NR} \\ \text{O}}}{\overset{\text{O}}{\diagup}} \qquad (5)
$$

which can be generated *in situ* from a 4-methoxyhydantoin[30] thermally or under acid catalysis.[31] Alternatively, the dehydrohydantoin can be produced by a chlorination–dyhydrochlorination route from a monosubstituted hydantoin.[32] Table 2-IV shows a number of representative examples of [4 + 2] cycloadditions using dehydrohydantoins. In general, these reactions show a high degree of stereo- and regioselectivity. Invariably, product stereochemistry derives from a transition state in which the two carbonyl groups of hydrohydantoin are *endo* to the diene, as can be seen in several cases contained in the table. The regiochemistry of these cycloadditions is again consistent with Scheme 2-I if one considers dipolar forms **A** and **C**.

The regiochemical features of this process can be seen in the results of a study by Kim and Weinreb using several highly substituted dienes and a dehydrohydantoin precursor (Scheme 2-IV).[34] It was found that as the substituent R becomes more electron donating, the ratio of regioisomeric adducts changes in the order one would expect based on the Scheme 2-I

TABLE 2-IV

Cycloadditions of Dehydrohydantoins

Dehydrohydantoin precursor	Diene	Conditions	Product	Yield	Re:
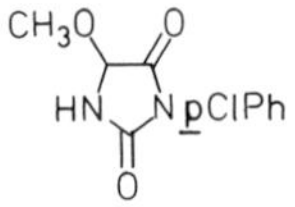		C_6H_6, 170°C, 3 days	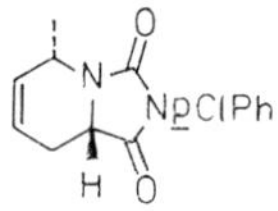	61%	3
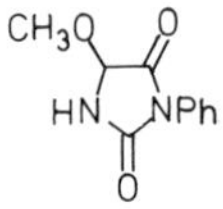	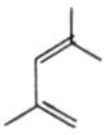	C_6H_6, 170°C, 3 days	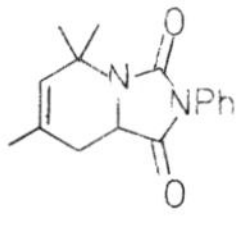	70%	3
		C_6H_6, ArSO$_3$H, Δ		75%	3
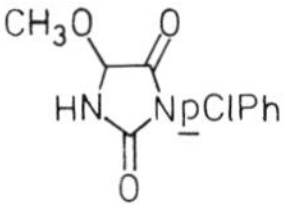		TFA, C_6H_6 C_6H_6, 170°C, 3 days	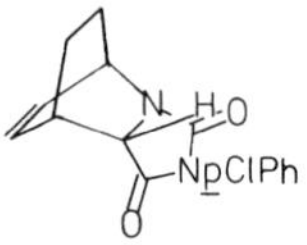	30% 74%	3 3
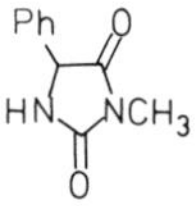		(1) tBuOCl, C_6H_6; (2) Et$_3$N, 25°C	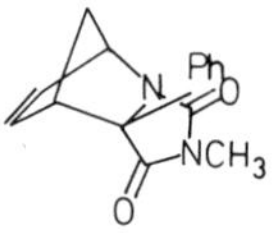	68%	3
		(1) tBuOCl, C_6H_6; (2) Et$_3$N, 25°C	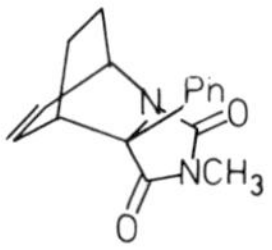	70%	3
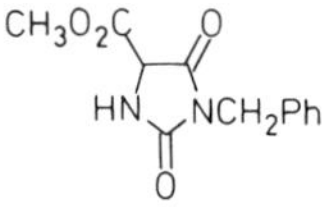		(1) tBuOCl, C_6H_6; (2) Et$_3$N, 25°C	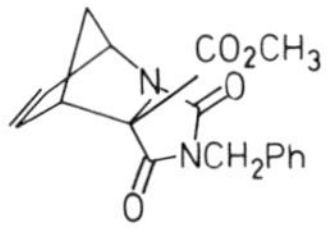	79%	3
	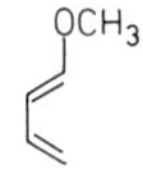	(1) tBuOCl, C_6H_6; (2) Et$_3$N, 25°C	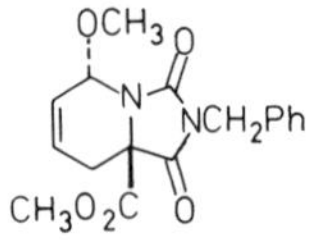	72%	3
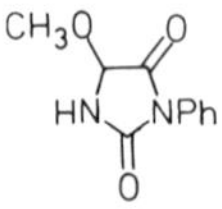	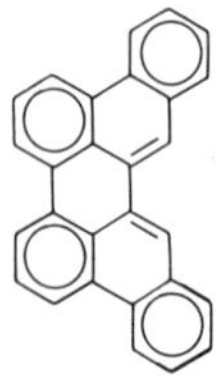	TFA, C_6H_6, Δ	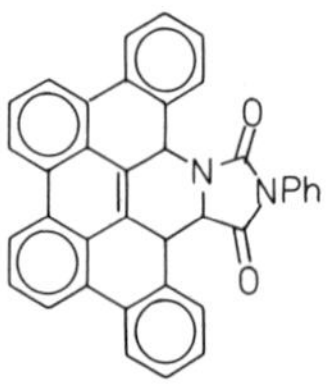	95%	3

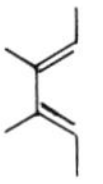

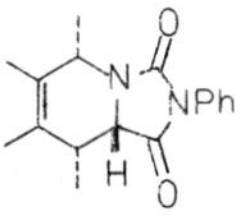

Ar = 4-ClC$_6$H$_4$		ratio		
R = CO$_2$CH$_3$		0	:	100
= H		8	:	92
= CH$_2$OAc		55	:	45
= CH$_3$		67	:	33
= CH$_2$CH$_3$		75	:	25

Scheme 2-IV

hypothesis. In all cases, *only* the stereoisomers derived from endo addition were produced.

This methodology has recently been applied to a total synthesis of the antitumor antibiotic streptonigrin (**24**).[35] On cycloaddition with a dehydrohydantoin, diene **23** afforded a 3 : 1 mixture of regioisomeric adducts (Scheme 2-V). The major isomer was aromatized to a pyridine ester, which was transformed in several steps to the natural product.

Ben-Ishai and co-workers have developed syntheses of a number of cyclic *N*-acylimines other than hydantoins.[36-40] Several of these compounds were shown to be reactive imino dienophiles, and some examples

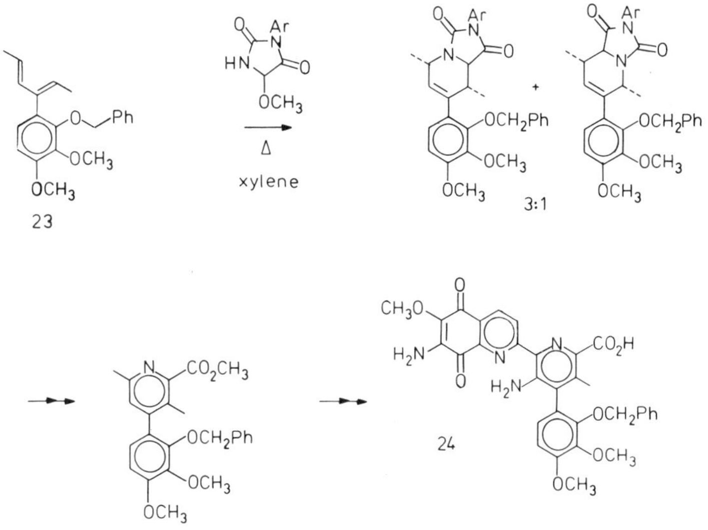

Scheme 2-V

TABLE 2-V

Cycloadditions of Various Cyclic *N*-Acylimines

Entry	Imine or precursor	Diene	Conditions	Product	Yield	Ref.
1	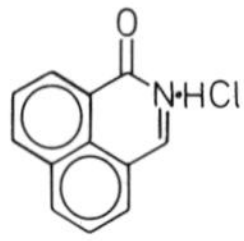	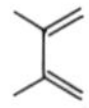	C_6H_6, RT	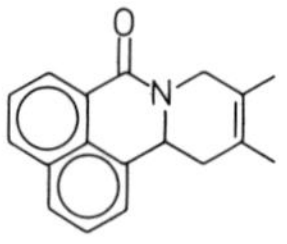	32%	36
2	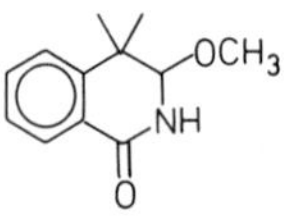		TFA, C_6H_6, Δ	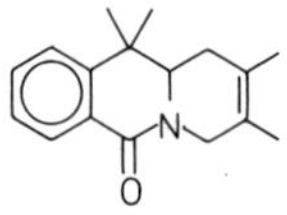	53%	37
3	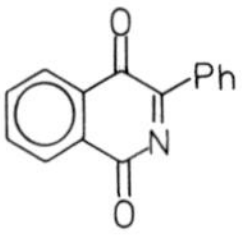		TFA, C_6H_6	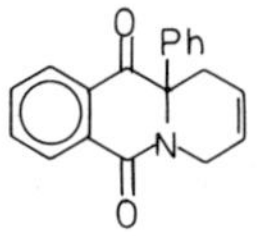	77%	37
4	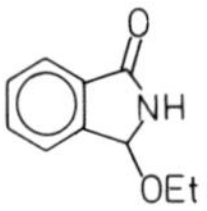	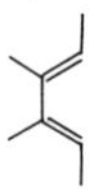	TFA, C_6H_6, Δ	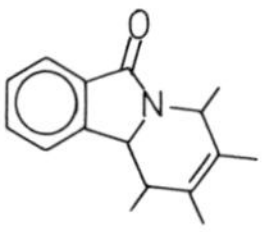	41%	38
5	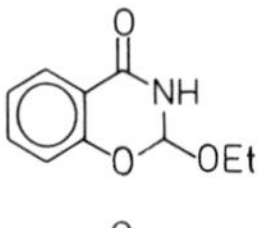		$BF_3 \cdot Et_2O$, Et_2O, RT	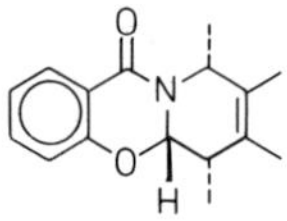	86%	39
6	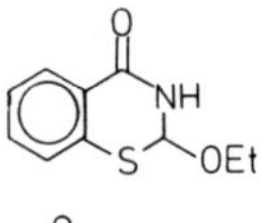	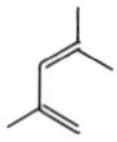	$BF_3 \cdot Et_2O$, Et_2O, Δ	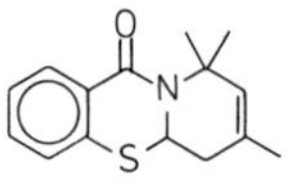	42%	40
7	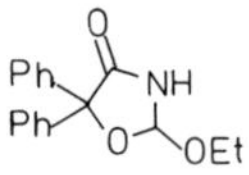		$BF_3 \cdot Et_2O$, Et_2O	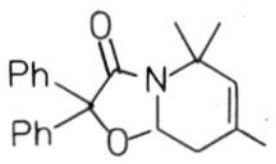	72%	40
8	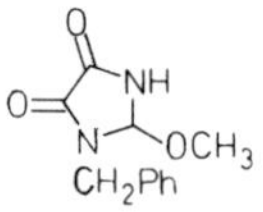	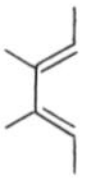	$ArSO_3H$, C_6H_6, Δ	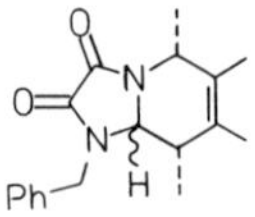	50%	40
9	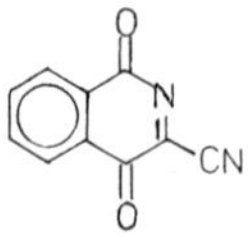		C_6H_6, 25°C	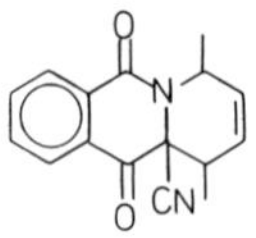	95%	41
10	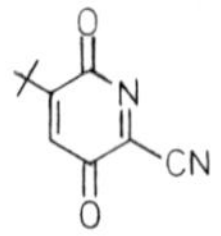		C_6H_6, 25°C	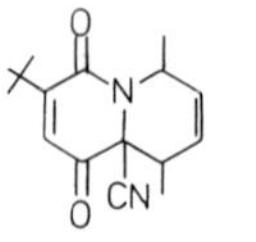	98%	41

TABLE 2-V (Continued)

Entry	Imine or precursor	Diene	Conditions	Product	Yield	Ref.
11			C_6H_6, 25°C	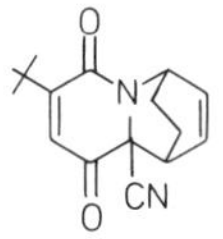	82%	41
12			C_6H_6, 25°C	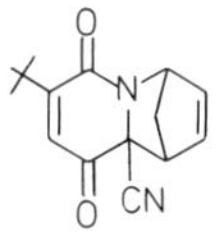	98%	41

are listed in Table 2-V (entries 1–8). It should be noted that stereochemical and regiochemical preferences have not been established for most of these dienophiles, and they have found little use to date.

Moore *et al.* have prepared some 2-cyanoaza-*p*-quinones via an interesting thermal rearrangement of *o*-diazido-*p*-quinones [Eq. (6)].[41]

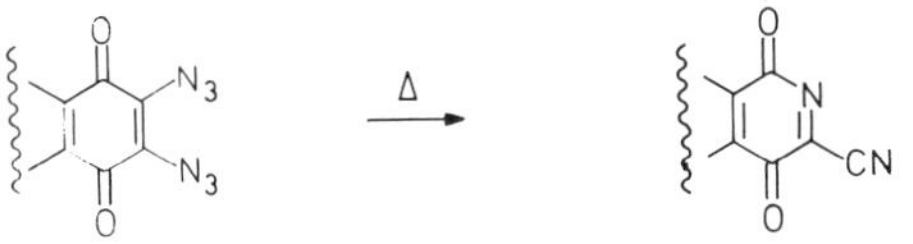

$$\tag{6}$$

These azaquinones are very reactive imino dienophiles. Some examples of cycloadditions of this type are shown in Table 2-V (entries 9–12).

4. C-ACYLIMINES

Only a few scattered examples of Diels–Alder cycloadditions of this type of dienophile have been described to date, and not all *C*-acylimines appear to be reactive. The indolone shown in Eq. (7)

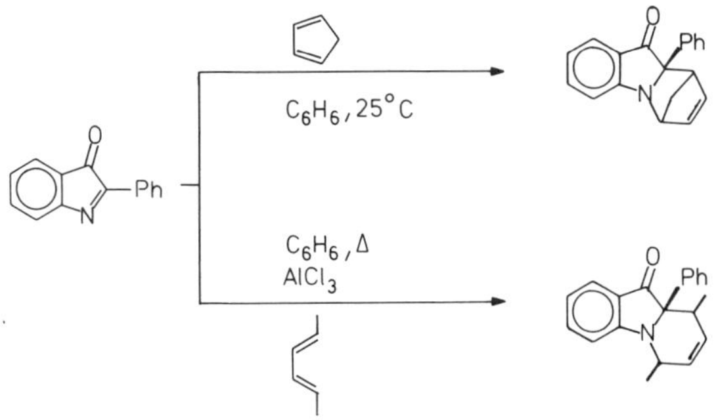

$$\tag{7}$$

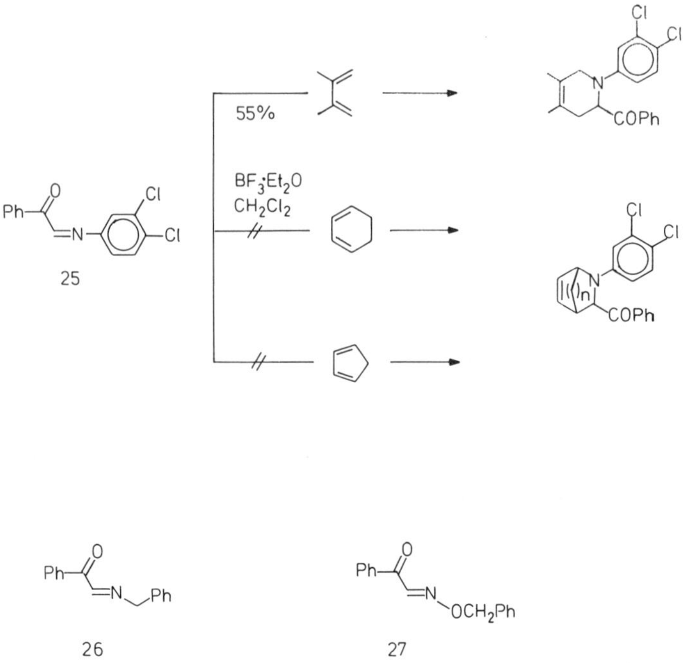

Scheme 2-VI

acts as an imino dienophile, but yields for these reactions were not reported, nor was product stereochemistry proven.[42]

McKay and Proctor prepared imine **25** and reported that under BF$_3$ catalysis (but not thermally) it combines with 1,3-dienes to give adducts as shown in Scheme 2-VI.[43] However, a recent reinvestigation of these reactions suggests that while the adduct from 2,3-dimethylbutadiene does have the assigned structure, the proposed structures from the cyclic dienes are *not* correct.[44] Rather, products of electrophilic substitution are obtained. McKay and Proctor noted that imines **26** and **27** were unreactive as dienophiles both thermally and under Lewis acid catalysis.[43]

Trost and Whitman have described the transient formation of a diimine [Eq. (8)].

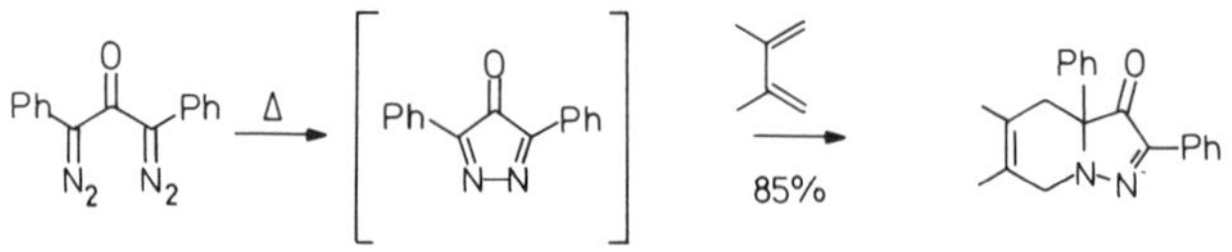

$$(8)$$

They found that it reacted in Diels–Alder fashion with 2,3-dimethylbutadiene.[45]

5. IMINIUM SALTS AND NEUTRAL IMINES

Cycloadditions of iminium compounds with 1,3-dienes to afford Diels–Alder adducts seems to be a reaction of general applicability.[46] On the other hand, simple electron-rich imino species are less reliable dienophiles. Some electron-rich imines will undergo Diels–Alder reactions under acid catalysis and thus are probably actually reacting as iminium salts. However, inverse electron demand [4 + 2] cycloadditions of neutral imines with electron-deficient dienes have been reported, as have additions to exceptionally reactive dienes (vide infra).

Böhme et al. described the initial example of a [4 + 2] cycloaddition of an iminium salt [Eq. (9)].[47]

$$\text{Et}_2\text{NCH}_2\text{X} \rightleftharpoons \left[\overset{+}{\text{Et}_2\text{N}} = \text{CH}_2 \quad \text{X}^- \right] \xrightarrow{69\%} \quad (9)$$

$$X = \text{Br, Cl}$$

In this case, an α-haloamine was used to generate the iminium dienophile in situ.

In a related series of reactions, Babayan et al.[48] found that aminals, on treatment with acetyl chloride, generated iminium chlorides which could be trapped with isoprene to give Diels–Alder adducts [Eq. (10)].

$$\text{R}_2\text{NCH}_2\text{NR}_2 \xrightarrow[\text{CH}_2\text{Cl}_2]{\text{CH}_3\text{COCl}} \left[\overset{\text{H}}{\underset{\text{H}}{}} \overset{+}{\text{NR}_2} \quad \text{Cl}^- \right] \longrightarrow \xrightarrow{^-\text{OH}} \quad (10)$$

$$R = CH_3, \text{Et, Bu, } (CH_2)_5$$

The adducts themselves were not isolated but were fragmented with base to afford amino-substituted dienes in mediocre yields.

Electron-rich oxygenated dienes react readily with iminium salts and with imines under Lewis acid catalysis.[49–51] Some examples are shown in Table 2-VI. In all cases, only one regioisomer was detected in the reaction, and in every instance product structure is in agreement with predictions based on Scheme 2-I.

The Danishefsky group has examined the uncatalyzed and Lewis adic-catalyzed cycloadditions of highly oxygenated dienes with simple imines in preparation of yohimbine[50b] analogs and the indolizidine alkaloid ipalbidine (Scheme 2-VII).[50c] What is quite interesting is that some of these dienes are sufficiently reactive that catalysis was not necessary with neutral imines. Also, the reaction seems to show excellent facial stereoselectivity with respect to the imine.

TABLE 2-VI

Cycloadditions of Electron-Rich Dienes with Imines and Iminium Salts

Imino dienophile	Diene	Conditions	Product(s)	Yield	Ref.
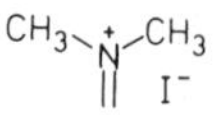	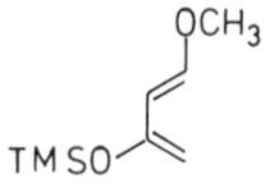	RT	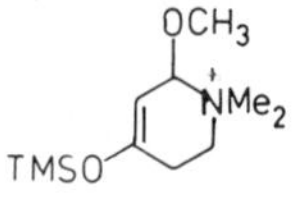	95%	49
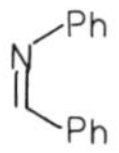	4.3 equiv	ZnCl₂, THF, RT	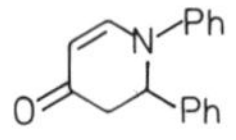	62%	50a
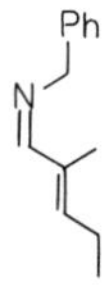	2.3 equiv	ZnCl₂, THF, RT	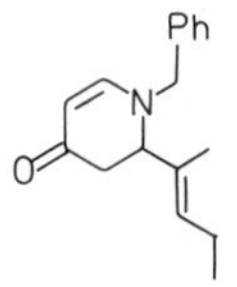	47%	50a
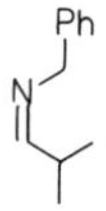	3.8 equiv	ZnCl₂, THF, RT	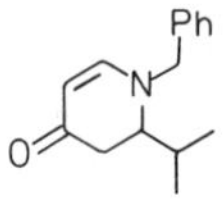	69%	50a
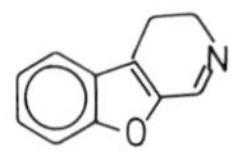	1.3 equiv	ZnCl₂, THF, RT	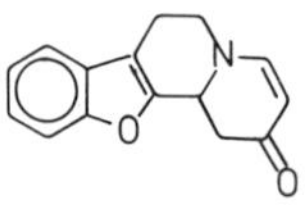	73%	51a,b
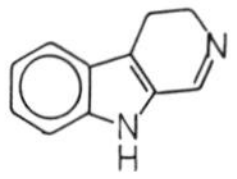	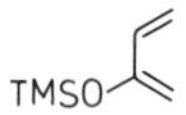	ZnCl₂, CH₃CN, 50°C	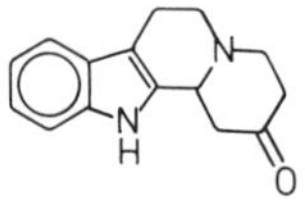	55%	51a,b
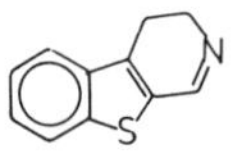		ZnCl₂, CH₃CN, 50°C	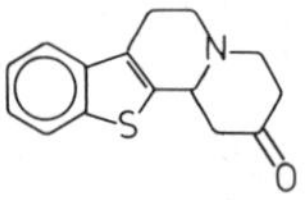	30%	51a,b
	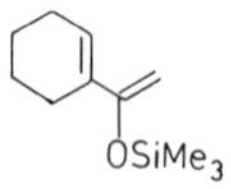	BF₃·Et₂O, CH₂Cl₂, −78°C, 5 min	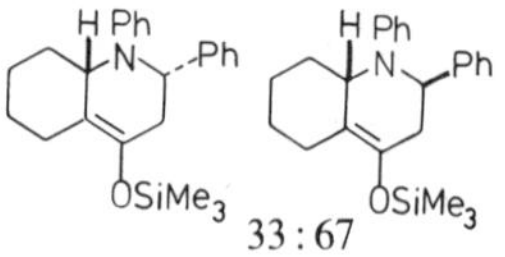 33 : 67		51c

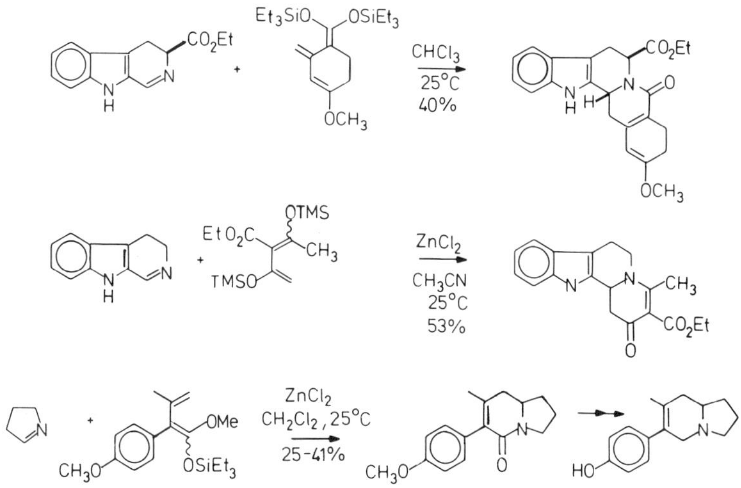

Scheme 2-VII

Recently, Larsen and Grieco discovered that iminium salts generated under aqueous Mannich-like conditions will undergo Diels–Alder reactions.[52] Thus, if one combines an amine hydrochloride, a 1,3-diene, and an electrophilic aldehyde at temperatures ranging from 25 to 55°C, [4 + 2] cycloadducts can be isolated in reasonable yields. Several examples of this process are shown in Table 2-VII. With optically active α-methylbenzylamine hydrochloride a 4 : 1 mixture of diastereomeric adducts was obtained (entry 5). The reaction could also be effected intramolecularly, as noted in entries 6 and 7 of Table 2-VII. Although acetaldehyde could be used successfully, it did not work as cleanly as formaldehyde, and other carbonyl components such as acetone were not at all useful in the intermolecular reaction. However, α-ketoaldehydes are useful substrates. Interestingly, the cycloaddition with (*E,E*)-hexadiene gave only one stereoisomeric product, which seems to point to a concerted pericyclic mechanism, rather than a stepwise ionic one, for the cycloaddition. Furthermore, the regiochemistry of the cycloaddition is again in accord with the predictions of Scheme 2-I.

A few additional imino Diels–Alder reactions have been reported which involve iminium salts. Böhme and Ahrens found that the perchlorate salt in Eq. (11)

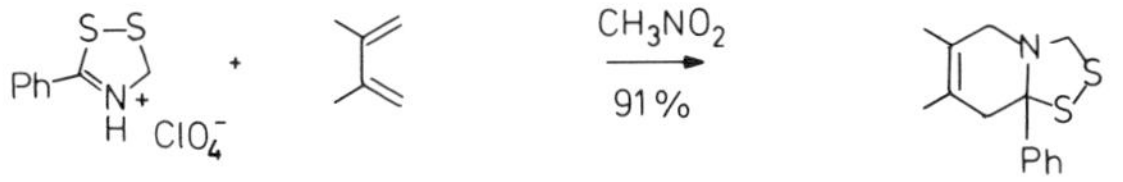

$$\tag{11}$$

TABLE 2-VII

Cycloadditions of Iminium Salts Generated under Aqueous Mannich Conditions[a]

Entry	Amine	Diene	Conditions	Product(s)	Yield
1	PhCH$_2$NH$_2$·HCl		37% HCHO, H$_2$O	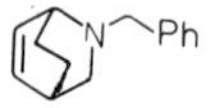	35%
2		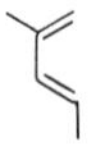	37% HCHO, H$_2$O	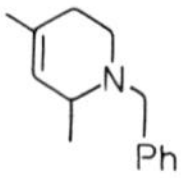	45%
3		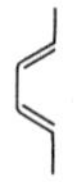	37% HCHO, H$_2$O	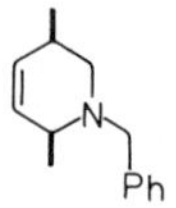	62%
4	NH$_4$Cl		37% HCHO, H$_2$O		44%
5	HCl·H$_2$N	37% HCHO, H$_2$O	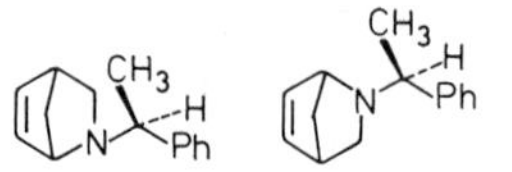 4:1	86%
6	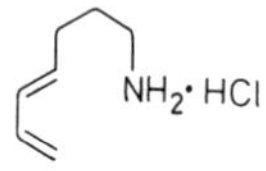	—	37% HCHO, H$_2$O, 50°C	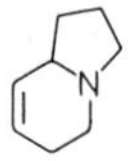	95%
7	PhCH$_2$NH$_2$·HCl		H$_2$O, 70°C	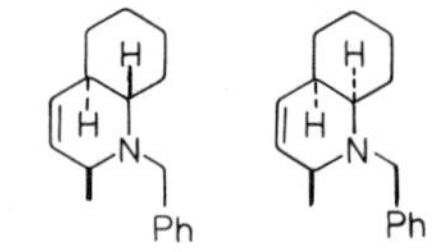 2.5:1	63%
8	CH$_3$NH$_2$·HCl		PhCOCHO, H$_2$O, RT, 22 hr	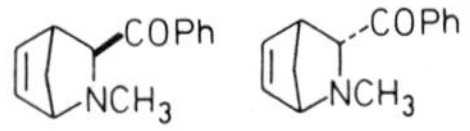 1:4.2	82%
9	PhCH$_2$NH$_2$·HCl		CH$_3$COCHO, H$_2$O, RT, 18 hr	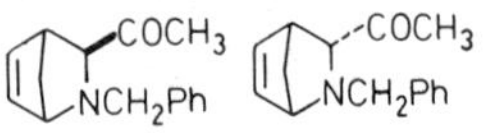 1:1.9	86%

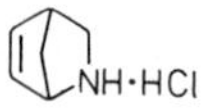

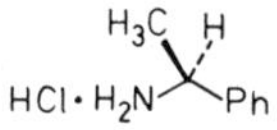

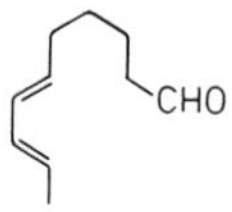

[a] Adapted from Ref. 52.

reacts with 2,3-dimethylbutadiene to give a [4 + 2] cycloadduct.[53] The free base derived from this salt was unreactive as a dienophile.

An imino cumulene acts as an imino dienophile with a few cyclic dienes as shown in Scheme 2-VIII.[54] However, with 2,3-dimethylbutadiene a mixture of [4 + 2] and [2 + 2] cycloadducts were produced. With butadi-

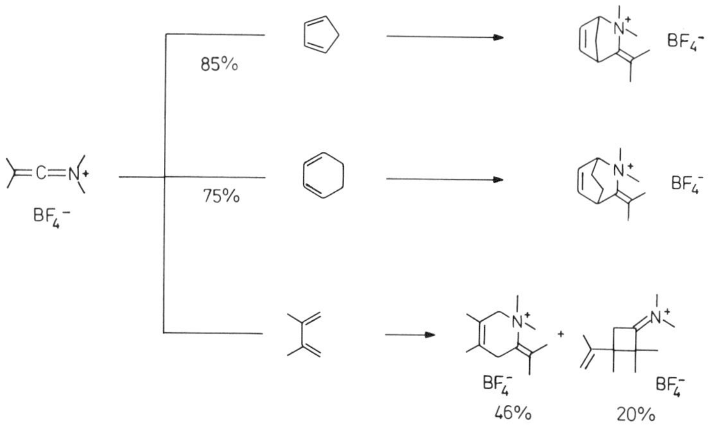

Scheme 2-VIII

ene, isoprene, and piperylene, *only* [2 + 2] cycloadducts were formed. Thus, this iminium salt does not seem to be a generally useful dienophile.

Several examples are known of neutral, electron-rich imines reacting in [4 + 2] cyclodditions with electron-deficient dienes. Bohlmann *et al.* found that simple imines react thermally with butadienecarboxylates to afford Diels–Alder-type adducts [Eq. (12)].[55]

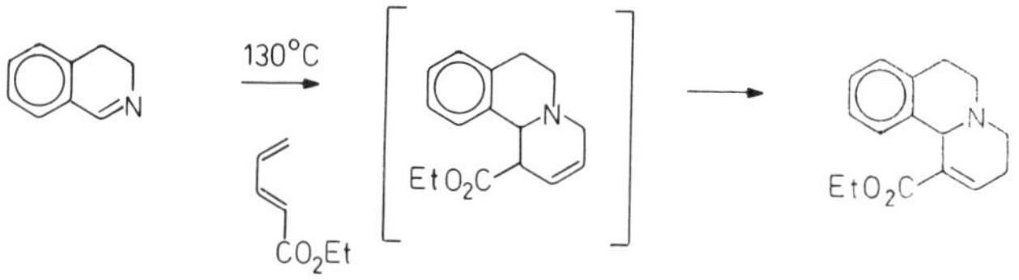

In the cases reported, only the conjugated esters, presumably formed by isomerization of the initial unconjugated [4 + 2] adducts, were isolated.

More recently, Langlois and co-workers have adapted this methodology to total synthesis of some complex indole alkaloids.[56] Cyclic imines (Scheme 2-IX) reacted thermally in good yields with methylbutadiene-1-carboxylate to afford a mixture of Diels–Alder adducts. Alkylation of the enolate derived from the mixture of isomers gave a β, γ-unsaturated ester stereospecifically. This type of compound was coverted to vindoline (**28**, R = OMe) and vindorosine (**28**, R = H). These workers found that 1-cyanobutadiene is also useful in this sort of cycloaddition [Eq. (13)]

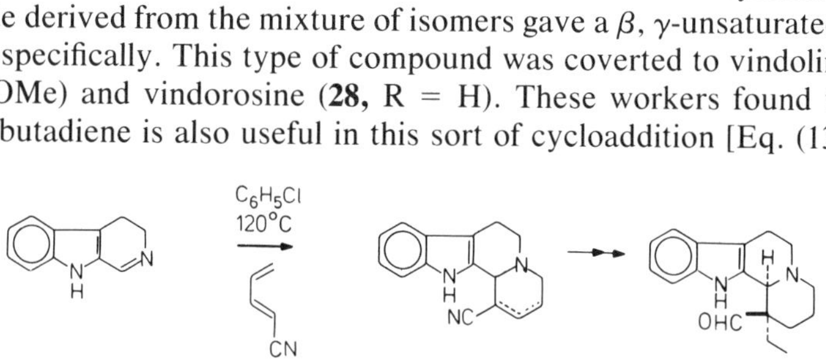

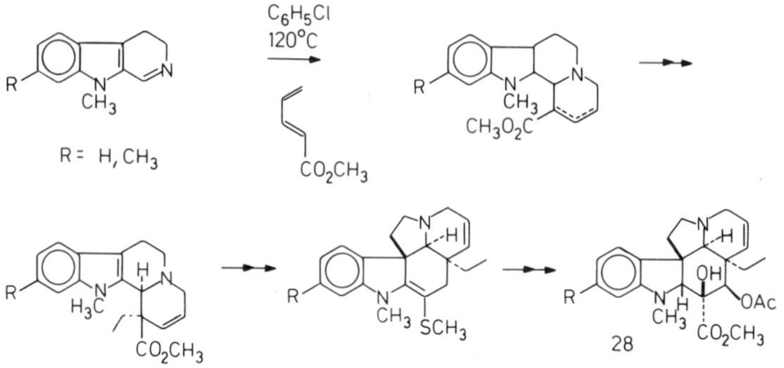

Scheme 2-IX

to generate intermediates for *Eburnane* alkaloid synthesis.[56b]

Several examples exist of inverse electron demand Diels–Alder reactions involving electron-deficient *sym*-tetrazines acting as dienes. However, the neutral, electon-rich imino compounds involved in these cycloadditions do not generally appear to be useful dienophiles with other types of dienes.

As shown in Eq. (14),

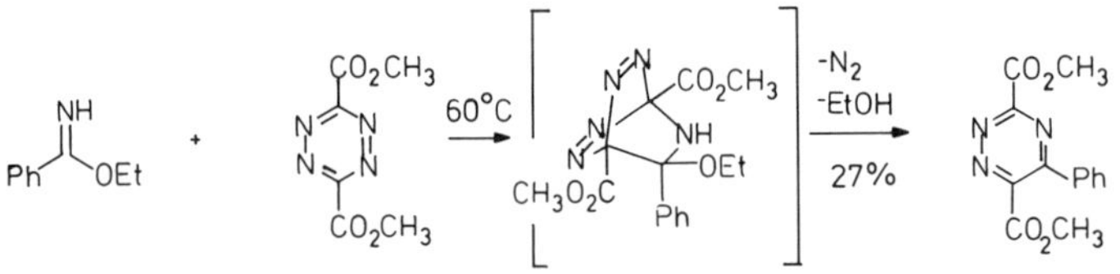

$$(14)$$

an *sym*-tetrazine dicarboxylate will react with an imino ether, presumably via a [4 + 2] cycloadduct, to produce a triazine, although the yield in this case was poor.[57] In a related series of reactions, amidine **29** was found to combine with several disubstituted tetrazines to yield a variety of heterocyclic products, depending on the substitution pattern of the 4π component.[58] The results of this work are outlined in Table 2-VIII. *asym*-Triazines **30** and **31** arise via a Diels–Alder addition producing an unstable bicyclic intermediate. The *sym*-triazines **32** and **33** have been shown to arise by further Diels–Alder reactions of excess amidine **29** with *asym*-triazines **30** and **31** [Eq. (15)].

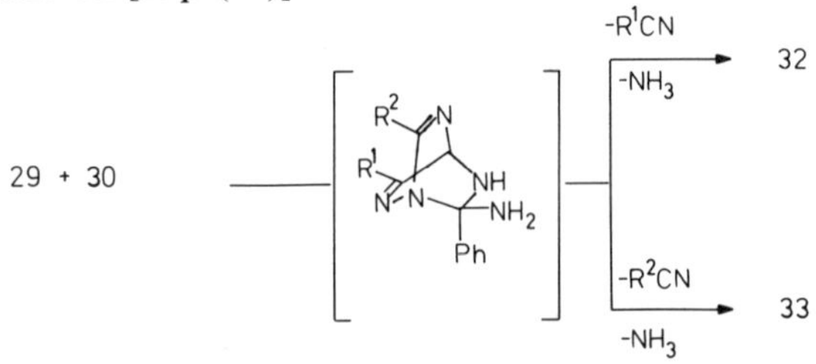

$$(15)$$

TABLE 2-VIII
Synthesis of *asym*-Triazines and *sym*-Triazines

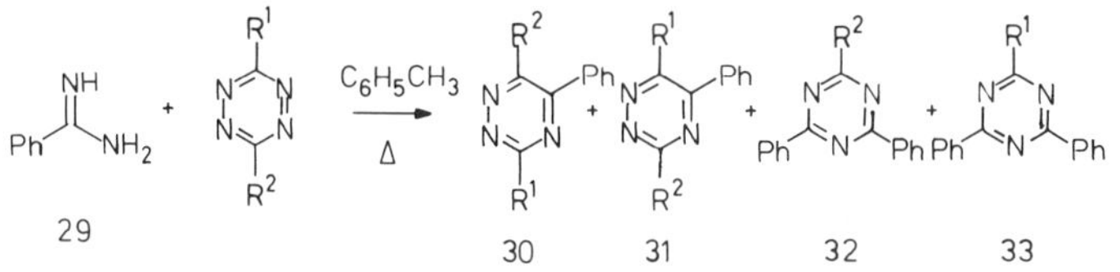

R^1	R^2	30	31	32	33
2-Pyridyl	2-Pyridyl	33%		5%	
Ph	Ph	34%		—	
Ph	2-Pyridyl	7.5%	30%	28%	6%
Ph	Me	—	70%	—	7%
2-Pyridyl	Me	—	35%	—	—

Several aldehyde N,N-dialkylhydrazones and some related imino compounds have been reported to add thermally in [4 + 2] fashion to *sym*-tetrazines to give adducts in good yields [Eq. (16)].[59-63]

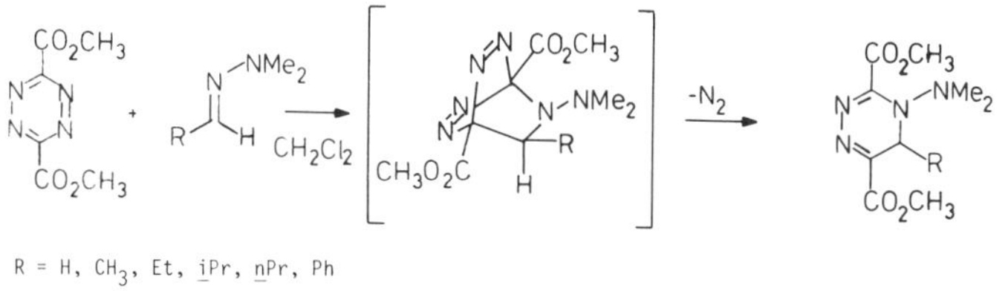

A bicyclic Diels–Alder adduct is presumably in intermediate in these reactions although it was not actually observed. Interestingly, ketone hydrazones react with the tetrazine via their enamine tautomers, giving diazenes as products.

In a series of papers, Kametani and co-workers reported that *o*-quinodimethanes are sufficiently reactive dienes to combine with neutral, electon-rich cyclic and acyclic imines to produce Diels–Alder adducts.[64-67] In all cases, the *o*-quinodimethane was generated by thermal opening of a substituted benzocyclobutene. Several examples of the transformations which could be effected are shown in Scheme 2-X.

In an analogous system, it was found that a phenanthrene *o*-quinodimethane, generated by a reductive process, acted as a diene in [4 + 2] cycloadditions with two neutral imines to give pentacyclic adducts, although yields were rather low [Eq. (17)].[68]

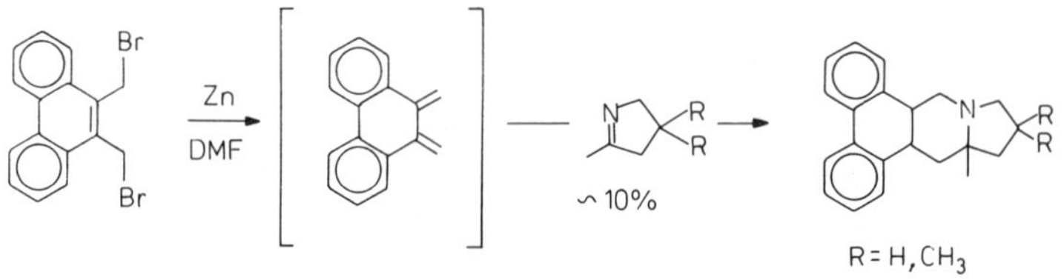

$$(17)$$

$$R = H, CH_3$$

One other isolated example of a neutral imine which was reported to undergo a [4 + 2] cycloaddition is shown in Eq. (18).[69]

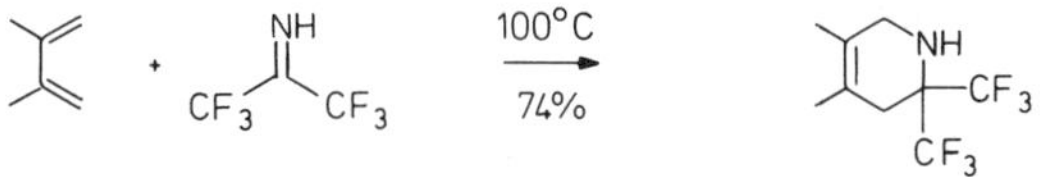

$$(18)$$

Presumably, there is some activation by the electron-withdrawing trifluoromethyl groups in this case.

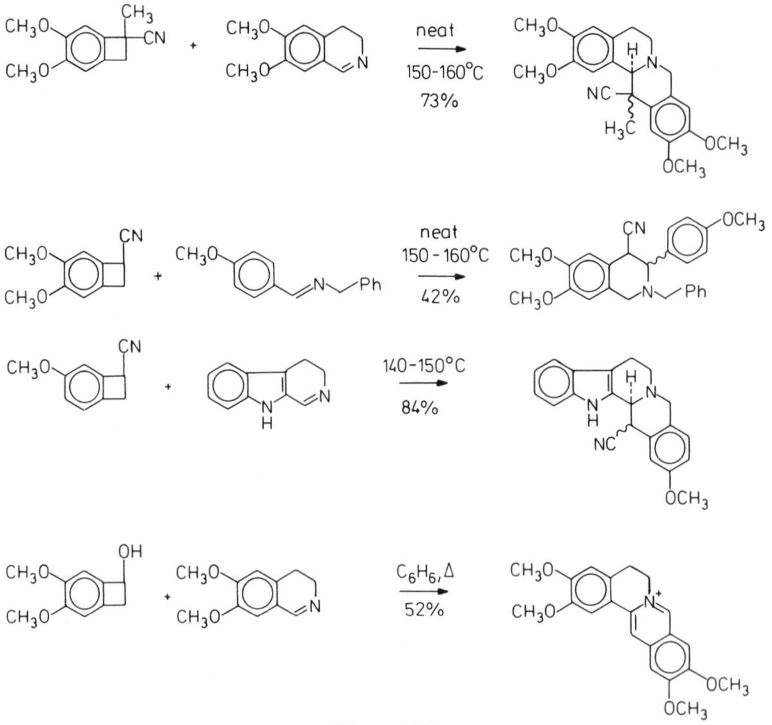

Scheme 2-X

6. AZIRINES

1-Azirines constitute another class of neutral, electon-rich imines which will undergo [4 + 2] cycloadditions with certain types of dienes. As can be seen in the following discussion, the initial adducts of these reactions are often not isolable and that a Diels–Alder reaction has, in fact, occurred is sometimes inferred by mechanistic reasoning.

Several examples have been published of thermal cycloadditions of aryl- and alkyl-substituted 1-azirines to symmetrically substituted cyclopentadienones, as outlined in [Eq. (19)].[70–74]

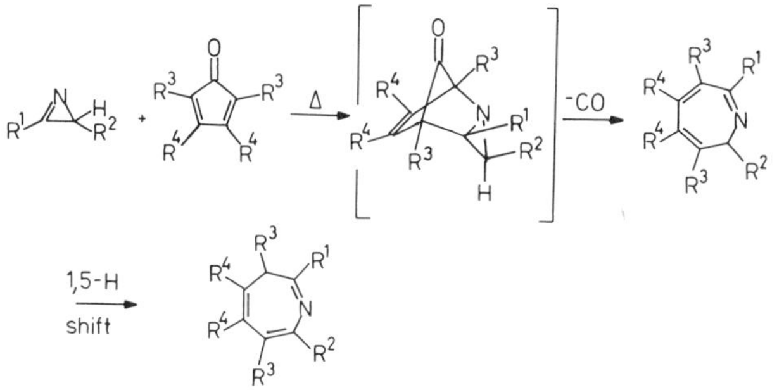

(19)

TABLE 2-IX
Thermal Cycloadditions of Substituted 1-Azirines and Cyclopentadienones[a]

Azirine	Cyclopentadienone	Conditions	Product	Yield
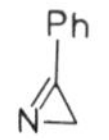	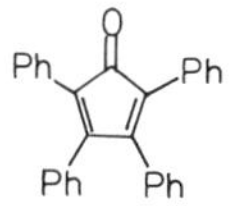	Toluene, Δ	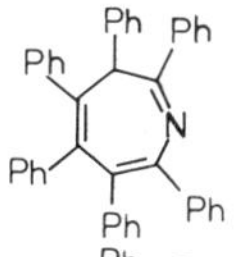	61%
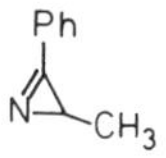	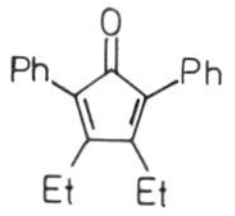	Toluene, Δ	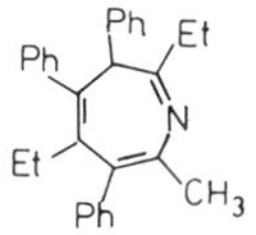	66%
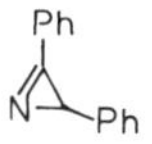	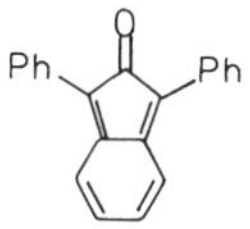	Toluene, Δ	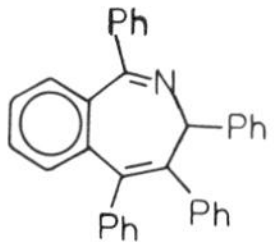	63%
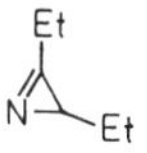	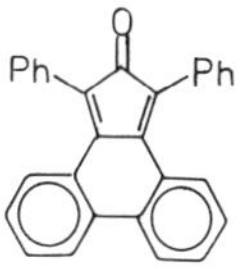	Toluene, Δ	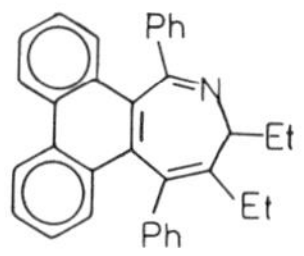	90%

[a] From Refs. 71, 72.

The initial bridged Diels–Alder adducts were never actually isolated but under the reaction conditions lost carbon monoxide to initially give $2H$-azepines. These heterocycles can undergo a thermally allowed 1,5-hydrogen shift, affording the isomeric $3H$-azepines. Some representative examples of this reaction are shown in Table 2-IX.[71,72] Hassner and Anderson[71] also examined the regiochemistry of addition of substituted 1-azirines to unsymmetrical cyclopentadienones. These reactions gave varying mixtures of isomeric $3H$-azepines depending on the nature of the substituents on both diene and dienophile. The results were rationalized in terms of electronic and steric factors that might possibly influence the cycloaddition.

1-Azirines add stereoselectively to diphenylisobenzofuran to afford exo bicyclic adducts [Eq. (20)].[73,74]

$$R^1 = CH_3, Ph, H; R^2 = CH_3, Ph, H, \underline{t}Bu$$

(20)

The exo selectivity of this process was explained on the basis of unfavorable secondary orbital interactions that destabilize the endo transition state.[73] Some of the products of this cycloaddition tend to rearrange to benzazepine derivatives.[74]

Several research groups have reported cycloaddition reactions of 1-azirines with electron-poor *sym*-tetrazines.[75] The nature of the products of this type of reaction is highly dependent on the structures of the azirine and tetrazine and on the reaction conditions. A secondary product of the [4 + 2] cycloaddition (Scheme 2-XI) is the triazepine **34**, presumably formed through the intermediacy of a bridged adduct. However, compound **34** can rearrange by successive 1,5-hydrogen shifts to triazepines **35** and **36**. In addition, these triazepines can also rearrange to give pyrimidines and/or pyrazoles.

1-Azirines bearing alkyl or aryl substituents apparently do not add to simple aliphatic 1,3-dienes. However, it has been reported that a benzoylazirine does combine rapidly with cyclopentadiene, affording an endo Diels–Alder adduct [Eq. (21)].[76]

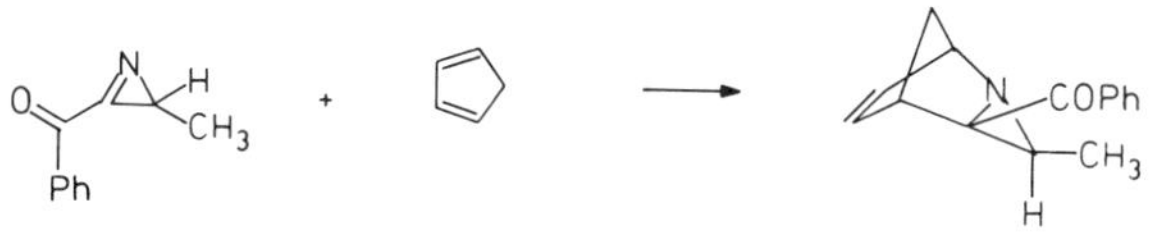

$$(21)$$

Finally, Nair and Kim have found that substituted 1-azirines will undergo [4 + 2] cycloaddition reactions with thiobenzoyl isocyanate, which acts as a heterodiene, yielding cycloadducts [Eq. (22)].[77]

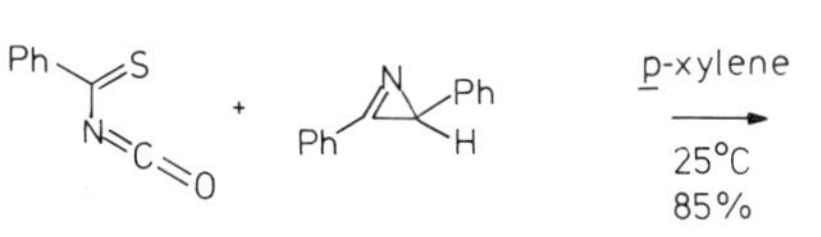

$$(22)$$

7. OXIMINO COMPOUNDS

Oximino compounds such as **37,** which bear two strongly electon-withdrawing groups on carbon and an acyl or sulfonyl group on oxygen, are reactive dienophiles with certain types of dienes. In an extensive study, Fleury and co-workers have examined reactions of variously substituted imino compounds **37** with cyclopentadiene to give adducts **38.**[78] The results of this research are detailed in a Table 2-X. Dienophile reactivity imparted by the carbon substituents R^1 and R^2 is CN $\gg$ CO$_2$R > CONH$_2$ and that imparted by the oxygen substituent R^3 is Ts > Ms > COpNO$_2$Ph

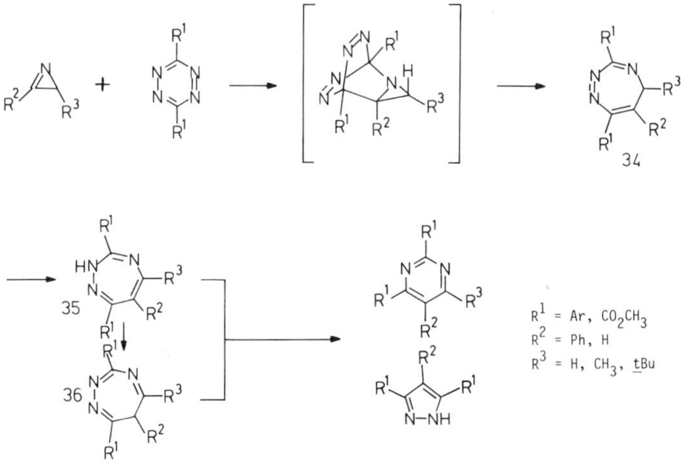

Scheme 2-XI

TABLE 2-X

Cycloadditions of Isonitrosomalonate Derivatives (37) with Cyclopentadiene[a]

R^1	R^2	R^3	Conditions	Yield of adduct **38**
CN	CN	Ts	Et$_2$O, 20°C, 12 hr	88%
CN	CN	COpNO$_2$Ph	Et$_2$O, 20°C, 24 hr	90%
CN	CN	COPh	Et$_2$O, 20°C, 48 hr	80%
CN	CO$_2$Et	Ts	C$_6$H$_6$, 80°C, 3 hr	55%
CN	CO$_2$Et	Ms	C$_6$H$_6$, 80°C, 3 hr	51%
CN	CO$_2$Et	COpNO$_2$Ph	C$_6$H$_6$, 80°C, 4 hr	65%
CN	CO$_2$Et	COPh	C$_6$H$_6$, 80°C, 8 hr	0%
CN	CO$_2$CH$_3$	Ts	C$_6$H$_6$, 80°C, 4 hr	57%
CN	CO$_2$CH$_3$	Ms	C$_6$H$_6$, 80°C, 4 hr	50%
CN	CO$_2$CH$_3$	COpNO$_2$Ph	C$_6$H$_6$, 80°C, 4 hr	62%
CN	CO$_2$CH$_3$	COPh	C$_6$H$_6$, 80°C, 8 hr	0%
CN	CONH$_2$	Ts	Acetone, 60°C, 20 hr	42%
CN	CONH$_2$	Ms	Dioxane, 70°C, 36 hr	39%
CN	CONH$_2$	COpNO$_2$Ph	Dioxane, 70°C, 48 hr	20%
CN	CONH$_2$	COPh	Dioxane, 80°C, 48 hr	0%
CO$_2$Et	CO$_2$Et	Ts	Acetone, 60°C, 72 hr	0%
CO$_2$Et	CO$_2$Et	Ms	Acetone, 60°C, 72 hr	0%
CO$_2$Et	CO$_2$Et	COpNO$_2$Ph	Acetone, 60°C, 72 hr	0%

> COPh. In general, it appears that carbonyl groups as substituent R^2 are better endo directors than cyano, but the situation is sometimes complicated by subsequent endo/exo equilibration of the adducts.

Several other dienes also react with the dicyano-O-tosyl compound **37**.[79] A few specific examples are shown in Scheme 2-XII. In some cases, the initial Diels–Alder adduct is unstable and on heating decomposes to a 2-cyanopyridine. This elimination can be effected cleanly in refluxing ethanol and provides a potentially useful pyridine synthesis. However, the methodology is severely limited by the fact that only very reactive dienes undergo the initial cycloaddition. For example, butadiene, 1,4-diphenylbutadiene, and phellandrene are unreactive.[79]

The reaction seems to show good regioselectivity in cycloadditions with most unsymmetrical dienes (Scheme 2-XII). This selectivity is in accord with FMO theory if one considers the diene HOMO and dienophile LUMO. The results can also be qualitatively rationalized based on Scheme 2-I if dipolar forms **B** and **D** are considered.

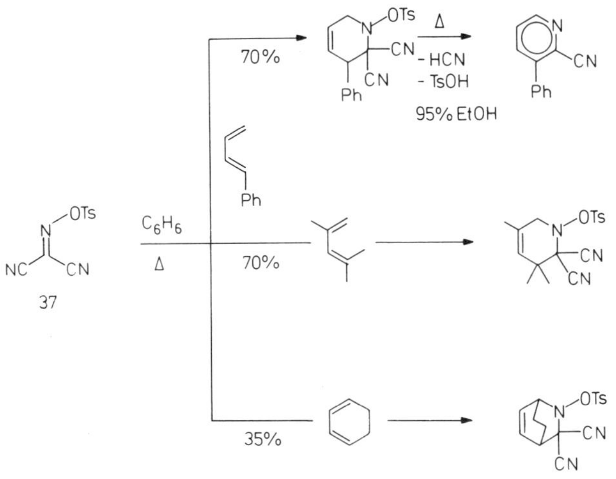

Scheme 2-XII

8. INTRAMOLECULAR CYCLOADDITIONS

The first example of an intramolecular imino Diels-Alder reaction was described by Oppolzer, who found that on heating a benzocyclobutene derivative a mixture of epimeric tricyclic adducts was formed [Eq. (23)].[80]

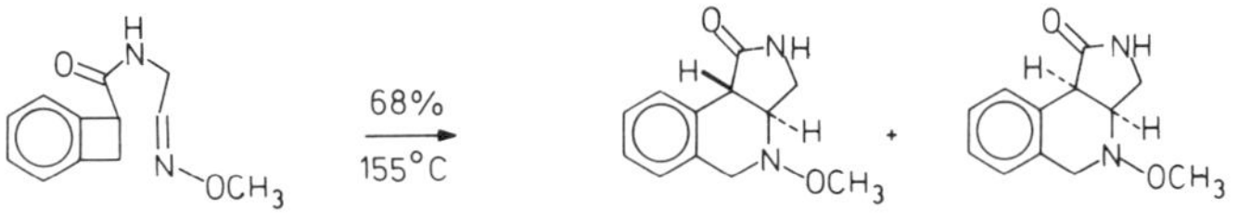

(23)

This cycloaddition proceeds through a thermally generated o-quinodimethane, which is a reactive enough diene to add to an oximino group.

An elegant modification of this sort of intrmolecular cycloaddition strategy was reported by Funk and Vollhardt in 1980.[81] Cooligomerization of bis(trimethylsilyl)acetylene with a substituted diyne under mild reaction conditions catalyzed by cyclopentadienylcobalt dicarbonyl gave a trans-fused tetracyclic adduct [Eq. (24)].

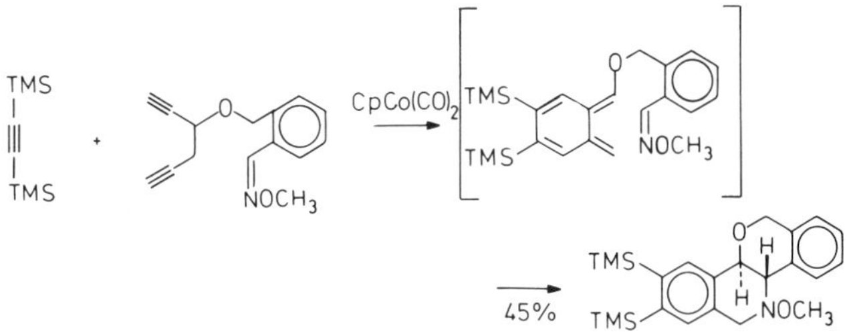

(24)

It appears that a *o*-quinodimethane is also an intermediate in this transformation, and the cycloaddition must have occurred through an exo transition state which leads to the product stereochemistry observed. A similar stereochemical situation probably applies in Oppolzer's case [Eq. (23)], but the trans isomer apparently partially epimerized thermally to the cis isomer under the cyclization conditions.

More recently, Berkowitz and John have used an intramolecular cycloaddition of a *o*-quinodimethane with an *N*-acylimine to synthesize a tricyclic precursor to the antischistosomiasis drug praziquantel [Eq. (25)].[82]

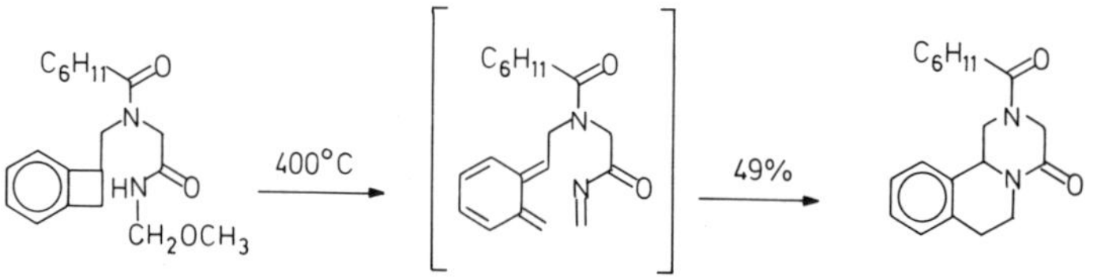

$$(25)$$

Both the unstable diene and the dienophile were generated simultaneously by pyrolysis of appropriate precursor functionality.

Oppolzer *et al.* devised a clever total synthesis of lysergic acid (**40**) that has as its key step an intramolecular imino Diels–Alder reaction (Scheme 2-XIII).[83] A thermal retro-Diels–Alder reaction of **39** liberating cyclopentadiene was used to afford a diene oxime ether which cyclized to give a tetracyclic indole as a 3:2 mixture of diastereomers. Three additional steps served to convert this adduct to (±)-lysergic acid (**40**). Of particular interest here is the fact that simple oximino compounds are not normally reactive dienophiles, and the intramolecularity of the conversion is apparently crucial to the success of this transformation.

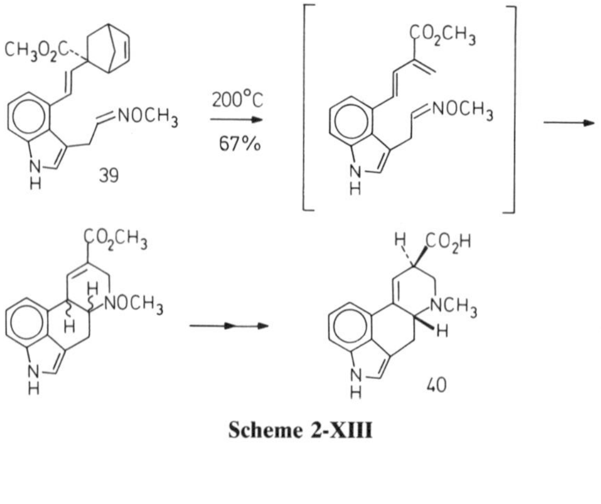

Scheme 2-XIII

During the past few years, Weinreb and co-workers have published a series of papers on studies of intramolecular imino Diels–Alder cycloadditions and applications of this methodology to alkaloid total synthesis.[84] These cycloadditions have involved N-acylimines as the dienophiles, generated thermally (hot tube pyrolysis) by acetic acid elimination of methylol acetates [Eq. (26)].[85]

$$\text{(26)}$$

This route was used in a synthesis of the simple indolizidine alkaloid δ-coniceine (**41**). N-Acylimines are only transient intermediates in these Diels–Alder reactions and are not usually detected. However, Lasne *et al.*,[87] using a flash vacuum pyrolysis (FVP) technique, have generated neutral N-acylimines and showed that they are indeed involved in Diels–Alder cycloadditions.

In a related cycloaddition, Earl and Vollhardt used an interesting retro-ene reaction [Eq. (27)]

$$\text{(27)}$$

to produce a transient N-acylimine/diene which underwent an intramolecular [4 + 2] cycloaddition.[88] Unfortunately, the yield of product was rather poor in this case, and the methodology does not appear to be synthetically useful.

The Weinreb group has also applied intramolecular imino Diels–Alder chemistry to the synthesis of the phenanthroindolizidine antitumor alkaloid tylophorine (**42**) (Scheme 2-XIV).[85,86] In this example, a diene incorporated into a phenanthrene unit proved sufficiently reactive in an intramolecular cyclization to provide the pentacyclic alkaloid ring system. A similar strategy was later used to synthesize the phenanthroquinolizidine alkaloid cryptopleurine.[89]

In a further application of this methodology to the *Elaeocarpus* alkaloid elaeokanine A, Weinreb *et al.* prepared Diels–Alder precursor **43** (Scheme 2-XV).[86,90] On hot tube pyrolysis, compound **43** underwent a cheletropic loss of sulfur dioxide as well as acetic acid elimination to generate an intermediate diene/N-acylimine, which cyclized to an indoli-

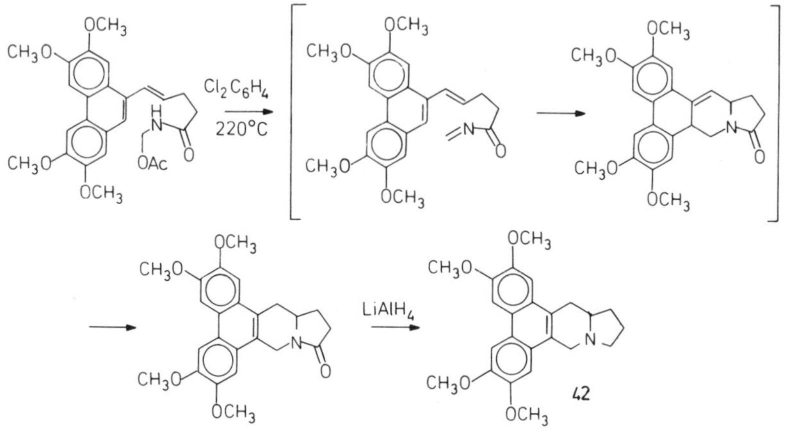

Scheme 2-XIV

zidinone. This Diels–Alder adduct was transformed in a few simple steps to elaeokanine A (**44**).

An intramolecular imino Diels–Alder reaction has also been used as the key step in a total synthesis of the neurotoxic fungal metabolite slaframine (**49**).[91] The [4 + 2] cycloaddition of an (*E,E*)-diene was used to establish the cis stereochemistry in the six-membered ring of slaframine (Scheme 2-XVI), although relative stereochemistry in the five-membered ring could not be controlled as efficiently. Thus, on heating, compound **45** provided a 1 : 1.8 mixture of epimeric indolizidinones **46** and **47,** respectively, in good yield. The desired isomer **46** was converted in a few steps, including a Curtius rearrangement, to carbamate **48,** which by straightforward functional group manipulations afforded racemic slaframine (**49**). The incorrect stereoisomer **47** could also be used to prepare the natural product via an epimerization sequence.

The quinolizidine alkaloid epilupinine (**50**) has been synthesized by a

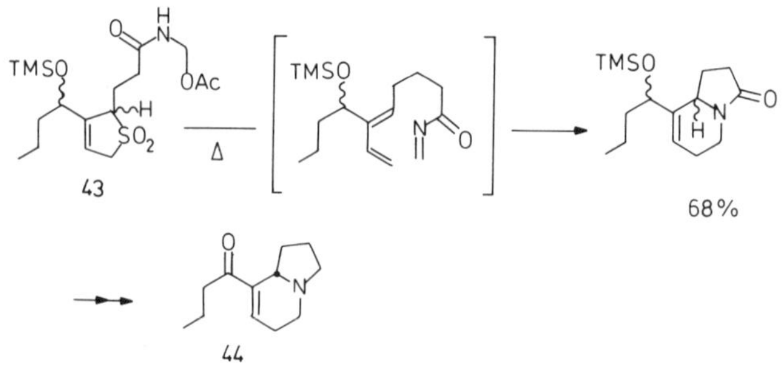

Scheme 2-XV

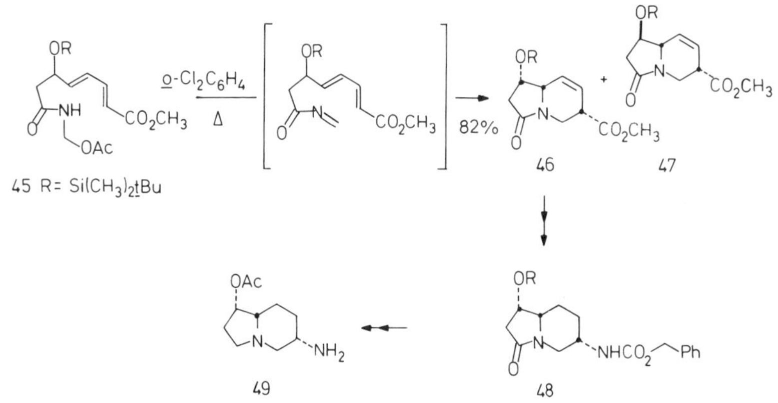

Scheme 2-XVI

stereoselective intramolecular Diels–Alder route utilizing a formalde-hyde-derived N-acylimine as the dienophile [Eq. (28)].[89,92]

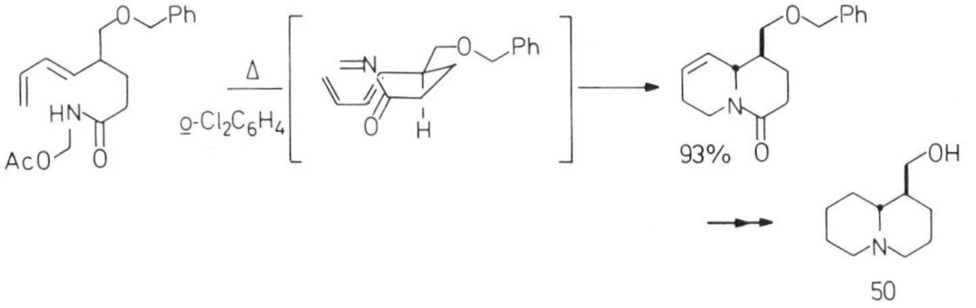

(28)

A methylol acetate was heated to afford a *single* bicyclic Diels–Alder adduct in 93% yield. The stereoselectivity in this process was rationalized by assuming a transition state in which the carbonyl group of the acyl imine is endo to the diene and the bridging chain has a chair-like confor-mation with the large benzyloxy methyl group quasi-equatorial. Such a conformation, which minimizes nonbonded and maximized secondary or-bital interactions, would afford the observed product.

Stereochemical aspects of thermal intramolecular [4 + 2] cycloaddi-tions of disubstituted N-acylimines have also been investigated by Weinreb *et al.*[93] The cyclization experiments in Eqs. (29) and (30)

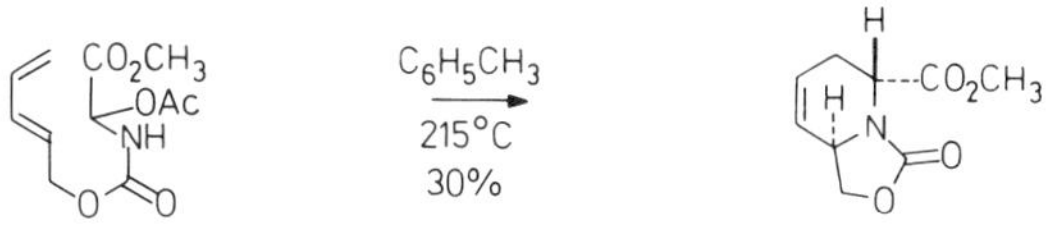

(29)

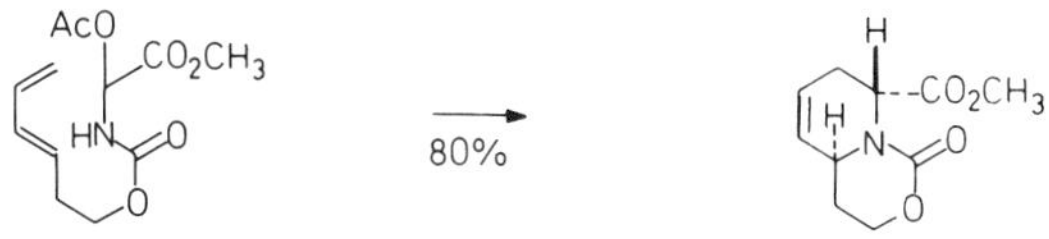

(30)

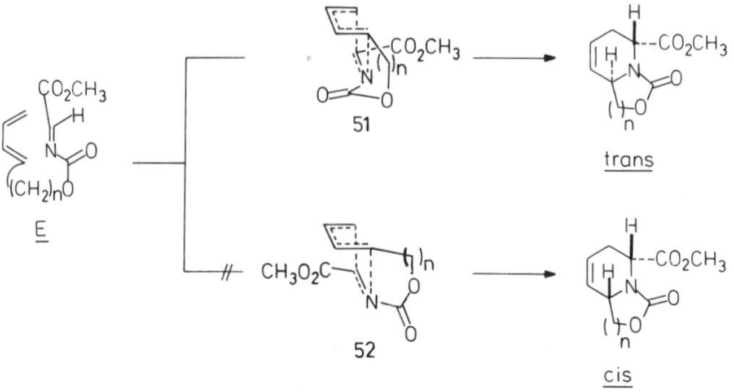

Scheme 2-XVII

were shown to give *single* bicyclic products possessing the trans relative stereochemistry adjacent to the ring of nitrogen.

To rationalize the stereochemical outcome of these cycloadditions, the involvement of an intermediate (*E*)-acylimine was postulated (Scheme 2-XVII). The corresponding (*Z*)-acylimine was discounted, primarily because it would require a Diels–Alder transition state having both carbonyl groups of the dienophile exo to produce the observed trans products. The *N*-acylimine can cyclize through transition states like **51** or **52**. Transition state **51,** which has the nitrogen carbonyl endo and the carbomethoxyl group exo, leads to the observed trans products. On the other hand, transition state **52,** having the *N*-acyl group exo and the carbomethoxyl group endo, would have produced the cis stereoisomers, which were not found. From inspection of molecular models there seems to be little difference between **51** and **52** sterically or in terms of strain. However, as the

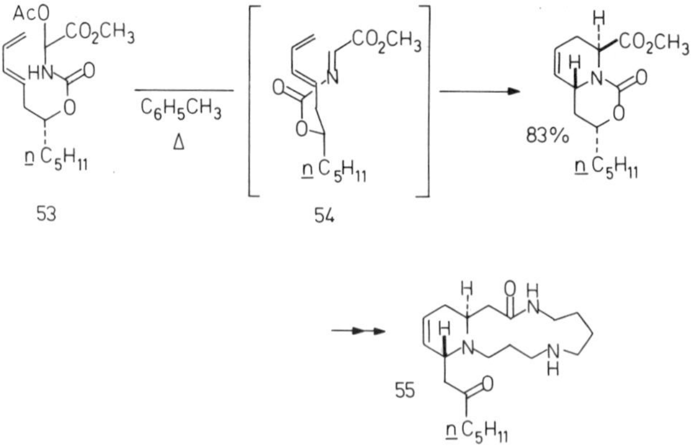

Scheme 2-XVIII

work of Krow indicates (Table 2-III), N-acyl groups prefer to be endo in intermolecular imino Diels–Alder reactions, and this preference seems to hold in intramolecular cases as well. The findings of this stereochemical study were used to design a total synthesis of the macrocyclic spermidine alkaloid anhydrocannabisativene (55).[94] Diels–Alder precursor 53 (Scheme 2-XVIII) was found to cyclize thermally to afford a single adduct in good yield. This transformation presumably involves an N-acylimine, which cyclizes via a transition state conformation as shown in 54. As in the cases shown in Eqs. (29) and (30), the adduct has the protons adjacent to nitrogen in a trans relationship. Interestingly, the remote chiral center in the adduct was also controlled. This result could be explained by invoking a bridging chain conformation as shown in N-acylimine 54. Such a conformation was invoked in the cyclization leading to epilupinine [Eq. (28)]. A series of steps was subsequently used to convert this adduct to anhydrocannabisativene (55).

REFERENCES

1. Alder, K. (1943). "Neuer Methoden der Praperative Organischen Chemie." Verlag Chemie, Weinheim.
2. Previous comprehensive reviews: a. Hamer, J. (ed.) (1967). "1,4-Cycloaddition Reactions, the Diels–Alder Reaction in Heterocyclic Syntheses." Academic Press, New York. b. Weinreb, S.M., and Levin, J. I. (1979). *Heterocycles* 12, 949. c. Weinreb, S. M., and Staib, R. R. (1982). *Tetrahedron* 38, 3087.
3. Desimoni, G., Tacconi, G., Barco, A., and Pollini, G. P. (1983). "Natural Products Syntheses Through Pericyclic Reactions," pp. 229–235. American Chemical Society, Washington, D.C., Burnier, J. S., and Jorgensen, W. L. (1983). *J. Org. Chem.* 48, 3923; Fleming, I. (1976). "Frontier Orbitals and Organic Chemical Reactions." Wiley, New York.
4. Krow, G., Rodebaugh, R., Carmosin, R., Figures, W., Pannella, H., DeVicaris, G., and Grippi, M. (1973). *J. Am. Chem. Soc.* 95, 5273.
5. Albrecht, R., and Kresze, G. (1964). *Chem. Ber.* 97, 490.
6. a. Krow, G., Rodebaugh, R., Marakowski, J., and Ramey, K. C. (1973). *Tetrahedron Lett.* 1899. b. Krow, G. R., Pyun, C., Rodebaugh, R., and Marakowski, J. (1974). *Tetrahedron* 30, 2977.
7. Kresze, G., and Wagner, U. (1972). *Liebigs Ann. Chem.* 762, 106.
8. Rijsenbrij, P. P. M., Loven, R., Wijnberg, J. B. P. A., Speckamp, W. N., and Huisman, H. O. (1972). *Tetrahedron Lett.* 1425.
9. Loven, R.P., Zunnebeld, W. A., and Speckamp, W. N. (1975). *Tetrahedron* 31, 1723.
10. Albrecht, R., and Kresze, G. (1965). *Chem. Ber.* 98, 1431; Barco, A., Benetti, S., Baraldi, P. G., Moroder, F., Pollini, G. P., and Simoni, D. (1982). *Liebigs Ann. Chem.* 960.
11. Zunnebeld, W. A., and Speckamp, W. N. (1975). *Tetrahedron* 31, 1717.
12. Holmes, A. B., Thompson, J., Baxter, A. J. G., and Dixon, J. (1985). *J. Chem. Soc., Chem. Commun.* 37.
13. Abramovich, R. A., and Stovers, J. R. (1984). *Heterocycles* 22, 671.

14. a. Merten, R., and Muller, G. (1962). *Angew. Chem.* **74,** 866. b. Merten, R., and Muller, G. (1964). *Chem. Ber.* **97,** 682. c. Baldwin, J. E., Forrest, A. K., Monaco, S., and Young, R. J. (1985). *J. Chem. Soc., Chem. Commun.,* 1586. d. Fischer, G., Fritz, H., and Prinzbach, H. (1986). *Tetrahedron Lett.* **27,** 1269.
15. Cava, M. P., and Wilkens, C. K., Jr., (1964). *Chem. Ind.* (*London*), 1422; Cava, M. P., Wilkens, C. K., Jr., Dalton, D. R., and Bessho, K. (1965). *J. Org. Chem.* **30,** 3772.
16. Quan, P. M., Karns, T. K. B., and Quin, L. D. (1964). *Chem. Ind.* (*London*), 1553; Quan, P. M., Karns, T. K. B., and Quin, L. D. (1965). *J. Org. Chem.* **30,** 2769.
17. Heine, H. W., Barchiese, B. J., and Williams, E. A. (1984). *J. Org. Chem.* **49,** 2560.
18. Härter, H. P., and Liisbert, S. (1968). *Acta Chem. Scand.* **22,** 2685.
19. Krow, G. R., Pyun, C., Leitz, C., Marakowski, J., and Ramey, K. (1974). *J. Org. Chem.* **39,** 2449.
20. Baxter, A. J. G., and Holmes, A. B. (1977). *J. Chem. Soc., Perkin Trans. 1,* 2343.
21. Krow, G. R., Johnson, C., and Boyle, M. (1978). *Tetrahedron Lett.,* 1971; see also Gaitanopoulos, D. E.., and Weinstock, J. (1985). *J. Heterocycl. Chem.* **22,** 957.
22. Vor der Bruck, D., Buhler, R., and Plieninger, H. (1972). *Tetrahedron* **28,** 791.
23. Jung, M. E., Shishido, K., Light, L., and Davis, L. (1981). *Tetrahedron Lett.* **22,** 4607; see also Ueda, Y., and Maynard, S. C. (1985). *Tetrahedron Lett.* **26,** 6309.
24. a. Imagawa, T., Sisido, K., and Kawanisi, M. (1973). *Bull. Chem. Soc. Jpn.* **46,** 2922. b. Raasch, M. S. (1975). *J. Org. Chem.* **40,** 161.
25. Krow, G. R., Henz, K. J., and Szczepanski, S. W. (1985). *J. Org. Chem.* **50,** 1888, and references cited therein.
26. Krow, G. R., Johnson, C. A., Guare, J. P., Kubrak, D., Henz, K. J., Shaw, D. A., Szczepanski, S. W., and Carey, J. T. (1982). *J. Org. Chem.* **47,** 5238.
27. Jankowski, K. (1976). *Tetrahedron Lett.,* 3309.
28. Krow, G. R., Pyun, C., Rodebaugh, R., and Marakowski, J. (1974). *Tetrahedron* **30,** 2977.
29. Diels–Alder reactions of acyclic *N*-acylimines with other cyclic dienes and studies of the stereochemistry of these cycloadditions: Krow, G. R., Rodebaugh, R., Grippi, M., DeVicaris, G., Hyndman, C., and Marakowski, J. (1973). *J. Org. Chem.* **38,** 3094; Krow, G. R., Damodaran, K. M., Fan, D. M., Rodebaugh, R., Gaspari, A., and Nadir, U. K. (1977). *J. Org. Chem.* **42,** 2486.
30. Ben-Ishai, D., Ben-Et, G., and Warshawsky, A. (1970). *J. Heterocycl. Chem.* **7,** 1289.
31. a. Goldstein, E., and Ben-Ishai, D. (1969). *Tetrahedron Lett.,* 2631. b. Goldstein, E., and Ben-Ishai, D. (1971). *Tetrahedron* **27,** 3119.
32. Evnin, A. B., Lam, A., and Blyskal, J. (1970). *J. Org. Chem.* **35,** 3097.
33. Tokita, S., Hiruta, K., Yaginuma, Y., Ishikawa, S., and Nishi, H. (1984). *Synthesis,* 270.
34. Kim, D., and Weinreb, S. M. (1978). *J. Org. Chem.* **43,** 121.
35. Weinreb, S. M., Basha, F. Z., Hibino, S., Khatri, N. A., Kim, D., Pye, W. E., and Wu, T.-T. (1982). *J. Am. Chem. Soc.* **104,** 536.
36. Warshawsky, A., and Ben-Ishai, D. (1969). *J. Heterocycl. Chem.* **6,** 681.
37. Ben-Ishai, D., Inbal, Z., and Warshawsky, A. (1970). *J. Heterocycl. Chem.* **7,** 615.
38. Warshawsky, A., and Ben-Ishai, D. (1970). *J. Heterocycl. Chem.* **7,** 917.
39. Ben-Ishai, D., and Warshawsky, A. (1971). *J. Heterocycl. Chem.* **8,** 865.
40. Ben-Ishai, D., Gillon, I., and Warshawsky, A. (1973). *J. Heterocycl. Chem.* **10,** 149
41. Pearce, D. S., Locke, M. J., and Moore, H. W. (1975). *J. Am. Chem. Soc.* **97,** 6181.
42. Ch'ng, H. S., and Hooper, M. (1969). *Tetrahedron Lett.,* 1527.
43. McKay, W. R., and Proctor, G. R. (1981). *J. Chem. Soc., Perkin Trans. 1,* 2443.
44. Lucchini, V., Prato, M., Quintily, U., and Scorrano, G. (1984). *J. Chem. Soc., Chem. Commun.,* 48.

45. Trost, B. M., Whitman, P. J. (1974). *J. Am. Chem. Soc.* **96,** 7421.
46. For early work in this area, see Ref. 2a.
47. Böhme, H., Hartke, K., and Müller, A. (1963). *Chem. Ber.* **96,** 607.
48. Babayan, A. T., Martirosyan, G. T., and Grigoryan, D. V. (1968). *Zh. Org. Khim.* **4,** 984.
49. Danishefsky, S., Kitahara, T., McKee, R., and Schuda, P. (1976). *J. Am. Chem. Soc.* **98,** 6715.
50. a. Kerwin, J. F., Jr., and Danishefsky, S. (1982). *Tetrahedron Lett.* **23,** 3739. b. Danishefsky, S., Langer, M. E., and Vogel, C. (1985). *Tetrahedron Lett.* **26,** 5983. c. Danishefsky, S., and Vogel, C. (1986). *J. Org. Chem.* **51,** 3916.
51. a. Vacca, J. P. (1985). *Tetrahedron Lett.* **26,** 1277. b. Huff, J. R., Anderson, P. S., Baldwin, J. J., Clineschmidt, B. V., Guare, J. P., Lotti, V. J., Pettibone, D. J., Randall, W. C., and Vacca, J. P. (1985). *J. Med. Chem.* **28,** 1759. c. Veyrat, C., Wartski, L., and Seyden-Penne, J. (1986). *Tetrahedron Lett.* **27,** 2981.
52. Larsen, S. D., and Grieco, P. A. (1985). *J. Am. Chem. Soc.* **107,** 1768; Grieco, P. A., Larsen, S. D., and Fobare, W. F. (1986). *Tetrahedron Lett.* **27,** 1975; see also Grieco, P. A., and Larsen, S. D. (1986). *J. Org. Chem.* **51,** 3553.
53. Böhme, H., and Ahrens, K. H.(1971). *Tetrahedron Lett.,* 149; Böhme, H., and Ahrens, K. H.(1974). *Arch. Pharm.* **307,** 828.
54. Marchand-Brynaert, J., and Ghosez, L. (1974). *Tetrahedron Lett.,* 377.
55. Bohlmann, F., Habeck, D., Poetsch, E., and Schumann, D. (1967). *Chem. Ber.* **100,** 2742.
56. a. Andriamialisoa, R. Z., Langlois, N., and Langlois, Y. (1982). *J. Chem. Soc., Chem. Commun.,* 1118. b. Langlois, Y., Pouilhes, A., Genin, D., Andriamialisoa, R. Z., and Langlois, N. (1983). *Tetrahedron* **39,** 3755. c. Andriamialisoa, R. Z., Langlois, N., and Langlois, Y. (1985). *J. Org. Chem.* **50,** 961.
57. Roffey, P., and Verge, J. P. (1969). *J. Heterocycl. Chem.* **6,** 497; see also Seitz, G., and Overheu, W. (1977). *Arch. Pharm.* **310,** 936; Seitz, G., and Mohr, R. (1986). *Arch. Pharm.* **319,** 690.
58. Figeys, H. P., and Mathy, A. (1981). *Tetrahedron Lett.,* 1393; see also Boger, D. L., and Panek, J. S. (1985). *J. Am. Chem. Soc.* **107,** 5745.
59. Seitz, G., and Overheu, W. (1979). *Arch. Pharm.* **312,** 452.
60. Seitz, G., and Overheu, W. (1981). *Arch. Pharm.* **314,** 376.
61. Seitz, G., Dhar, R., and Hühnermann, W. (1982). *Chem. Zeit.,* 100.
62. Seitz, G., Shar, R., and Dietrich, S. (1983). *Arch. Pharm.* **316,** 472.
63. Seitz, G., Dhar, R., Mohr, R., and Overheu, W. (1984). *Arch. Pharm.* **317,** 237.
64. Kametani, T., Takahashi, T., Honda, T., Ogasawara, K., and Fukumoto, K. (1974). *J. Org. Chem.* **39,** 447.
65. Kametani, T., Takahashi, T., Ogasawara, K., and Fukumoto, K. (1974). *Tetrahedron* **30,** 1047.
66. Kametani, T., Katoh, Y., and Fukumoto, K. (1974). *J. Chem. Soc., Perkin Trans. 1,* 1712.
67. Kametani, T., Kajiwara, M., Takahashi, T., and Fukumoto, K. (1975). *J. Chem. Soc., Perkin Trans. 1,* 737.
68. Dannhardt, G., and Wiegrebe, W. (1977). *Arch. Pharm.* **310,** 802.
69. Middleton, W. J., and Krespan, C. G. (1965). *J. Org. Chem.* **30,** 1398.
70. Previous reviews: Anderson, D. J., and Hassner, A. (1975). *Synthesis,* 486; Hassner, A. (1980). *Heterocycles* **14,** 1517.
71. Hassner, A., and Anderson, D. J. (1974). *J. Org. Chem.* **39,** 3070.
72. Nair, V. (1972). *J. Org. Chem.* **37,** 802.
73. Nair, V. (1972). *J. Org. Chem.* **37,** 2508.

74. Anderson, D. J., and Hassner, A. (1974). *J. Org. Chem.* **39**, 2031.
75. a. Anderson, D. J., and Hassner, A. (1974). *J. Chem. Soc., Chem. Commun.*, 45. b. Johnson, G. C., and Levin, R. H, (1974). *Tetrahedron Lett.*, 2303. c. Takahashi, M., Suzuki, N., and Igari, Y. (1975). *Bull. Chem. Soc. Jpn.* **48**, 2605. d. Nair, V. (1975). *J. Heterocycl. Chem.* **12**, 183.
76. Hemetsberger, H., and Knittel, D. (1972). *Monatsh. Chem.* **103**, 205.
77. Nair, V., and Kim, K. H. (1974). *Tetrahedron Lett.*, 1487.
78. a. Biehler, J. M., Fleury, J. P., Perchais, J., and Regent, A. (1968). *Tetrahedron Lett.*, 4227. b. Fleury, J. P., Biehler, J. M., and Desbois, M. (1969). *Tetrahedron Lett.*, 4091. c. Biehler, J. M., Perchais, J., and Fleury, J. P. (1971). *Bull. Soc. Chim. Fr.*, 2711. d. Biehler, J. M., and Fleury, J. P. (1971). *J. Heterocycl. Chem.* **8**, 431. e. Biehler, J. M., and Fleury, J. P. (1971). *Tetrahedron* **27**, 3171. f. Fleury, J. P. (1977). *Chemia* **31**, 143.
79. Fleury, J. P., Desbois, M., and See, J. (1978). *Bull. Soc. Chim. Fr. II*, 147.
80. Oppolzer, W. (1972). *Angew. Chem. Int. Ed. Engl.* **11**, 1031.
81. Funk, R. L., and Vollhardt, K. P. C. (1980). *J. Am. Chem. Soc.* **102**, 5045.
82. Berkowitz, W. F., and John, T. V. (1984). *J. Org. Chem.* **49**, 5269.
83. Oppolzer, W., Francotte, E., and Battig, K. (1981). *Helv. Chim. Acta.* **64**, 478.
84. Weinreb, S. M. (1985). *Acc. Chem. Res.* **18**, 16.
85. Weinreb, S. M., Khatri, N. A., and Shringarpure, J. (1979). *J. Am. Chem. Soc.* **101**, 5073.
86. Khatri, N. A., Schmitthenner, H. F., Shringarpure, J., and Weinreb, S. M. (1981). *J. Am. Chem. Soc.* **103**, 6387.
87. Lasne, M. C., Ripoll, J. L., and Thuiller, A. (1982). *J. Chem. Res. Synop.*, 214.
88. Earl, R. A., and Vollhardt, K. P. C. (1982). *Heterocycles* **19**, 265.
89. Bremmer, M. L., Khatri, N. A., and Weinreb, S. M. (1983). *J. Org. Chem.* **48**, 3661.
90. Schmitthenner, H. F., and Weinreb, S. M. (1980). *J. Org. Chem.* **45**, 3672.
91. Gobao, R. A., Bremmer, M. L., and Weinreb, S. M. (1982). *J. Am. Chem. Soc.* **104**, 7065.
92. Bremmer, M. L., and Weinreb, S. M. (1983). *Tetrahedron Lett.* **24**, 261; see also Nomoto, T., and Takayama, H. (1985). *Heterocycles* **23**, 2913.
93. Nader, B., Franck, R. W., and Weinreb, S. M. (1980). *J. Am. Chem. Soc.* **102**, 1153; Nader, B., Bailey, T. R., Franck, R. W., and Weinreb, S. M. (1981). *J. Am. Chem. Soc.* **103**, 7573.
94. Bailey, T. R., Garigipati, R. S., Morton, J. A., and Weinreb, S. M. (1984). *J. Am. Chem. Soc.* **106**, 3240.

Chapter 3

Nitroso and Thionitroso Dienophiles

INTRODUCTION

A wide variety of nitroso compounds are known to react thermally with conjugated 1,3-dienes to afford 3,6-dihydro-1,2-oxazines [Eq. (1)].[1]

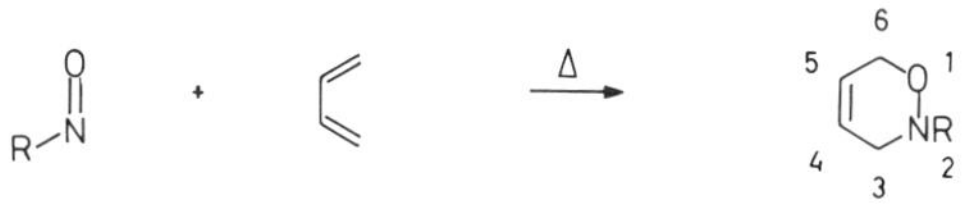

$$\underset{R^{\diagdown N}}{\overset{\overset{O}{\parallel}}{}} \quad + \quad \diagup\diagdown \quad \overset{\Delta}{\longrightarrow} \quad \underset{4\quad3\quad2}{\overset{6}{\underset{5}{\diagup}}\diagdown\overset{O}{\underset{NR}{}}^{1}} \qquad (1)$$

R = alkyl, aryl, acyl, sulfonyl

In general, most types of C-nitroso compounds appear to be reactive heterodienophiles. On the other hand, N-nitroso compounds and nitrosyl chloride do not usually undergo [4 + 2] cycloadditions.[1a] Much of the early work in this area can be found in the Hamer and Kresze reviews, which cover the literature through 1967.[1a,b] This chapter will emphasize newer aspects of [4 + 2] cycloadditions involving nitroso dienophiles.

1. MECHANISTIC AND REGIOCHEMICAL CONSIDERATIONS

A completely clear and totally consistent picture of the mechanistic course of nitroso Diels–Alder reactions is not yet available. It would appear that the mechanism of this sort of cycloaddition, and often the resulting product regiochemistry, is very much dependent on the electronic and sometimes steric nature of both the nitroso dienophile and the diene employed. The most thoroughly studied dienophiles of this class are the arylnitroso compounds. Kresze and co-workers,[1b,2] in an extensive series of papers involving kinetic studies and investigations of adduct regiochemistry, have proposed that these cycloadditions may vary widely from concerted pericyclic processes to those involving dipolar intermediates. The details of Kresze's mechanistic investigations can be found in an excellent previous review,[1b] and some typical examples of regiochemical results using arylnitroso compounds are given in the following section.

Interestingly, α-chloronitroso compounds seem to react with isoprene to given the *opposite* regioisomer from nitrosobenzene.[3,4] This phenomenon has been rationalized on steric grounds, for lack of a better explanation. More recently, Boger *et al.*[5] have studied the regiochemistry of addition of acylnitroso dienophiles to some unsymmetrical cyclic dienes and have suggested that these reactions involve either normal ($HOMO_{diene}$ controlled) or inverse electron demand ($LUMO_{diene}$ controlled) electrocyclic processes, depending on the diene (*vide infra*). Overall, nitroso Diels–Alder cycloadditions do show excellent regiochemical control, but further mechanistic studies are required to more firmly establish the basis for this selectivity and to explain some of the apparently anomalous results.

2. ARYLNITROSO COMPOUNDS

Arylnitroso compounds of various types react with both electron-rich and electron-deficient dienes under mild conditions (usually 0–100°C) to produce dihydrooxazines. As mentioned in the preceding section, Kresze's group has looked in considerable detail at orientational preferences and kinetics of [4 + 2] cycloadditions with various phenylnitroso compounds and substituted 1,3-dienes.[1b,2] Table 3-I contains some of the results of cycloadditions of aromatic nitroso dienophiles with representative unsymmetrical 1- and 2-substituted dienes.[1b] Scheme 3-I shows the subtle regiochemical trends that can be observed in a series of these reactions.[2b] These results were rationalized based on consideration of putative dipolar and nonpolar transition states and their relative stabilization by aromatic electron-donating and -withdrawing substituent groups.

TABLE 3-I

Cycloadditions of Aromatic Nitroso Compounds with Unsymmetrical Dienes[a]

Dienophile	Diene	Conditions	Product(s)	Yield
PhNO		CH_3OH, RT	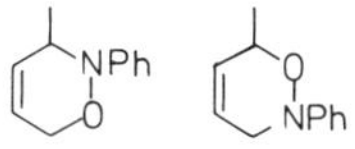	86%
pClPhNO	R = OAc	CH_3OH, RT	Ar = pClPh	35%
pClPhNO	R = CO_2CH_3	CH_3OH, RT	Ar = pClPh	83%
pClPhNO	R = Ph	CH_3OH, RT	Ar = pClPh	70%
pClPhNO	R = CH_3	CH_3OH, RT	Ar = pClPh	33%
pClPhNO	R = CN	CH_3OH, RT	Ar = pClPh	70%
PhNO	R = Ph	CH_3OH, 0°C	Ar = Ph	83%
pClPhNO	R = CF_3	CH_3OH, RT	Ar = pClPh	61%

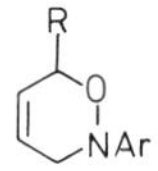

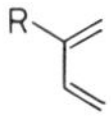

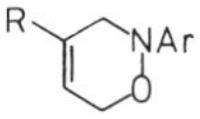

[a] From Ref. 1b.

A related study shown in Scheme 3-II indicates that substituents in the aromatic ring of the 2π nitroso component also affect the regioisomer ratio, again presumably by electronic stabilization or destabilization of polar transition states leading to each product.[6]

One useful reaction of N-aryldihydrooxazines is their conversion to N-

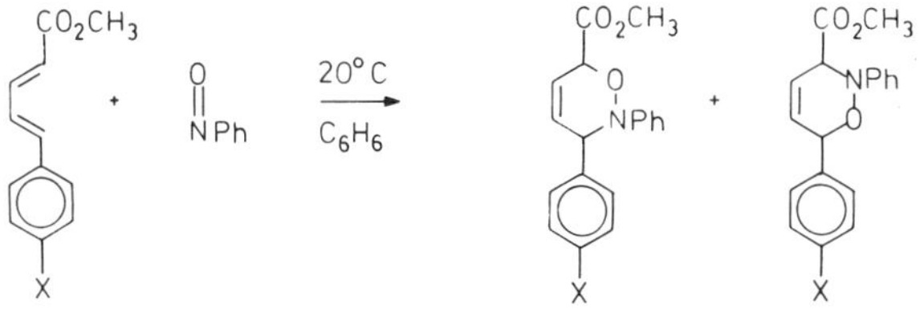

X	Isomer Ratio
CN	61 : 39
Cl	76 : 24
H	80 : 20
CH_3	81 : 19
OCH_3	78 : 22
NMe_2	66 : 24

Scheme 3-I

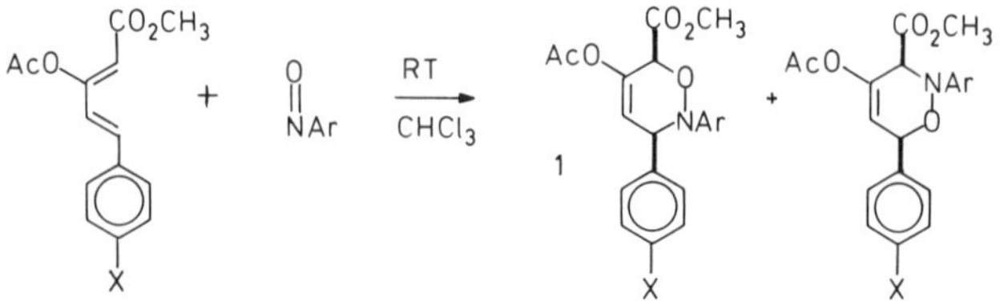

X	Ar	Isomer Ratio
H	Ph	76 : 24
H	pClPh	68 : 32
H	pNO$_2$Ph	54 : 46
H	2,4-(NO$_2$)$_2$Ph	31 : 69
CH$_3$O	Ph	62 : 38

Scheme 3-II

arylpyrroles. For example, Kresze and Hartner[6] found that adduct **1** (Scheme 3-II) can be converted to a pyrrole on treatment with silica gel [Eq. (2)].[1b]

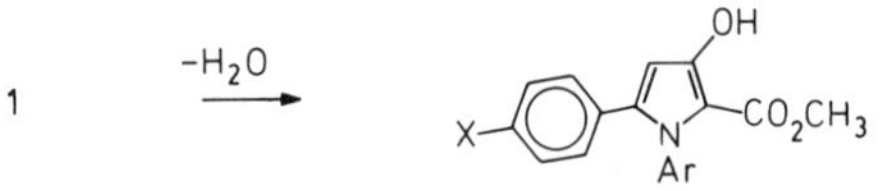

However, this transformation may only be general if there is an electron-withdrawing group at C-6 of the 3,6-dihydro-1,2-oxazine. More recently, Givens *et al.* reported that dihydrooxazines derived from *p*-carbomethoxynitrosobenzene on photolysis afford the *N*-arylpyrroles irrespective of whether an electron-withdrawing substituent is present at C-6 (Scheme 3-III).[7] The *N*-*p*-carbomethoxyphenyl group was chosen to provide a good chromophore for the photochemical extrusion of water and to allow subsequent facile Birch reduction to remove the *N*-aryl substituent. Interestingly, *p*-methoxycarbonylnitrosobenzene gives the *opposite* regioisomeric adduct from nitrosobenzene (see Table 3-I).

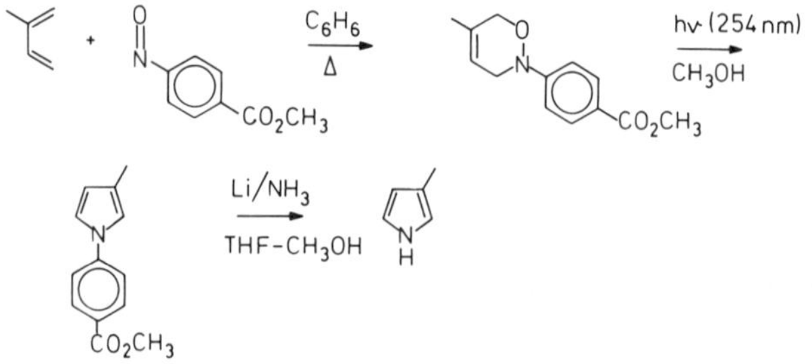

Scheme 3-III

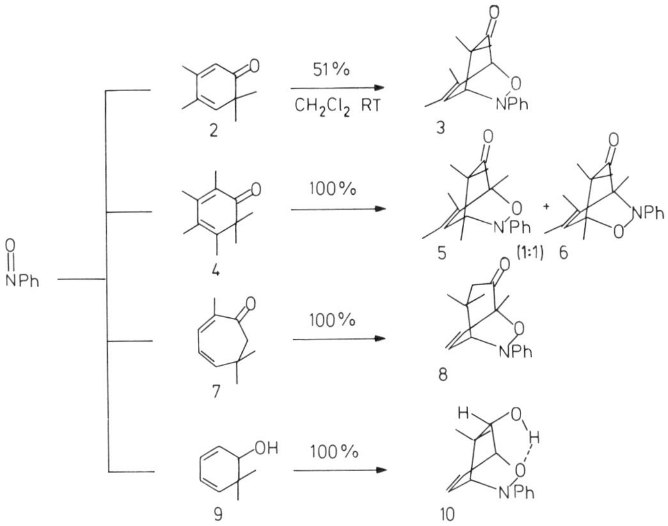

Scheme 3-IV

Hart *et al.* studied the cycloaddition of nitrosobenzene to several conjugated cyclic dienones and dienols, and their results are summarized in Scheme 3-IV.[8] Additions to dienones **2** and **7** produced *only* adducts **3** and **8**, respectively. This regioselectivity can be qualitatively rationalized by FMO theory if one considers the interaction of the dienone HOMO and the LUMO of nitrosobenzene. The regiochemistry and syn selectivity of the reaction with dienol **9** to give **10** may be due to hydrogen bonding in the transition state for cycloaddition. The lack of regioselectivity in the reaction of nitrosobenzene with permethylated dienone **4**, yielding **5** and **6**, is puzzling but is probably due to steric factors.

Substituted nitrosobenzenes have been reported to add both regioselectively and stereoselectively to thebaine at room temperature to afford Diels–Alder adducts in high yields [Eq. (3)].[9]

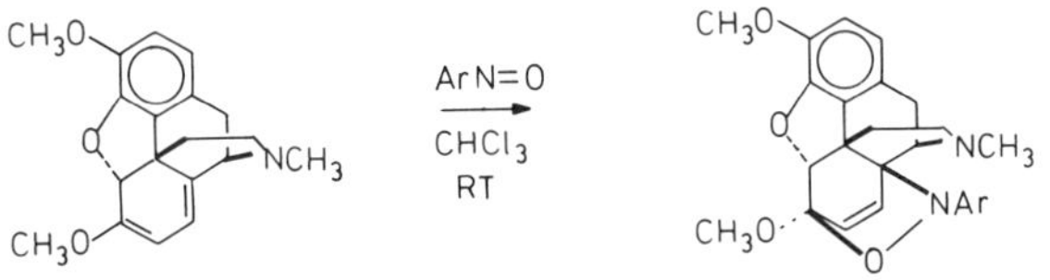

$$(3)$$

These reactions are generally reversible at ambient temperatures. Electron-withdrawing para substituents diminished dissociation, and the opposite effect was observed for electron-donating para substituents.

In recent work, Taylor and co-workers have developed a new procedure to synthesize aryl and heteroaromatic nitroso compounds from the parent amine [Eq. (4)].[10,11]

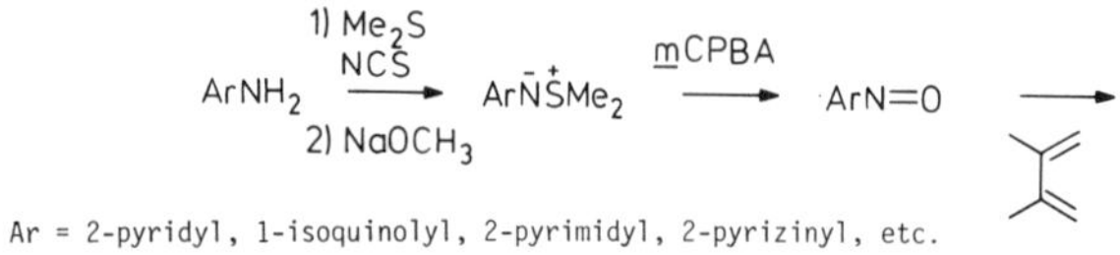
1) Me_2S
NCS
$ArNH_2$
$Ar\bar{N}\overset{+}{S}Me_2$
2) $NaOCH_3$
mCPBA
ArN=O
NAr
Ar = 2-pyridyl, 1-isoquinolyl, 2-pyrimidyl, 2-pyrizinyl, etc.

(4)

These nitroso compounds undergo [4 + 2] cycloadditions readily with 2,3-dimethylbutadiene to yield the expected adducts. This method was used to synthesize arylnitroso compound **11,** which was in turn used to prepare a 3-pyrrolidinone via a Diels–Alder strategy (Scheme 3-V). The initial cycloaddition was apparently regioselective, giving only the isomer shown (cf. Scheme 3-III).[11]

N-Acyl-1,2-dihydropyridines undergo regioselective Diels–Alder cycloadditions with nitrosobenzene [Eq. (5)].[12]

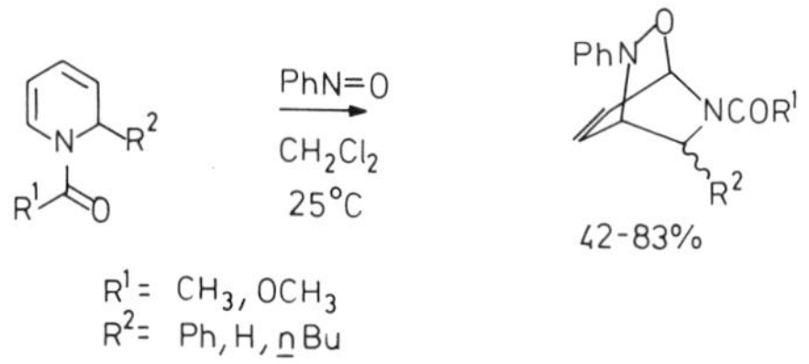
PhN=O
CH_2Cl_2
25°C
PhN
$NCOR^1$
R^2
42-83%
R^1 = CH_3, OCH_3
R^2 = Ph, H, nBu

(5)

The sole regioisomer produced is in accord with the predictions of FMO theory. 1,2-Dihydropyridines bearing a substituent at C-2 underwent cycloadditions to give *signals* stereoisomers, but the configuration was not established for these adducts. However, a recent reinvestigation showed the addition of nitrosobenzene to dihydropyridine **12** gives the anti adduct, as might be expected on steric grounds [Eq. (6)].[13]

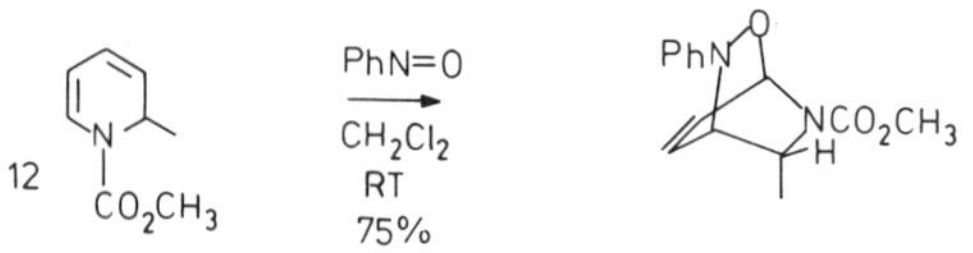
12
CO_2CH_3
PhN=O
CH_2Cl_2
RT
75%
PhN
NCO_2CH_3
H

(6)

The adducts in Eq. (5) presumably also have this anti stereochemistry.

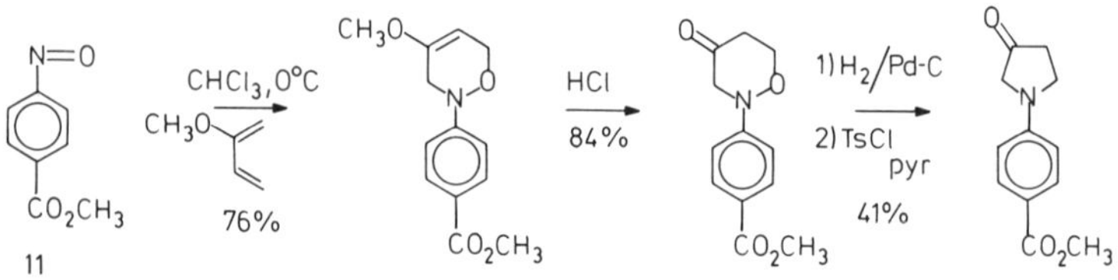
N=O
CO_2CH_3
11
$CHCl_3$, 0°C
CH_3O
76%
HCl
84%
1) H_2/Pd-C
2) TsCl
pyr
41%

Scheme 3-V

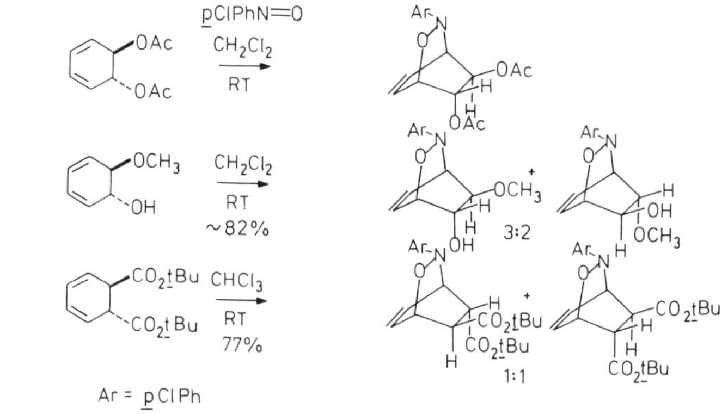

Scheme 3-VI

Cycloadditions of nitrosobenzenes with simple carbocyclic 1,3-dienes have been known for many years.[1b,4] In a recent investigation, Kresze *et al.* have added *p*-chloronitrosobenzene to some 5,6-difunctionalized 1,3-cyclohexadienes (Scheme 3-VI).[15] In the first example shown, only one cycloadduct was produced, although it is not obvious why this is the case. Also puzzling is the stereochemical result of the second example shown in the scheme. Based on Hart's work (Scheme 3-IV), one might have expected the *syn*-hydroxyl isomer to be the major product of this reaction.

Some studies have been reported recently that involve [4 + 2] cycloadditions of porphyrin derivatives with nitroso aromatics.[16] For instance, protoporphyrin II dimethyl ester reacts with *p*-nitronitrosobenzene to give primarily a mono Diels–Alder adducts [Eq. (7)].

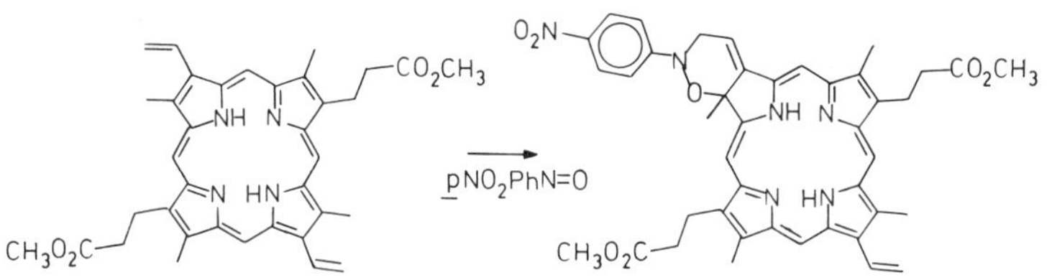

$$(7)$$

On treatment of this product with excess nitroso compound, an interesting vinyl to formyl group degradation occurred involving the presumed intermediacy of the bis Diels–Alder adduct.

The reaction of propellane **13** with nitrosobenzene was found to afford a 2:1 mixture of isomeric *anti/syn* Diels–Alder adducts [Eq. (8)].[17]

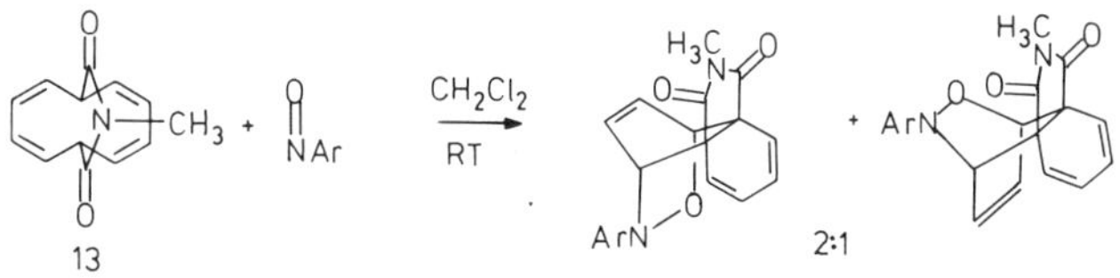

$$(8)$$

This result is of some theoretical interest since **13** reacts with *N*-methylmaleimide to give only the anti product and with *N*-phenyltriazolinedione to give exclusively the syn adduct.

The interesting organometallic 2-substituted 1,3-diene in Eq. (9)

$$
\text{Fp}\diagdown\diagup \quad + \quad \underset{\text{NPh}}{\overset{\text{O}}{\|}} \quad \xrightarrow[\substack{\text{RT}\\88\%}]{\text{CH}_2\text{Cl}_2} \quad \text{Fp}\diagdown\underset{\text{NPh}}{\overset{\text{O}}{\diagup}} \quad + \quad \text{Fp}\diagdown\underset{\text{O}}{\overset{\text{NPh}}{\diagup}} \qquad (9)
$$

$$\text{Fp} = \eta\text{-}\text{C}_5\text{H}_5(\text{CO})_2\text{Fe}$$

2:3

reacts with nitrosobenzene to give Diels–Alder adducts, but the cycloaddition is unfortunately not regioselective even though the yield is good.[18] It was noted that the ratio was independent of the solvent used, thus suggesting that the products are formed via a [4 + 2] cycloaddition rather than by some other reaction pathway.

Another reaction sequence involving organometallic compounds is shown in Eq. (10).[19]

$$
\diagup\!\!\!\diagdown_{\text{NPh}} \xrightarrow[\substack{\text{C}_6\text{H}_6\\40°\text{C}}]{\text{Fe}_2(\text{CO})_9} \quad \diagdown + \quad \diagdown + \quad \diagdown + \quad \diagdown \qquad (10)
$$

| 8% | 6% | 8% | 25% |

In general, treatment of dihydrooxazines with $\text{Fe}_2(\text{CO})_9$ gives mixtures of varying composition depending on the structure of the initial Diels-Alder adduct used. The mechanism of this reaction has been probed, and a plausible sequence for formation of the products shown has been offered. The synthetic potential of this type of transformation has not yet been pursued.

3. α-HALONITROSO COMPOUNDS

Several aliphatic nitroso compounds are known to act as heterodienophiles, but only the α-halonitroso type has been used to any significant extent.[1] It has been known for about 40 years[1,20] that these α-chloronitroso compounds react readily with 1,3-dienes to yield unstable adducts [Eq. (11)].

$$
\underset{\text{Cl}}{\overset{\text{R}}{\underset{\text{R}}{\diagup}}}\!\!-\!\text{N}=\text{O} \quad + \quad \diagup\!\!\!\diagdown \longrightarrow \left[\diagdown\underset{\text{O}}{\overset{\text{R}}{\diagup}}\!\!-\!\!\underset{\text{Cl}}{\overset{\text{R}}{\diagup}} \rightleftharpoons \diagdown\underset{\text{O}}{\overset{+}{\diagup}}\!\!-\!\!\overset{\text{R}}{\diagup}\text{R} \right] \xrightarrow{\text{R'OH}} \diagdown\underset{\text{O}}{\overset{\text{NH}\cdot\text{HCl}}{\diagup}} \qquad (11)
$$

$$+ \quad \text{R}_2\text{C}(\text{OR'})_2$$

If these cycloaddition reactions are either run in an alcoholic solvent or if an alcohol is subsequently added, the product that is isolated is the hydrochloride salt of a 3,6-dihydro-1,2-oxazine.

For example, addition of a chloronitroso compound to 1,3-cyclohexadiene [Eq. (12)][21]

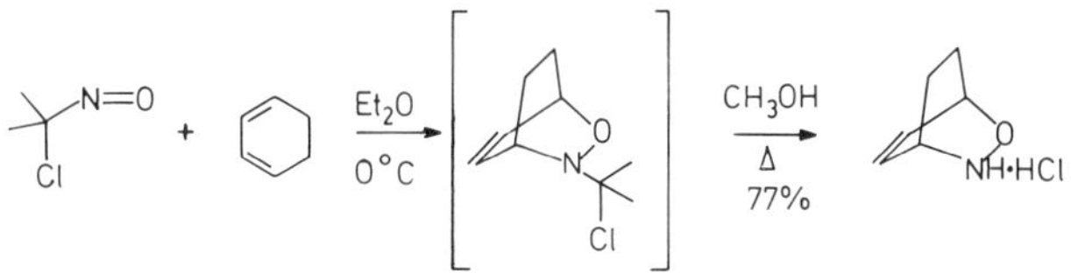

$$(12)$$

gives an unstable adduct, which when heated in methanol affords the dihydrooxazine hydrochloride. The adduct of cyclopentadiene can be similarly prepared (89% yield).[22]

The regiochemistry of the addition of α-chloronitroso compounds with unsymmetrical acyclic dienes has been probed, and some results are listed in Table 3-II.[1b,3a,23] In general, orientational preferences are in accord with those observed for arylnitroso compounds (cf. Table 3-I). As noted above, a glaring exception is the case of isoprene, which gives the opposite selectivities with chloronitroso compounds and nitrosobenzene. It was suggested that steric factors become important in cycloadditions

TABLE 3-II
Cycloadditions of 1-Chloro-1-nitrosocyclohexane to Unsymmetrical 1,3-Dienes[a]

Diene	Conditions[b]	Product(s)	Yield
 R = CH₃ R = Ph	0°C, 96 hr RT, 5 hr	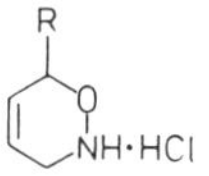	47% 53%
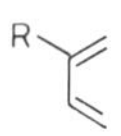 R = CH₃ R = Ph	RT, 48 hr RT, 1 hr	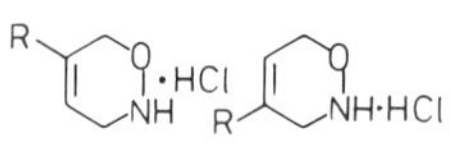	80 : 20 32% 30 : 70 58%
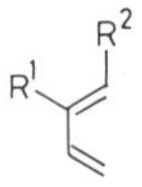 R¹ = Ph, R² = CH₃ R¹ = CH₃, R² = Ph	0°C, 72 hr 0°C, 24 hr	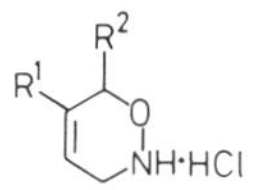	33% 73%
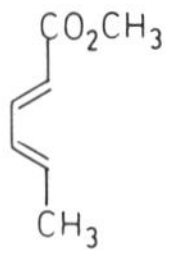	0°C, 14 days	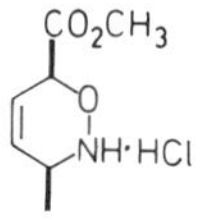	67%

[a] Adapted from Refs. 1b, 3a, and 23.
[b] Reactions were run in C₆H₆–EtOH or Et₂O–EtOH.

with chloronitroso compounds, and that the bulky alkyl group on nitrogen will approach the least substituted end of the 1,3-diene. However, such an argument does not satisfactorily explain the inconsistent results with 2-substituted butadienes such as isoprene and 2-phenylbutadiene.

Leonard *et al.* have used the major adduct from isoprene cycloaddition to a chloronitroso compound in an efficient total synthesis of the cell division stimulant *cis*-zeatin [Eq. (13)].[3c]

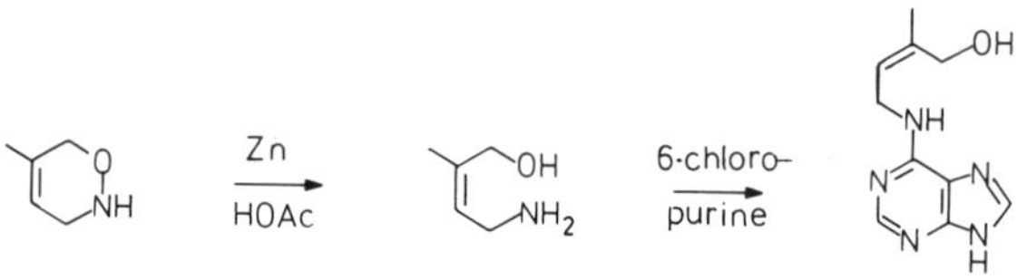

$$\underset{\text{HOAc}}{\overset{\text{Zn}}{\longrightarrow}} \qquad \underset{\text{purine}}{\overset{\text{6-chloro-}}{\longrightarrow}} \qquad (13)$$

A key point here is that the cycloaddition process allows one to control double bond geometry (Z) in the acyclic amino alcohol product.

Kresze and co-workers utilized several of the adducts of α-chloronitroso compounds and 5,6-disubstituted 1,3-cyclohexadienes in syntheses of some inosamine derivatives.[15,24–26] For example, konduramin-Fl has been synthesized as depicted in Eq. (14).[24]

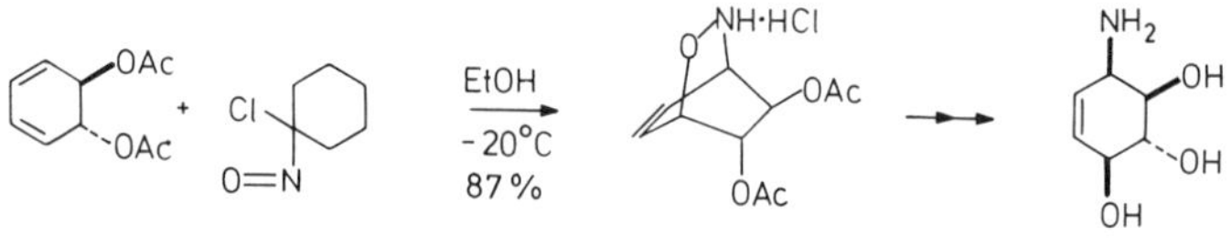

$$\underset{- 20°C}{\overset{\text{EtOH}}{\longrightarrow}} \quad \underset{87\%}{} \qquad \longrightarrow \qquad (14)$$

Interestingly, addition of l-chloro-l-nitrosocyclohexane to the oxygenated cyclohexadiene gave *only* the adduct shown (cf. Scheme 3-VI).

In a related study, Kresze *et al.* examined the [4 + 2] cycloadditions of the nitrogen-containing cyclohexadienes shown in Scheme 3-VII.[25] In the

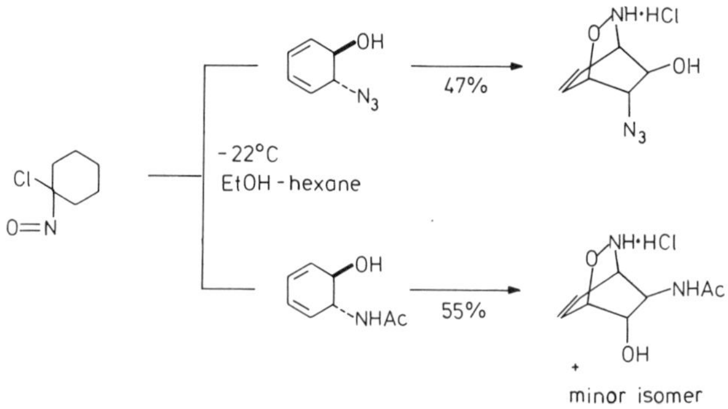

Scheme 3-VII

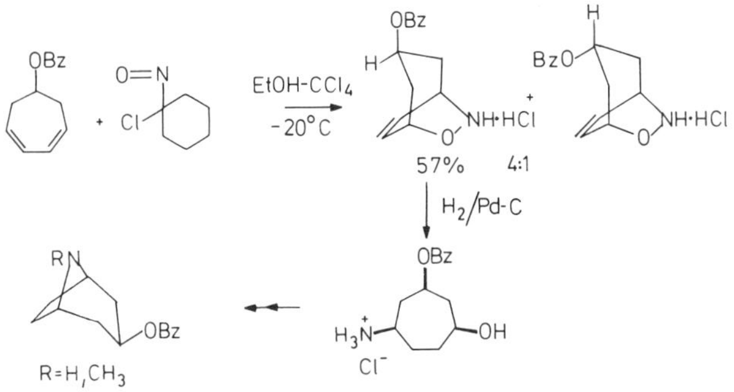

Scheme 3-VIII

first example, only a single detectable adduct was formed, which has the structure indicated. Interestingly, the acetamido derivative gave a different regio- and stereoisomeric major product along with a small amount of an uncharacterized minor adduct (ratio 4 : 1). Also rather interesting is that p-chloronitrosobenzene gave complex mixtures of adducts with both azido and acetamido dienes. Several of these adducts were subsequently transformed to inosamine analogs.[26]

In a recent investigation, Kibayashi et al. have used a nitroso Diels–Alder strategy in the synthesis of tropane alkaloids (Scheme 3-VIII).[27] The initial cycloaddition of α-chloronitrosocyclohexane with 6-benzyloxy-1,3-cycloheptadiene showed a 4 : 1 facial selectivity. The major isomer of this cycloaddition could be converted to pseudotropane and tropacocaine by a straightforward sequence of steps.

Kresze and co-workers have looked at the possibility of effecting nitroso Diels–Alder reactions enantioselectively. In a preliminary experiment, the optically active α-chloronitroso compound in Eq. (15)

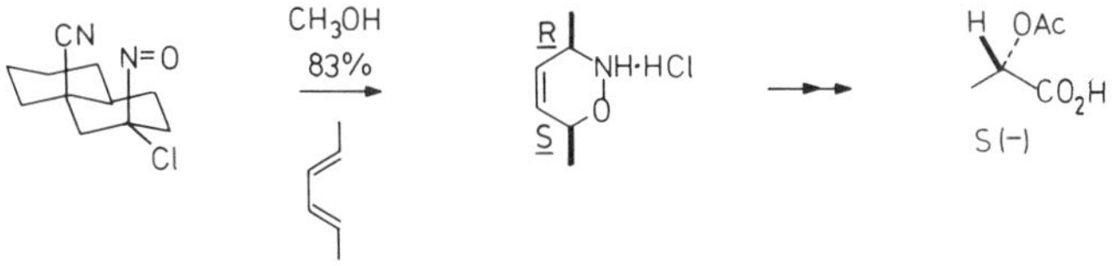

was combined with (E,E)-2,4-hexadiene to give an adduct in good yield. Degradation of this dihydrooxazine to acetyl lactic acid showed that the cycloaddition had proceeded to give the adduct having the 3R,6S configuration with only 39% optical purity.[28]

In more recent studies, Kresze has found that by using more effective chiral auxiliaries, high enantiomeric excesses can be achieved.[29] Thus, the α-chloronitroso dienophile derived from epiandrosterone [Eq. (16)]

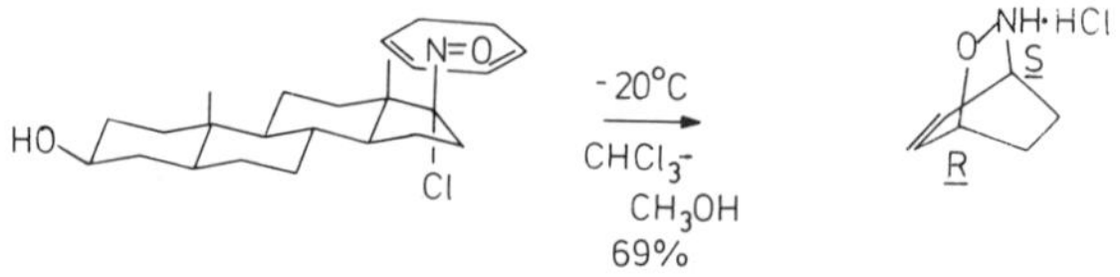

$$(16)$$

reacts with cyclohexadiene to afford an adduct with an enantiomeric excess of at least 95%.[29a] The configuration of this adduct is $1R,4S$, which would result if the diene approaches the dienophile from the most exposed face as shown in Eq. (16).

It is possible to prepare the enantiomeric series of adducts by using a chiral auxiliary derived from mannose [Eq. (17)].[29b]

$$(17)$$

In this case, the adduct formed from 1,3-cyclohexadiene had the $1S,4R$ configuration and again was produced in greater than 95% enantiomeric excess. It was proposed that the diene approaches the chloronitroso dienophile as shown in Eq. (17). It thus appears that methodology is now potentially available for synthesizing chiral dihydrooxazines of any absolute configuration.

4. ACYL- AND CYANONITROSO COMPOUNDS

The pioneering studies of Kirby et al. have established that acylnitroso compounds are useful heterodienophiles in [4 + 2] cycloadditions.[1c] These species, which can best be generated by periodate oxidation of a hydroxamic acid, are unstable and may undergo rapid solvolysis or dimerization.[30] However, in the presence of a diene such as 9,10-dimethylanthracene an isolable adduct such as 14 is produced [Eq. (18)].

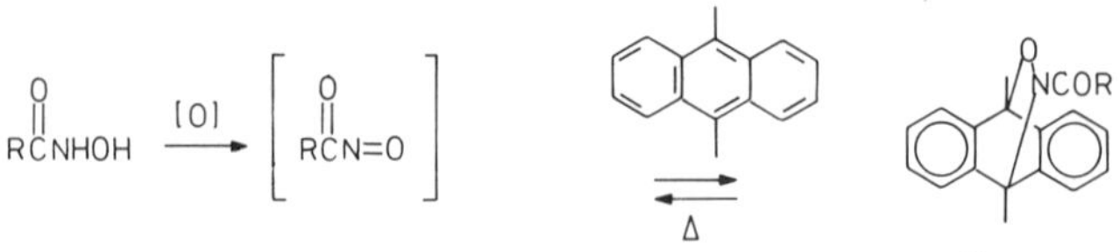

$$(18)$$

Compound 14 undergoes a retro-Diels–Alder reaction thermally at about 60°C to generate the acylnitroso intermediate which can be trapped by a different diene. Some adducts of acylnitroso compounds with cyclopentadiene also tend to undergo facile retro-[4 + 2] reactions readily and can

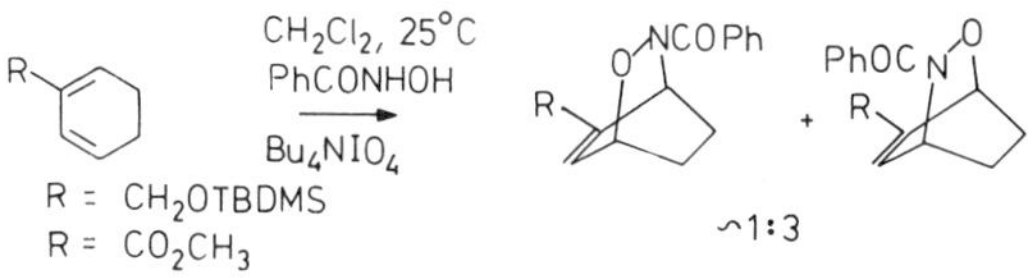
CH2Cl2, 25°C
PhCONHOH
Bu4NIO4
R = CH2OTBDMS
R = CO2CH3
NCOPh
PhOC
~1:3

Scheme 3-IX

often be used as an alternative source of acylnitroso dienophile.[30] One problem associated with acylnitroso compounds is their tendency to undergo ene reactions that are sometimes competitive with the Diels–Alder process [Eq. (19)].[30,31,34]

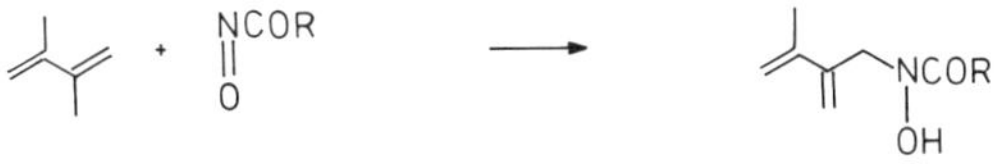
NCOR
O
NCOR
OH

(19)

However, [4 + 2] cycloadditions are usually more rapid than the corresponding ene reaction. Table 3-III shows some representative examples of Diels–Alder cycloadditions of acylnitroso dienophiles.[38]

Boger and co-workers have recently determined the regioselectivity of acylnitroso compound additions to some 2-substituted 1,3-cyclohexadienes.[5] In the cases of *both* electron-rich and electron-deficient substituted dienes the same regioisomer predominated by about 3 : 1 (Scheme 3-IX). It was suggested that the first case is consistent with a normal ($HOMO_{diene}$ controlled) Diels–Alder process, whereas the second reaction is consistent with either a $HOMO_{diene}$ or inverse electron demand ($LUMO_{diene}$ controlled) cycloaddition.

In approaches to diamino sugars, Defoin *et al.* have investigated additions of various acylnitroso compounds to 1,2-dihydropyridine **15** and found some rather interesting regiochemical preferences (Scheme 3-X).[39]

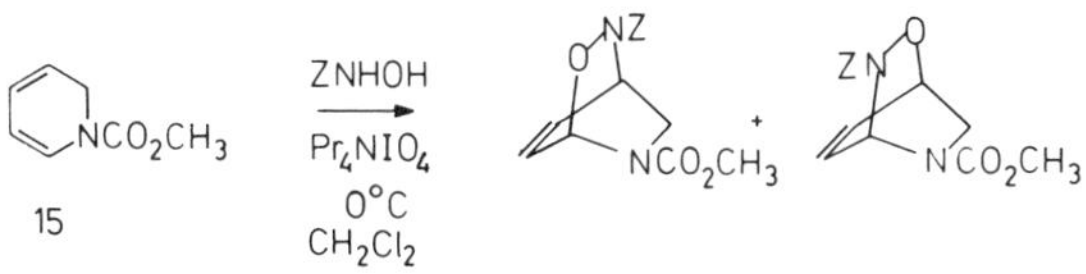
NCO2CH3
15
ZNHOH
Pr4NIO4
0°C
CH2Cl2
NZ
NCO2CH3
Z
NCO2CH3

	Ratio
Z = PhCO	0 : 100
= $PhCH_2CO$	0 : 100
= CH_3OCO	50 : 50
= $PhCH_2OCO$	50 : 50
= Me_2NCO	75 : 25

Scheme 3-X

TABLE 3-III

Cycloadditions of Acylnitroso Compounds with 1,3-Dienes

Nitroso dienophile precursor	Diene	Conditions	Product	Yield	Ref.
PhCONHOH	Thebaine	Et$_4$NIO$_4$, EtOAc	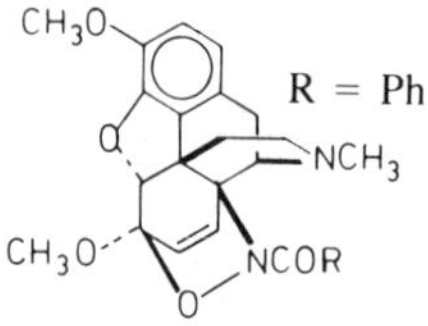 R = Ph	97%	32
CCl$_3$CH$_2$OCONHOH	Thebaine	NaIO$_4$, H$_2$O– EtOAc, 0°C	R = CCl$_3$CH$_2$O	82%	36, 3
MeCONHOH		Et$_4$NIO$_4$, CH$_2$Cl$_2$, 0°C	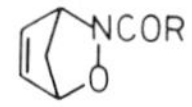 R = CH$_3$	70%	30
CF$_3$CONHOH		BnMe$_3$NIO$_4$, CH$_2$Cl$_2$, −70°C	R = CF$_3$	78%	35
2,4-(NO$_2$)$_2$C$_6$H$_3$ NHOH		BnMe$_3$NIO$_4$, CH$_2$Cl$_2$, −20°C	R = 2,4-(NO$_2$)$_2$C$_6$H$_3$	82%	35
tBuOCONHOH		NaIO$_4$, H$_2$O– EtOAc, 0°C	R = tBuO	70%	36
PhCONHOH		Et$_4$NIO$_4$, CH$_2$Cl$_2$, 0°C	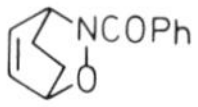	84%	30a,
14, R = Ph		C$_6$H$_6$, 80°C	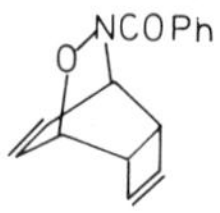	61%	30a
Me$_2$NCONHOH		Et$_4$NIO$_4$, CH$_2$Cl$_2$, 0°C	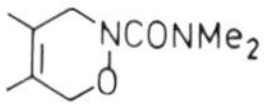	60%	34
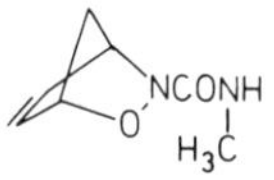	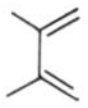	C$_6$H$_6$, 80°C		71%	34
EtOCONHOH		Pr$_4$NIO$_4$, CHCl$_3$, RT	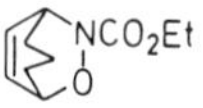	71%	29
PhCH$_2$OCONHOH		Et$_4$NIO$_4$, CH$_3$NO$_2$, −10°C	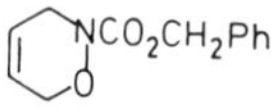	72%	36

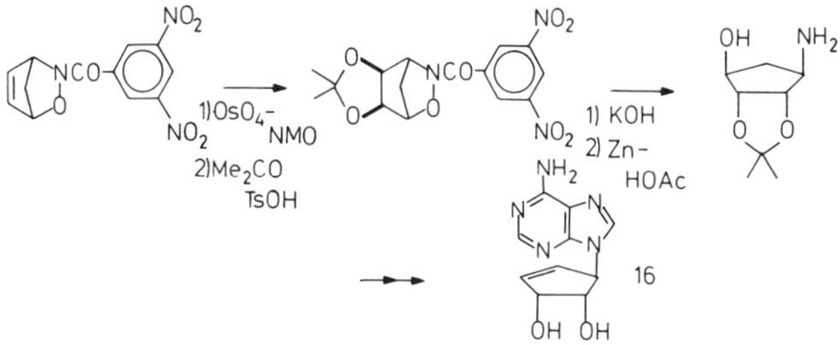

Scheme 3-XI

These results, which seem rather surprising, have not been satisfactorily rationalized.

During the past few years, acylnitroso Diels–Alder cycloadditions have been used as the key step in several natural product total syntheses. Retey *et al.*[40] have used the adduct from cyclopentadiene and the acylnitroso compound derived from 3,5-dinitrobenzoic acid[35] to synthesize the antitumor compound neplanocin A (**16**) (Scheme 3-XI).

Baldwin and co-workers have used a nitroso Diels–Alder reaction as a key step in total synthesis of tabtoxin (**17**), a metabolite causing leafspot disease in tobacco.[41] This group has also prepared tabtoxinine β-lactam (**18**), the active principle generated by *in vivo* enzymatic hydrolysis of **17**.

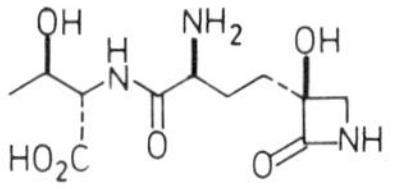

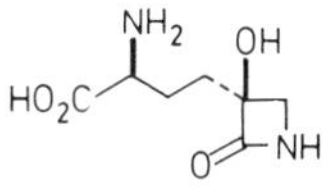

The synthesis of tabtoxinine β-lactam is briefly outlined in Scheme 3-XII. The initial nitroso formate cycloaddition was totally regioselective and provided the basic functionality needed to ultimately prepare **18**.

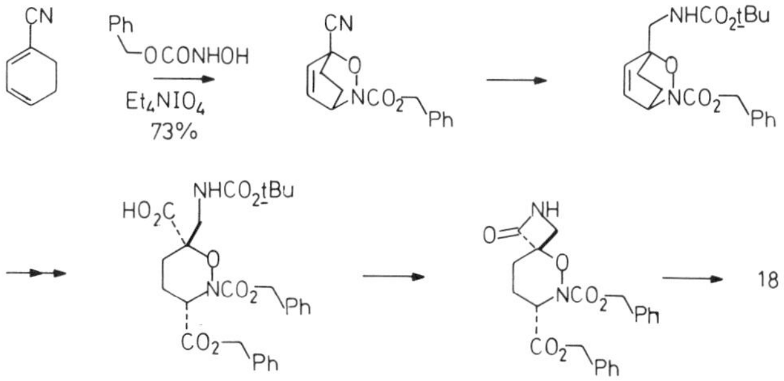

Scheme 3-XII

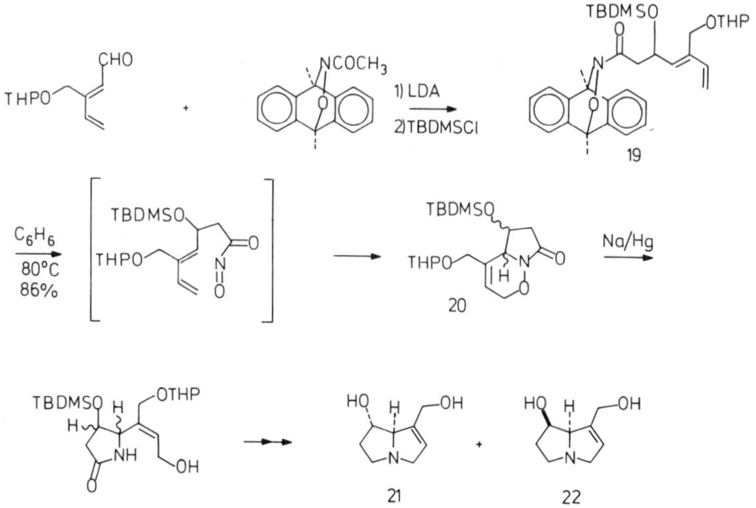

Scheme 3-XIII

A few examples have recently appeared involving intramolecular cy-cloadditions of acylnitroso dienophiles in alkaloid total synthesis. Keck has utilized such an intramolecular reaction as the key step in total syntheses of the necine bases heliotridine (**21**) and retronecine (**22**) (Scheme 3-XIII).[42] In this work, the dimethylanthracene adduct **19** was used as a precursor for the acylnitroso compound. The Diels–Alder cyclization afforded a 1.3 : 1 mixture[42a] of epimeric adducts **20**, which was converted to the individual alkaloids.

An intramolecular nitroso Diels–Alder approach to the frog neurotoxin gephyrotoxin 223AB (**25**) has recently been reported (Scheme 3-XIV).[43] Cyclization of the nitroso compound derived from hydroxamic acid **23** afforded the bicyclic adduct **24** in good yield, directly establishing two of

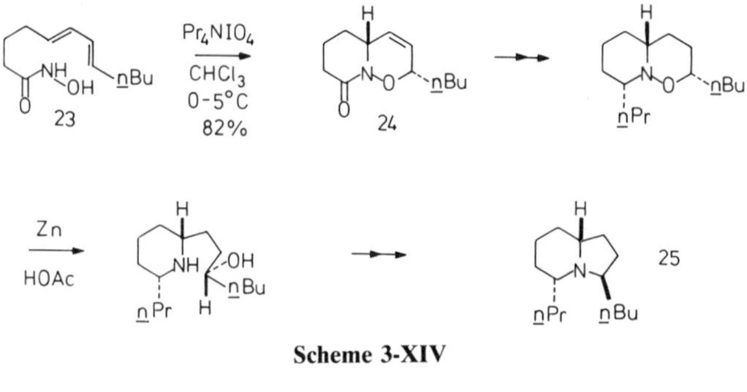

Scheme 3-XIV

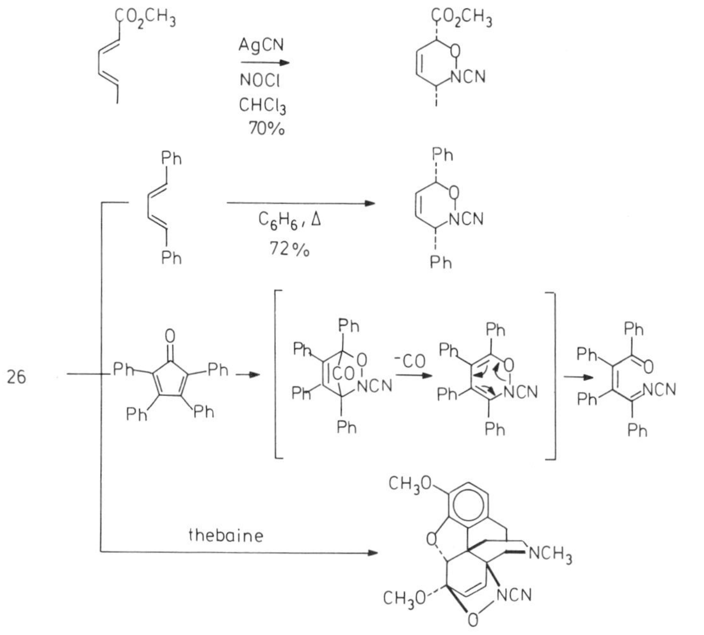

Scheme 3-XV

the three chiral centers of the alkaloid. A few additional steps served to convert **24** to the natural product.

Kirby *et al.* have examined the [4 + 2] cycloaddition chemistry of nitrosyl cyanide.[1c,44] This unstable species, which can be generated from nitrosyl chloride and silver cyanide, can be trapped *in situ* by conjugated dienes. However, as with acylnitroso compounds it is often cleaner and more convenient to use the 9,10-dimethylanthracene adduct **26** as a latent source of nitrosyl cyanide [Eq. (20)].

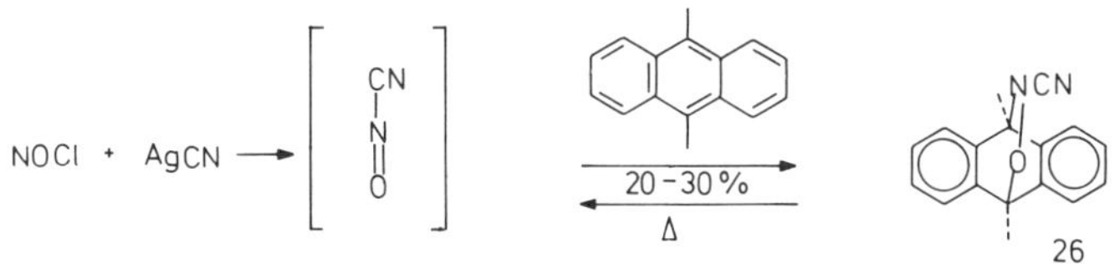

Several examples of hetero Diels–Alder reactions with nitrosyl cyanide are presented in Scheme 3-XV. Orientation of addition to unsymmetrical dienes generally corresponds to that of aryl- and chloroalkylnitroso compounds (*vide supra*).

One final type of electron-deficient nitroso dienophile worthy of mention are sulfonylnitroso compounds.[35,45] Surprisingly, these compounds have found only scant use in Diels–Alder reactions. The sulfonylnitroso compounds can be easily generated from an alkyl nitrite and a sulfinic acid [Eq. (21)]

$$
ArSO_2H + RONO \xrightarrow{Et_2O} \left[\begin{array}{c} NSO_2Ar \\ \parallel \\ O \end{array} \right] \xrightarrow{30\%} \quad (21)
$$

and trapped *in situ* with a diene to give a dihydrooxazine.

5. VINYLNITROSO AND RELATED COMPOUNDS

Work has recently appeared on the synthesis and cycloaddition reactions of vinylnitroso compounds.[46,47] Although these compounds are capable of acting either as heterodienes or as nitroso dienophiles, their reactivity is critically dependent on their structure. In general, if the vinylnitroso system contains any β substituents, the compounds will react as dienophiles; if there is no β substituent, the system prefers to act as the 4π component of a Diels–Alder reaction and will react with dienophiles.

Vinylnitroso compounds are generally unstable and are generated *in situ* by base-promoted elimination of α-haloximes [Eq. (22)].

$$
\begin{array}{c} NOH \\ R^1 \diagdown \diagup \diagdown_{R^3} \\ R^2 \diagup \quad \diagup \\ X \end{array} \xrightarrow{base} \begin{array}{c} R^3 \\ R^1 \diagdown \diagup \diagdown_{N\diagdown} \diagup^{O} \\ R^2 \end{array} \quad (22)
$$

These dienophiles react with various dienes to initially produce the usual dihydrooxazine adduct, but in many cases these compounds rearrange to epoxyaziridines. The examples depicted in Scheme 3-XVI are indicative of the various types of reactions these systems can undergo.

The iminonitroso compound of the type shown in Eq. (23),

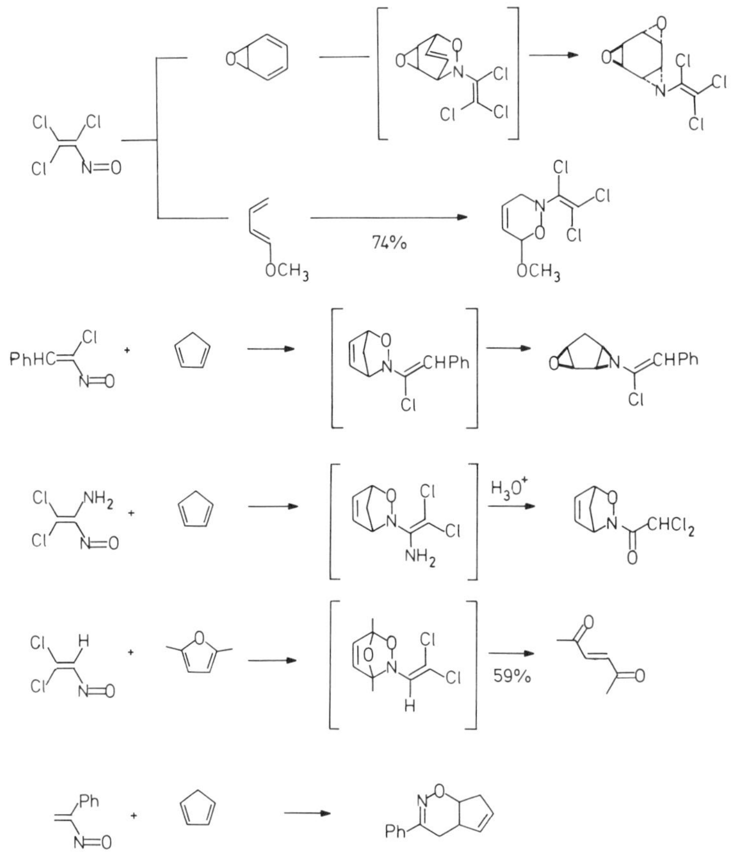

Scheme 3-XVI

which can be generated by oxidation of the corresponding oxime, reversibly cyclizes. However, the nitroso intermediate can be trapped by thebaine to give a stable Diels–Alder adduct.[48,49]

6. THIONITROSO COMPOUNDS

Thionitroso compounds have found very little synthetic use as dienophiles.[50] In general, these species are quite unstable and are often trapped *in situ* by conjugated dienes as a method of procf for their fleeting existence.

Some early methods of generation of thionitrosobenzenes are shown in Eqs. (24)[51] and (25).[52]

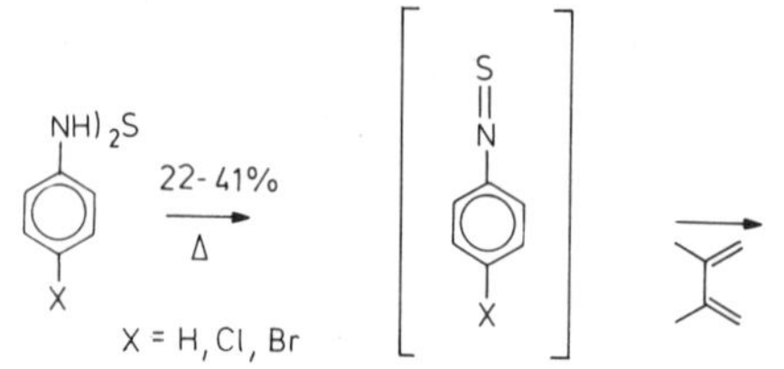

(24)

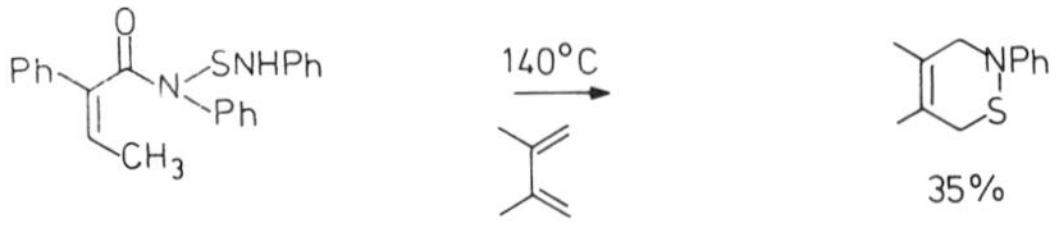

(25)

A more recent approach to aromatic thionitroso compounds was reported by Joucla and Rees,[53] which makes use of a photolytic decomposition of an azidobenzisothiazole ring system [Eq. (26)].

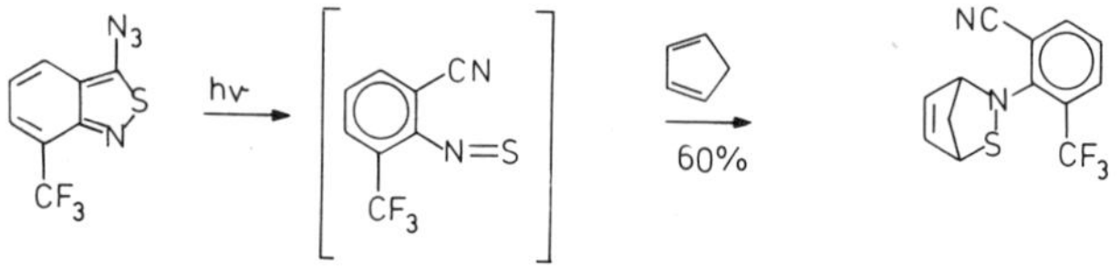

(26)

A very interesting method for generation of simple alkyl thionitroso compounds has been described [Eq. (27)].[54]

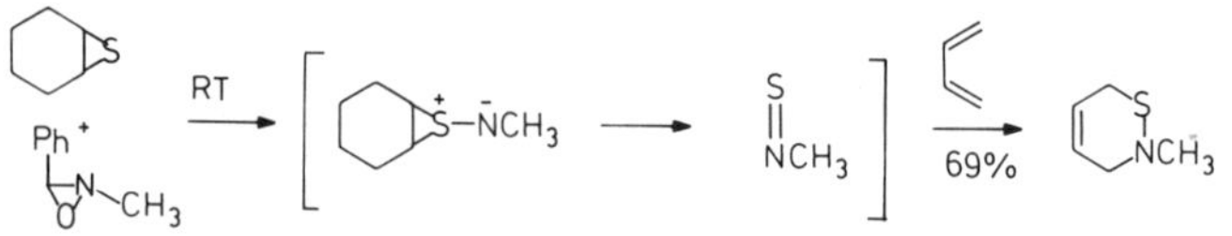

(27)

The thionitroso species could be trapped with butadiene to give the expected Diels–Alder adducts. An S,N-ylid is probably an initial intermediate in this sequence.

Acyl- and sulfonyl-substituted thionitroso compounds can also be prepared using an analogous strategy.[55] The S,N-ylid derived from tetrachlorothiophene reacts with alkenes in Diels–Alder fashion to give an adduct which spontaneously extrudes a thionitroso compound at ambient temperature. These species can be efficiently trapped with 1,3-dienes [Eq. (28)].

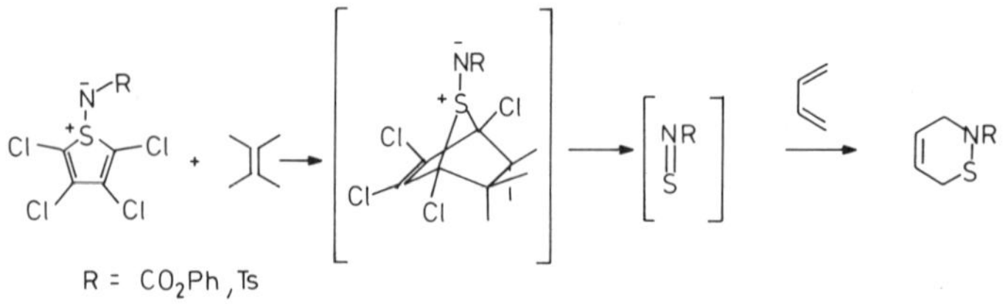

(28)

R = CO_2Ph, Ts

One example of what is perhaps an inverse electron demand Diels–Alder reaction of a thionitroso compound has appeared [Eq. (29)].[56]

$$\text{(29)}$$

Interestingly, N-thionitrosodimethylamine is a relatively stable compound,[57] unlike the C-thionitroso compounds mentioned above.

REFERENCES

1. Previous reviews: a. Hamer, J. (ed.) (1967). "1,4-Cycloaddition Reactions, the Diels–Alder Reaction in Heterocyclic Syntheses." Academic Press, New York. b. Kresze, G., and Firl, J. (1969). *Fort. Chem. Forsch.* **11**, 245. c. Kirby, G. W. (1977). *Chem. Soc. Rev.* **6**, 1. d. Weinreb, S. M., and Staib, R. R. (1982). *Tetrahedron* **38**, 3087.
2. a. Kresze, G., and Koshahn, W. (1971). *Tetrahedron* **27**, 1931. b. Kresze, G., Saitner, H., Firl, J., and Kosbahn, W. (1971). *Tetrahedron* **27**, 1941. c. Haussinger, P., and Kresze, G. (1978). *Tetrahedron* **34**, 689. d. Jacob, D., Niederman, H. P., and Meier, H. (1986). *Tetrahedron Lett.* **27**, 5703.
3. a. Labaziewicz, H., and Riddell, F. G. (1979). *J. Chem. Soc., Perkin Trans. 1*, 2926. b. Riddell, F. G. (1975). *Tetrahedron* **31**, 523. c. Leonard, N. J., Playtis, A. J., Skoog, F., and Schmitz, R. Y. (1971). *J. Am. Chem. Soc.* **93**, 3056; Leonard, N. J., and Playtis, A. J. (1972). *J. Chem. Soc., Chem. Commun.*, 133.
4. Sasaki, T., Eguchi, S., Ishii, T., and Yamada, H. (1970). *J. Org. Chem.* **35**, 4273; see also Ohno, M., Mori, K., and Eguchi, S. (1986). *Tetrahedron Lett.* **27**, 3381.
5. Boger, D. L., and Patel, M. (1984). *J. Org. Chem.* **49**, 4099; Boger, D. L., Patel, M., and Takusagawa, F. (1985). *J. Org. Chem.* **50**, 1911.
6. Kresze, G., and Hartner, H. (1973). *Liebigs Ann. Chem.*, 650; see also Defoin, A., Fritz, H., Geffroy, G., and Streith, J. (1986). *Tetrahedron Lett.* **27**, 3135.
7. Givens, R. S., Chou, D. J., Merchant, S. N., Stitt, R. P., and Matuszewski, B. (1982). *Tetrahedron Lett.* **23**, 1327.
8. Hart, H., Ramaswami, S. K., and Willer, R. (1979). *J. Org. Chem.* **44**, 1.
9. Kirby, G. W., Bentley, K. W., Horsewood, P., and Singh, S. (1979). *J. Chem. Soc., Perkin Trans. 1*, 3064.
10. Taylor, E. C., Tseng, C.-P., and Rampal, J. B. (1982). *J. Org. Chem.* **47**, 552.
11. Taylor, E. C., McDaniel, K., and Skotnicki, J. S. (1984). *J. Org. Chem.* **49**, 2500.
12. Knaus, E. E., Avasthi, K., and Giam, C. S. (1980). *Can. J. Chem.* **58**, 2447.
13. Streith, J., Augelmann, G., Fritz, H., and Strub, H. (1982). *Tetrahedron Lett.* **23**, 1909; Augelmann, G., Streith, J., and Fritz, H. (1985). *Helv. Chim. Acta* **68**, 95.
14. Kresze, G., and Rubner, R. (1969). *Chem. Ber.* **102**, 1280; Kresze, G., Heidegger, P., and Asbergs, A. (1970). *Liebigs Ann. Chem.* **738**, 113.
15. a. Kresze, G., Dittel, W., and Melzer, H. (1981). *Liebigs Ann. Chem.*, 224. b. Kresze, G., Kysela, E., and Dittel, W. (1981). *Liebigs Ann. Chem.*, 210. c. Kresze, G., and Kysela, E. (1981). *Liebigs Ann. Chem.*, 202.

16. Cavaleiro, J. A. S., Jackson, A. H., Neves, M. G. P. M. S., and Rao, K. R. N. (1985). *J. Chem. Soc., Chem. Commun.*, 776.
17. Ashkenazi, P., Gleiter, R., von Philipsborn, W., Bigler, P., and Ginsburg, D. (1981). *Tetrahedron* **37**, 127.
18. Waterman, P. S., Belmonte, J. E., Bauch, T. E., Belmonte, P. A., and Giering, W. P. (1985). *J. Organomet. Chem.* **294**, 235.
19. Becker, Y., Eisenstadt, A., and Shvo, Y. (1978). *Tetrahedron* **34**, 799, and references cited therein.
20. Wichterle, O. (1947). *Coll. Czech. Chem. Commun.* **12**, 292.
21. Kesler, E. (1980). *J. Heterocycl. Chem.* **17**, 1113.
22. Ranganathan, D., Ranganathan, S., Rao, C. B., and Raman, K. (1981). *Tetrahedron* **37**, 629.
23. Belleau, B., and Au-Young, Y.-K. (1963). *J. Am. Chem. Soc.* **85**, 64.
24. Kresze, G., and Dittel, W. (1981). *Liebigs Ann. Chem.*, 610.
25. Kresze, G., Weiss, M. M., and Dittel, W. (1984). *Liebigs Ann. Chem.*, 203.
26. See also Kresze, G., and Melzer, H. (1981). *Liebigs Ann. Chem.*, 1874.
27. Iida, H., Watanabe, Y., and Kibayashi, C. (1985). *J. Org. Chem.* **50**, 1818.
28. Nitsch, H., and Kresze, G. (1976). *Angew. Chem. Int. Ed. Engl.* **15**, 760.
29. a. Sabuni, M., Kresze, G., and Braun, H. (1984). *Tetrahedron Lett.* **25**, 5377. b. Felber, H., Kresze, G., Braun, H., and Vasella, A. (1984). *Tetrahedron Lett.* **25**, 5381. c. Felber, H., Kresze, G., Prewo, R., and Vasella, A. (1986). *Helv. Chim. Acta* **69**, 1137. d. Braun, H., Klier, K., Kresze, G., Sabuni, M., Werbitzky, O., and Winkler, J. (1986). *Liebigs Ann. Chem.* 1360.
30. a. Corrie, J. E., Kirby, G. W., and MacKinnon, J. W. M. (1985). *J. Chem. Soc., Perkin Trans. 1*, 883. b. Dao, L. H., Dust, J. M., McKay, D., and Watson, K. N. (1979). *Can. J. Chem.* **57**, 1712.
31. Keck, G. E., and Webb, R. R. (1982). *J. Org. Chem.* **47**, 1302, and references cited therein.
32. Kirby, G. W., and Sweeny, J. G. (1973). *J. Chem. Soc., Chem. Commun.*, 704; Kirby, G. W., and Sweeny, J. G. (1981). *J. Chem. Soc., Perkin Trans. 1*, 3250.
33. Keck, G. E., and Fleming, S. A. (1978). *Tetrahedron Lett.*, 4763.
34. Christie, C. C., Kirby, G. W., McGuigan, H., and MacKinnon, J. W. M. (1985). *J. Chem. Soc., Perkin Trans. 1*, 2469.
35. Just, G., and Cutrone, L. (1976). *Can. J. Chem.* **54**, 867.
36. Kirby, G. W., McGuigan, H., MacKinnon, J. W. M., McLean, D., and Sharma, R. P. (1985). *J. Chem. Soc., Perkin Trans. 1*, 1437.
37. Kirby, G. W., and McLean, D. (1985). *J. Chem. Soc., Perkin Trans. 1*, 1443.
38. For some interesting rearrangements of these heterocycles, see a. Kirby, G. W., and MacKinnon, J. W. M. (1985). *J. Chem. Soc., Perkin Trans. 1*, 887. b. Ranganathan, D., Ranganathan, S., and Rao, C. B. (1981). *Tetrahedron* **37**, 637. c. MacKay, D., Dao, L. H., and Dust, J. M. (1980). *J. Chem. Soc., Perkin Trans. 1*, 2408.
39. Defoin, A., Schmidlin, C., and Streith, J. (1984). *Tetrahedron Lett.* **25**, 4515; see also Defoin, A., Fritz, H., Geffroy, G., and Streith, J. (1986). *Tetrahedron Lett.* **27**, 4727.
40. Jung, M., Offenbächer, G., and Retey, J. (1983). *Helv. Chim. Acta* **66**, 1915.
41. a. Baldwin, J. E., Bailey, P. D., Gallacher, G., Singleton, K. A., and Wallace, P. M. (1983). *J. Chem. Soc., Chem. Commun.*, 1049. b. Baldwin, J. E., Otsuka, M., and Wallace, P. M. (1985). *J. Chem. Soc., Chem. Commun.*, 1549. c. Baldwin, J. E., Otsuka, M., and Wallace, P. M. (1986). *Tetrahedron* **42**, 3097.
42. a. Keck, G. E. (1978). *Tetrahedron Lett.*, 4767. b. Keck, G. E., and Nickell, D. G. (1980). *J. Am. Chem. Soc.* **102**, 3632.

43. Iida, H., Watanabe, Y., and Kibayashi, C. (1985). *J. Am. Chem. Soc.* **107**, 5535.
44. Horsewood, P., and Kirby, G. W. (1980). *J. Chem. Soc., Perkin Trans. 1*, 1587; Horsewood, P., Kirby, G. W., Sharma, R. P., and Sweeny, J. G. (1981). *J. Chem. Soc., Perkin Trans. 1*, 1802.
45. Kresze, G., and Kort, W. (1961). *Chem. Ber.* **94**, 2624.
46. Faragher, R., and Gilchrist, T. L. (1979). *J. Chem. Soc., Perkin Trans. 1*, 249.
47. Viehe, H. G., Merényi, R., Francotte, E., Van Meerssche, M., Germain, G., Reclercq, J. P., and Bodart-Gilmont, J. (1977). *J. Am. Chem. Soc.* **99**, 2340; Francotte, E., Merényi, R., and Viehe, H. G. (1978). *Angew. Chem. Int. Ed. Engl.* **17**, 936; Francotte, E., Merényi, R., Vandenbulcke-Coyette, B., and Viehe, H. G. (1981). *Helv. Chim. Acta* **64**, 1208.
48. Gilchrist, T. L., Peek, M. E., and Rees, C. W. (1975). *J. Chem. Soc., Chem. Commun.*, 913, 915.
49. See also Ref. 38a.
50. For a brief review, see Mayer, R., Bleisch, S., and Domschke, G. (1978). *Zeit. Chem.* **18**, 323.
51. Tavs, P. (1966). *Angew. Chem. Int. Ed. Engl.* **5**, 1048; Davis, F. A., and Skibo, E. R. (1976). *J. Org. Chem.* **41**, 1333.
52. Minami, T., Yamataka, K., Ohshiro, Y., Agawa, T., Yasuoka, N., and Kasai, N. (1972). *J. Org. Chem.* **37**, 3810.
53. Joucla, M. F., and Rees, C. W. (1984). *J. Chem. Soc., Chem. Commun.*, 374.
54. Hata, Y., and Watanabe, M. (1980). *J. Org. Chem.* **45**, 1691.
55. a. Meth-Cohn, O., and van Vuuren, G. (1984). *J. Chem. Soc., Chem. Commun.*, 1144.
    b. Meth-Cohn, O., and van Vuuren, G. (1986). *J. Chem. Soc., Perkin Trans. 1*, 245.
56. Seitz, G., and Overheu, W. (1979). *Chem. Zeit.* **103**, 230.
57. Middleton, W. J. (1966). *J. Am. Chem. Soc.* **88**, 3842.

Chapter **4**

Carbonyl Dienophiles

INTRODUCTION

Cycloaddition of a 1,3-diene with a carbonyl compound provides a 5,6-dihydro-2H-pyran derivative [Eq. (1)].[1]

$$
\underset{R}{\overset{O}{\underset{\,}{\bigwedge}}}\!\!R' \quad + \quad \Big\langle\!\!\!\Big\langle \quad \longrightarrow \quad {}^{3}\overset{2}{\underset{4}{\bigotimes}}\overset{1}{\underset{5\ R}{\bigg|}}\overset{R'}{_{6}} \tag{1}
$$

In general, reactions of this type proceed poorly with aliphatic and aromatic aldehydes and ketones unless highly reactive dienes and/or Lewis acid catalysts are used. For example, acetaldehyde and isoprene give only a trace of adduct at 100°C.[2] Aromatic aldehydes do not react with 2,3-dimethylbutadiene under thermal conditions, although [4 + 2] adducts can be obtained in mediocre yields if p-tolunesulfonic acid is used as a catalyst.[3,4]

Electron-deficient aldehydes and ketones, on the other hand, are generally more reliable dienophiles in both thermal and acid-catalyzed cycloadditions with alkyl-substituted 1,3-dienes. These reactions can also be promoted by high pressure techniques. Interestingly, carbonyl compounds other than aldehydes and ketones (i.e., amides, esters, etc.) do not appear to be reactive as dienophiles.

It has been suggested that cycloadditions of carbonyl dienophiles with 1,3-dienes are in agreement with FMO theory.[5] However, aside from some very recent qualitative mechanistic work on cycloadditions of aldehydes and highly oxygenated dienes under Lewis acid catalysis (*vide infra*), there are no rigorous mechanistic studies in this area. As outlined in specific examples in the following sections, carbonyl dienophile Diels–Alder reactions show good regioselectivity usually in accord with the predictions of FMO theory. These reactions are syn stereoselective with respect to the 4π component, but relatively little information is available on endo/exo selectivity. What is known about the stereochemistry of the process is included in some of the following sections.

1. ELECTRON-DEFICIENT ALDEHYDES

It has been recognized for many years that chloral will react with simple 1,3-dienes to afford reasonable yields of [4 + 2] adducts.[6,7] The stereochemistry of the Diels–Alder reaction of cyclohexadiene and chloral[7] has been reinvestigated and was unambiguously established to give the endo adduct [Eq. (2)].[8]

$$\text{(2)}$$

In order for this cycloaddition to be reproducible, it was necessary to exclude all moisture and conduct the reaction with highly purified reagents under vacuum. The [4 + 2] cycloadduct is extremely acid labile, rearranging rapidly to the [3.2.1] isomer.

[4 + 2] Cycloadditions of 1,3-dienes with glyoxylates have also been known for quite a long time.[1,6b,9,10] In recent years this reaction has been developed into a powerful method for carbohydrate synthesis (*vide infra*).[1c,d]

Two groups have studied the stereochemistry of addition of (*E*)-1-alkoxybutadienes with simple glyoxylate esters.[9b,e,10] In virtually all of the cases reported, approximately a 2:1 mixture of cis (endo) to trans (exo) adducts were obtained if the reaction was conducted near room tempera ture and under neutral conditions, although there was some variation depending on the solvent [Eq. (3)].

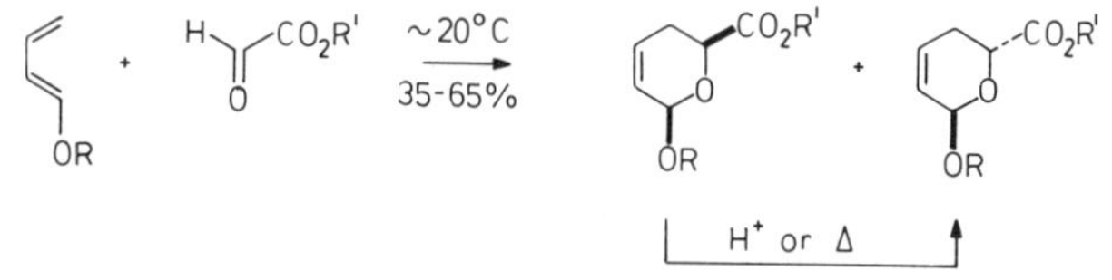

(3)

R = CH₃, Et, nPr, nBu; R' = Et, nBu

However, if the cycloaddition was run at higher temperatures ($\sim 150°C$) or if an acid was added, the trans isomer predominated ($>4:1$ ratio). It could be demonstrated that the cis isomer is the kinetically favored product but tends to readily isomerize to the trans under thermal or acidic conditions, and this fact has been used in a number of synthetic applications (vide infra).

Jung et al. recently found that methyl glyoxylate adds to Danishefsky's diene to afford, after hydrolysis, a good yield of a pyranone as an equal mixture of cis and trans stereoisomers [Eq. (4)].[11a]

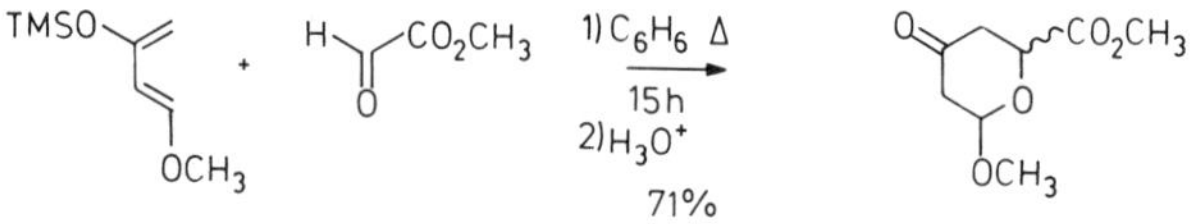

(4)

At high pressure the initial adducts themselves can be isolated.[11b] Other highly oxygenated dienes also undergo this cycloaddition.[11c]

The additions of glyoxylates with unsymmetrical dienes appear to be completely regioselective. As has been pointed out by Fleming,[5] FMO theory rationalizes these results, and the regiochemistry of cycloaddition is presumably controlled by the interaction of the HOMO of the diene and the LUMO of the carbonyl compound.

More recently, Hosomi et al.[12] reported some regioselective cycloadditions of silyl-substituted 1,3-dienes with n-butyl glyoxylate to afford dihydropyrans in good yields (Scheme 4-I). As in the examples cited above, products in the last two cases were stereoisomer mixtures due to ready epimerization at C-2. n-Butyl glyoxylate also reacts smoothly with 1,3-cyclohexadiene under thermal conditions, affording a $9:1$ mixture of the endo and exo bicyclic adducts, respectively [Eq. (5)].[10b]

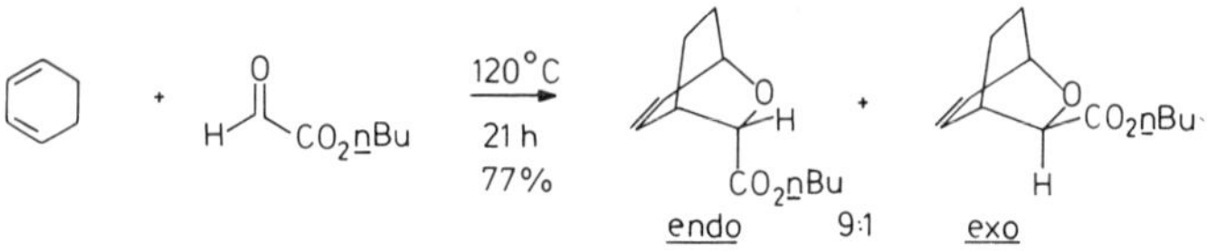

(5)

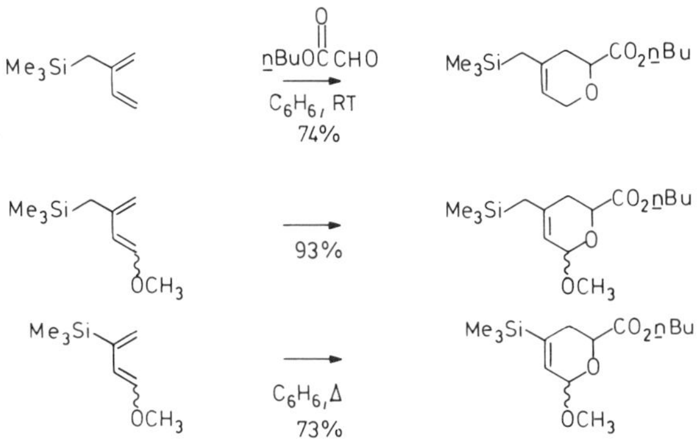

Scheme 4-I

Diels–Alder reactions of glyoxylates have recently found application in natural product total synthesis. For example, Schmidt and Wagner used such a cycloaddition as a key step in synthesis of uracinine derivative **3**, a component of the nucleoside antibiotic blasticidin S (Scheme 4-II).[13] Interestingly, when unsymmetrical diene **1** undergoes [4 + 2] cycloaddition with a glyoxylate, only one series of regioisomers is produced, which has the incorrect orientation for synthesis of **3**.[14] However, if the o-benzoyl

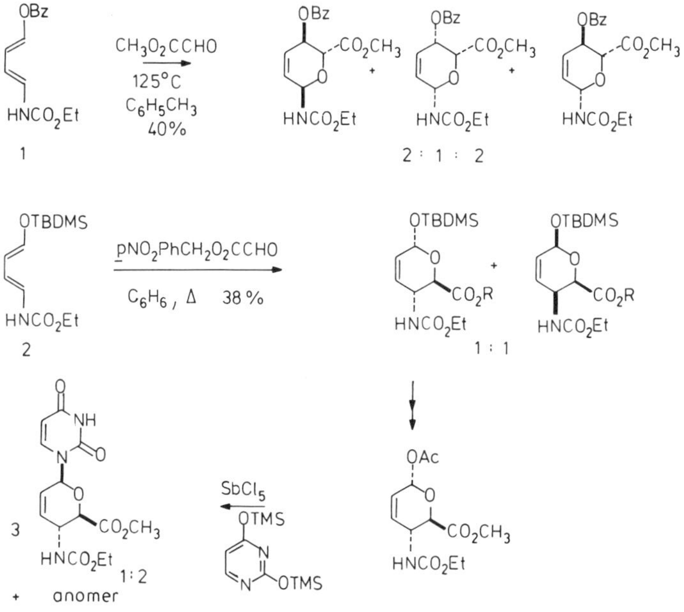

Scheme 4-II

group was replaced by *tert*-butyldimethylsilyl, the regiochemistry of cycloaddition totally reversed. It was suggested that these processes are $HOMO_{diene}$–$LUMO_{dienophile}$ controlled, and that by substituting an electron-donating group on oxygen for an electron-withdrawing one, the sizes of the coefficients at C-4 and C-1 of the diene would be reversed. This effect should thus change the regiochemistry of addition with a glyoxylate, which was actually observed experimentally. The appropriate stereoisomer formed from diene **2** was converted to uracinine **3** as outlined in the scheme.

Schmidt and Angerbauer have used the Diels–Alder adducts of (*E,E*)-1,4-diacetoxybutadiene (**4**) and *n*-butyl glyoxylate in synthesis of carbohydrate pseudoglycals.[15] Cycloaddition of **4** gave a mixture of three isomeric adducts **5–7** (Scheme 4-III). The two compounds **6** and **7** probably arise by rearrangement of the all-cis (endo) adduct, presumably caused by a trace contaminant of acetic acid in the glyoxylate ester which was used. These adducts were found to react with various alcohols to cleanly yield α-glycosides. For instance, treatment of adduct **5** with the ribose derivative shown gave greater than 95% of the α-linked disaccharide. Similarly, adducts **6** and **7** also produced only the α anomers.

In an extensive study Jurczak *et al.* examined asymmetric induction in the Diels–Alder reactions of a number of achiral dienes with several optically active esters of glyoxylic acid.[16,17] Cycloadditions were run both at atmospheric pressure[16a,c] and at high pressures (6–10 kbar).[16b,d,e] To summarize these studies, asymmetric induction using glyoxylate esters of

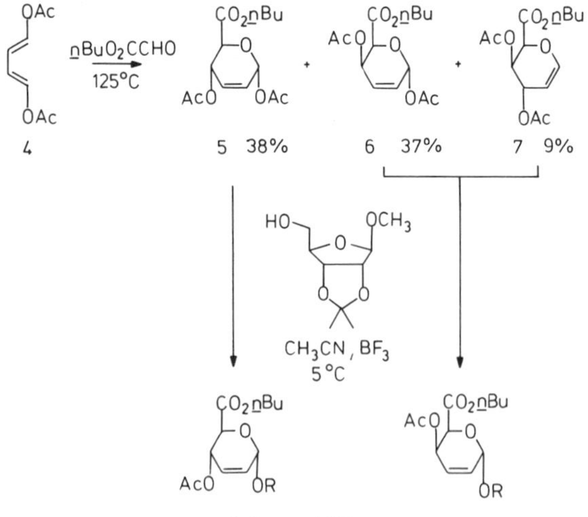

Scheme 4-III

optically active menthol, borneol, 2-octanol, and 2,2-dimethyl-3-heptanol gave rather poor optical yields in the range of 0.4–13% at atmospheric pressure, depending on the solvent used. At high pressures, optical yields of adducts with $(R)(-)$-menthyl glyoxylate esters were somewhat better (1–21%) depending on diene, solvent, and pressure. Jurczak has proposed a mechanistic model to account for these results.[16e] Unfortunately, since the enantiomeric excesses here are so low, this process would not seem to have much synthetic utility.

In a series of publications, David and co-workers have utilized cycloadditions of glyoxylates in an elegant strategy for synthesis of disaccharides and more complex carbohydrates.[18] The basic approach involves use of E-1-oxygenated butadienes attached to various sugar derivatives in Diels–Alder reactions with glyoxylates. The power of this method is best exemplified by the total synthesis of the antigenic determinant of blood group A (**9**), a trisaccharide of considerable complexity (Scheme 4-IV).[18c] A primary feature of this particular route is that the chiral diene and chiral dienophile reinforce each other in controlling absolute stereochemistry at C-5 in adduct **8.** An exo/endo mixture (i.e., trans/cis) was initially formed in the cycloaddition but, as noted above, could be equilibrated by BF_3 to the desired trans (α,D) configuration.

Another important feature of this synthetic approach is that even with achiral glyoxylates cycloadditions generally show good diastereofacial selectivity, even though endo/exo mixtures (which are subsequently equilibratable) are usually formed. Table 4-I shows some stereochemical

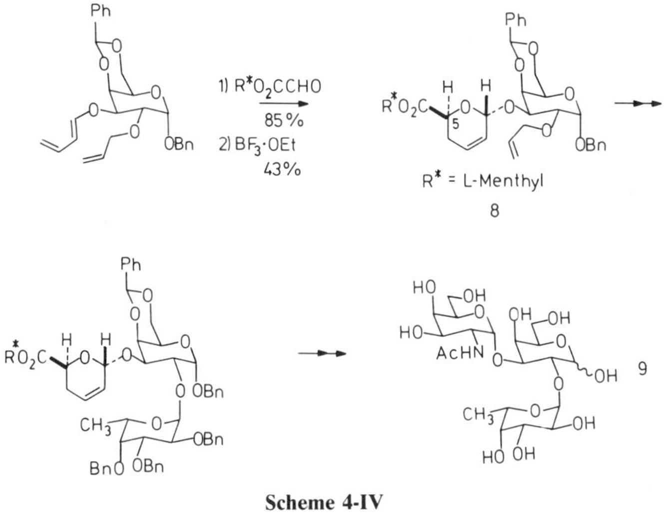

Scheme 4-IV

TABLE 4-I

Stereochemistry of Cycloadditions of Chiral Dienes with *n*-Butyl Glyoxylate[a]

Diene	α-D	β-D	α-L	β-L	Endo	Face (+)	Face(−)
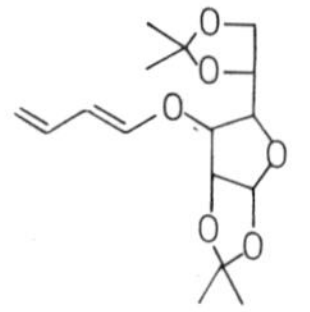	13%	18%	9%	60%	78%	27%	73%
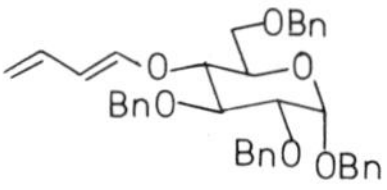	44%	4%	0%	52%	56%	4%	96%
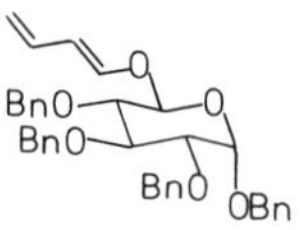	48%	18%	0%	34%	52%	18%	82%
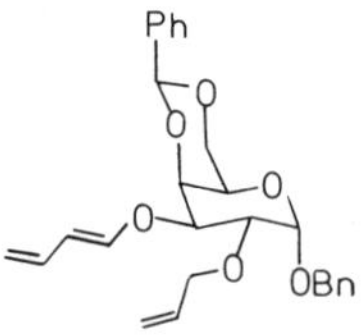	1.1%	50.3%	46.5%	2.1%	52.4%	96.8%	3.2%
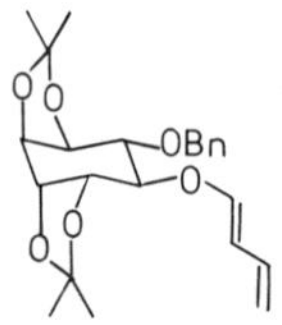	20%	25%	47%	8%	33%	72%	28%

[a] From Ref. 18d.

results of cycloadditions of dienes bearing chiral sugar appendages with a simple, achiral glyoxylate ester.[18d]

2. ELECTRON-DEFICIENT KETONES

Several electron-deficient ketones are known to participate in Diels–Alder reactions as heterodienophiles. Carbonyl cyanide[19] and α-halogenated ketones[11b,20] have been used, but relatively little has been done with these dienophiles in recent years.[1] Diethyl ketomalonate is also a reactive 2π component,[9b] and this ketone has been applied widely in Diels–Alder reactions.[21–24] Some examples of cycloadditions of this type are listed in Table 4-II. As in the case of glyoxylates, these reactions show excellent regioselectivity and the usual cis stereoselectivity with respect to the 1,3-diene.

Ruden and Bonjouklian have suggested that ketomalonate might serve as the synthetic equivalent of carbon dioxide, which itself does not react as a dienophile.[21] They have found that adducts can, in fact, be transformed via a Curtius sequence into β,γ-unsaturated lactones [Eq. (6)].

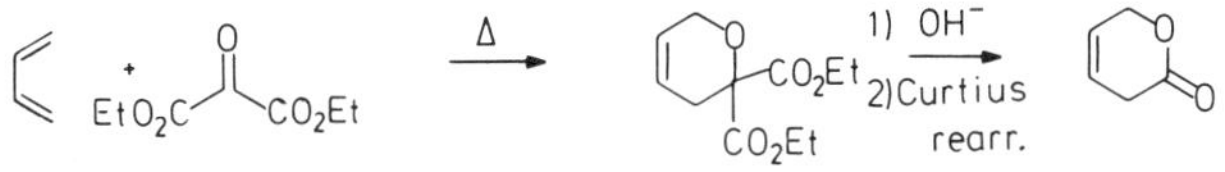

$$\tag{6}$$

A cycloaddition of diethyl ketomalonate has been used in total synthesis of the perfumery product (+)-ambreinolide (12) and its C-8 epimer 14 (Scheme 4-V).[25] The highly congested conjugated diene 10 did not react thermally with diethyl ketomalonate, nor did Lewis acids help. However, at high pressure a mixture of epimeric [4 + 2] adducts 11 and 13 were produced in a 65 : 35 ratio. The major product results from attack of the dienophile on the face of the diene opposite to the angular methyl group. The reaction was totally regioselective, giving only the orientation products shown. These adducts could be converted to 12 and 14 using the methodology of Ruden and Bonjouklian described above [Eq. (6)].[21]

Some other electron-deficient ketones have been used as dienophiles in Diels–Alder reactions. Various types of 1,2,3-tricarbonyl compounds seem to be quite reactive in this type of process, and some selected examples are listed in Table 4-III.[26–28]

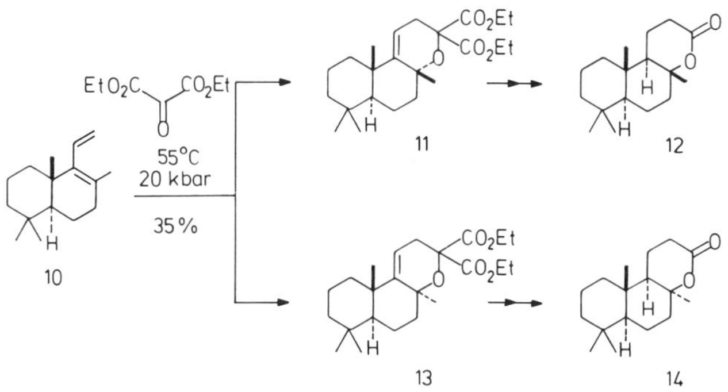

Scheme 4-V

TABLE 4-II

Thermal Cycloadditions of Diethyl Ketomalonate

Diene	Conditions	Product(s)	Yield	Ref.
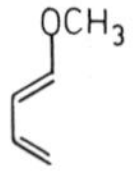	80°C, 1.5 hr	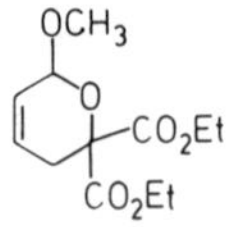	52%	9b
	CH₃CN, 130°C, 4 hr	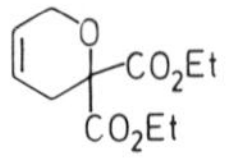	63%	21
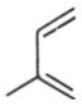	CH₃CN, 130°C, 4 hr	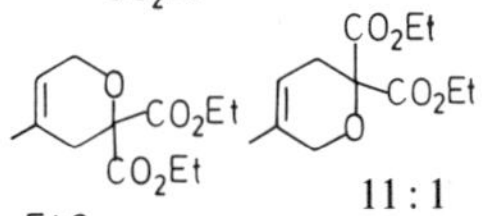 11:1	64%	21
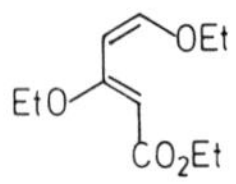	50°C, 6 hr	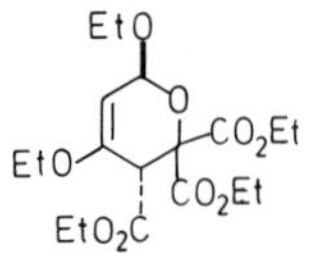	54%	22
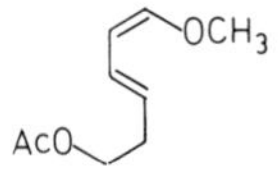	C₆H₆, 115°C, 60 hr 60°C, 8.5 kbar, 48 hr	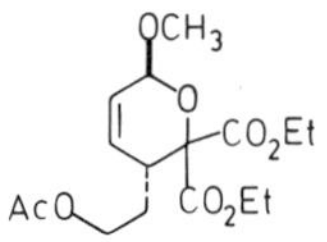	51% 98%	22
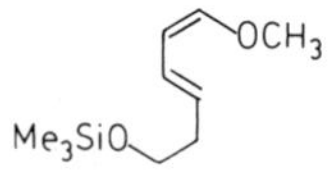	80°C, 72 hr	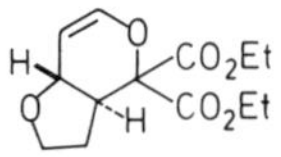	43%	22a
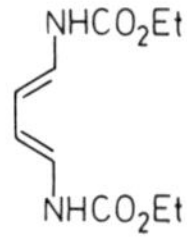	80°C, 44 hr	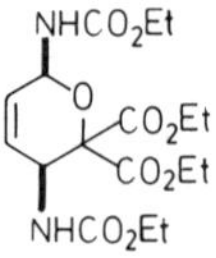	34%	22b
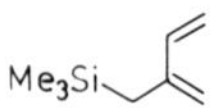	C₆H₆, Δ, 10 hr	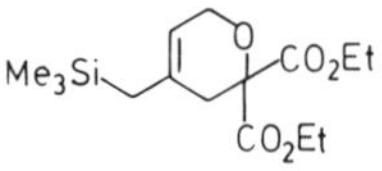	71%	12
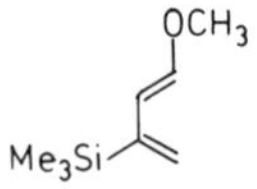	C₆H₆, 50°C, 10 hr	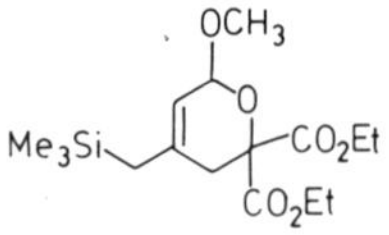	100%	12
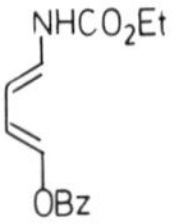	Toluene, 110°C, 80 hr	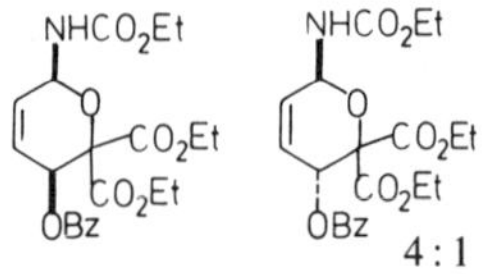 4:1	56%	14

TABLE 4-II (*Continued*)

Diene	Conditions	Product(s)	Yield	Ref.
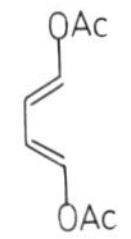 [structure: diene with OAc groups]	C_6H_6, 140°C, 20 hr	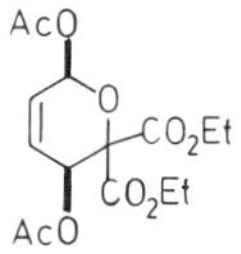 [structure: AcO ... CO_2Et, CO_2Et, AcO]	87%	15b
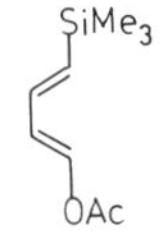 [structure: diene with SiMe$_3$ and OAc]	Xylene, Δ, 20 hr	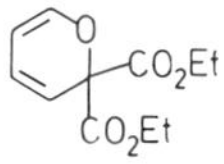 [structure: product with CO_2Et, CO_2Et]	12%	24
 [cyclohexadiene structure]	CH_3CN, 130°C, 4 hr	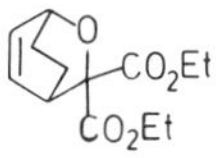 [bicyclic product with CO_2Et, CO_2Et]	74%	21
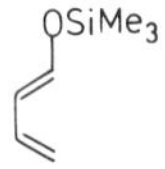 [structure with OSiMe$_3$]	150–160°C, 24 hr	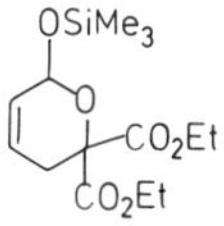 [OSiMe$_3$ product with CO_2Et, CO_2Et]	55%	23
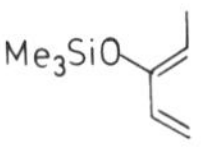 [Me$_3$SiO diene structure]	150–160°C, 24 hr	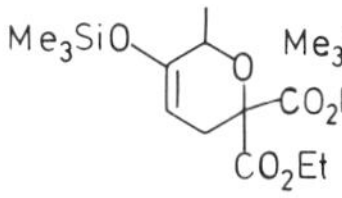 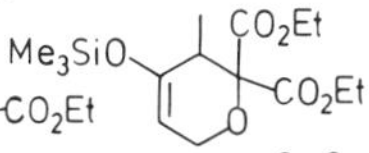 [Me$_3$SiO ... CO_2Et, CO_2Et and Me$_3$SiO ... CO_2Et products] 3:2	24%	23
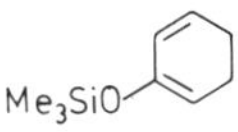 [Me$_3$SiO cyclohexadiene]	150–160°C, 24 hr	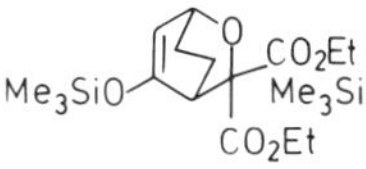 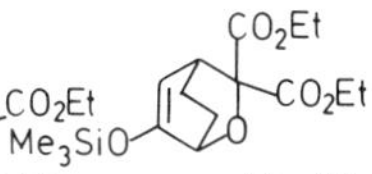 [Me$_3$SiO bicyclic products with CO_2Et] 55:45	41%	23
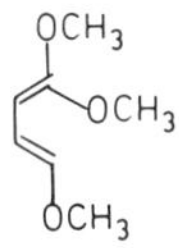 [structure with OCH$_3$, OCH$_3$, OCH$_3$]	CCl_4, Δ, 2 hr	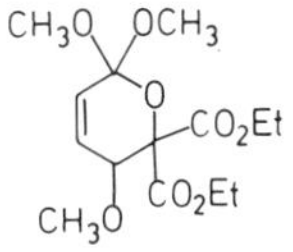 [CH$_3$O, OCH$_3$ product, CH$_3$O, CO_2Et, CO_2Et]	70%	11c
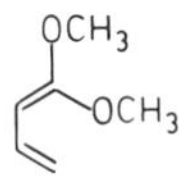 [structure with OCH$_3$, OCH$_3$]	CCl_4, RT, 1 hr	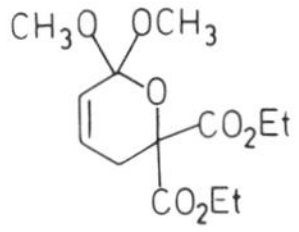 [CH$_3$O, OCH$_3$ product with CO_2Et, CO_2Et]	95%	11c

Although *o*- and *p*-quinones generally react with 1,3-dienes to give carbocycles, a few scattered examples of quinones acting as carbonyl dienophiles have been reported. For example, *o*-chloranil and *o*-bromanil react with 2,3-dimethylbutadiene to give [4 + 2] adducts [Eq. (7)].[29]

TABLE 4-III
Cycloadditions of Triones

Trione	Diene	Conditions	Product	Yield	Ref.
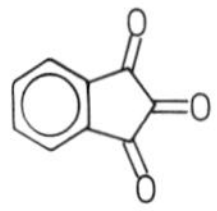	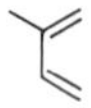	RT, 28 days	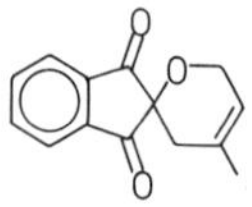	76%	26
	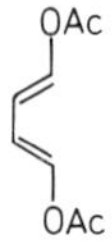	CH$_2$Cl$_2$, 25°C, 11 days	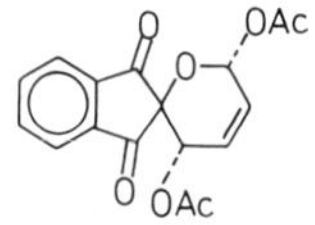	71%	27
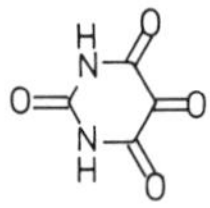	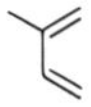	RT, 5 days	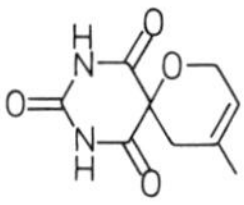	95%	26
	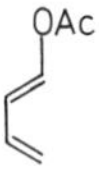	DME, 25°C, 5 days	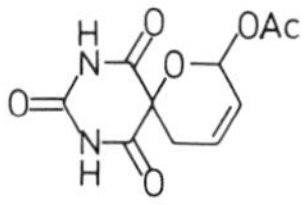	50%	27
	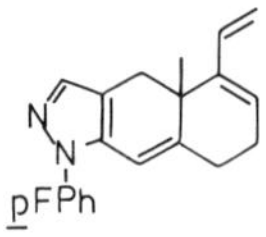	Toluene, Δ, 24 hr	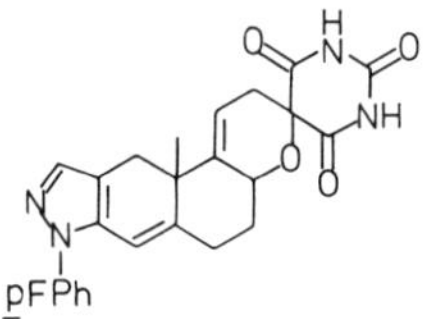	92%	28
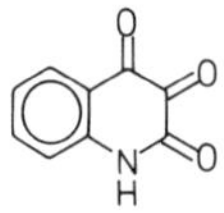	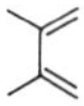	RT, 14 days	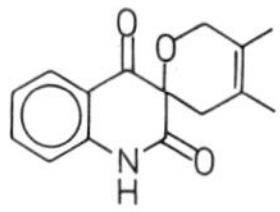	84%	26
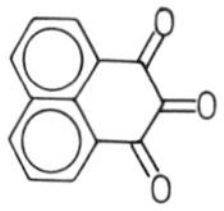	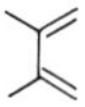	RT	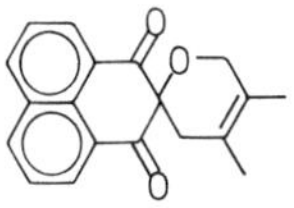	84%	26
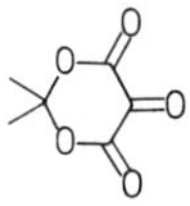	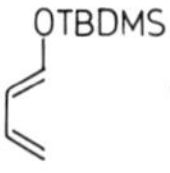	CH$_2$Cl$_2$, 25°C, 9 days	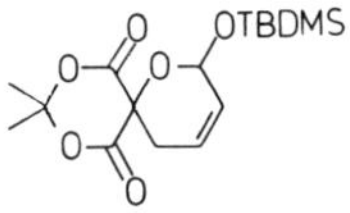	59%	27
	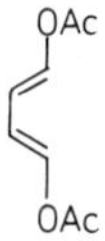	CH$_2$Cl$_2$, 25°C, 18 days	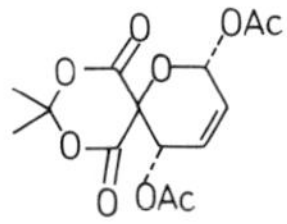	56%	27

$$X = Cl, Br \tag{7}$$

However, this is not a general reaction of o-quinones. Barltrop and Hesp described a few photochemical [4 + 2] cycloadditions of p-benzoquinone [Eq. (8)].[30]

$$R^1, R^2 = H, CH_3 \tag{8}$$

It was suggested that a stepwise mechanism involving a diradical intermediate was operating in this process.

3. ALIPHATIC AND AROMATIC ALDEHYDES AND KETONES

As noted in the Introduction [4 + 2] cycloadditions of simple aliphatic and aromatic carbonyl compounds with dienes which are not particularly electron-rich are generally unreliable.[1,3,4,31] However, recent papers by Jurczak and co-workers give hope that the scope of carbonyl dienophile cycloadditions of this type might be broadened if reactions are carried out under neutral conditions at ultra-high pressures. (E)-1-Methoxybutadiene was found to add smoothly at ~20 kbar to several aldehydes and ketones that are normally not good dienophiles (Table 4-IV).[32a] In view of the observations of the Danishefsky group that lanthanide reagents are mild Lewis acid catalysts for hetero Diels–Alder reactions (*vide infra*), Jurczak *et al.* have reported that such catalysts allow one to conduct cycloadditions with carbonyl dienophiles and 1-methoxybutadiene at lower pressures (~10 kbar).[32b] Some data are given in Table 4-IV. It is yet to be demonstrated that other dienes will react as well with simple aldehydes. Some of the adducts listed in the table have been used in short total syntheses of two δ-lactone insect pheromones.[33]

In an extension of this methodology, Jurczak *et al.* found that under high pressure (E)-1-methoxybutadiene reacts with (R)-glyceraldehyde acetonide to give adducts with a high degree of stereoinduction (Scheme

TABLE 4-IV

High Pressure Cycloadditions of (E)-1-Methoxybutadiene and Carbonyl Compounds[a]

Carbonyl compound	Conditions	Products	Yield
$CH_3COCO_2CH_3$	50°C, 20.0 kbar	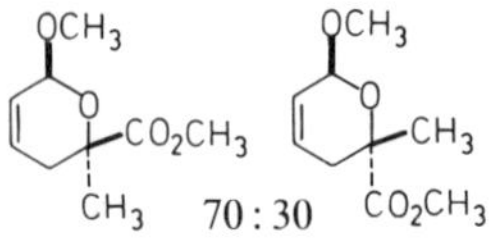 (OCH$_3$ ring, –CO$_2$CH$_3$, CH$_3$) : (OCH$_3$ ring, –CH$_3$, CO$_2$CH$_3$) 70 : 30	85%
CF_3COPh	20°C, 19.5 kbar	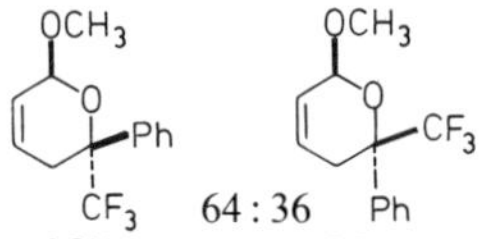 (OCH$_3$ ring, –Ph, CF$_3$) : (OCH$_3$ ring, –CF$_3$, Ph) 64 : 36	81%
$PhCHO$	50°C, 19.5 kbar	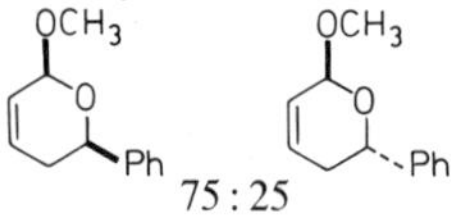 (OCH$_3$ ring, –Ph) : (OCH$_3$ ring, ⋯Ph) 75 : 25	80%
CH_3CHO	65°C, 20.0 kbar	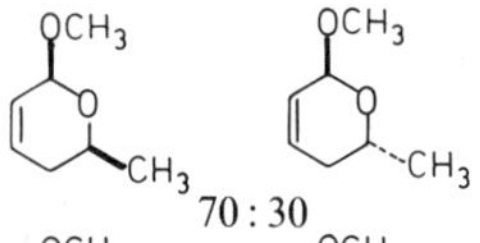 (OCH$_3$ ring, –CH$_3$) : (OCH$_3$ ring, ⋯CH$_3$) 70 : 30	62%
$nC_5H_{11}CHO$	65°C, 20.0 kbar 20°C, 23.5 kbar	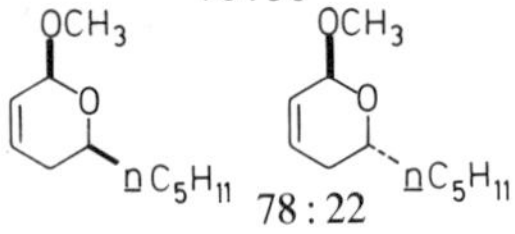 (OCH$_3$ ring, –nC$_5$H$_{11}$) : (OCH$_3$ ring, ⋯nC$_5$H$_{11}$) 78 : 22	28% 16%
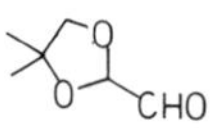 (2,2-dimethyl-1,3-dioxolane)CHO	Eu(fod)$_3$ cat., 50°C, 10 kbar	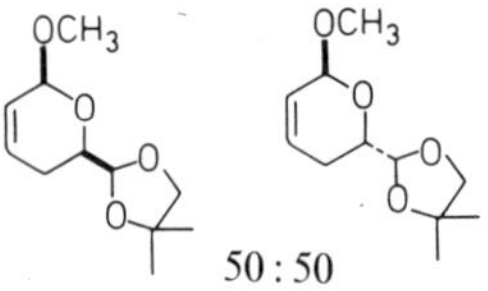 50 : 50	53%
	Eu(fod)$_3$ cat., 50°C, 10 kbar	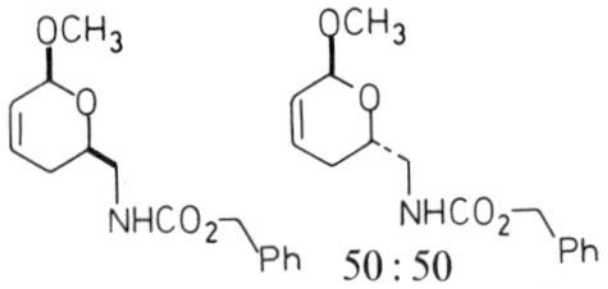 (OCH$_3$ ring, NHCO$_2$–Ph) : (OCH$_3$ ring, NHCO$_2$–Ph) 50 : 50	50%

[a] Adapted from Refs. 32a,b.

4-VI).[34a] Significantly, the cycloaddition shows a 82 : 18 facial selectivity with respect to the diene (i.e., anti : syn) which can be rationalized based on a "Felkin-like" model of approach.[1d,35] This cycloaddition also demonstrated reasonably good endo selectivity. It was found that the ratio of stereoisomeric products was affected by the pressure under which the reaction was run and by the temperature. These adducts have been used in synthesis of some 4-deoxyheptoses.[34c]

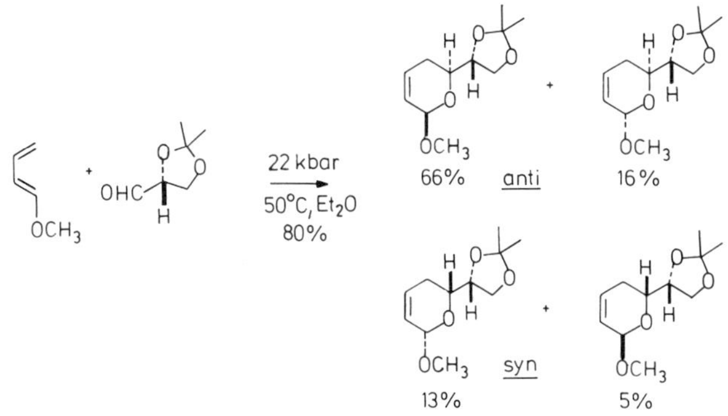

Scheme 4-VI

Another application of this high pressure cyclization technique to stereochemically more complex aldehydes has recently been described (Scheme 4-VII).[34b] For example, aldehyde **15** reacts with (*E*)-1-methoxybutadiene to give *exclusively* adduct **16**, which can be epimerized to **17**. A Felkin model for cycloaddition can again be used to rationalize these results (cf. **18**). It is the greater rigidity and steric bulk of this system that imparts the high facial selectivity relative to the glyceraldehyde derivative in Scheme 4-VI.

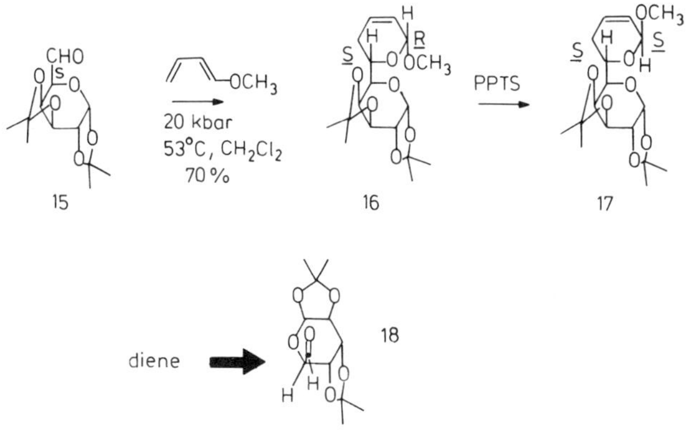

Scheme 4-VII

4. HIGHLY OXYGENATED AND REACTIVE DIENES

A major advance in the Diels–Alder chemistry of carbonyl compounds has come with the recognition that simple aldehydes will react readily with many electron-rich oxygenated dienes[36] under Lewis acid cataly-

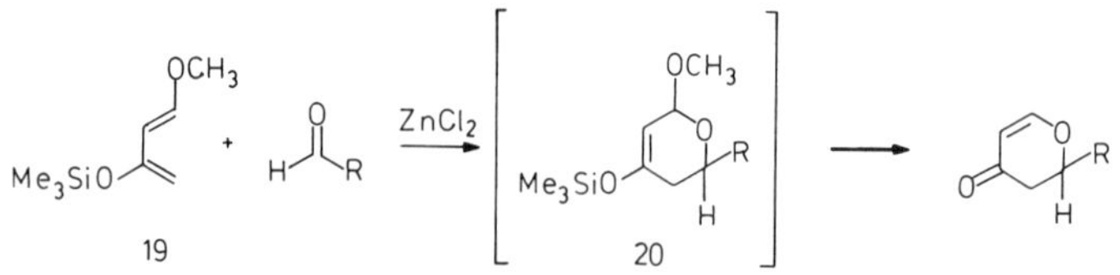

R	% Yield of γ-Pyrone
H	55
CH$_2$OCH$_2$Ph	87
CH$_2$SPh	70
CH$_2$NHCbz	80
Ph	65
pNO$_2$Ph	58
oCH$_3$OPh	58
CH$_3$	17
Et	48
iPr	43
iBu	37
CH=CH$_2$	50
C(CH$_3$)=CH$_2$	62
E-CH=CHPh	75

Scheme 4-VIII

sis.[11b] In their initial reports,[37,45c] the Danishefsky group described the reactions of oxygenated diene **19** with a wide range of unactivated aldehydes using zinc chloride or boron trifluoride etherate as catalyst to directly afford a 2,3-dihydro-δ-pyrone (Scheme 4-VIII), presumably via the intermediacy of a transient [4 + 2] adduct like **20**. Interestingly, under these conditions, α, β-unsaturated aldehydes gave γ-pyrones rather than carbocyclic adducts.[4,38,39] An application of this basic cycloaddition reaction to unsaturated aldehyde **21** has led to a simple total synthesis of the antibiotic LL-Z1120 (**22**) [Eq. (9)].[40]

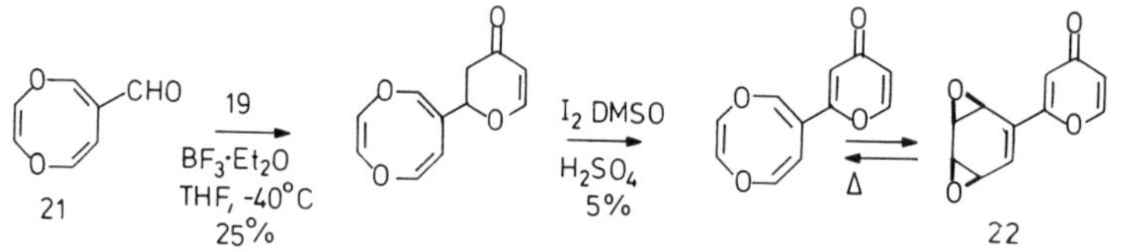

Danishefsky *et al.* have also discovered that Eu(fod)$_3$ is a very mild Lewis acid catalyst that is highly effective in promoting this cycloaddition process.[41] A major advantage of using this catalyst is that one can usually isolate the initial [4 + 2] adduct (Scheme 4-IX), which can subsequently be converted to the γ-pyrone, cleaved to the ketone, or transformed in other ways. It might be noted that the particular cycloaddition shown in this scheme was totally stereoselective (*vide infra*), giving the C-5,6 cis

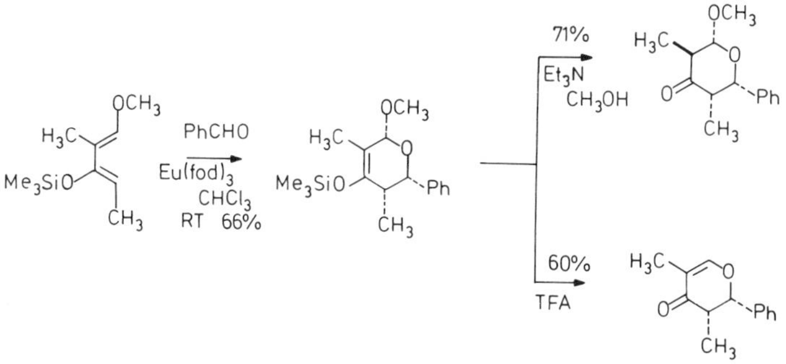

Scheme 4-IX

product. In an application of this lanthanide catalyst method, Castellino and Sims found that 3-oxo-δ-lactones can be efficiently prepared from aldehydes and diene **23** (Scheme 4-X).[42]

The ready availability of optically active NMR lanthanide shift reagents led Danishefsky to investigate the possibility of inducing absolute chirality in these cycloadditions.[41] Using Eu(hfc)$_3$ as catalyst, it was possible to induce modest (20–50%) enantiomeric excesses, depending on the oxygenated diene used and on the reaction conditions.[41b] Equation (10) shows one typical result.

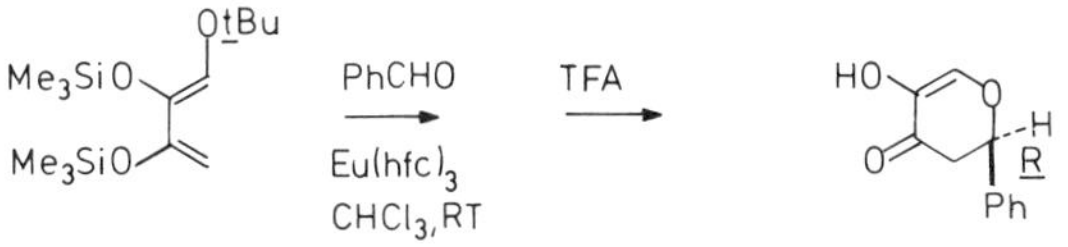

$$\text{(10)}$$

In an exceptionally elegant series of papers, the Danishefsky group has described investigations into the mechanism,[43] stereochemistry,[44] and synthetic applications[45] of this Diels–Alder process. Unfortunately, an in-

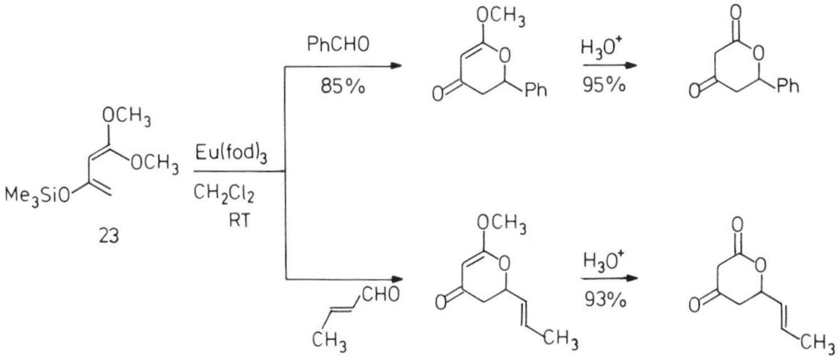

Scheme 4-X

depth analysis of this work is beyond the scope of this chapter. The material presented below is intended to highlight some of the important features of this research.

Mechanistically, the cycloaddition reaction is rather complex. Depending on the catalyst or solvent used and the reaction substrates, pericyclic and/or Mukaiyama aldol-like pathways may be involved.[43] The pericyclic mechanism, generally favored by zinc chloride and the lanthanide catalysts, tends to produce adducts having the cis relative stereochemistry at C-5,6. It is assumed that chelation of the aldehyde with the Lewis acid occurs in an anti fashion and that the steric bulk of R is less than that of the Lewis acid–solvent complex L [Eq. (11)], thus favoring a Diels–Alder transition state with R endo.

$$(11)$$

$$(12)$$

Interestingly, when boron trifluoride etherate was used as catalyst, a reversal of C-5,6 stereochemistry was observed, affording primarily trans compounds. It was postulated that these reactions involve an aldol-like mechanism [Eq. (12)] that is trans (threo) selective. Combinations of the pericyclic and aldol mechanisms are probably operative in some cases.

Some interesting stereochemical results have been obtained with α-substituted aldehydes.[44] These cycloadditions can show excellent facial selectivity depending on the nature of the α substituent and the Lewis acid catalyst used. For example, Scheme 4-XI indicates the facial control obtained with α-methylphenylacetaldehyde and a highly oxygenated diene using two different catalysts.[44a,b] In both cases shown in the scheme the facial approach was the same (i.e., the so-called Cram type), but, as noted above, C-5,6 stereochemistry was reversed as expected.

Good facial selectivity also was found in the case of α-alkoxyaldehydes but again was highly dependent on the Lewis acid and substrates used. Equation (13) shows an example of this type of process.[44e]

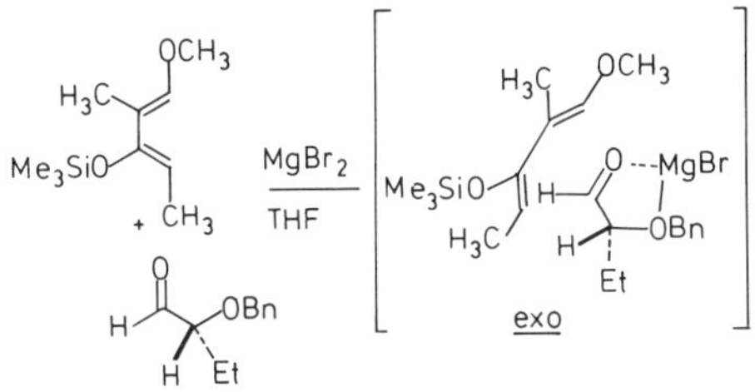

(13)

Using magnesium bromide as catalyst, cycloaddition apparently occurs via a *chelated* aldehyde with diene attack from the least congested face. For steric reasons, an exo transition state in a pericyclic process is favored here, leading to the trans C-5,6 stereochemistry. With other Lewis acids, much lower facial selectivity was found in systems like this.[44e]

Midland and Graham recently discovered that europium(III) complexes give very high facial selectivity explicable by a chelation controlled model as in Eq. (13) using diene 24 [Eq. (14)].[46]

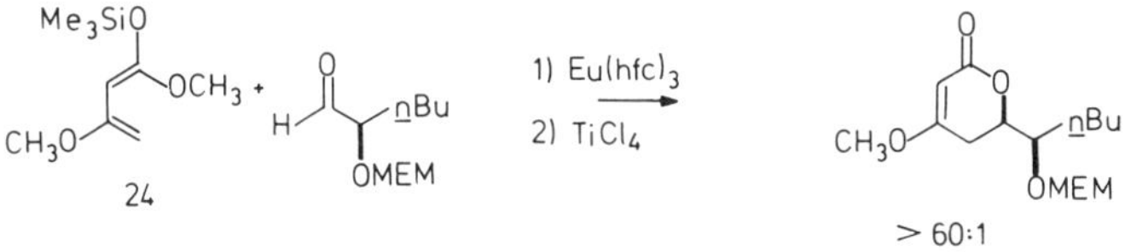

(14)

> 60:1

Another interesting point is that diene 24 also condenses with simple unactivated **ketones** under zinc chloride catalysis to afford δ-lactones like those shown in Scheme 4-X.

Danishefsky has looked at chiral induction in [4 + 2] cycloadditions of aldehydes to oxygenated dienes bearing chiral auxiliaries.[44d] Optically inactive europium(III) catalysts proved effective in promoting the cycloaddition reaction but did not give favorable diastereomer mixtures (Scheme 4-XII). However, combination of the appropriate chiral diene

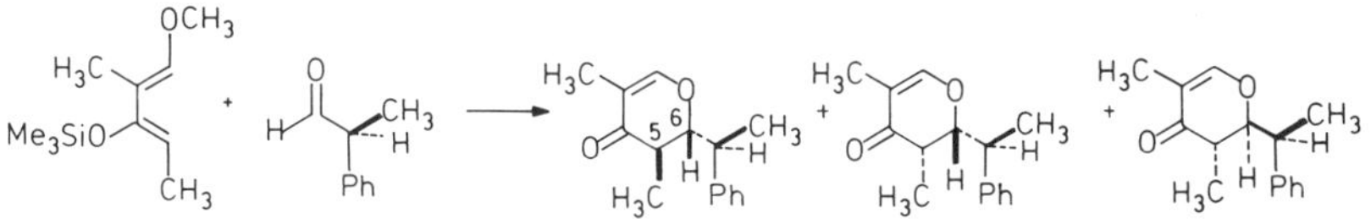

| $BF_3 \cdot Et_2O$, CH_2Cl_2; | 90% | 81 | 19 | 0 |
| $ZnCl_2$, THF; | 92% | 0 | 89 | 11 |

Scheme 4-XI

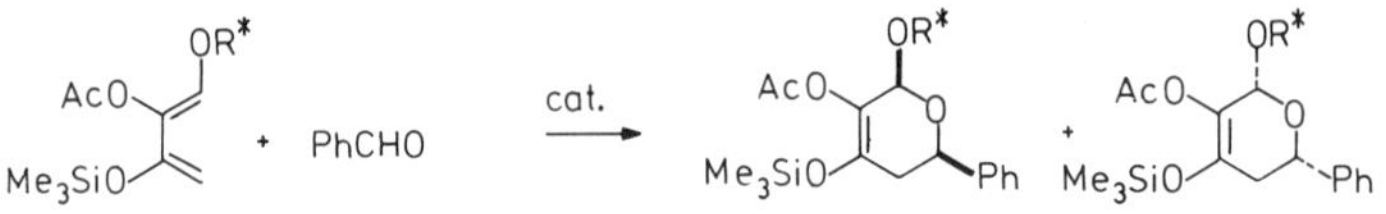

	cat.	Ratio
R* = D-Menthyl	Eu(fod)$_3$	45 : 55
	Eu(hfc)$_3$	41 : 59
R* = L-Menthyl	Eu(fod)$_3$	55 : 45
	Eu(hfc)$_3$	7 : 93

Scheme 4-XII

with a chiral lanthanide catalyst provided the reinforcement needed to generate one diastereomer in good excess. This is a nice example of a reaction involving double diastereoselection.

The Danishefsky group has extensively applied this methodology to total synthesis of a number of natural products.[45] In particular, the method has provided entry to several types of carbohydrate, both simple and complex. For example, the monosaccharide L-talose could be prepared in just a few steps using a hetero Diels–Alder reaction [Eq. (15)].[44a]

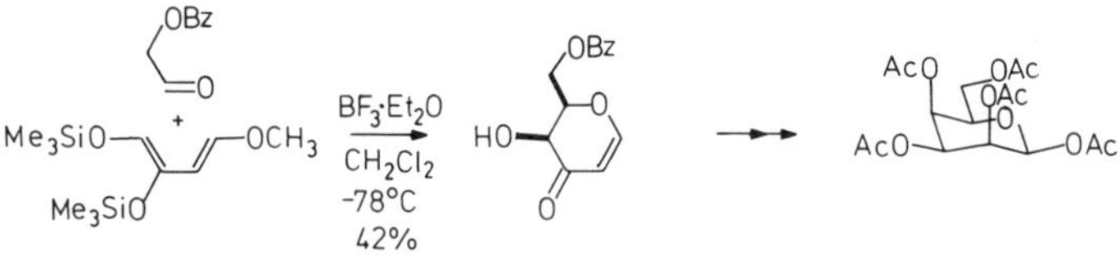

In a much more complicated piece of work recently disclosed, two stereospecific aldehyde Diels–Alder reactions have been used in preparation of hikosamine derivative **29,** a component of the antibacterial compound hizikimycin (Scheme 4-XIII).[45k] Diene **25** reacted with furfural to give *cis*-γ-pyrone **26,** which was transformed in several steps to aldehyde **27.** Condensation of **27** with **25** using magnesium bromide as catalyst afforded *only* adduct **28,** presumably via chelated intermediate **27A.** Compound **28** was converted in a series of steps to acetylhikosamine **29.** The methodlogy described here allowed total synthesis of this unusual sugar having 10 contiguous chiral centers with complete stereocontrol.

Recently, an α-oxoquinodimethane has been used as a highly reactive diene in cycloadditions with simple aromatic aldehydes [Eq. (16)].[47]

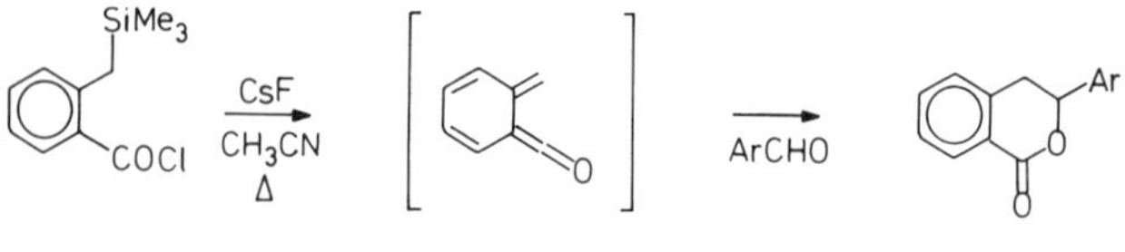

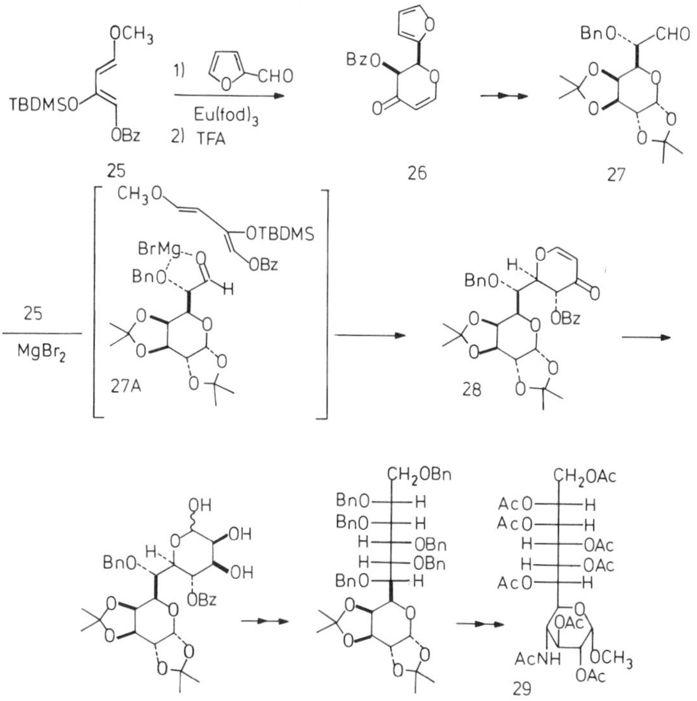

Scheme 4-XIII

5. KETENES

On rare occasion, Ketene carbonyl groups can act as heterodienophiles with all-carbon dienes. Other cycloaddition pathways, such as [2 + 2] addition, are usually preferred, and this Diels–Alder reaction is certainly not of general utility.[48–49] Equations (17) and (18) indicate how relatively small structural changes in reactants can affect the nature of the cycloaddition products.[48]

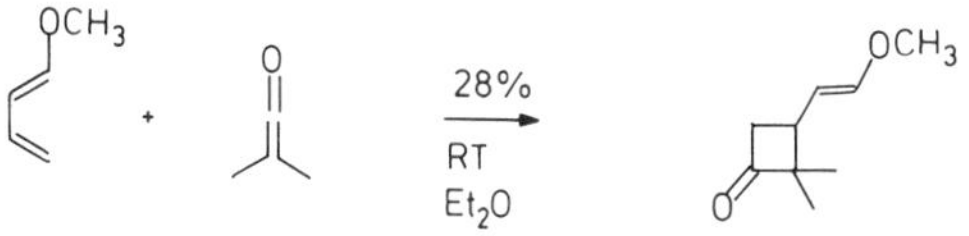

(17)

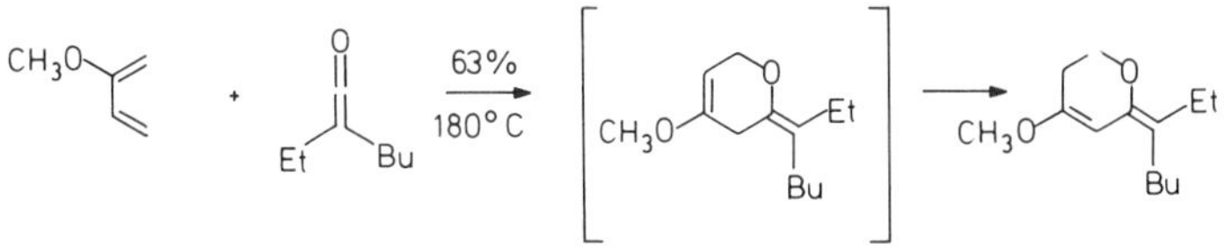

(18)

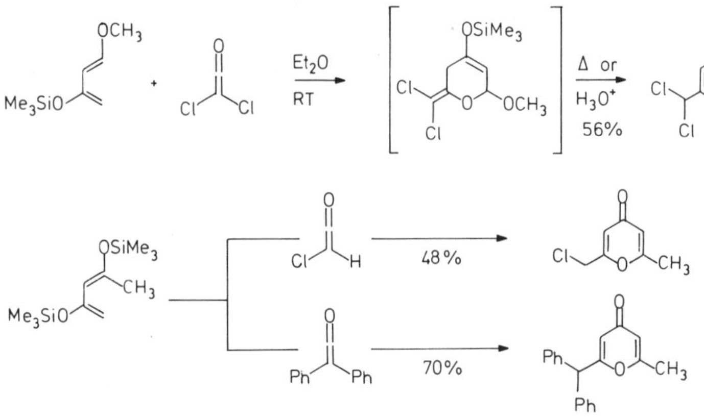

Scheme 4-XIV

A recent report by Brady and Agho has shown that highly oxygenated dienes will give [4 + 2] cycloadducts with ketenes (Scheme 4-XIV).[50] The initial adducts were not actually isolated since they readily rearranged to 4-pyrones in good yields.

6. INTRAMOLECULAR REACTIONS

A few instances of intramolecular carbonyl dienophile Diels–Alder cyclizations are known. In the first reported example, Oppolzer found that an aldehyde benzocyclobutene on heating afforded a 25% yield of two epimeric cycloadducts [Eq. (19)].[51]

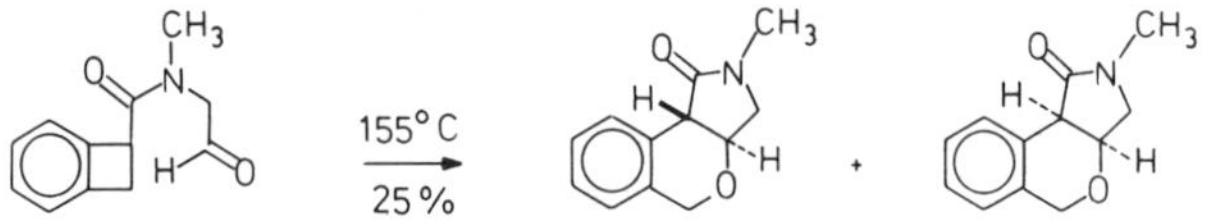

(19)

An o-quinodimethane is presumably the diene in this reaction.

Using their novel cobalt-catalyzed oligomerization method, Funk and Vollhardt discovered that condensation of **30** and **31** afforded exclusively a bridged adduct [Eq. (20)].[52]

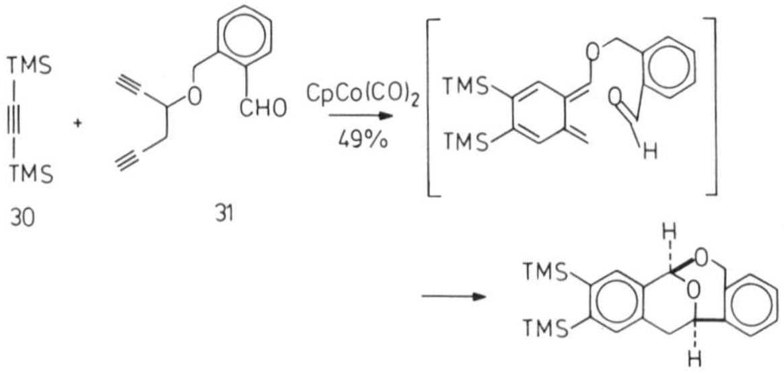

(20)

Interestingly, the putative *o*-quinodimethane intermediate in this case undergoes cyclization with regiochemistry opposite to Oppolzer's system. Whether this result is due to the nature of the bridge between diene and dienophile or to some sort of polar interaction is not clear. From the product stereochemistry it is evident that cyclizations must have occurred via an *endo*-phenyl transition state.

The intramolecular cycloaddition of a furan ester was reported several years ago to yield a tricyclic adduct [Eq. (21)].[53]

$$ (21) $$

This is apparently the *only* example reported to date of a simple ester carbonyl group acting as a heterodienophile.

Snider, Phillips, and Cordova very cleverly combined two consecutive ene reactions with a formaldehyde Diels–Alder cycloaddition to produce a dihydropyran that has previously been used in a total synthesis of pseudomonic acid (**39**).[54] Thus, 1,5-diene **32** underwent dimethylaluminum chloride-catalyzed ene reaction with formaldehyde to afford **33** (Scheme 4-XV) as a 8:1 mixture of trans:cis isomers (80%). Isomers were not separated since the cis compound did not undergo the subsequent Diels–Alder reaction. Treatment of acetate **34** with ethylaluminum dichloride and formaldehyde in CH_2Cl_2–CH_3NO_2 (25°C, 12 hr) gave a 37% yield of adduct **38**. This transformation presumably involves initial ene reaction of **34** to give **35**, which reacted with formaldehyde to produce complex **36**. A quasi-intramolecular Diels–Alder cycloaddition then ensued that led to **37**. Hydrolysis of aluminum complex **37** gave the desired pyran **38**.

In a more recent example of an intramolecular carbonyl Diels–alder process, Trost *et al.* prepared a diene aldehyde [Eq. (22)],

$$ (22) $$

which under Lewis acid catalysis cyclized via an endo transition state to give only the stereoisomer shown.[55]

Rigby and co-workers have also looked at some intramolecular Lewis acid-catalyzed cyclizations of dienic carbonyl compounds.[56] The systems examined to date by this group have involved the cycloheptadienes shown in Eq. (23),

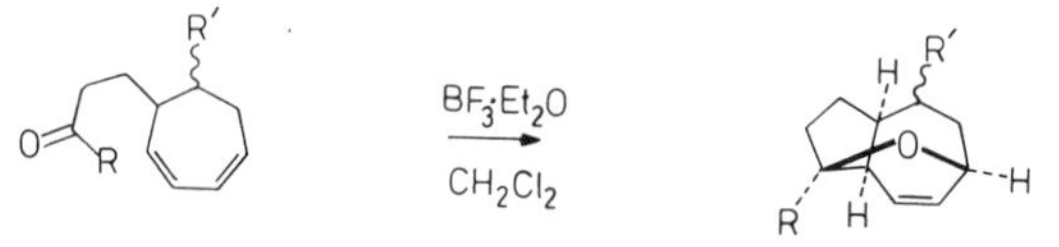

$$\text{(23)}$$

R = H, CH$_3$, iPr
R' = CH$_3$CO, CH$_3$, CH$_2$Ph

which give bridged Diels–Alder adducts in good yields.

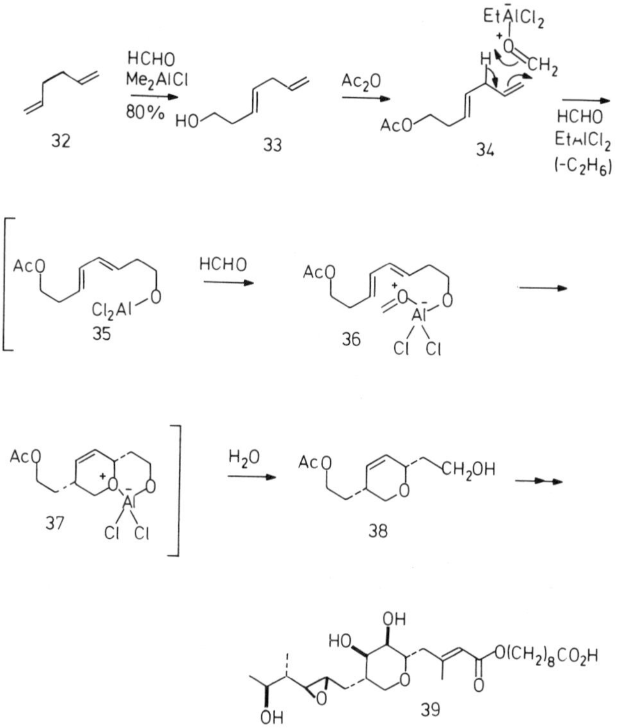

Scheme 4-XV

REFERENCES

1. Previous reviews: a. Hamer, J. (ed.) (1967). "1,4-Cycloaddition Reactions, the Diels–Alder Reaction in Heterocyclic Syntheses." Academic Press, New York. b. Weinreb, S. M., and Staib, R. R. (1982). *Tetrahedron* **38**, 3087. c. Zamojski, A., Banaszek, A., and Grynkiewicz, G. (1982). *Adv. Carbohydr. Chem. Biochem.* **40**, 36–38, 123–128. d. McGarvey, G. J., Kimura, M., Oh, T., and Williams, J. M. (1984). *J. Carbohydr. Chem.* **3**, 125. e. Schmidt, R. R. (1986). *Acct. Chem. Res.* **19**, 250.

2. Dale, W. J., and Sisti, A. J. (1954). *J. Am. Chem. Soc.* **76**, 81.
3. Ansell, M. F., and Charalambides, A. A. (1972). *J. Chem. Soc., Chem. Commun.*, 739.
4. Gramenitskaya, V. N., Vodka, V. S., Golovkina, L. S., and Vulfson, N. S. (1977). *Zh. Org. Khim.* **13**, 2329.
5. Fleming, I. (1976). "Frontier Orbitals and Organic Chemical Reactions." Wiley, New York.
6. a. Dale, W. J., and Sisti, A. J. (1957). *J. Org. Chem.* **22**, 449. b. Shavrygina, O. A., and Makin, S. M. (1966). *Zh. Org. Khim.* **2**, 1354.
7. Smushkevich, Y. I., Belov, V. N., Kleev, B. V., and Akimova, A. Y. (1967). *Zh. Org. Khim.* **3**, 1036.
8. Begley, M. J., Benner, J. P., and Gill, G. B. (1981). *J. Chem. Soc., Perkin Trans. 1*, 1112.
9. Inter alia, see a. Arbuzov, Y. A., Klimov, E. M., and Klimova, E. J. (1962). *Dokl. Akad. Nauk SSSR* **142**, 341. b. Kanowal, A., Jurczak, J., and Zamojski, A. (1968). *Rocz. Chem.* **42**, 2045. c. Shavrygina, O. A., Jablonovskaya, S. D., and Makin, S. M. (1969). *Zh. Org. Khim.* **5**, 775. d. Mochalin, V. B., Porshnev, Y. N., and Samokhvalov, G. I. (1969). *Zh. Obshch. Khim.* **39**, 109. e. Yablonovskaya, S. D., Shekhtman, N. M., Antonova, N. D., Bogatkov, S. V., Makin, S. M., and Zefirov, N. S. (1970). *Zh. Org. Khim.* **6**, 871.
10. a. Zamojski, A., Konowal, A., and Jurczak, J. (1970). *Rocz. Chem.* **44**, 1981. b. Achmatowicz, O., Jurczak, J., and Pyrek, J. S. (1975). *Rocz. Chem.* **49**, 1831.
11. a. Jung, M. E., Shishido, K., Light, L., and Davis, L. (1981). *Tetrahedron Lett.*, 4607. b. Jurczak, J., Golebiowski, A., and Rahm, A. (1986). *Tetrahedron Lett.* **27**, 853. c. van Balen, H. C. J. G., Brockhuis, A. A., Scheeren, J. W., and Nivard, R. J. F. (1979). *Recl. Trav. Chim. Pays-Bas* **98**, 36. See also Ref. 42.
12. a. Hosomi, A., Sakata, Y., and Sakurai, H. (1985). *Tetrahedron Lett.* **26**, 5175. b. Hosomi, A., Otaka, K., and Sakurai, H. (1986). *Tetrahedron Lett.* **27**, 2881.
13. Schmidt, R. R., and Wagner, A. (1983). *Tetrahedron Lett.* **24**, 4661.
14. Schmidt, R. R., and Wagner, A. (1982). *Synthesis*, 958.
15. a. Schmidt, R. R., and Angerbauer, R. (1977). *Angew. Chem. Int. Ed. Engl.* **16**, 783. b. Angerbauer, R., and Schmidt, R. R. (1981). *Carbohydr. Res.* **89**, 193. c. Schmidt, R. R., and Abele, W. (1982). *Angew. Chem. Int. Ed. Engl.* **21**, 302.
16. a. Jurczak, J., and Zamojski, A. (1972). *Tetrahedron* **28**, 1505. b. Jurczak, J., and Baranowski, B. (1978). *Pol. J. Chem.* **52**, 1857. c. Jurczak, J. (1979). *Pol. J. Chem.* **53**, 209. d. Jurczak, J. (1979). *Pol. J. Chem.* **53**, 2539. e. Jurczak, J., and Tkacz, M. (1979). *J. Org. Chem.* **44**, 3347.
17. See also Kanowal, A., Jurczak, J., and Zamojski, A. (1976). *Tetrahedron* **32**, 2957; Golebiowski, A., Izdebski, J., Jacobsen, U., and Jurczak, J. (1986). *Heterocycles* **24**, 1205.
18. Inter alia, see a. David, S., Eustache, J., and Lubineau, A. (1974). *J. Chem. Soc., Perkin Trans. 1*, 2274. b. David, S., and Lubineau, A. (1977). *Nouv. J. Chim.* **1**, 375. c. David, S., Lubineau, A., and Vatele, J. M. (1978). *J. Chem. Soc., Chem. Commun.*, 535. d. David, S., Lubineau, A., and Thieffry, A. (1978). *Tetrahedron* **34**, 299; David, S., and Eustache, J. (1979). *J. Chem. Soc., Perkin Trans. 1*, 2230; David, S., and Eustache, J. (1979). *J. Chem. Soc., Perkin Trans. 1*, 2521; David, S., Eustache, J., and Lubineau, A. (1979). *J. Chem. Soc., Perkin Trans. 1*, 1795.
19. Achmatowicz, O., and Zamojski, A. (1961). *Rocz. Chem.* **35**, 1251.
20. Linn, W. J. (1964). *J. Org. Chem.* **29**, 3111.
21. a. Ruden, R. A., and Bonjouklian, R. (1975). *J. Am. Chem. Soc.* **97**, 6892. b. Bonjouklian, R., and Ruden, R. A. (1977). *J. Org. Chem.* **42**, 4095.

22. a. Abele, W., and Schmidt, R. R. (1981). *Tetrahedron Lett.* **22**, 4807. b. Schmidt, R. R., and Wagner, A. (1981). *Synthesis*, 273.
23. a. Belanger, J., Landry, N. L., Pare, J. R. J., and Jankowski, K. (1982). *J. Org. Chem.* **47**, 3649. b. Potthoff, B., and Breitmaier, E. (1986). *Chem. Ber.* **119**, 2059, 3204.
24. Koreeda, M., and Ciufolini, M. A. (1982). *J. Am. Chem. Soc.* **104**, 2308.
25. Daniewski, W. M., Kubak, E., and Jurczak, J. (1985). *J. Org. Chem.* **50**, 3963.
26. Schonberg, A., and Singer, E. (1971). *Chem. Ber.* **104**, 160.
27. Schmidt, R. R., and Vogt, K. (1983). *Synthesis*, 799.
28. Bellm, M. R., and Herrmann, J. L. (1982). U.S. Patent Appl. 4,349,559.
29. Ansell, M. F., and Leslie, V. J. (1971). *J. Chem. Soc.* (*C*), 1423.
30. Barltrop, J. A., and Hesp, B. (1965). *J. Chem. Soc.* (*C*), 5182.
31. Kubler, D. G. (1962). *J. Org. Chem.* **27**, 1435.
32. a. Jurczak, J., Chmielewski, M., and Filipek, S. (1979). *Synthesis*, 41. b. Jurczak, J., Golebiowski, A., and Bauer, T. (1985). *Synthesis*, 928.
33. Chmielewski, M., and Jurczak, J. (1981). *J. Org. Chem.* **46**, 2230.
34. a. Jurczak, J., Bauer, T., Filipek, S., Tkacz, M., and Zygo, K. (1983). *J. Chem. Soc., Chem. Commun.*, 540. b. Jurczak, J., Bauer, T., and Jarosz, S. (1984). *Tetrahedron Lett.* **25**, 4809. c. Jurczak, J., and Bauer, T. (1985). *J. Carbohydr. Chem.* **4**, 447. d. Jurczak, J., and Bauer, T. (1986). *Tetrahedron* **42**, 5045.
35. Cherest, M., Felkin, H., and Prudent, N. (1968). *Tetrahedron Lett.*, 2199.
36. Danishefsky, S. (1981). *Acc. Chem. Res.* **14**, 400. Petrzilka, M., and Grayson, J. I. (1981). *Synthesis*, 753.
37. a. Danishefsky, S., Kerwin, J. F., Jr., and Kobayashi, S. (1982). *J. Am. Chem. Soc.* **104**, 358. b. Danishefsky, S., and Kerwin, J. F., Jr. (1982). *J. Org. Chem.* **47**, 3183.
38. Danishefsky, S., Yan, C. F., Singh, R. K., Gammill, R. B., McCurry, P. M., Jr., Fritsch, N., and Clardy, J. (1979). *J. Am. Chem. Soc.* **101**, 7001.
39. See also Scheeren, H. W., and Aben, R. W. (1982). *Synthesis*, 779.
40. Altenbach, H.-J., Lex, J., Linkenheil, D., Voss, B., and Vogel, E. (1984). *Angew. Chem. Int. Ed. Engl.* **23**, 966.
41. a. Bednarsky, M., and Danishefsky, S. (1983). *J. Am. Chem. Soc.* **105**, 3716. b. Bednarski, M., Maring, C., and Danishefsky, S. (1983). *Tetrahedron Lett.* **24**, 3451.
42. Castellino, S., and Sims, J. J. (1984). *Tetrahedron Lett.* **25**, 2307.
43. a. Larsen, E. R., and Danishefsky, S. (1982). *J. Am. Chem. Soc.* **104**, 6458. b. Larson, E. R., and Danishefsky, S. (1982). *Tetrahedron Lett.* **23**, 1975. c. Danishefsky, S., Larson, E., Askin, D., and Kato, N. (1985). *J. Am. Chem. Soc.* **107**, 1246.
44. a. Danishefsky, S., Kato, N., Askin, D., and Kerwin, J. F., Jr. (1982). *J. Am. Chem. Soc.* **104**, 360. b. Danishefsky, S., Larson, E. R., and Askin, D. (1982). *J. Am. Chem. Soc.* **104**, 6457. c. Danishefsky, S., Kobayashi, S., and Kerwin, J. F., Jr. (1982). *J. Org. Chem.* **47**, 1981. d. Bednarski, M., and Danishefsky, S. (1983). *J. Am. Chem. Soc.* **105**, 6968. e. Danishefsky, S., Pearson, W. H., and Harvey, D. F. (1984). *J. Am. Chem. Soc.* **106**, 2455. f. Danishefsky, S., Pearson, W. H., Harvey, D. F., Maring, C. J., and Springer, J. P. (1985). *J. Am. Chem. Soc.* **107**, 1256; see also Garner, P. (1984). *Tetrahedron Lett.* **25**, 5855; Garner, P., and Ramakanth, S. (1986). *J. Org. Chem.* **51**, 2609.
45. a. Danishefsky, S., and Kerwin, J. F., Jr. (1982). *J. Org. Chem.* **47**, 1597. b. Larson, E. R., and Danishefsky, S. (1983). *J. Am. Chem. Soc.* **105**, 6715. c. Danishefsky, S., and Webb, R. R. (1984). *J. Org. Chem.* **49**, 1955. d. Danishefsky, S. J., Maring, C. J., Barbachyn, M. R., and Segmuller, B. E. (1984). *J. Org. Chem.* **49**, 4565. e. Danishefsky, S. J., and Maring, C. J. (1985). *J. Am. Chem. Soc.* **107**, 1269. f. Danishefsky, S. J., Larson, E., and Springer, J. P. (1985). *J. Am. Chem. Soc.* **107**, 1274. g. Danishefsky, S. J., Pearson, W. H., and Segmuller, B. E. (1985). *J. Am. Chem. Soc.* **107**, 1280. h.

Danishefsky, S. J., Wang, B. J., and Quallich, G. (1985). *J. Am. Chem. Soc.* **107**, 1285.
i. Danishefsky, S., and Harvey, D. F. (1985). *J. Am. Chem. Soc.* **107**, 6647. j.
Danishefsky, S., and Barbachyn, M. (1985). *J. Am. Chem. Soc.* **107**, 7761. k.
Danishefsky, S., and Maring, C. (1985). *J. Am. Chem. Soc.* **107**, 7762. l. Danishefsky,
S., and Hungate, R. (1986). *J. Am. Chem. Soc.* **108**, 2486. m. Bednarski, M., and
Danishefsky, S. (1986). *J. Am. Chem. Soc.* **108**, 7060. n. Danishefsky, S., Aube, J., and
Bednarski, M. (1986). *J. Am. Chem. Soc.* **108**, 4145. o. Danishefsky, S., and DeNinno,
M. P. (1986). *J. Org. Chem.* **51**, 2617.
46. Midland, M. M., and Graham, R. S. (1984). *J. Am. Chem. Soc.* **106**, 4294.
47. Kessar, S. V., Singh, P., and Venugopal, D. (1985). *J. Chem. Soc., Chem. Commun.*,
    1258.
48. Martin, J. C., Gott, P. G., Goodlett, V. W., and Hasek, R. H. (1965). *J. Org. Chem.* **30**,
    4175; see also Gouesnard, J. P. (1974). *Tetrahedron* **30**, 3113.
49. [4 + 2] Cycloadditions of ketenes with heterodienes: Bargagna, A., Evangalisti, F., and
    Schenone, P. (1979). *J. Heterocycl. Chem.* **16**, 93; Mosti, L., Schenone, P., and
    Menozzi, G. (1978). *J. Heterocycl. Chem.* **15**, 181; Brady, W. T., and Watts, R. D.
    (1981). *J. Org. Chem.* **46**, 4047; Gotthardt, H., and Schenk, K.-H. (1985). *Angew.
    Chem. Int. Ed. Engl.* **24**, 608.
50. Brady, W. T., and Agho, M. O. (1982). *Synthesis*, 500.
51. Oppolzer, W. (1972). *Angew. Chem. Int. Ed. Engl.* **11**, 1031.
52. Funk, R. L., and Vollhardt, K. P. C. (1980). *J. Am. Chem. Soc.* **102**, 5245.
53. Tagmazyan, K. T., Mkrtchyan, R. S., and Babayan, A. T. (1974). *Zh. Org. Khim.* **10**,
    1657.
54. a. Snider, B. B., Phillips, G. B., and Cordova, R. (1983). *J. Org. Chem.* **48**, 3003. b.
    Snider, B. B., and Phillips, G. B. (1982). *J. Am. Chem. Soc.* **104**, 1113.
55. Trost, B. M., Lautens, M., Hung, M.-H., and Carmichael, C. S. (1984). *J. Am. Chem.
    Soc.* **106**, 7641.
56. a. Rigby, J. H. (1982). *Tetrahedron Lett.* **23**, 1863. b. Rigby, J. H., and Wilson, J. Z.
    (1984). *J. Am. Chem. Soc.* **106**, 8217. c. Rigby, J. H., Wilson, J. Z., and Senanayake, C.
    (1986). *Tetrahedron Lett.* **27**, 3329. d. Rigby, J. H., and Wilson, J. A. Z. (1987). *J. Org.
    Chem.* **52**, 34.

Chapter **5**

Thiocarbonyl and Selenocarbonyl Dienophiles

Introduction
1. Thioketones
2. Thioaldehydes
3. Thioesters, Dithioesters, and Related Compounds
4. Thiophosgene and Related Compounds
5. Thienium Salts
6. Thioketenes
7. Sulfines and Related Compounds
8. Selenoaldehydes
References

INTRODUCTION

Cycloaddition of a thiocarbonyl compound and a 1,3-diene affords a 5,6-dihydro-2H-thiapyran [Eq. (1)].[1]

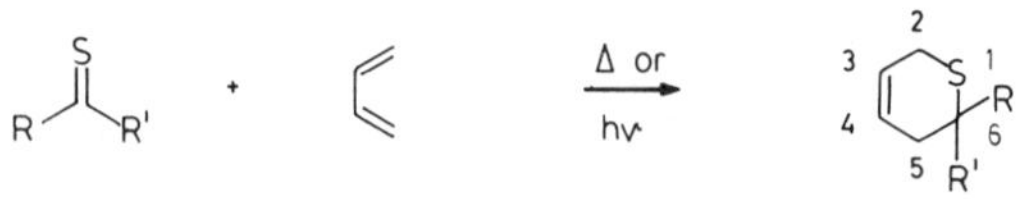

$$\underset{R}{\overset{S}{\Vert}}\underset{R'}{} \quad + \quad \text{(diene)} \quad \xrightarrow[h\nu]{\Delta \text{ or}} \quad \text{(thiapyran)} \tag{1}$$

Virtually all types of thiocarbonyl compounds have been found to react as heterodienophiles. In general, thiocarbonyl compounds are more reactive, versatile dienophiles than are the corresponding carbonyl compounds. One major difference between the two types of dienophile is that a dihydrothiapyran formed from a thiocarbonyl group often has a tendency to undergo a thermal retro-Diels–Alder reaction, whereas the corresponding dihydropyran shows less of a proclivity to do so (see Chapter 4). As noted in Section 8, selenoaldehydes are rare species that have only recently been used as heterodienophiles (*vide infra*).

Mechanistic studies in this area are essentially nonexistent. Recently, the regiochemistry of some [4 + 2] cycloadditions of thioaldehydes has been rationalized based on FMO theory.[2] It is difficult to generalize the resulting stereochemistry of this type of Diels–Alder reaction since systematic studies are lacking and since many structurally diverse kinds of thiocarbonyl dienophiles have been used as 2π components.

1. THIOKETONES

Thioketones of various types are readily available[1c,d] and are well documented as effective dienophiles.[3–12] Representative thioketone cycloadditions are listed in Table 5-I. In general, it appears that thioketones usually add to most dienes in high yield at exceptionally low temperatures to afford stable adducts, although some of these adducts tend to undergo retro-Diels–Alder reactions.[5,6] Very little has been done toward establishing the regiochemical selectivity of thioketone additions to unsymmetrical 1,3-dienes, and the few such entries in Table 5-I indicate that mixtures were obtained. The exo/endo stereochemistry of [4 + 2] cycloadditions with unsymmetrical thioketones has not been probed to date. It has been reported that Diels–Alder cycloadditions of thioketones can also be photochemically induced.[13,14]

Relatively little in the way of further chemistry has been performed on these adducts,[6] although Biellmann has found an interesting rearrangement of some 6,6-disubstituted thiapyrans [Eq. (2)].[10,15]

$$\tag{2}$$

The carbanion derived from such a system rearranges to a cyclopropane derivative which can be S-alkylated to give the *cis*-thioether. Several possible mechanisms have been offered for this rearrangement.[10]

2. THIOALDEHYDES

Until recently, thioaldehydes found little application in Diels–Alder cycloadditions[16] owing to their relative unavailability and pronounced

TABLE 5-I
Cycloadditions of Thioketones

Thioketone	Diene	Conditions	Product(s)	Yield	Ref.
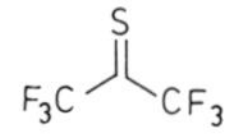 F_3C—C(=S)—CF_3		Neat, −78°C	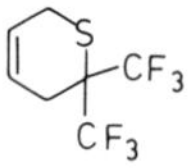 —CF₃, CF₃	90%	3a
		Neat, −78°C	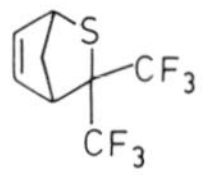 —CF₃, CF₃	85%	3a
		Neat, −78°C	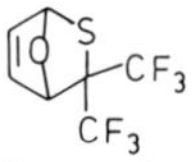 —CF₃, CF₃	73%	3a
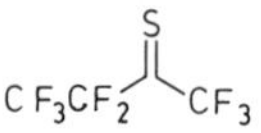 CF_3CF_2—C(=S)—CF_3	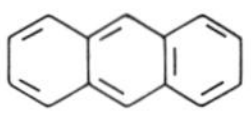	0°C	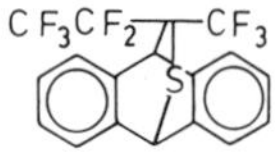 CF_3CF_2—CF_3	95%	3a
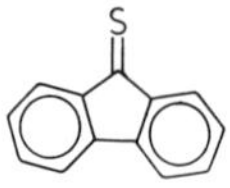	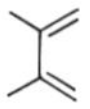	Neat, 200°C	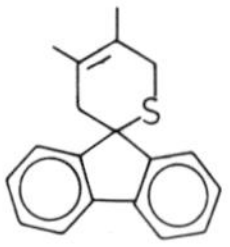	76%	4
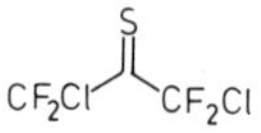 CF_2Cl—C(=S)—CF_2Cl	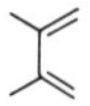	Pentane, RT	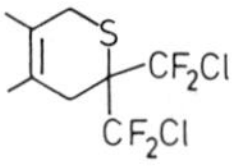 —CF₂Cl, CF₂Cl	84%	3a, 5b, 6
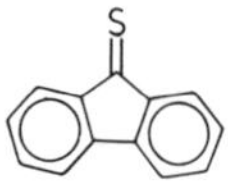	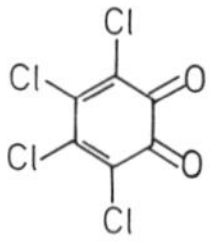	Et₂O, −5°C	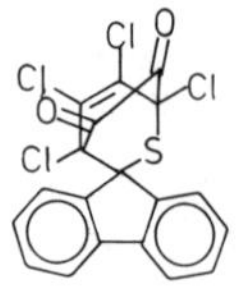	82%	5
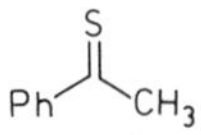 Ph—C(=S)—CH_3	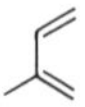	Neat, RT, 2 days	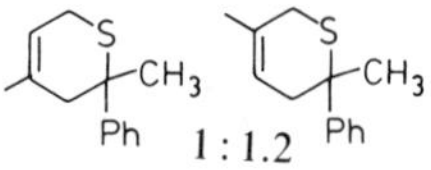 —CH₃, Ph 1:1.2 —CH₃, Ph	100%	7
Ph—C(=S)—Ph R = CH₃	R—	Neat, RT, 10 hr	R——Ph, Ph 1.5:1 R——Ph, Ph	100%	7
R = Cl		Neat, RT, 10 hr	1:1.8	100%	7
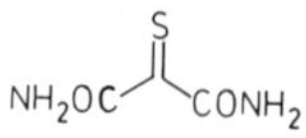 NH_2OC—C(=S)—$CONH_2$		DMSO, 25°C	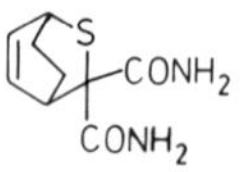 —CONH₂, CONH₂	60%	8a

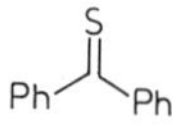

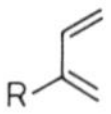

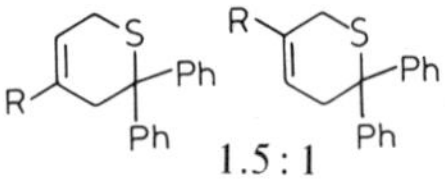

TABLE 5-I (*Continued*)

Thioketone	Diene	Conditions	Product(s)	Yield	Ref.
	 Ph ... Ph	DMSO, 25°C	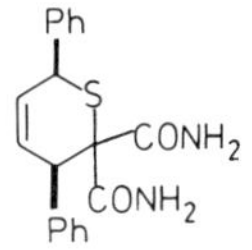 Ph ... S, CONH₂, CONH₂, Ph	40%	8a
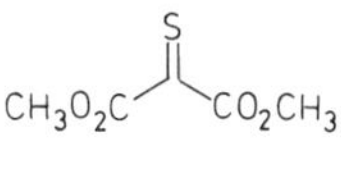 CH₃O₂C–C(=S)–CO₂CH₃	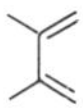	Acetone, 0°C	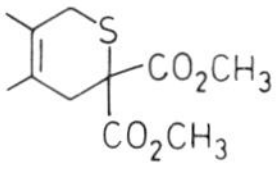 S, CO₂CH₃, CO₂CH₃	70%	8b
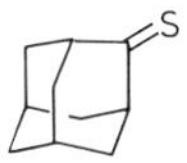 (adamantanethione)		Neat, 100°C, 24 hr	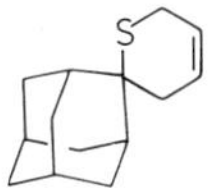 (spiro product)	60%	9
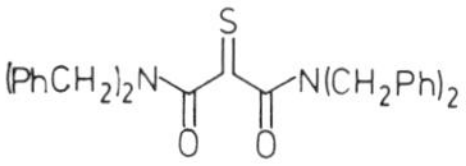 (PhCH₂)₂N–C(=O)–C(=S)–C(=O)–N(CH₂Ph)₂	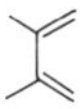	CH₂Cl₂, RT	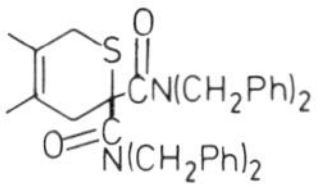 S, O, C N(CH₂Ph)₂, O=C N(CH₂Ph)₂	87%	10
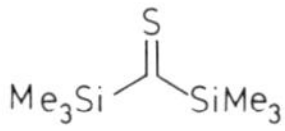 Me₃Si–C(=S)–SiMe₃	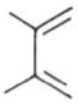	Neat, Δ	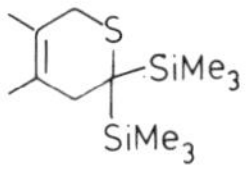 S, SiMe₃, SiMe₃	46%	11
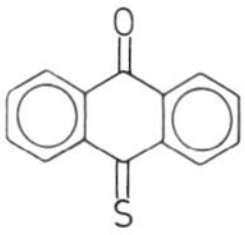 (anthraquinone thioketone)		CH₂Cl₂, RT	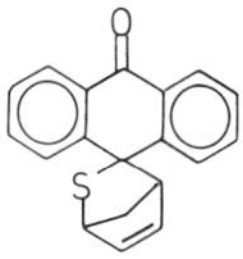 (product)	90%	12

tendency toward polymerization. However, during the past few years improved synthetic methods for preparation of these compounds have been developed, and thioaldehydes have proved to be useful hetero-dienophiles.

Vedejs *et al.* have used a versatile photochemically initiated fragmentation of α-thioacetophenone derivatives to generate a variety of thioaldehydes which have been trapped *in situ* by substituted 1,3-dienes [Eq. (3)].[2,17]

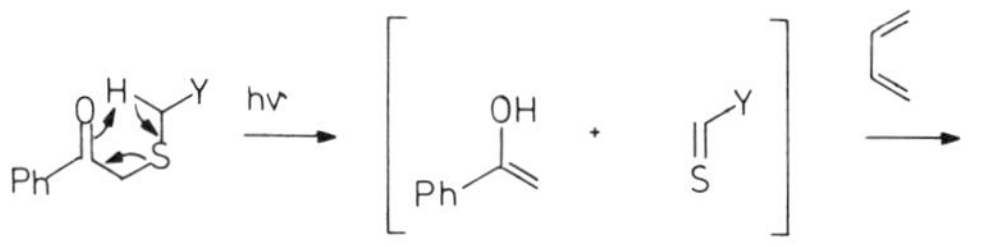

$$\tag{3}$$

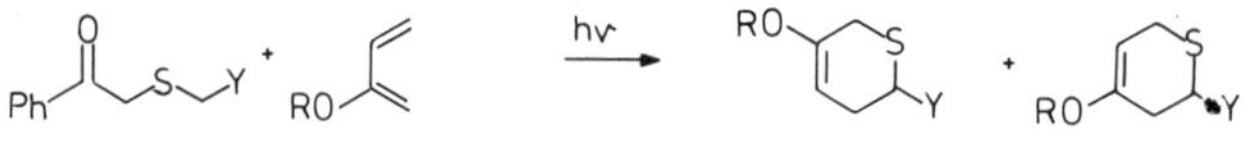

Y	R	Ratio (% Yield)
CN	Et	13 : 1 (76)
CN	TBDMS	18 : 1 (74)
COPh	TBDMS	12 : 1 (64)
CO$_2$iPr	TBDMS	6 : 1 (62)
COCH$_3$	TBDMS	3.5 : 1 (55)
COCH$_3$	Et	7 : 1 (38)
CH$_2$CH$_2$Ph	TBDMS	0 : 100 (23)
H	TBDMS	1 : 11 (72)
Ph	TBDMS	1 : 2 (15)
SiMe$_3$	TBDMS	1 : 10 (33)
SO$_2$Ph	TBDMS	9.5 : 1 (63)
POPh$_2$	TBDMS	100 : 0 (75)

Scheme 5-I

Several such reactions were conducted using 2-alkoxy- and 2-siloxybuta-dienes, and product regiochemistry was determined. Some results are shown in Scheme 5-I.[2,17] Moreover, additions to 1,3-dioxygenated dienes were conducted and found to be completely regioselective (Scheme 5-II).[2] As can be seen from these data, a *reversal* of adduct regiochemistry occurs as thioaldehyde substituent Y becomes more electron rich. This phenomenon has been rationalized by FMO theory.[2] The orientation in these reactions was suggested to be related to the size of the LUMO coefficients at C versus S of the thioaldehyde. These coefficients are strongly affected by the nature of substituent Y. Strong electron-with-drawing groups enlarge the sulfur LUMO coefficient and thus make the sulfur terminus more electrophilic. Electron-donating groups have the opposite effect.

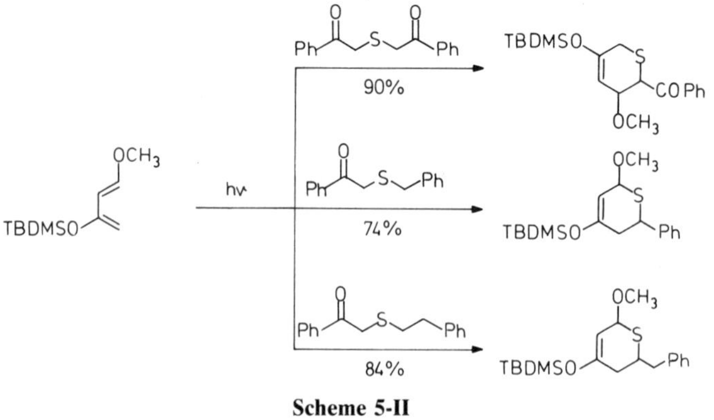

Scheme 5-II

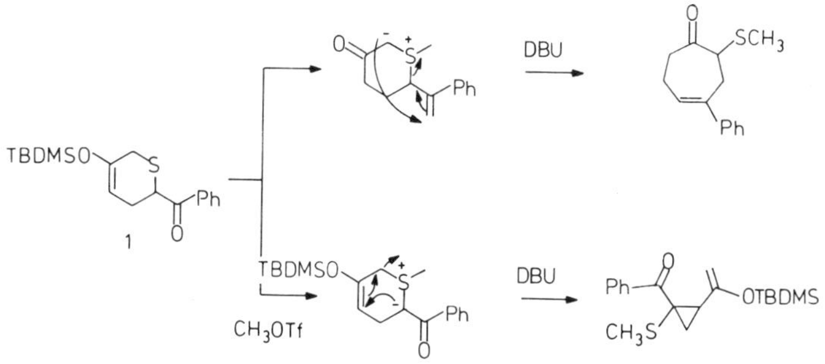

Scheme 5-III

Several interesting and potentially useful transformations of these adducts have been described.[17a] For example, Scheme 5-III indicates some of the chemistry possible for derivatives of adduct **1**, in these cases leading to cycloheptenones and cyclopropane derivatives via [2,3]-sigmatropic rearrangements.

Kirby *et al.* have recently described some new fragmentation methods for generation of thioaldehydes and their subsequent trapping by conjugated dienes.[18-21] One approach involves base-promoted elimination of HCl from a sulfenyl chloride [Eq. (4)].[18]

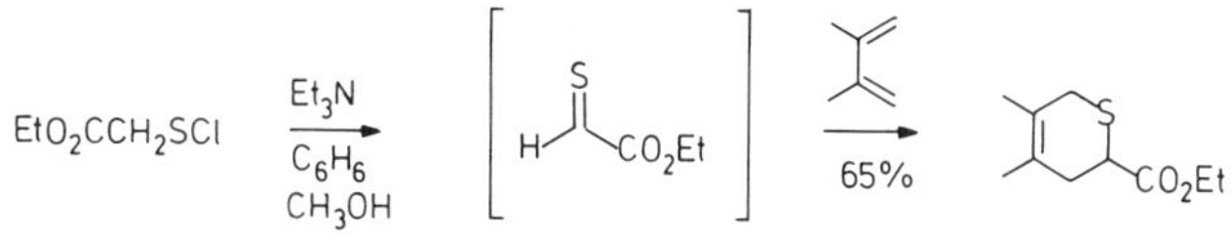

Cycloaddition of this thioaldehyde with thebaine gave predominantly one regioisomeric adduct as the kinetic product (**2**) [Eq. (5)].

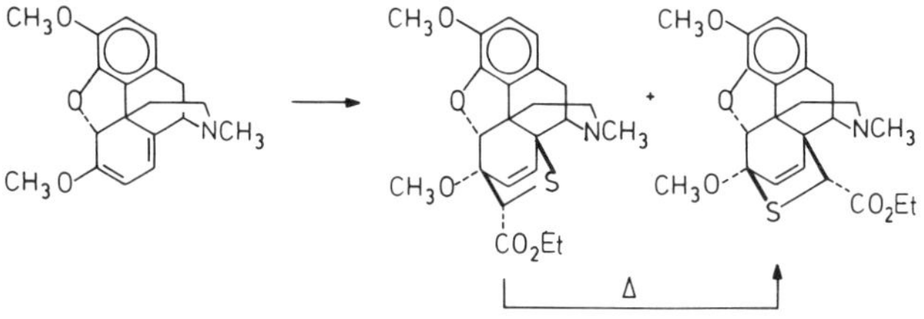

Interestingly, isomer **2** is opposite to what one obtains from thebaine and glyoxylate esters (see Chapter 4). If isomer **2** is warmed, it isomerizes to **3** via a retro-Diels–Alder process. One major drawback of this methodology is that the starting sulfenyl chloride has a tendency to react directly with the 1,3-diene. These workers have therefore looked at several other routes for generating thioaldehydes that do not suffer from this limitation.

One such method shown in Eq. (6)[19] makes use of a fragmentation of phthalimide derivative **4** to generate ethyl thioxoacetate.

$$EtO_2CCH_2S-N \quad \xrightarrow[\substack{C_6H_6 \\ RT}]{Et_3N} \quad \left[EtO_2C \overset{S}{\underset{}{\overset{\|}{C}}} H \right] \quad \xrightarrow{100\%} \quad \underset{exo}{CO_2Et} \quad + \quad \underset{endo}{CO_2Et} \qquad (6)$$

4

This thioaldehyde could be trapped by 1,3-dienes in good yields. With cyclopentadiene, a 7:3 endo:exo kinetic stereoisomer mixture was obtained. Heating either pure isomer in toluene afforded a 3:7 endo:exo equilibrium mixture via a retro-Diels–Alder pathway.

Another efficient procedure developed by the Kirby group to produce thioaldehydes, using Bunte salts **5,** is outlined in Scheme 5-IV.[20] Stereochemical results of cycloadditions with cyclopentadiene are also shown in the scheme. As one might have anticipated, endo adducts predominate as the kinetically favored products. Thiotosylates **6** could also be used to generate thioaldehydes, and trapping results with cyclopentadiene were similar to those with the Bunte salts.[21]

$$ZCH_2SX \quad \xrightarrow[\substack{CH_3OH \\ CoCl_2 \\ RT}]{Et_3N} \quad \left[Z \overset{S}{\underset{}{\overset{\|}{C}}} H \right] \quad \longrightarrow \quad \underset{}{Z} \quad + \quad \underset{}{Z}$$

5 X = SO₃Na
6 X = SO₂Ts

Z	Ratio (% Yield)
CO₂Et	7 : 3 (67)
PhNHCO	6 : 1 (66)
PhCO	7 : 3 (65)
CN	2 : 1 (77)
pNO₂Ph	7 : 1 (88)

Scheme 5-IV

Baldwin and Lopez have formed both aromatic and aliphatic thioaldehydes by pyrolysis of thiosulfinates like **7,** and they were able to trap these species as Diels–Alder adducts [Eq. (7)].[22]

$$Ph\overset{-O}{\underset{S}{\overset{+}{S}}}Ph \quad \xrightarrow[\substack{C_6H_5CH_3 \\ \Delta \\ 87\%}]{9,10-DMA} \quad \underset{}{} \quad \xrightarrow[\substack{C_6H_5CH_3 \\ \Delta}]{- 9,10-DMA} \quad \underset{92\%}{Ph} \qquad (7)$$

7

R	Ratio	(Yield)
CH$_3$	7 : 2	(89%)
Et	6 : 1	(92%)
nPr	9 : 2	(94%)
nBu	9 : 2	(91%)
iPr	7 : 1	(66%)
PhCH$_2$	7 : 2	(80%)

Scheme 5-V

Addition of thiobenzaldehyde to 9,10-dimethylanthracene gave a [4 + 2] cycloadduct, which on heating in the presence of another diene gave a new dihydrothiapyran, again via a facile retro-Diels–Alder reaction.[18-21]

An interesting method for preparation and trapping of alkyl and aryl thioaldehydes was recently disclosed.[23] Fluoride-induced fragmentation of α-silyldisulfides such as **8** gave thioaldehydes which were trapped *in situ* by cyclopentadiene in high yields (Scheme 5-V). In all these systems, endo adducts were the predominant stereoisomers.

Recently, Lee *et al.* found that compound **10**, prepared from **9** and sulfur dichloride via a complex mechanistic pathway, on thermolysis in the presence of a 1,3-diene afforded dihydrothiapyrans [Eq. (8)].[24]

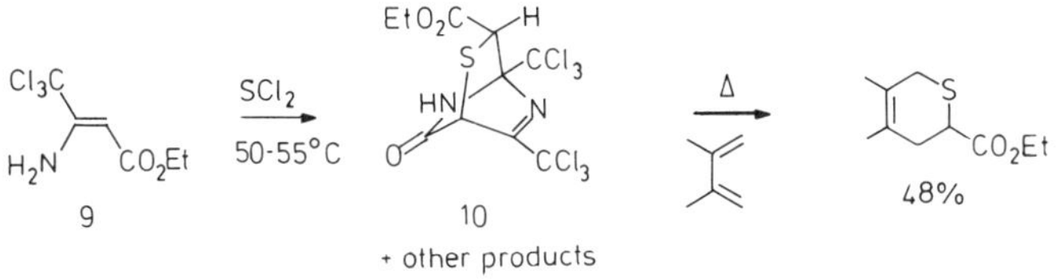

$$(8)$$

This transformation undoubtedly occurs via a retro-Diels–Alder reaction of **10** to give a thioaldehyde.

Baldwin and Lopez reported the first example of an intramolecular thiocarbonyl dienophile [4 + 2] cycloaddition.[22b] Thermolysis of diene thiosulfinate **11** afforded a 1 : 1 mixture of bicyclic adducts that are the result of exo and endo intramolecular cycloadditions [Eq. (9)].

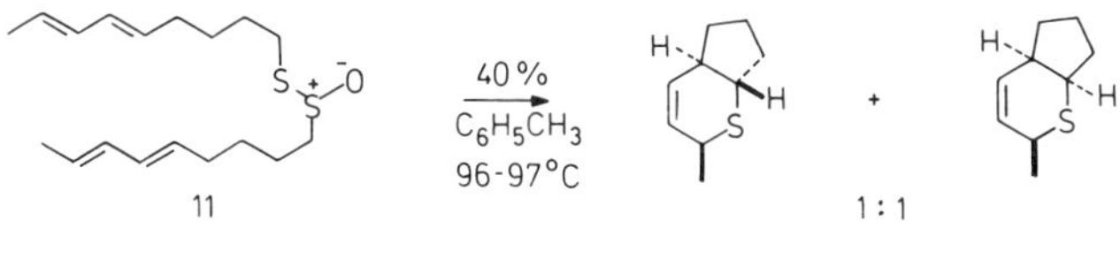

$$(9)$$

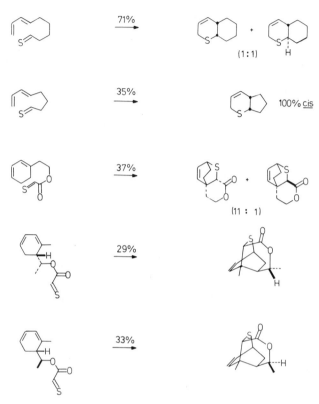

Scheme 5-VI

Vedejs and Eberlein have observed that several diene thioaldehydes generated by their photochemical process [cf. Eq. (3)] undergo stereoselective intramolecular Diels–Alder cycloadditions.[17d] Some examples are shown in Scheme 5-VI.

3. THIOESTERS, DITHIOESTERS, AND RELATED COMPOUNDS

A number of different types of thio- and dithioesters have proved to be reactive heterodienophiles. The majority of the examples of this type of cycloaddition involve systems bearing an electron-withdrawing group on the thiocarbonyl carbon. For example, the thiooxalates in Scheme 5-VII give Diels–Alder adducts in high yields.[25] A rather interesting feature of these thiooxalates is that they will also react as heterodienes with some simple alkenes.[25]

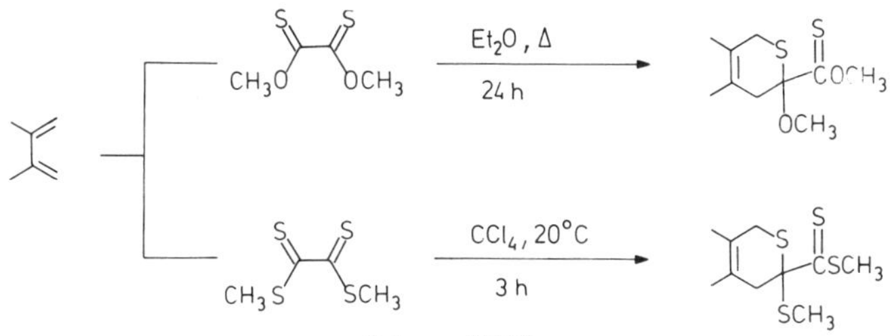

Scheme 5-VII

Vedejs and co-workers have developed a clever method for generation of monothiooxalate derivatives by the dithiolanium ylide cycloreversion shown in Eq. (10).[26]

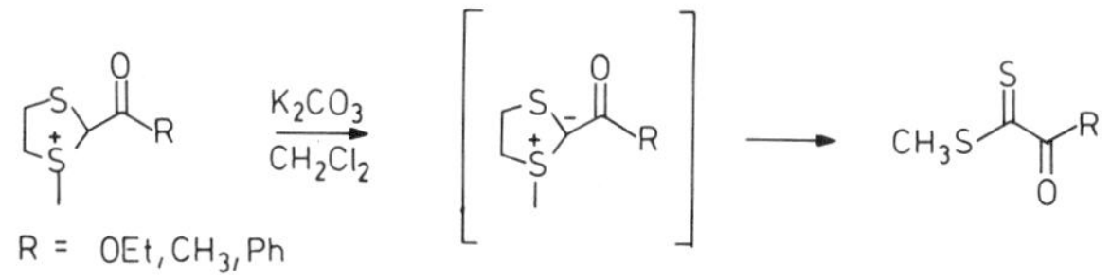

$$R = OEt, CH_3, Ph \tag{10}$$

These thiocarbonyl compounds were not isolated but were trapped as formed by 1,3-dienes to afford good yields of Diels–Alder adducts (Scheme 5-VIII). 1,3-Dimethylbutadiene gave only a single regioisomer (>98%), having the orientation shown. This regiochemistry is in line with that found previously in cycloadditions of some cyanothioformamides with unsymmetrical dienes (*vide infra*).

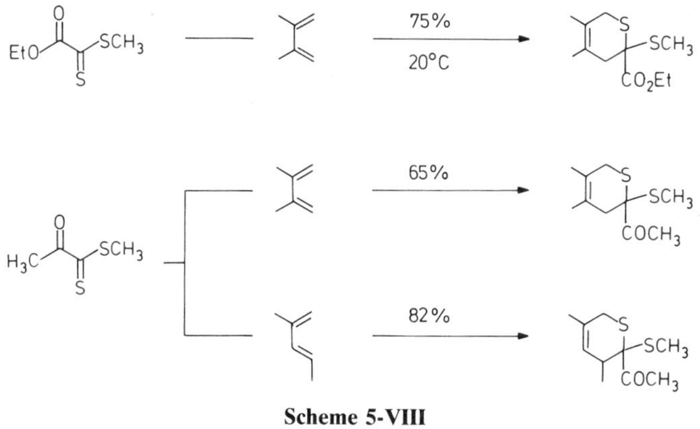

Scheme 5-VIII

Several years ago Middleton found that highly fluorinated dithioesters and -lactones are good dienophiles [Eq. (11)].[3a]

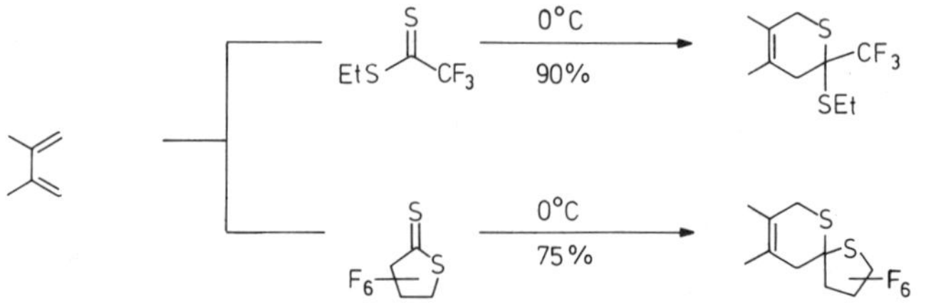

$$(11)$$

These reactions proceed in good yield at very low temperatures.

Some simple alkyl and aryl thioesters have been used as dienophiles.[27] These cycloadditions require much more vigorous conditions than do related systems bearing electron-withdrawing substituents (Scheme 5-IX). The regiochemistry and stereochemistry of these reactions were also briefly investigated (Scheme 5-X).[27] No regioselectivity was found in the cycloaddition of methyl dithioacetate and isoprene. Slightly more regioselectivity was seen with *trans*-piperylene, but little in the way of stereoselectivity was observed.

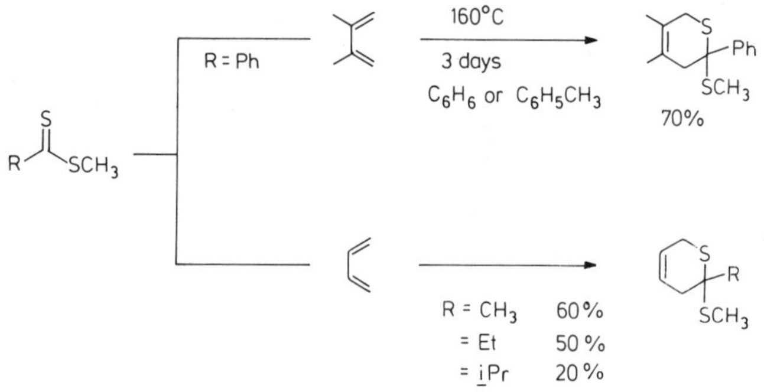

Scheme 5-IX

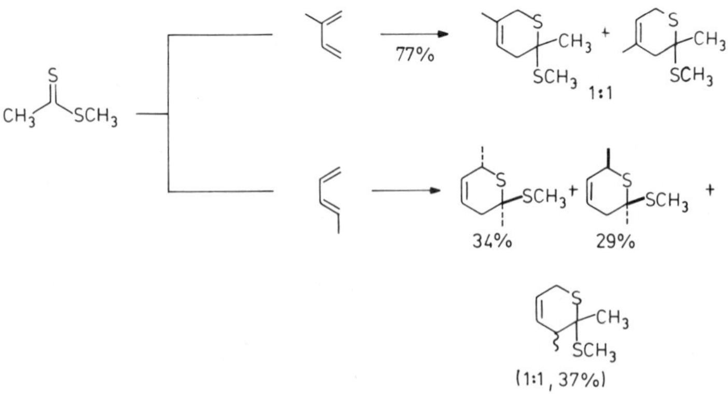

Scheme 5-X

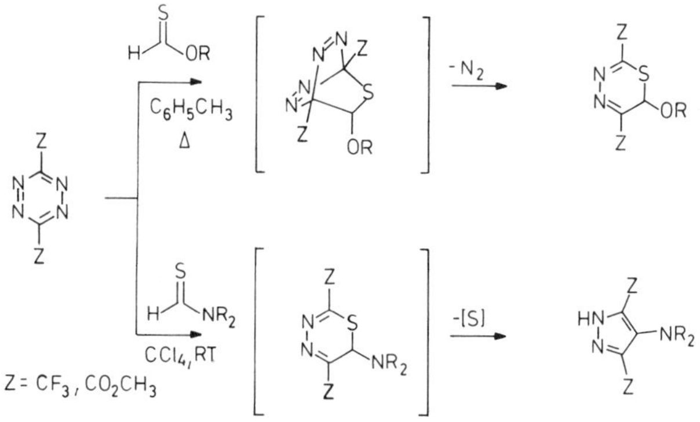

Scheme 5-XI

In what may be inverse electron demand Diels–Alder reactions, Seitz and co-workers found that electron-deficient tetrazines act as dienes with thioformates and thioamides to give heterocyclic products (Scheme 5-XI).[28] In neither case was the initial Diels–Alder adduct detected. Thiobenzaldehyde was also found to react in a similar manner with these tetrazines.

In a recent report Hungarian workers found that sugar-derived thioformates undergo cycloadditions thermally or under high pressure (2.5 kbar).[29] In general, high pressure reactions showed better facial selectivity than did thermal ones. An example is shown in Eq. (12).

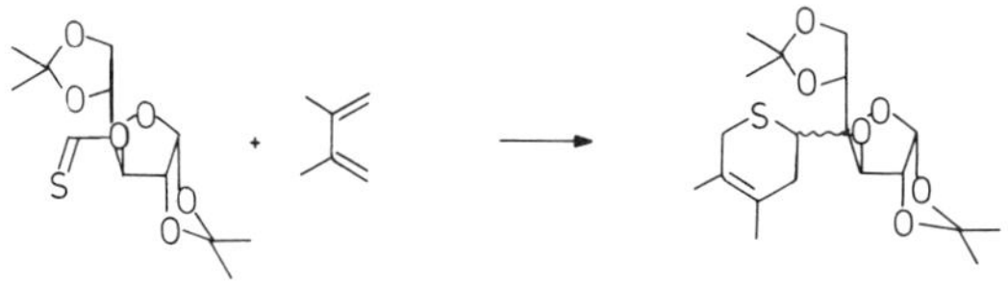

(12)

	isomer ratio
toluene, 150°C, 10 – 16h	1 : 1
2.5 kbar, 6 d	5 : 2

Unfortunately the stereostructures of the adducts were not determined, and thus a mechanistic rationale for the stereoinduction was not offered.

Dithiophthalimide **12** is also known to be a reactive thiocarbonyl dienophile.[30] A few examples of cycloadditions of **12** with unsymmetrical dienes are shown in Scheme 5-XII. In general, these cycloadditions show excellent regioselectivity with dienes other than isoprene. In addition, they proceed with fair stereoselectivity. These reactions, like many of the thiocarbonyl additions discussed in this chapter, are reversible on heating at about 90°C.

Some isolated examples of thioacyl halides acting as dienophiles have

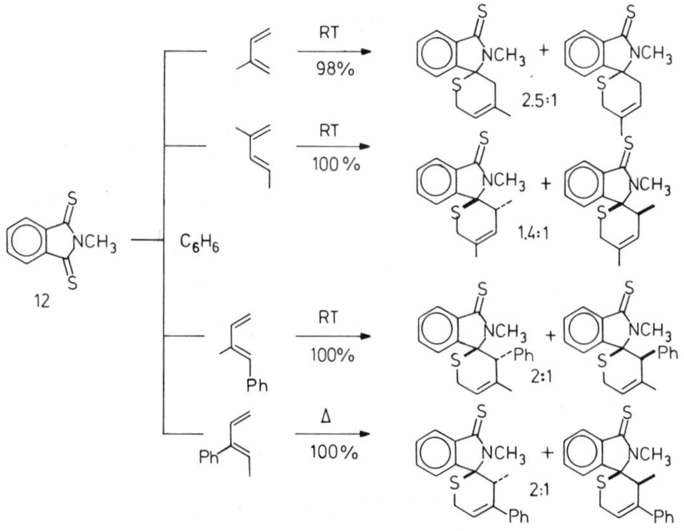

Scheme 5-XII

been described. Middleton found that thioacyl fluorides undergo rapid cycloadditions [Eq. (13)].[3a]

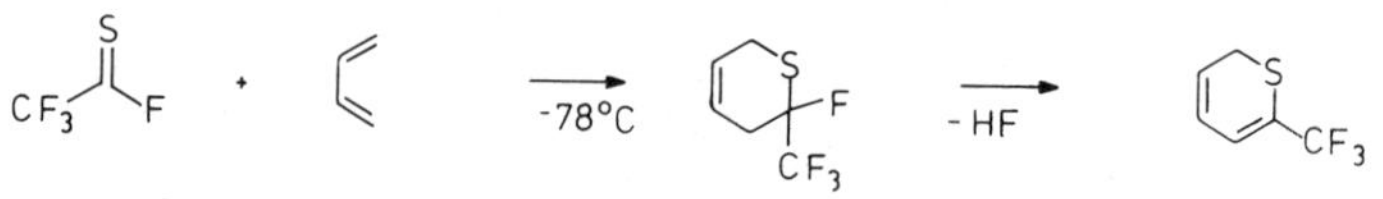

$$\qquad\qquad (13)$$

With butadiene, the adduct tends to lose HF on standing. With cyclic dienes, however, more stable products are formed.

In 1981 Martin *et al.* reported that thiophosgene reacts with dimethylketene to yield a bis(thioacid chloride) that combines rapidly in Diels–Alder fashion with cyclopentadiene, leading to a [4 + 2] cycloadduct [Eq. (14)].[31]

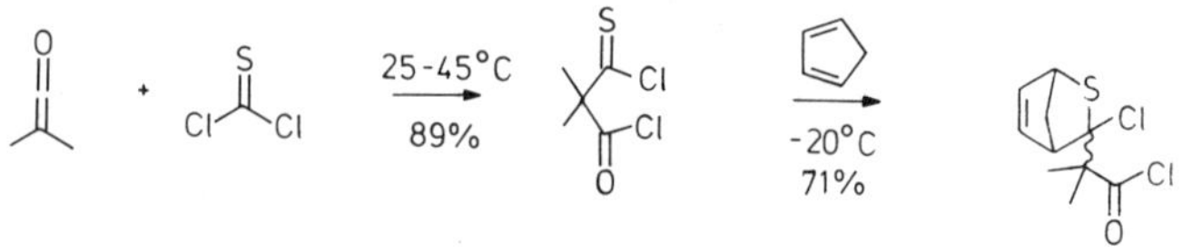

$$\qquad\qquad (14)$$

4. THIOPHOSGENE AND RELATED COMPOUNDS

Although carbon disulfide does not normally react with 1,3-dienes,[32] other thiocarbonyl compounds in this oxidation state are generally reactive dienophiles. Thiophosgene has been added to several different

TABLE 5-II

Cycloadditions of Thiophosgene

Entry	Diene	Conditions	Product	Yield	Ref.
1		Neat, 0°C	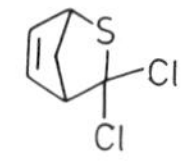	84%	3a, 33
2		CH_2Cl_2, 0°C	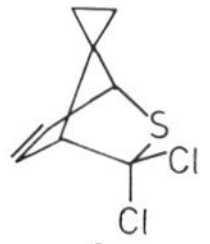	74%	34
3	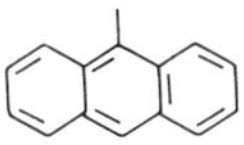	Xylene, 25°C, 2 hr	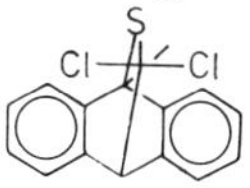	65%	35
4	R = CH₃	THF, 35°C		—	36
5	R = H	Pentane, Δ		50%	36
6		C_6H_6, Δ, 2 hr	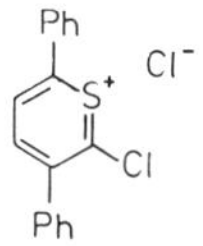	85%	37

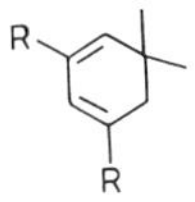

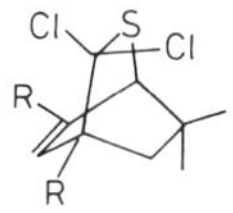

dienes, and some examples are listed in Table 5-II.[33–37] Adducts of this type are usually unstable and undergo hydrolysis in the presence of water to a thiolactone [Eq. (15)],

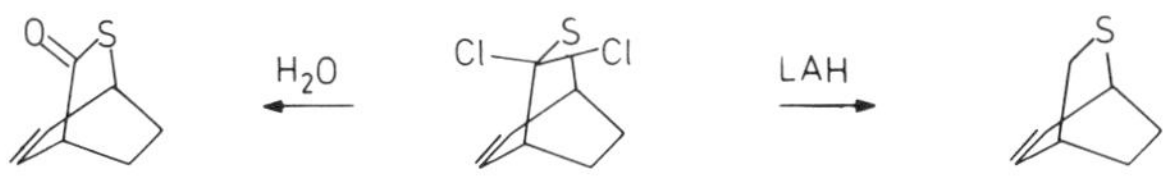

$$\text{(15)}$$

or they may oxidize to a thiopyrrylium salt (Table 5-II, entry 6). Reduction of these adducts with lithium aluminum hydride will remove the chlorines, giving compounds equivalent to thioformaldehyde Diels–Alder products.[36] Thiocarbonyl fluoride has also been used as a dienophile.[3a]

Larson and Harpp have prepared the bis(triazole) **13** and found that it undergoes cycloaddition with 1,3-dienes to afford stable adducts in high yields (Scheme 5-XIII).[38] These adducts, on treatment with dry methanolic HCl, can be opened to mercaptoesters [Eq. (16)].

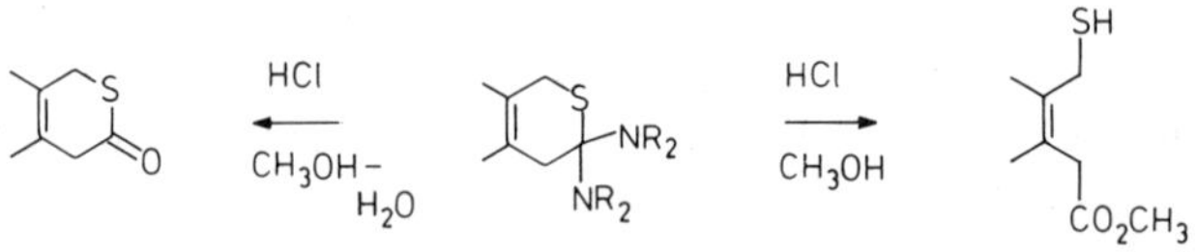

$$(16)$$

Under slightly different conditions, the thiolactone can be obtained.

Trithiocarbonate S,S-dioxides **14** are also reactive thiocarbonyl dienophiles (Scheme 5-XIV).[39] Addition of **14** to cyclopentadiene afforded a high yield of a 3:1 mixture of the endo and exo adducts. The corre-

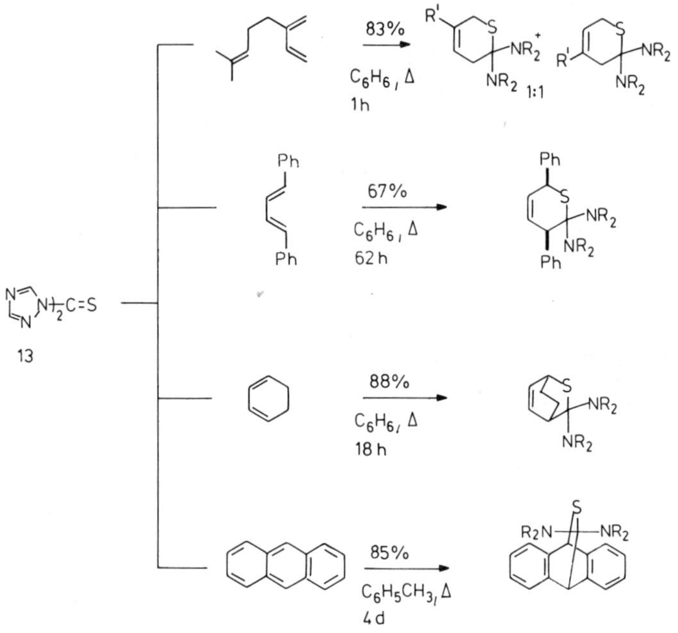

Scheme 5-XIII

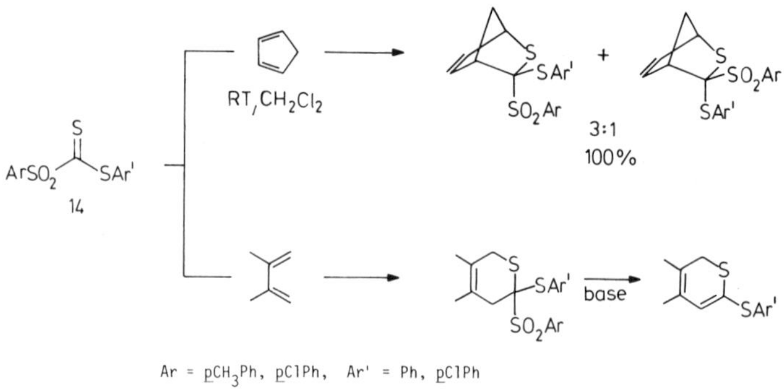

Ar = pCH$_3$Ph, pClPh,   Ar' = Ph, pClPh

Scheme 5-XIV

sponding adducts with acyclic dienes are unstable and readily eliminate arylsulfinic acids to provide thiopyrans. Similarly, a trithiocarbonate derivative is a good dienophile and reacts with cyclopentadiene to give a mixture of exo and endo adducts [Eq. (17)].[40]

$$PhS \overset{\overset{\displaystyle S}{\|}}{\diagup} SSTs \quad + \quad \diamond \quad \longrightarrow \quad \diamond\!\!\!\overset{S}{\underset{SPh}{\diagdown}}\!-SSTs \qquad (17)$$

Vyas and Hay reported that methyl cyanodithioformate (15) will function as a dienophile (Scheme 5-XV).[41] Addition of 15 to cyclopentadiene gave a quantitative yield of exo and endo adducts in a 40:60 ratio. When 1-methoxybutadiene was used, dihydrothiapyran 16 was the major reaction product along with traces (<20%) of regioisomer 17 which was not characterized fully. The *endo*-cyano stereochemistry was established for 16 by extensive NMR studies. An interesting point here is that the regiochemistry of the major product of addition of 1-methoxybutadiene and 15 is like that found with some types of thioesters (Scheme 5-VIII) and thioimides (Scheme 5-XII). Similarly, cyanothioformamides of type 18

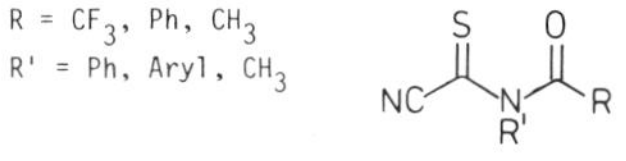

R = CF$_3$, Ph, CH$_3$
R' = Ph, Aryl, CH$_3$

$$NC \overset{\overset{\displaystyle S}{\|}}{\diagup} \underset{R'}{N} \overset{\overset{\displaystyle O}{\|}}{\diagup} R$$

18

were shown to be reactive thiocarbonyl dienophiles, and additions to several cyclic and acyclic 1,3-dienes have been described.[42] With cyclopentadiene, exo/endo mixtures were obtained.

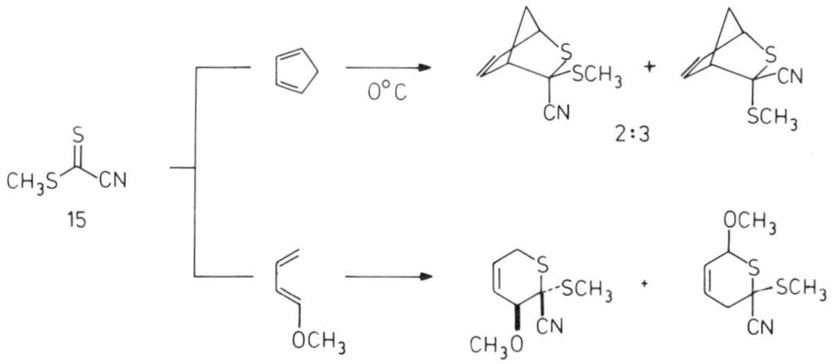

Scheme 5-XV

5. THIENIUM SALTS

A rather interesting route to 3-cyclopentenones has been developed by Corey and Walinsky (Scheme 5-XVI), which uses a dithienium salt as a heterodienophile in the initial step.[43] This species adds regioselectively to 2,3-dimethylbutadiene to give an adduct that can be transformed as shown via a vinyl cyclopropane to the unsaturated ketone in good overall yield. In related work, it was found that several substituted acyclic thienium salts will also provide vinyl cyclopropanes [Eq. (18)].[44,45]

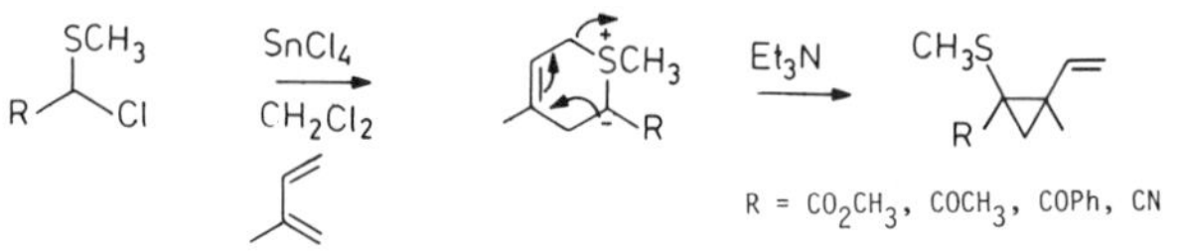

$$ R = CO_2CH_3, COCH_3, COPh, CN \tag{18} $$

In these cases, the intermediate Diels–Alder adducts were not isolated or characterized.

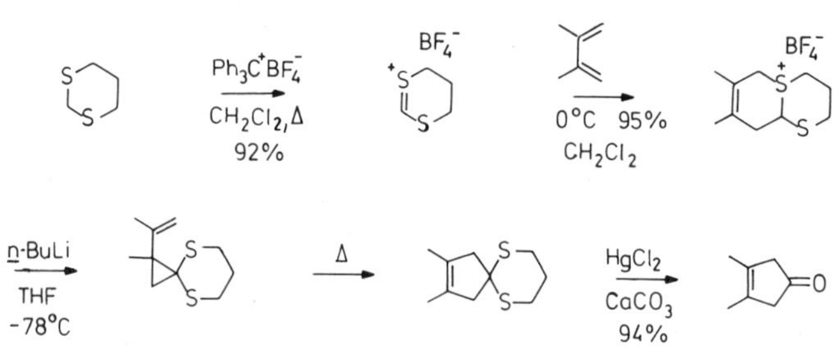

Scheme 5-XVI

A clever route to thienium salts from a nitro olefin like **19** has recently been described.[46] Treatment of **19** with aluminum chloride provides a thienium intermediate that adds to various 1,3-dienes. The product of [4 + 2] cycloaddition was not detected, rather a ring-opened compound was produced (Scheme 5-XVII). Some additional examples of this reaction are

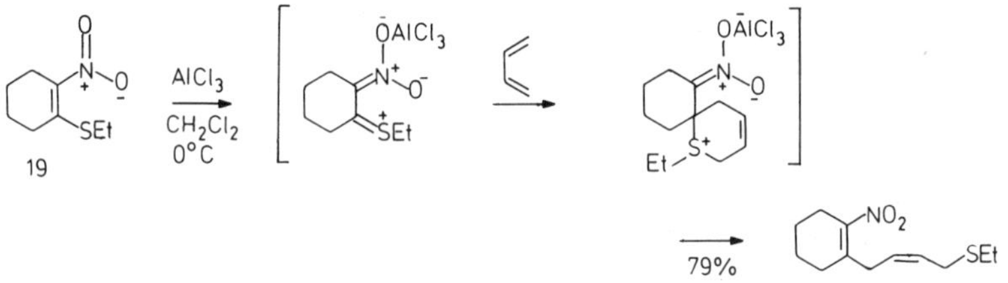

Scheme 5-XVII

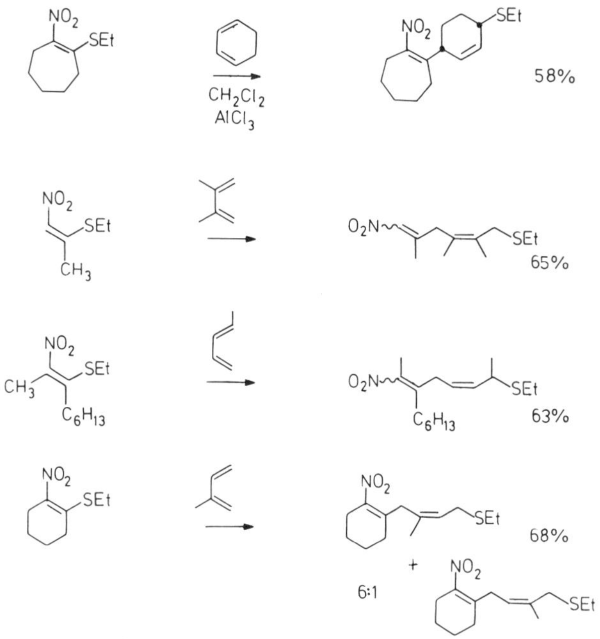

Scheme 5-XVIII

shown in Scheme 5-XVIII. The cycloaddition with *trans*-piperylene was totally regioselective, whereas with isoprene a 6:1 product mixture was obtained. In all cases, only the (Z)-alkene was produced. Cyclohexadiene affords the cis product exclusively. These results are all circumstantial evidence for a pericyclic process having occurred, although no firm mechanistic data are available.

6. THIOKETENES

Although monomeric thioketenes are generally highly unstable, Raasch has found that bis(trifluoromethyl) thioketene (**20**) is stable enough to handle and that it is excellent dienophile. Some examples of its reactions are compiled in Scheme 5-XIX.[34,47]

7. SULFINES AND RELATED COMPOUNDS

Sulfines, which are thiocarbonyl mono-S-oxides, represent another type of reactive C=S heterodienophile.[48] This functional group is nonlinear and can exist as (Z)-**21** or (E)-**22** isomers which have good configura-

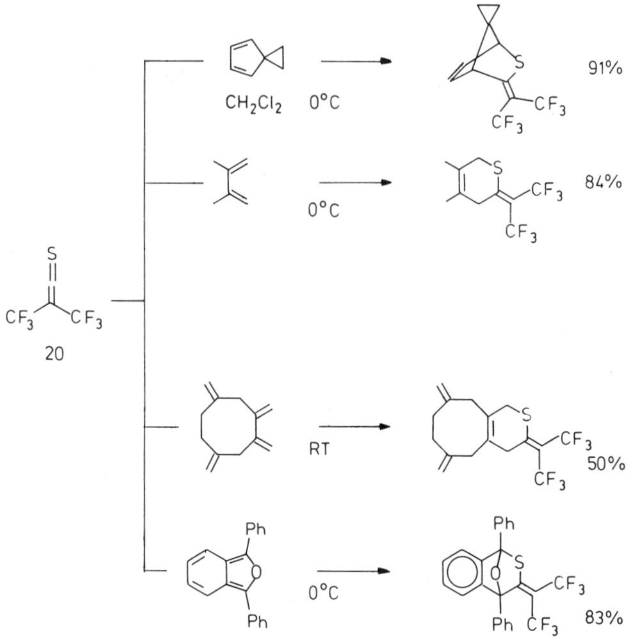

Scheme 5-XIX

tional stability. [4 + 2] Cycloadditions of sulfines are stereospecific syn processes.[48] Zwanenburg's group has reported several Diels–Alder cycloadditions of sulfines,[48–52] as have others,[53–55] and some results are listed in Table 5-III. Entries 5–8 demonstrate the syn selectivity of the cycloaddition. The product of addition of thiofluoxene S-oxide and cyclopentadiene (entry 1) is unstable, and the exo:endo ratio was found to change with time. It is therefore not clear which is the kinetic product of the reaction. The mixtures of isomeric products in entries 3 and 4 are probably due to an equilibration either of (E)- and (Z)-sulfines or of adduct stereoisomers during workup. The $\Delta S\dagger$ values were measured for these two reactions and were found to be −15 and −20 e.u. (electron units), respectively. Such parameters are indicative of a concerted [4 + 2] cycloaddition mechanism.[50]

ABLE 5-III

ycloadditions of Sulfines

ntry	Sulfine	Diene	Conditions	Adduct(s)	Yield	Ref.
1	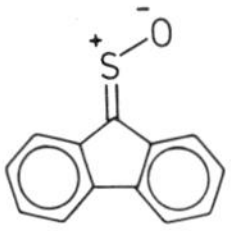		C_6H_6, RT	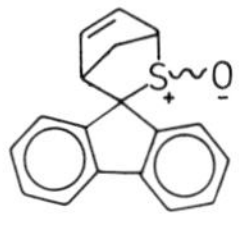 exo : endo = 4 : 1	48%	49
2	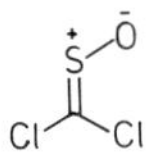	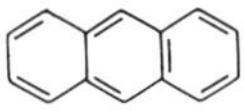	C_6H_6, 36 hr	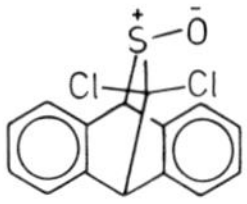	75%	49, 51
3	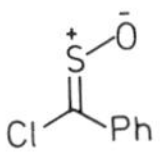	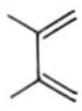	RT, 2 days	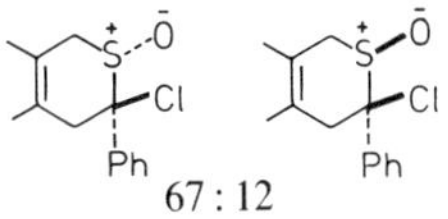 67 : 12	79%	49, 50
4	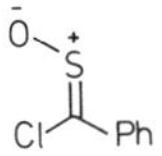		RT, 7 days	18 : 70	88%	49, 50
5	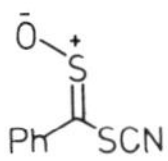		20°C, 4 days	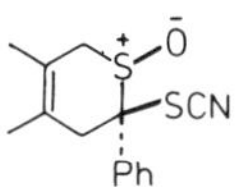	79%	49
6	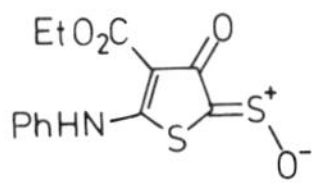		CH_2Cl_2, 25°C	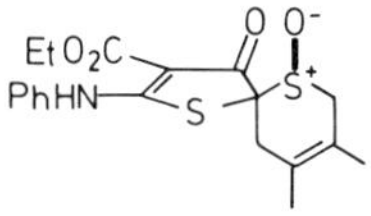	59%	52a
7	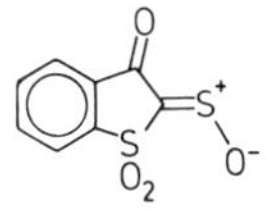		CH_2Cl_2, 0°C, 1 hr	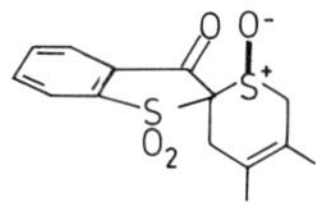	55%	52b
8	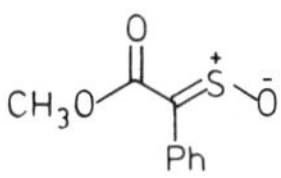		CH_2Cl_2, 20°C, 24 hr	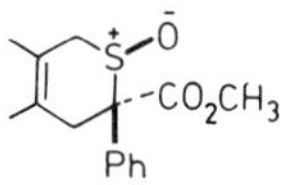	59%	52b
9	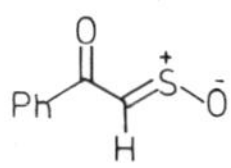		CH_2Cl_2, −78°C, 0.5 hr	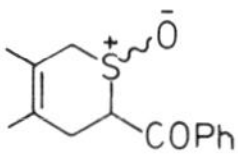	84%	52b
10	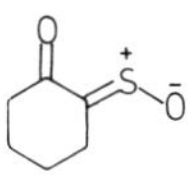		CH_2Cl_2, 0°C, 8 hr	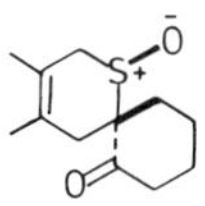	68%	52b

(*continued*)

TABLE 5-III (Continued)

Entry	Sulfine	Diene	Conditions	Adduct(s)	Yield	Ref.
11	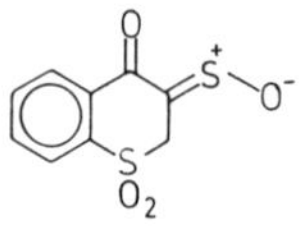		Et$_2$O, 25°C, 6 hr	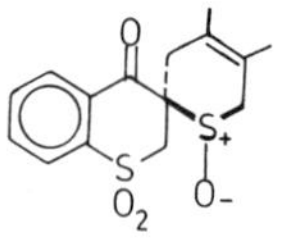	45%	53
12	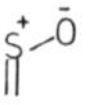		CH$_3$CN, −20°C	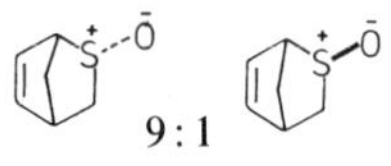 9 : 1	100%	54b
13	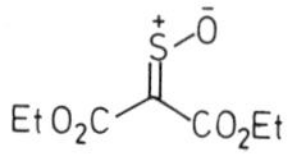		CH$_2$Cl$_2$, −20°C, 1 hr	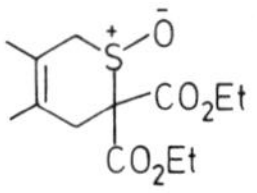	53%	52b, 5

In recent studies, Zwanenburg and co-workers looked at reactions of several sulfines bearing chiral auxiliaries.[56,57] (Z)-Sulfine **23** prepared from camphor was found to undergo totally stereospecific Diels–Alder cycloaddition with 2,3-dimethylbutadiene giving only one diastereomeric product [Eq. (19)].

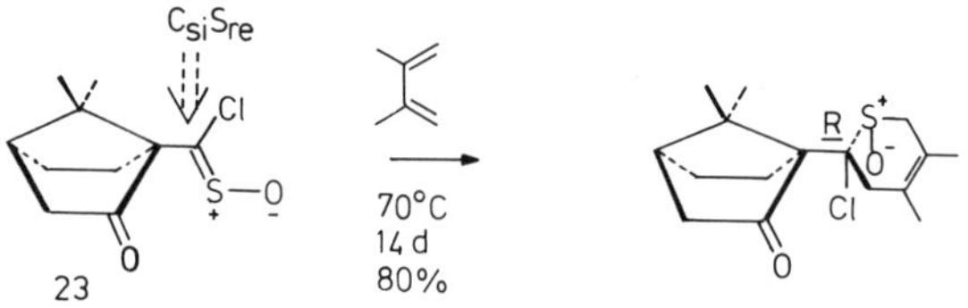

$$(19)$$

It was assumed that cycloaddition occurs via the sulfine conformation shown in **23** and that the diene approaches from the least congested C$_{si}$S$_{re}$ face, giving the observed adduct stereochemistry.[56] It was also found that the relative stereochemistry between chlorine and the sulfur bearing oxygen is retained (syn) in the product.

Another type of chiral sulfine **24** was prepared from an optically active sulfoximine [Eq. (20)].[56]

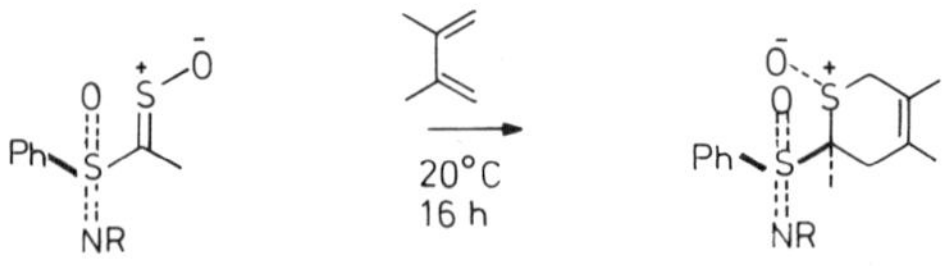

$$(20)$$

24   R=CH$_3$,Ts   **25**

Cyclization of either the *N*-methyl or *N*-tosyl sulfine with 2,3-dimethylbutadiene gave a single diastereomeric adduct, assumed to have structure **25.**

Some additional chiral sulfines prepared from (*S*)-proline have also been prepared and their reactions with 2,3-dimethylbutadiene studied.[57] (*E*)-Sulfines of type **26** all reacted to afford mixtures of diastereomeric adducts [Eq. (21)].

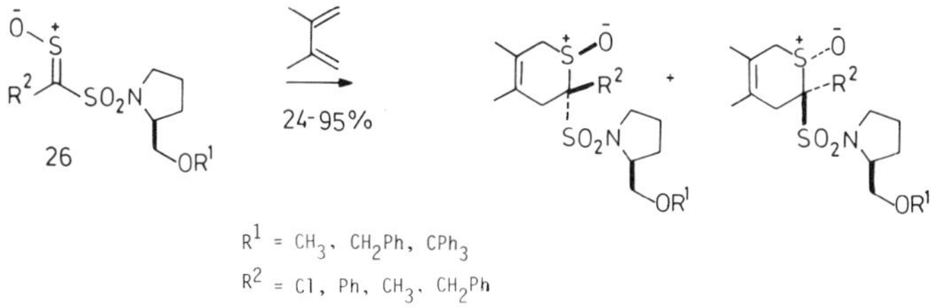
26
24-95%
R1 = CH3, CH2Ph, CPh3
R2 = Cl, Ph, CH3, CH2Ph

(21)

Unfortunately, diastereomeric excesses in this system ranged from 0% to a maximum of only 60%. A mechanistic model was offered to rationalize the selectivities observed here.

Thione *S*-imide **27,** which is the nitrogen analog of a sulfine, reacts with acyclic dienes to afford [4 + 2] cycloaddition products. With isoprene and 1,3-dimethylbutadiene the reactions were totally regioselective (Scheme 5-XX). However, cyclopentadiene afforded only a product of a 1,3-dipolar cycloaddition, *not* a Diels–Alder adduct.

Block *et al.* have discovered that sulfenes **28** can be generated by a desilylation route and that these reactive intermediates act as Diels–Alder dienophiles with cyclopentadiene (Scheme 5-XXI).[54a,b] It is of interest to note that other methods of generating sulfenes did not produce [4 + 2] adducts.[54a] Moreover, with substituted sulfenes endo stereoisomers predominated.

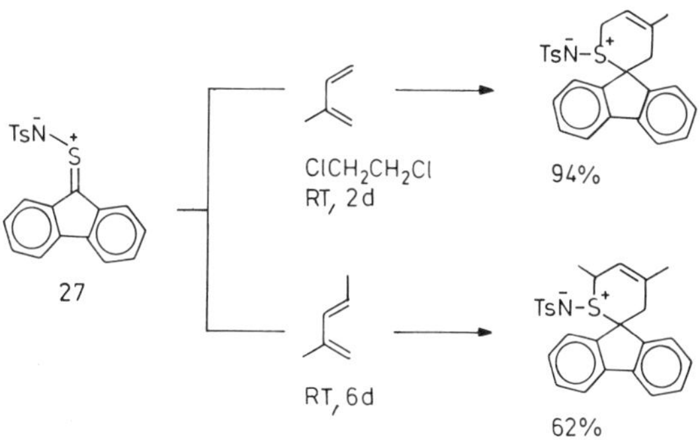
27
ClCH2CH2Cl
RT, 2d
94%
RT, 6d
62%

Scheme 5-XX

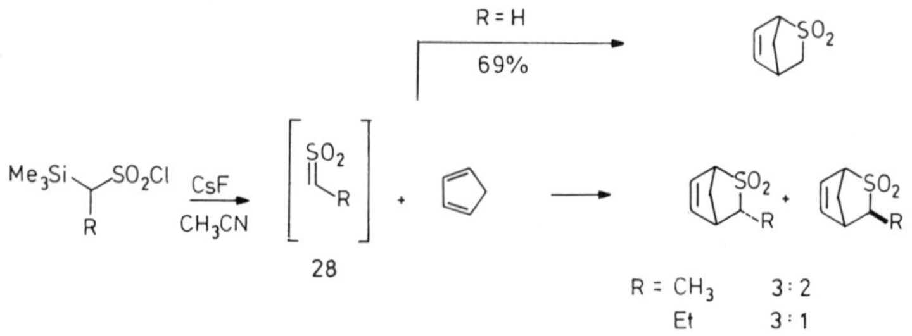

Scheme 5-XXI

8. SELENOALDEHYDES

Selenocarbonyl compounds have been virtually ignored as hetero-dienophiles. Recently, Krafft and Meinke have found a novel method to generate various selenoaldehydes (Scheme 5-XXII).[59] These compounds are reactive dienophiles and can be trapped *in situ* with cyclopentadiene to give Diels–Alder adducts. As can be seen in the scheme, endo adducts predominate in these cycloadditions. Similar results were found in thioaldehyde additions to cyclopentadiene (cf. Scheme 5-IV).

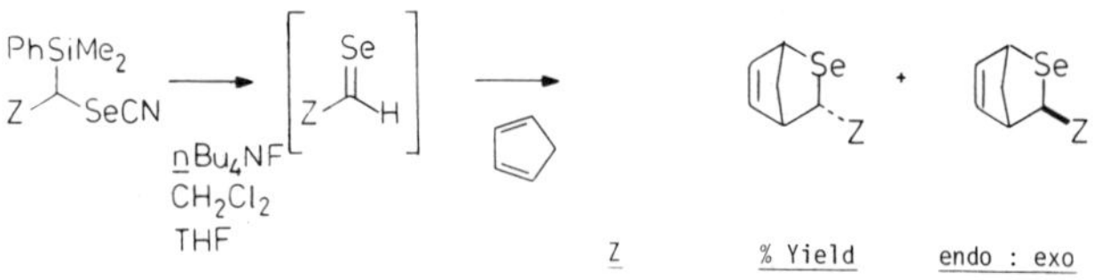

Z	% Yield	endo : exo
H	66%	
CH_3	83%	2.3 : 1
Et	78%	4.3 : 1
nPr	76%	3.4 : 1
Ph	81%	2.6 : 1
CH_2Ph	89%	3.5 : 1
tBu	39%	9.0 : 1

Scheme 5-XXII

REFERENCES

1. Previous reviews: a. Hamer, J. (ed.) (1967). "1,4-Cycloaddition Reactions, the Diels–Alder Reaction in Heterocyclic Syntheses." Academic Press, New York. b. Weinreb, S. M., and Staib, R. R. (1982). *Tetrahedron* **38,** 3087. c. Paquer, D. (1972). *Int. J. Sulf. Chem.* **7,** 269. d. Paquer, D. (1973). *Int. J. Sulf. Chem.* **8,** 173.

2. Vedejs, E., Perry, D. A., Houk, K. N., and Rondan, N. G. (1983). *J. Am. Chem. Soc.* **105,** 6999.
3. a. Middleton, W. J. (1965). *J. Org. Chem.* **30,** 1390. b. Middleton, W. J. (1965). *J. Org. Chem.* **30,** 1395.
4. Linn, W. J. (1964). *J. Org. Chem.* **29,** 3111.
5. a. Schönberg, A., and König, B. (1965). *Tetrahedron Lett.,* 3361. b. Schönberg, A., and König, B. (1968). *Chem. Ber.* **101,** 725.
6. König, B., Martens, J., Praefcke, K., Schönberg, A., Schwarz, H., and Zeisberg, R. (1974). *Chem. Ber.* **107,** 2931; Praefcke, K., and Weichsel, C. (1979). *Liebigs Ann. Chem.,* 784; Praefcke, K., and Weichsel, C. (1980). *Liebigs Ann. Chem.,* 1604.
7. Ohno, A., Ohnishi, Y., and Tsuchihashi, G. (1969). *Tetrahedron* **25,** 871.
8. a. Friedrich, K., and Gallmeier, H. J. (1981). *Tetrahedron Lett.* **22,** 2971 b. Beelitz, K., Höhne, G., and Praefcke, K. (1978). *Z. Naturforsch.* **33b,** 417.
9. Biellmann, J. F., Ducep, J. B., and Vicens, J. J. (1976). *Tetrahedron* **32,** 1801.
10. Malmberg, W.-D., Voss, J., and Weinschneider, S. (1983). *Liebigs Ann. Chem.,* 1694.
11. Block, E., and Aslam, M. (1985). *Tetrahedron Lett.* **26,** 2259.
12. a. Raasch, M. S. (1979). *J. Org. Chem.* **44,** 632. b. Lakshmikantham, M. V., Levinson, M., Menachery, M., and Cava, M. P. (1986). *J. Org. Chem.* **51,** 412.
13. Yamada, K., Yoshioka, M., and Sugiyama, N. (1968). *J. Org. Chem.* **33,** 1240; Omote, Y., Yoshioka, M., Yamada, K., and Sugiyama, N. (1967). *J. Org. Chem.* **32,** 3676; Sugiyama, N., Yoshioka, M., Aoyama, H., and Nishio, T. (1971). *J. Chem. Soc., Chem. Commun.,* 1063; Nishio, T., Yoshioka, M., Aoyama, H., and Sugiyama, N. (1973). *Bull. Soc. Chem. Jpn.* **46,** 2253; Ohno, A. (1971). *Int. J. Sulf. Chem.* **6,** 183.
14. Gotthardt, H. (1972). *Chem. Ber.* **105,** 2008.
15. Biellmann, J. F., and Ducep, J. B. (1970). *Tetrahedron Lett.,* 2899.
16. For some earlier trappings of thioaldehydes as Diels–Alder adducts, see Dice, D. R., and Steer, R. P. (1974). *Can. J. Chem.* **52,** 3518; Anastassiou, A. G., Wetzel, J. C., and Chao, B. (1976). *J. Am. Chem. Soc.* **98,** 6405.
17. a. Vedejs, E., Eberlein, T. H., and Vanle, D. L. (1982). *J. Am. Chem. Soc.* **104,** 1445. b. Vedejs, E., and Perry, D. A. (1983). *J. Am. Chem. Soc.* **105,** 1683. c. Vedejs, E., Eberlein, T. H., Mazur, D. J., McClure, C. K., Perry, D. A., Ruggeri, R., Schwartz, E., Stults, J. S., Varie, D. L., Wilde, R. G., and Wittenberger, S. (1986). *J. Org. Chem.* **51,** 1556. d. Vedejs, E., and Eberlein, T. H., Unpublished results.
18. Bladon, C. M., Ferguson, I. E. G., Kirby, G. W., Lochead, A. W., and McDougall, D. C. (1983). *J. Chem. Soc., Chem. Commun.,* 423; Bladon, C. M., Ferguson, I. E. G., Kirby, G. W., Lochead, A. W., and McDougall, D. C. (1985). *J. Chem. Soc., Perkin Trans. 1,* 1541.
19. Kirby, G. W., and Lochead, A. W. (1983). *J. Chem. Soc., Chem. Commun.,* 1325.
20. Kirby, G. W., Lochead, A. W., and Sheldrake, G. N. (1984). *J. Chem. Soc., Chem. Commun.,* 922.
21. Kirby, G. W., Lochead, A. W., and Sheldrake, G. N. (1984). *J. Chem. Soc., Chem. Commun.,* 1469.
22. Baldwin, J. E., and Lopez, R. C. G. (1982). *J. Chem. Soc., Chem. Commun.,* 1029; Baldwin, J. E., and Lopez, R. C. G. (1983). *Tetrahedron* **39,** 1487.
23. Krafft, G. A., and Meinke, P. T. (1985). *Tetrahedron Lett.* **26,** 1947.
24. Lee, L. F., Dolson, M. G., Howe, R. K., and Stults, B. R. (1985). *J. Org. Chem.* **50,** 3216.
25. Hartke, K., Kissel, T., Quante, J., and Henssen, G. (1978). *Angew. Chem. Int. Ed. Engl.* **17,** 953; Hartke, K., Quante, J., and Kampchen, T. (1980). *Liebigs Ann. Chem.,* 1482; Gillmann, T., and Hartke, K. (1986). *Chem. Ber.* **119,** 2859.

26. Vedejs, E., Arnost, M. J., Dolphin, J. M., and Eustache, J. (1980). *J. Org. Chem.* **45,** 2601.
27. Beslin, P., and Metzner, P. (1980). *Tetrahedron Lett.* **21,** 4657; interestingly, α,β-unsaturated dithioesters are apparently *not* heterodienophiles: Lawson, K. R., Singleton, A., and Whitham, G. H. (1984). *J. Chem. Soc., Perkin Trans. 1,* 859.
28. Seitz, G., Mohr, R., Overheu, W., Allmann, R., and Nagel, M. (1984). *Angew. Chem. Int. Ed. Engl.* **23,** 890.
29. Herczegh, P., Zsély, M., Bognár, R., and Szilágyi, L. (1986). *Tetrahedron Lett.* **27,** 1509.
30. Tamaru, Y., Satomi, H., Kitao, O., and Yoshida, Z. (1984). *Tetrahedron Lett.* **25,** 2561.
31. Martin, J. C., Gott, P. G., Meen, R. H., and Raynolds, P. W. (1981). *J. Org. Chem.* **46,** 3911.
32. See, for example, Mayer, R., Laban, G., and Wirth, M. (1967). *Liebigs Ann. Chem.* **703,** 140.
33. Johnson, C. R., Keiser, J. E., and Sharp, J. C. (1969). *J. Org. Chem.* **34,** 860; Benassi, R., Folli, U., and Iarossi, D. (1974). *Synthesis,* 735.
34. Raasch, M. S. (1975). *J. Org. Chem.* **40,** 161.
35. Allgeier, H., and Winkler, T. (1976). *Tetrahedron Lett.,* 215.
36. Reich, H. J., and Trend, J. E. (1973). *J. Org. Chem.* **38,** 2637.
37. Laban, G., and Mayer, R. (1967). *Z. Chem.* **7,** 227.
38. a. Larsen, C., and Harpp, D. N. (1980). *J. Org. Chem.* **45,** 3713. b. Harpp, D. N., MacDonald, J. G., and Larsen, C. (1985). *Can. J. Chem.* **63,** 951.
39. Boerma, J. A., Nilsson, N. H., and Senning, A. (1974). *Tetrahedron* **30,** 2735.
40. Hansen, H. C., and Senning, A. (1979). *J. Chem. Soc., Chem. Commun.,* 1135.
41. Vyas, D. M., and Hay, G. W. (1971). *Can. J. Chem.* **49,** 3755; Vyas, D. M., and Hay, G. W. (1975). *J. Chem. Soc., Perkin Trans. 1,* 180.
42. Friedrich, K., and Zamkanei, M. (1977). *Tetrahedron Lett.,* 2139; Friedrich, K., and Zamkanei, M. (1979). *Chem. Ber.* **112,** 1867.
43. Corey, E. J., and Walinsky, S. W. (1972). *J. Am. Chem. Soc.* **94,** 8932.
44. Ishibashi, H., Kitano, Y., Nakatani, H., Okado, M., Ikeda, M., Okura, M., and Tamura, Y. (1984). *Tetrahedron Lett.* **25,** 4231.
45. See also Tamura, Y., Ishiyama, K., Mizuki, Y., Maeda, H., and Ishibashi, H. (1981). *Tetrahedron Lett.* **22,** 3773.
46. Fuji, K., Khanapure, S. P., Node, M., Kawabata, T., and Ito, A. (1985). *Tetrahedron Lett.* **26,** 779. See also Ishibashi, H., Okada, M., Nakatani, H., Ikeda, M., and Tamura, Y. (1986). *J. Chem. Soc. Perkin Trans. 1,* 1763.
47. Raasch, M. (1970). *J. Org. Chem.* **35,** 3470; Raasch, M. (1978). *J. Org. Chem.* **43,** 2500.
48. For a review of the chemistry of sulfines see Zwanenburg, B. (1982). *Recl. Trav. Chim. Pays-Bas* **101,** 1; see also Bonini, B. F., Mazzanti, G., Zani, P., Maccagnani, G., Barbaro, G., Battaglia, A., and Giorgianni, P. (1986). *J. Chem. Soc., Chem. Commun.* 964; Schwab, M., and Sundermeyer, W. (1986). *Chem. Ber.* **119,** 2458.
49. Strating, J., Thijs, L., and Zwanenburg, B. (1967). *Recl. Trav. Chim. Pays-Bas* **86,** 641.
50. Zwanenburg, B., Thijs, L., Broens, J. B., and Strating, J. (1972). *Recl. Trav. Chim. Pays-Bas* **91,** 443.
51. Zwanenburg, B., Thijs, L., and Strating, J. (1969). *Tetrahedron Lett.,* 4461.
52. a. Lenz, B. G., Haltiwanger, R. C., and Zwanenburg, B. (1984). *J. Chem. Soc., Chem. Commun.,* 502. b. Lenz, B. G., Regeling, H., van Rozendaal, H. L. M., and Zwanenburg, B. (1985). *J. Org. Chem.* **50,** 2930. c. Lenz, B. G., Regeling, H., and Zwanenburg, B. (1984). *Tetrahedron Lett.* **25,** 5947.
53. Still, I. W. J., and Ablenas, F. J. (1985). *J. Chem. Soc., Chem. Commun.,* 524.

54. a. Block, E., and Aslam, M. (1982). *Tetrahedron Lett.* **23**, 4203. b. Block, E., and Wall, A. (1985). *Tetrahedron Lett.* **26**, 1425. c. Block, E., Wall, A., and Zubieta, J. (1985). *J. Am. Chem. Soc.* **107**, 1783.

55. Saalfrank, R. W., and Rost, W. (1985). *Angew. Chem. Int. Ed. Engl.* **24**, 855.

56. Porskamp, P. A. T. W., Haltiwanger, R. C., and Zwanenburg, B. (1983). *Tetrahedron Lett.* **24**, 2035.

57. van den Broek, L. A. G. M., Porskamp, P. A. T. W., Haltiwanger, R. C., and Zwanenburg, B. (1984). *J. Org. Chem.* **49**, 1691.

58. a. Saito, T., and Motoki, S. (1978). *Chem. Lett.,* 591. b. Saito, T., and Motoki, S. (1979). *J. Org. Chem.* **44**, 2493.

59. Krafft, G. A., and Meinke, P. T. (1986). *J. Am. Chem. Soc.* **108**, 1314; see also Fischer, H., Gerbing, U., Riede, J., and Benn, R. (1986). *Angew. Chem. Int. Ed. Engl.* **25**, 78; Kirby, G. W., and Tretheway, A. N. (1986). *J. Chem. Soc., Chem. Commun.* 1152.

Chapter 6

Miscellaneous Dienophiles

1. NITRILES

Introduction

The ability of a nitrile to act as a Diels–Alder dienophile has been known since 1935 when Dilthey *et al.* discovered that tetracyclone reacts with nitriles at high temperatures to afford substituted pyridines [Eq. (1)].[1-3]

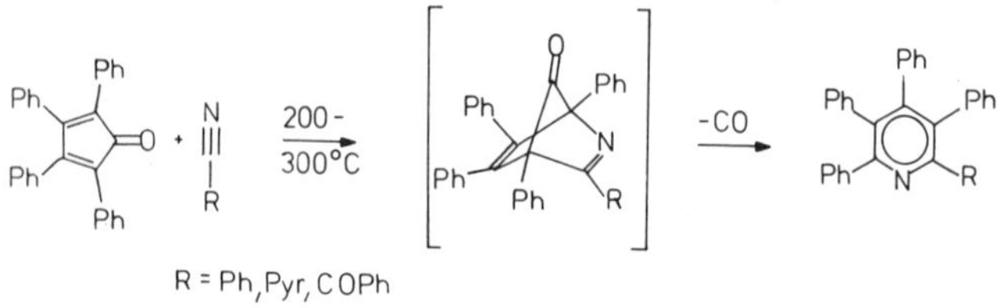

R = Ph, Pyr, C OPh

(1)

One of the features of Diels–Alder reactions with most alkyl and aryl nitriles as well as cyanamides that has made them rather unattractive as heterodienophiles is the requirement of very high reaction temperatures for cycloaddition.[1] Under these extreme conditions the initial 3,6-dihydropyridine reaction products of nitriles and simple acyclic 1,3-dienes invariably oxidize to the corresponding pyridines [Eq. (2)].

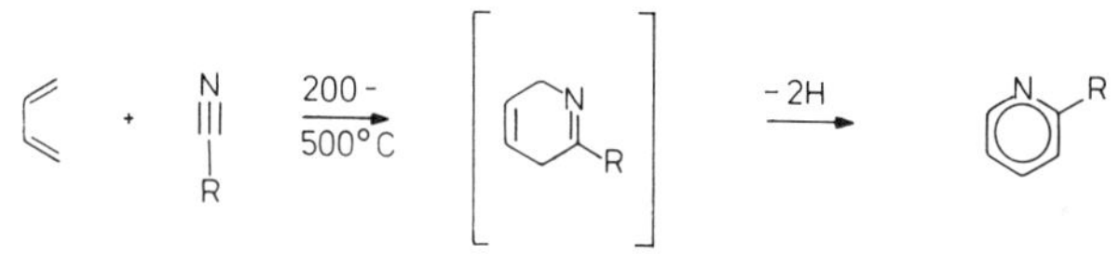

$$\text{(2)}$$

Although perhaps useful in industrial processes, the methodology is not well suited to general laboratory-scale preparations.

The various aspects, including kinetic and thermodynamic parameters, of this type of reaction have been thoroughly reviewed[1] and need not be reiterated here. Only more recent events of potentially wider synthetic utility are included in the following discussion.

Sulfonyl Cyanides

Sulfonyl cyanides, which are readily prepared by reaction of an alkyl or aryl sodium sulfinate with cyanogen chloride [Eq. (3)],

$$RSO_2^-Na^+ \; + \; ClCN \; \longrightarrow \; RSO_2CN \; + \; NaCl$$

$$\text{(3)}$$

R = aryl, benzyl, alkyl

are exceptionally reactive dienophiles.[4,5] Cycloaddition of a sulfonyl cyanide with a diene at reasonably low temperatures affords an initil 3,6-dihydropyridine, which under the reaction conditions is oxidized to a 2-sulfonylpyridine or is hydrolyzed *in situ,* probably by a trace of water, to give a 2-pyridone [Eq. (4)].

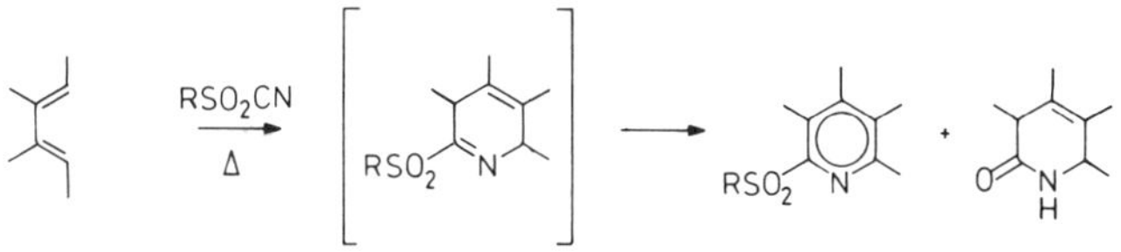

$$\text{(4)}$$

Table 6-I contains some typical examples of these cycloadditions. A clever total synthesis of an unusual isonitrile fungal metabolite has been achieved using the bicyclic adduct prepared from tosyl cyanide and cyclopentadiene (Scheme 6-I).[6]

TABLE 6-I

Cycloadditions of Sulfonyl Cyanides (RSO₂CN)

R	Diene	Conditions	Product(s)	Yield	Ref.
pCH₃Ph	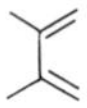	20°C, 60 hr	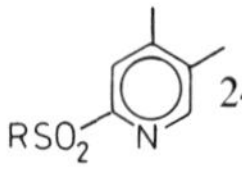 24% 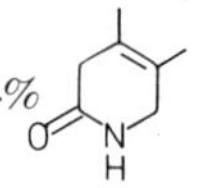	55%	4b
pCH₃Ph		72°C, 2 hr	28%	38%	4b
PhCH₂		20°C, 20 hr	20%	22%	4b
pCH₃Ph	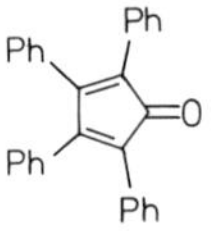	175°C, 45 min	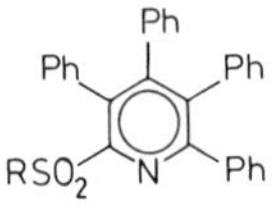	89%	4b
pCH₃Ph		RT, 30 min	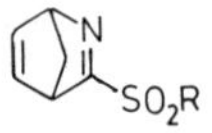	95%	4c
pClPh	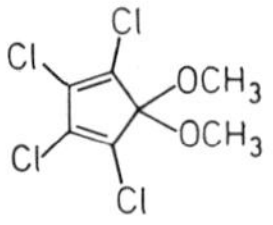	150°C, 10 hr	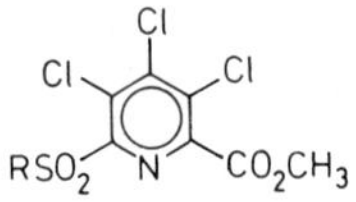	76%	5a
pCH₃Ph	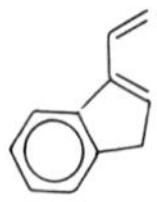	C₆H₆, 80°C	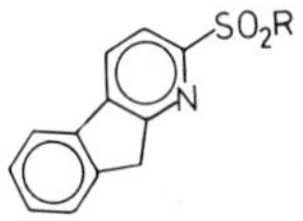	49%	5b

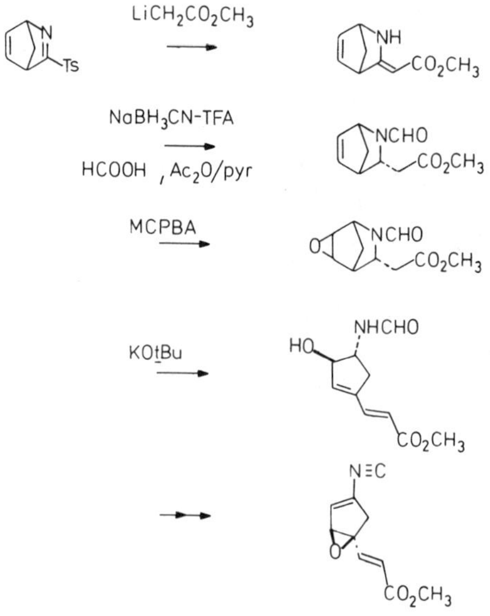

Scheme 6-I

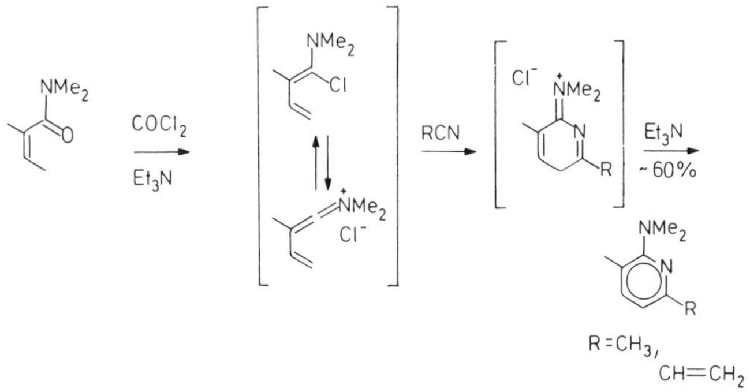

Scheme 6-II

Cycloadditions with Reactive Dienes

Ghosez *et al.* have recently prepared so-called push–pull dienes (Scheme 6-II).[7] These dienes are sufficiently electrophilic to react at relatively low temperatures with unactivated, electron-rich nitriles, affording initial [4 + 2] adducts which tautomerize to 2-aminopyridines. This method would seem to offer a potentially efficient approach to synthesis of certain substituted pyridines. An *s*-tetrazine dicarboxylate is also a sufficiently electrophilic diene to combine with an N-substituted cyanamide to afford a triazene in good yield [Eq. (5)].[8]

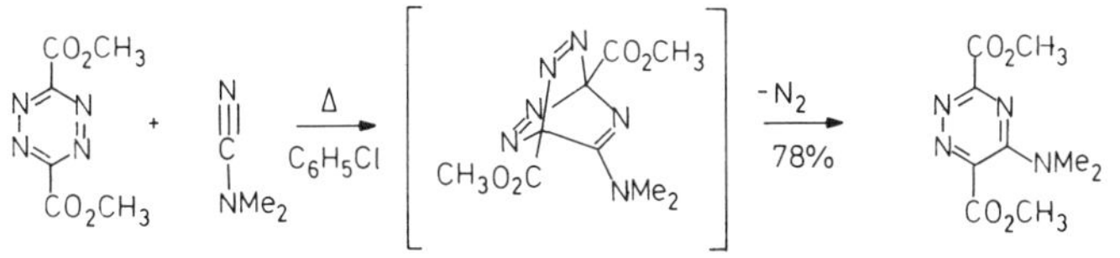

$$(5)$$

Intramolecular Cycloadditions[9]

A few scattered examples of intramolecular Diels–Alder cycloadditions of nitriles have been documented. Oppolzer has synthesized some benzocyclobutene nitriles (Scheme 6-III) and found that on pyrolysis they were transformed into isoquinoline derivatives in excellent yields.[10] *o*-Quinodimethanes are assumed to be transient dienes in all of these cycloadditions.

A brief report has appeared describing an interesting synthesis of an annulated pyridine by thermal intramolecular cycloaddition of a diene nitrile. Unfortunately, neither yields nor a detailed account of reaction conditions were provided [Eq. (6)].[11]

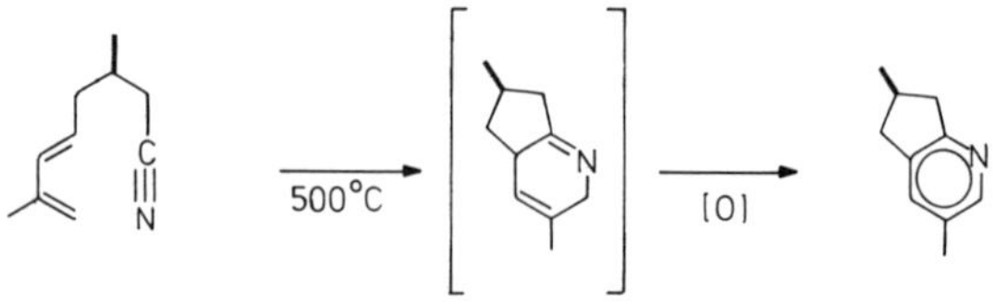

$$ (6) $$

A furan ring has been reported as the diene in some intramolecular [4 + 2] cycloadditions of nitriles [Eq. (7)].[12]

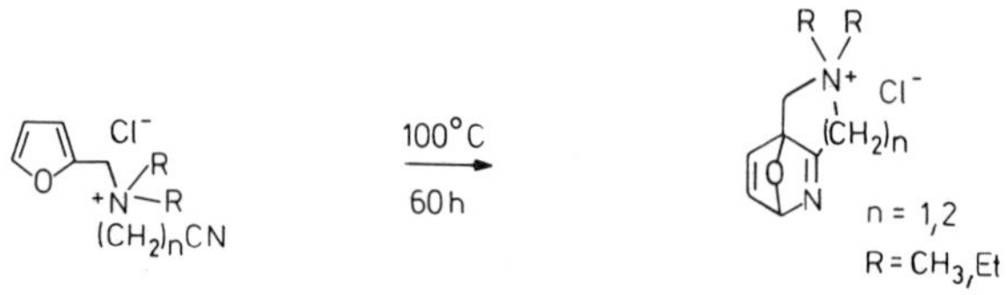

$$ (7) $$

These are the only examples to date of a furan reacting as a diene with a cyano dienophile. Taylor and French have recently reported two intramolecular cyclizations of nitrile dienophiles with triazines to give fused heterocycles [Eq. (8)] in modest yields.[12b]

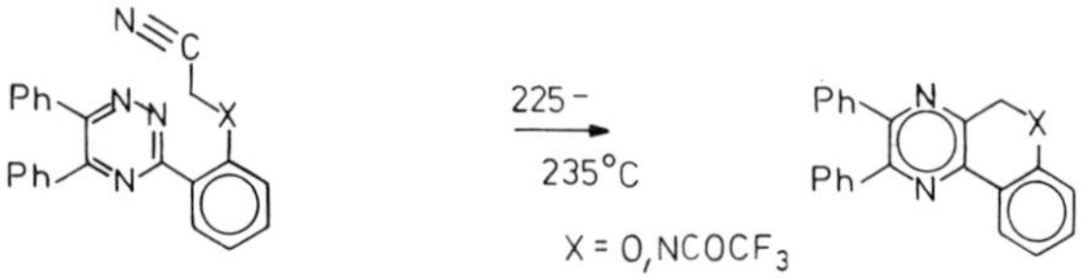

$$ (8) $$

X = O, NCOCF$_3$

2. PHOSPHORUS-CONTAINING DIENOPHILES

Phosphoalkenes and Related Compounds

Several structurally different types of phosphaalkenes containing a phosphorous–carbon double bond have been found to undergo [4 + 2]

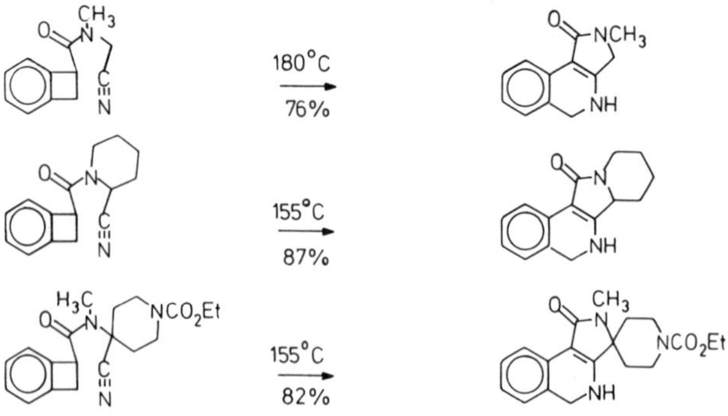

Scheme 6-III

TABLE 6-II

Cycloadditions with Phosphorus-Containing Dienophiles

Entry	Dienophile	Diene	Conditions	Product(s)	Yield	Ref.
1	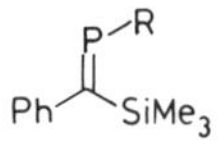	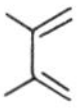	C_6H_6, 60°C	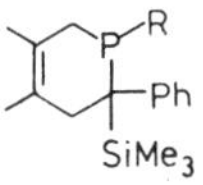	82%	14
	R = OCH$_3$					
2	R = N=CPh$_2$		C_6H_6, 60°C		53%	14
3	R = nBu		C_6H_6, RT, 2 hr		84%	14
4	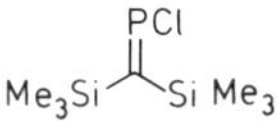		RT, 6 hr	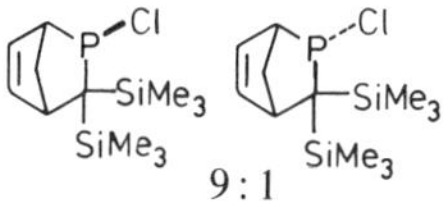 9:1	—	15
5	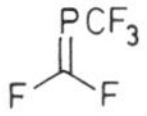		CH$_2$Cl$_2$, −15°C, 15 hr	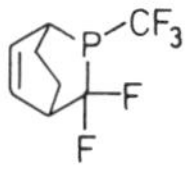	85%	16, 17
6	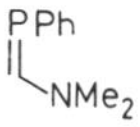	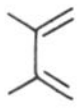	$h\nu$, CH$_3$CN	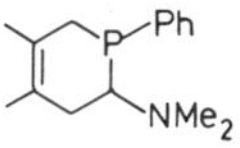	80%	18
7	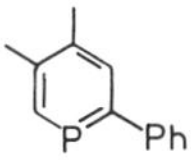		S$_8$, neat, 100°C, 24 hr	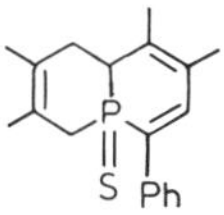	85%	19b
8	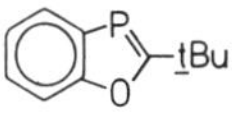	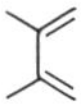	120°C, 120 hr	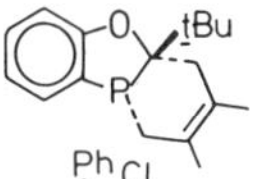	55%	20
9	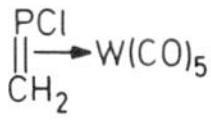	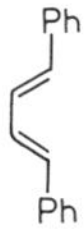	Toluene, 65°C, 4 hr	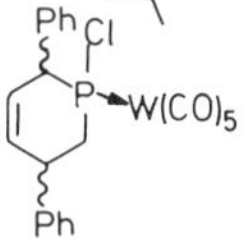	—	19a
10	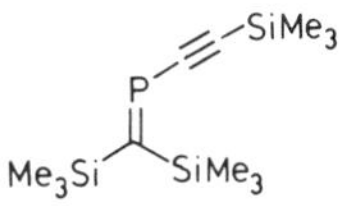	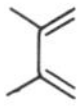	C_6H_6, 2 hr	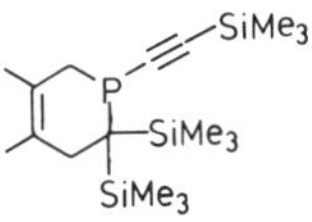	80%	21
11	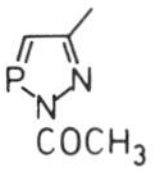		15°C, 5 min	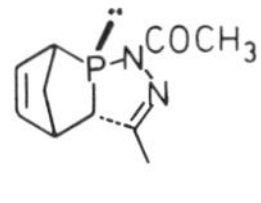	65%	22d

cycloadditions with 1,3-dienes. The electron demand of such dienophiles has been calculated using FMO theory,[13] and, at least for simple phosphaalkenes, these appear to be normal ($HOMO_{diene}$ controlled) Diels–Alder reactions. Table 6-II lists a number of Diels–Alder reactions involving this type of dienophile. It should be noted that these cycloadditions are often readily reversible, and diastereomer mixtures can be formed depending on whether the reactions are run under conditions of kinetic or thermodynamic control (cf. entries 1–5).[22d]

It was recently reported that the phosphole in Eq. (9)

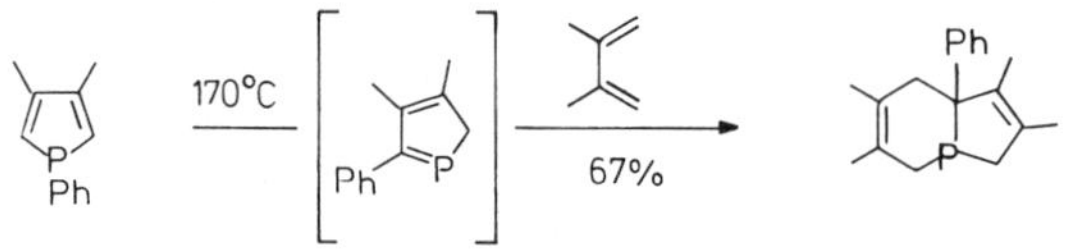

$$ \text{(9)} $$

rearranges thermally in the presence of excess 2,3-dimethylbutadiene to give a [4 + 2] adduct which could be isolated in excellent yield.[22] It was postulated that a transient 2H-phosphole was the dienophile in this case. Similarly, on heating an arsole [Eq. (10)]

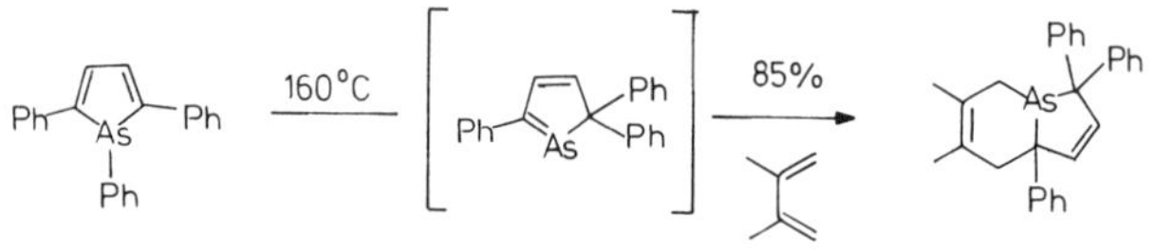

$$ \text{(10)} $$

in the presence of 2,3-dimethylbutadiene, a Diels–Alder adduct was isolated.[23] In this case the 2H-arsole is believed to be the reacting species.

Other Dienophiles

Phosphonodithioic anhydrides react regioselectively with unsymmetrical dienes to give Diels–Alder adducts.[24] For example, isoprene and *trans*-piperylene react to form adducts shown in Scheme 6-IV. Yields

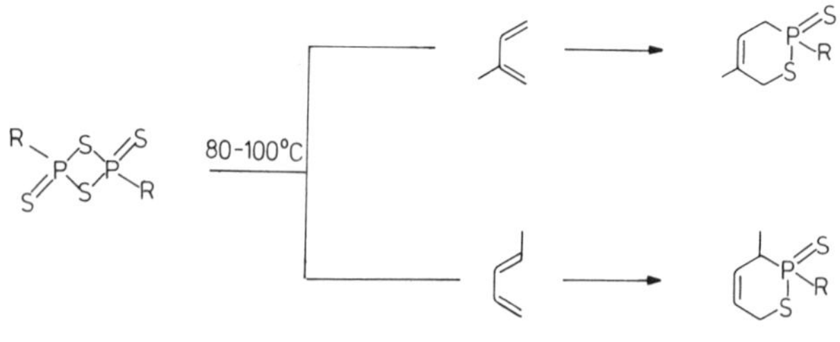

Scheme 6-IV

with most dienes are in the 70–90% range. It was suggested that a biradical mechanism is most consistent with the regiochemical results.[24c]

It has been proposed that a phosphinidene sulfide, formed by magnesium metal reduction of a dichloro precursor, can be trapped by 2,3-dimethylbutadiene to yield an adduct [Eq. (11)].[25]

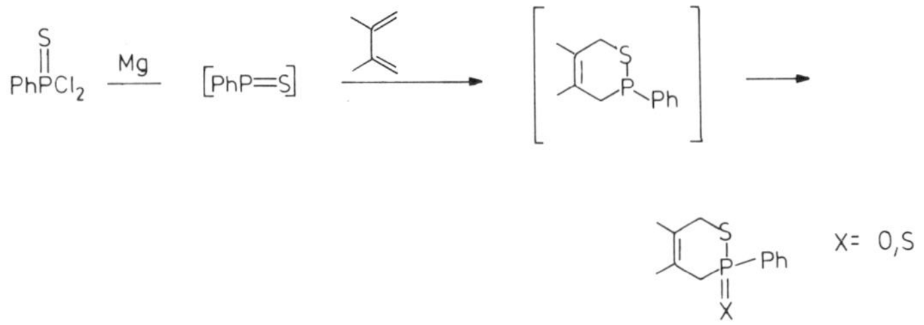

$$
\underset{\substack{\| \\ PhPCl_2}}{\overset{S}{}} \quad \xrightarrow{Mg} \quad [PhP{=}S] \quad + \quad \ce{<diene>} \quad \longrightarrow \quad \left[\ce{<adduct>} \right] \quad \longrightarrow
$$

(11)

X= O,S

However, the initial Diels–Alder product was not isolated, but rather two further transformation products were. It was postulated that these species are artifacts produced during the workup procedure; however, this interpretation has been questioned.[24b]

It has recently been found that a diphosphene [Eq. (12)]

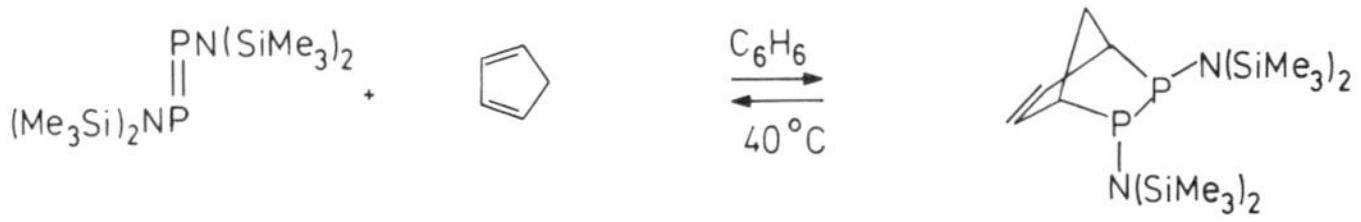

$$
\underset{(Me_3Si)_2NP}{\overset{PN(SiMe_3)_2}{\|}} \quad + \quad \ce{<cyclopentadiene>} \quad \underset{40\,^\circ C}{\overset{C_6H_6}{\rightleftarrows}} \quad \ce{<adduct>}
$$

(12)

can act as a dienophile.[26] Reaction with cyclopentadiene forms exclusively the trans adduct, but at 40°C the adduct undergoes a retro cycloaddition. In what appears to be a similar type of transformation, it was found that cyclophosphanes on irradiation in the presence of a diene give Diels–Alder adducts [Eq. (13)].[24b,27]

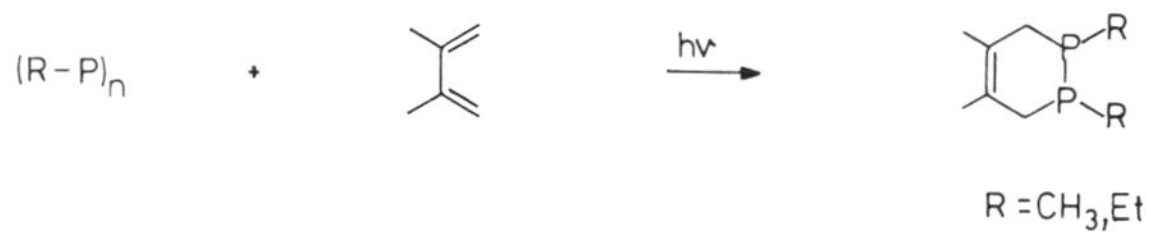

$$
(R{-}P)_n \quad + \quad \ce{<diene>} \quad \xrightarrow{h\nu} \quad \ce{<adduct>}
$$

(13)

$R = CH_3, Et$

The corresponding cycloarsanes undergo the same reaction in good yields [Eq. (14)].

$$
(CH_3As)_5 \quad + \quad \ce{<diene>} \quad \underset{95\%}{\overset{h\nu}{\longrightarrow}} \quad \ce{<adduct>}
$$

(14)

3. AZO COMPOUNDS

Introduction

The first [4 + 2] cycloaddition of an azo dienophile described by Diels *et al.* in 1925 [Eq. (15)]

is of historical importance since it was one of the earliest examples of a Diels–Alder reaction.[28] Since then, acyclic and cyclic azodicarbonyl compounds have proved to be among the most reactive dienophiles and have been used extensively in synthesis.[1,29] In general, [4 + 2] cycloadditions with azodicarbonyl compounds appear to be concerted, stereospecific processes that can be rationalized nicely by FMO theory.[30] The high reactivity of these systems is due to the lower LUMO of the N=N bond relative to that of a C=C bond. Incorporation of the azo system into a ring lowers the LUMO further, and consequently cyclic compounds such as 1,2,4-triazoline-3,5-diones are among the most reactive dienophiles known.[31] Azodicarboxylates also undergo facile ene reactions,[29b,32] which can sometimes lead to undesired side products.

Space limitations make it impossible to thoroughly survey this extensive area of hetero Diels–Alder chemistry. The following sections will highlight some of the more recent advances and previous reviews which should be consulted for additional information.[1,29]

Acyclic Azodicarboxylates and Related Compounds

Various azodicarboxylate esters have been used extensively as dienophiles.[1,29] Several representative examples of this type of cycloaddition are listed in Table 6-III.[33–40] As mentioned above, these reactions appear to be concerted processes that retain the 1,3-diene configuration in the adducts.[30a] The more bulky esters tend to react more slowly with dienes. Thus, dimethylazodicarboxylate reacts 5–6 times more rapidly with cyclopentadiene than does the ethyl ester,[42a] whereas the *tert*-butyl compound reacts sluggishly.[42b]

Kresze *et al.* have described an interesting pyrrole synthesis based on an azo Diels–Alder reaction [Eq. (16)].[43]

TABLE 6-III

Cycloadditions of Acyclic Azodicarboxylates

Azodicarboxylate	Diene	Conditions	Product	Yield	Ref.
NCO_2R RO_2CN $R = Et$		Toluene, 80°C, 100 min		89%	33
$R = Et$		Neat, 160°C, 6 hr		56%	34
$R = Et$		C_6H_{12}, 25°C, 24 hr		43%	35
$R = Et$		Toluene, Δ, 48 hr		65%	36
$R = CH_3$		$h\nu$, C_6H_{12}, 46–50°C		75%	37
$R = CH_3$		C_6H_6, RT		95%	38
$R = CH_2CCl_3$				79%	·39
$R = CH_2Ph$		RT, 1 hr		—	40
$R = Et$		C_6H_6, 150°C, 3 hr		82%	41

(16)

In this particular example, reduction of the initial adduct with zinc gave a trisubstituted pyrrole in 33% overall yield. The mechanism of this transformation is not immediately obvious.

Diels–Alder adducts of various azodicarboxylates have been used extensively in synthesis of cyclic azoalkanes[44] and pyridazine derivatives.[1,29] Since previous reviews have adequately dealt with this material, it will not be reiterated here.

Although simple azobenzenes do not usually undergo [4 + 2] cycloadditions,[41,45] Ahern, Gokel, and co-workers have found that arenediazocyanides are effective dienophiles.[46] These compounds are prepared by addition of potassium cyanide to diazonium salts in the presence of 18-crown-6 [Eq. (17)].

(17)

Addition of these dienophiles to a number of dienes gave good yields of cycloadducts. Four representative examples of this cycloaddition are shown in Scheme 6-V. In general, the *p*-methoxy-substituted aromatics were less reactive than those bearing electron-withdrawing groups. Kinetic data have indicated that these cycloadditions probably proceed via a concerted mechanism with the transition state having dipolar character.[47] The regioselectivity of the process is also in accord with this mechanistic postulate.

Cyclic Azodicarboxylates and Related Compounds

Since their introduction in 1962 by Cookson, 4-substituted 1,2,4-triazoline-3,5-diones have become some of the most widely used dienophiles in organic chemistry.[48] This type of dienophile has found such extensive use that it is not possible to more than touch on its applications here.[49] Previous reviews contain more detailed information.[1,29,30b] Table 6-IV contains a few recent representative examples of Diels–Alder reactions with

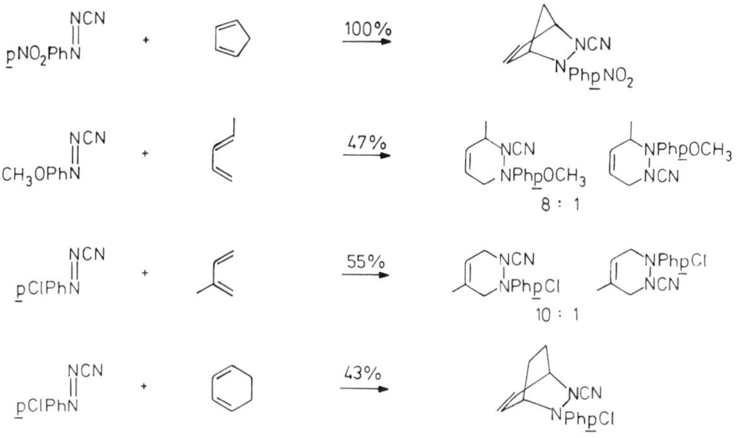

Scheme 6-V

triazolinediones.[50–57] In a thorough study, Stevens and co-workers[30b] have measured the kinetics of addition of a number of 4-substituted triazoline-3,5-diones with some dienes and concluded that the reaction is best rationalized by an FMO model.

One important use of triazolinediones as dienophiles has been to protect conjugated dienes, particularly in steroids and related compounds.[59] Several methods have been developed for conversion of the adducts back to the starting diene functionality [Eq. (18)],

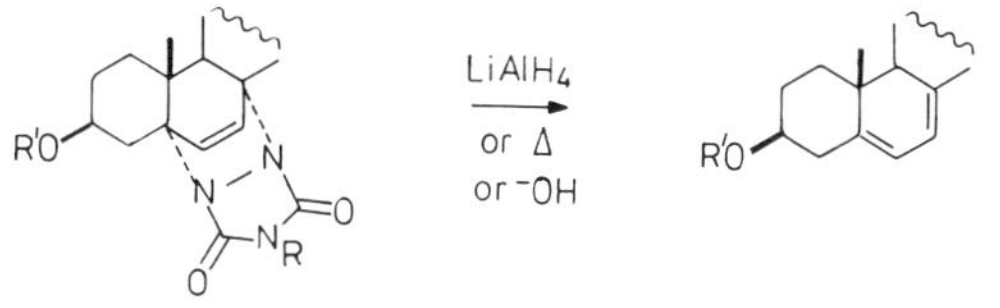

$$(18)$$

including reduction, thermolysis, and basic hydrolysis.

In a recent elegant synthetic application of this methodology, Vandewalle and co-workers have prepared some vitamin D_3 derivatives.[60] Scheme 6-VI shows a synthesis of 12-hydroxy vitamin D_3 from 7-dehydrocholesterol. The key feature of this route is the protection of the sensitive triene moiety as its Diels–Alder adduct with N-phenyltriazoline-3,5-dione.

In general, hydrolysis of adducts of triazolinediones to the hydrazines requires rather vigorous conditions. To solve this problem, Corey[61a] and Beak[61b] have independently generated 1,3,4-thiadiazole-2,5-dione, which can be trapped with reactive 1,3-dienes at 0°C or below [Eq. (19)].

TABLE 6-IV

Cycloadditions of *N*-Substituted 1,2,4-Triazoline-3,5-diones

Triazolinedione	Diene	Conditions	Product	Yield	Ref.
		CH$_2$Cl$_2$, 0°C		67%	50
		Neat, RT		65%	51
		CH$_2$Cl$_2$, −50°C		60%	52
		Me$_2$CO, 0°C		70%	53
		Me$_2$CO, RT, 0.5 hr		88%	55
		CHCl$_3$, −35°C		96%	56
		hν, 18°C		40%	57
		—		78%	54
		CH$_3$CN, RT		71%	58

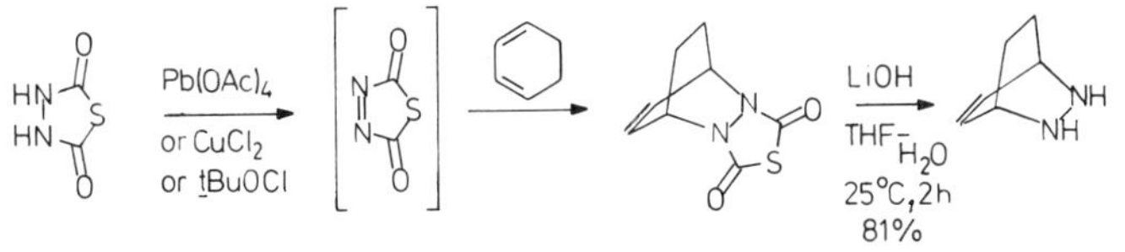

$$\text{(19)}$$

Since this dienophile is thermally unstable, it cannot be used effectively with relatively unreactive dienes that require heating for cycloaddition to occur.

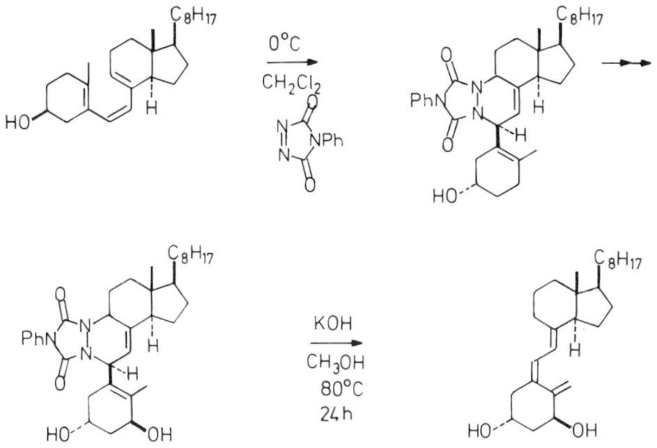

Scheme 6-VI

Keana and co-workers have recently developed some sulfonated 4-aryltriazolinediones which are sufficiently stable to hydrolysis that they can be used for aqueous Diels–Alder reactions.[62a] Thus, the triazolinedione sodium sulfonate shown in Eq. (20)

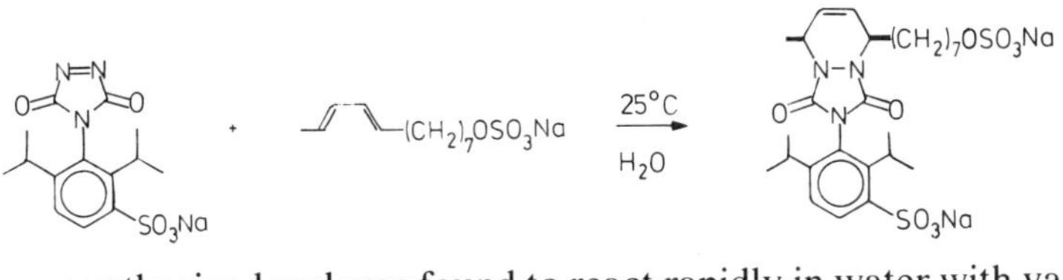

$$\text{(20)}$$

has been synthesized and was found to react rapidly in water with various dienic detergents to give [4 + 2] cycloadducts. This reagent allows one to

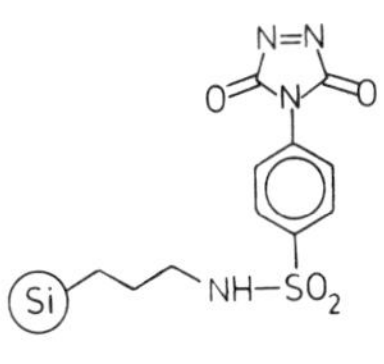

A

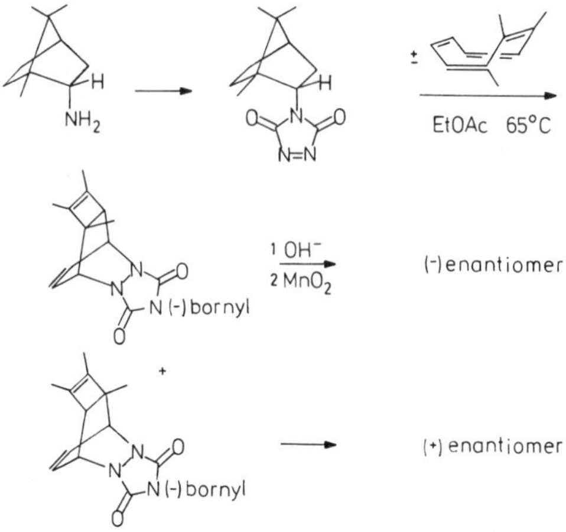

Scheme 6-VII

remove this and other new 1,3-diene-containing detergents[62b] and phase transfer catalysts[62c] since the adducts are highly water soluble. In addition, these workers have prepared a silica gel immobilized reagent (**A**), which can be used to remove 1,3-diene from solution.[62a,c]

Paquette's group has recently synthesized and evaluated chiral 4-substituted triazolinediones in Diels–Alder cycloadditions.[63] For example, *endo*-(−)-bornylamine has been converted to an azo dienophile, which has been used to resolve 1,2,3-trimethylcyclooctatetraene (Scheme 6-VII). In addition, triazolinediones derived from α-methylbenzylamine and dehydroabietylamine have been prepared and their reactions with achiral 1,3-dienes were investigated.[63b] Unfortunately, these dienophiles did not show good kinetic enantioselection, but they are useful for the type of resolution shown in Scheme 6-VII.[63c]

Other types of cyclic azo compounds have also been used as dienophiles, and a few representative systems are shown in Scheme 6-VIII.[64–68] Previous reviews should be consulted for additional examples.[1,29]

Aromatic Diazonium Salts and Other Charged Azo Dienophiles

The observation that diazodicyanoimidazole added to butadiene to afford a Diels–Alder adduct [Eq. (21)][69]

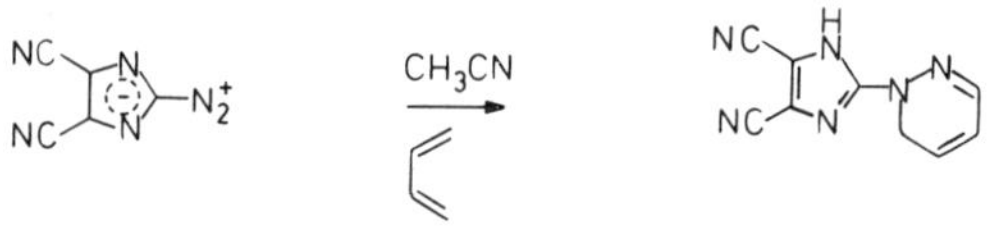

$$(21)$$

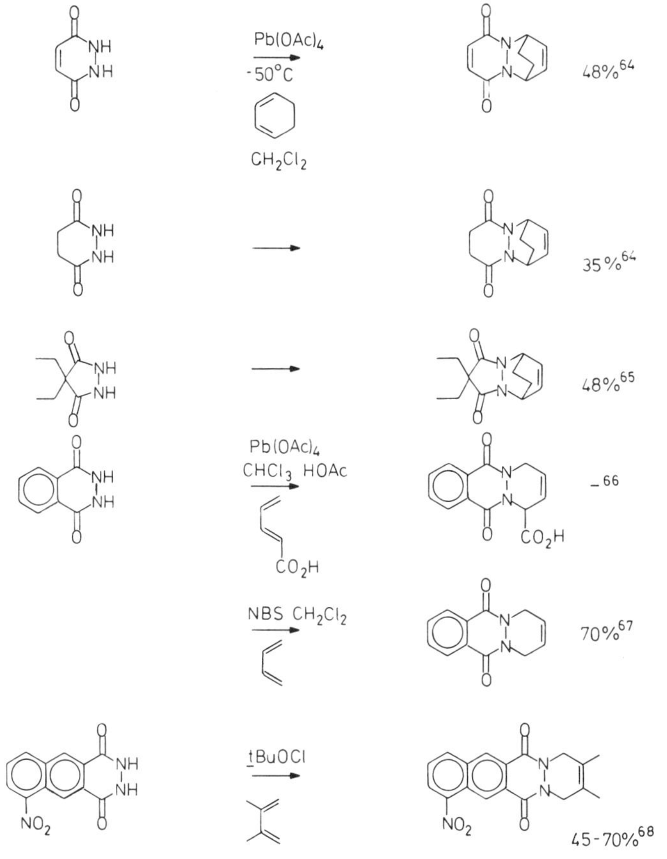

Scheme 6-VIII

led Sheppard and co-workers[70] to reexamine the reaction of aryldiazonium compounds and 1,3-dienes first reported by Meyer in 1919 to afford
simple acyclic substitution products.[71] In fact, aryldiazonium salts react
with electron-rich 1,3-dienes to initially give Diels–Alder adducts, which
tautomerize to 1,6-dihydropyridazines [Eq. (22)].

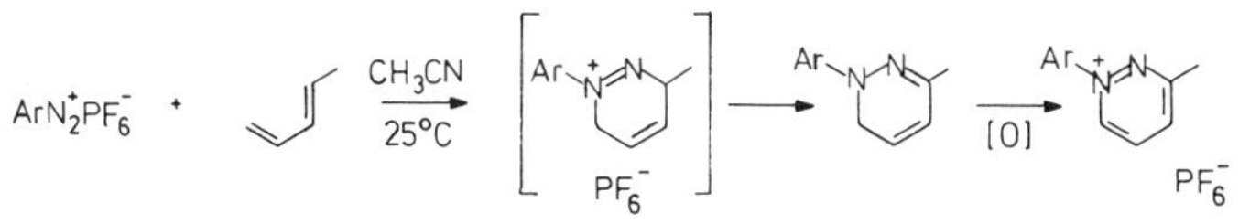

In those cases where Ar is p-nitrophenyl, these compounds can be isolated. However, with other diazonium compounds *in situ* oxidation occurs to afford pyridazinium salts. Several examples of this cycloaddition

are compiled in Table 6-V.[70–72] More recently, Bronberger and Huisgen have reinvestigated this cycloaddition from a mechanistic viewpoint and have concluded that the cycloaddition probably proceeds via a concerted process.[72]

Although the [4 + 2] cycloaddition of a simple azoalkane is not an energetically favorable process, Nelsen and co-workers have discovered that such a reaction can be catalyzed by HBF_4.[73] The sequence shown in Eq. (23) exemplifies this methodology.

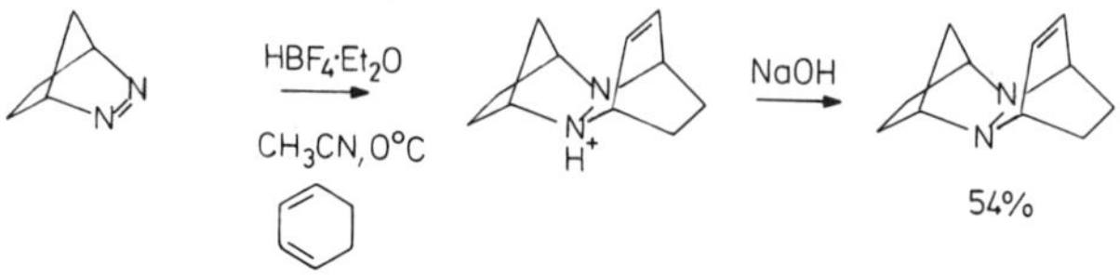

(23)

TABLE 6-V
Cycloadditions of Substituted Aryldiazonium Salts

Diene	Product	Yield: H	*p*Cl	*m*F	*p*F	*p*NO_2	Ref.
	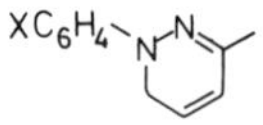	—	—	—	—	73%	70
	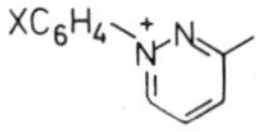 PF_6^-	Trace	26%	—	—	—	70
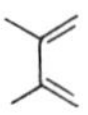	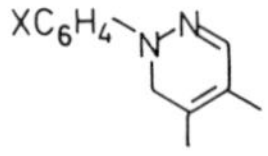	—	—	—	—	79%	70
	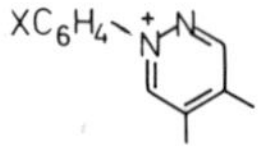 PF_6^-	42%	60%	52%	72%	—	70
	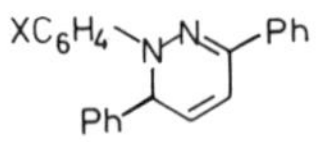	—	—	—	—	69%	72
	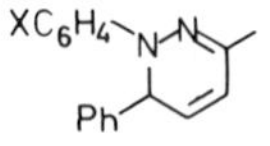	—	—	—	—	39%	72

Interestingly, only the stereoisomer shown was produced. This may, in fact, be the thermodynamically most stable product and may not reflect a kinetic preference.

4. OTHER DIENOPHILES

Steliou *et al.* have described a method for generation of what may be S_2 from silyl and germanyl trisulfides [Eq. (24)].[74]

$$R_3MSSSMR_3 \; + \; Ph_3PBr_2 \xrightarrow[25°C]{CH_2Cl_2} \left[\; S_2 \; \right] \longrightarrow \bigtimes\longrightarrow \bigcirc\!\!\!\!\!_S^S \quad 35\% \qquad (24)$$

$$R = Ph, \; C_6H_{11}, \; \underline{p}CH_3Ph; \; M = Si, \; Ge$$

In the presence of a 1,3-diene this species produces [4 + 2] cycloadducts. The exact nature of this reactive intermediate has not yet been determined. It was found that a boron imide undergoes a Diels–Alder cycloaddition with cyclopentadiene under mild conditions [Eq. (25)].[75]

$$C_6F_5\,B{=}N\underline{t}Bu \longrightarrow \bigcirc \xrightarrow[CHCl_3]{-30°C} \bigcirc\!\!\!\!\!\!{}_{B\,C_6F_5}^{N\underline{t}Bu} \qquad (25)$$

REFERENCES

1. Previous reviews: a. Hamer, J. (ed.) (1967). "1,4-Cycloaddition Reactions, the Diels–Alder Reaction in Heterocyclic Syntheses." Academic Press, New York. b. Weinreb, S. M., and Staib, R. R. (1982). *Tetrahedron* **38**, 3087. c. Wollweber, H. (1972). "Diels–Alder Reaktion." G. Thieme, Stuttgart.
2. Dilthey, W., Schommer, W., Hoschen, W., and Dierichs, H. (1935). *Chem. Ber.* **68**, 1159.
3. See also Martin, D., and Bauer, M. (1980). *Z. Chem.* **20**, 53; Arnold, Z., and Dvorak, D. (1985). *Coll. Czech. Chem. Commun.* **50**, 2265; Potthoff, B., and Breitmaier, E. (1986). *Synthesis* 584.
4. a. van Leusen, A. M., and Jagt, J. C. (1970). *Tetrahedron Lett.,* 971. b. Jagt, J. C., and van Leusen, A. M. (1973). *Recl. Trav. Chim. Pays-Bas* **92**, 1343. c. Jagt, J. C., and van Leusen, A. M. (1974). *J. Org. Chem.* **39**, 564.
5. a. Pews, R. G., Nyquist, E. B., and Corson, F. P. (1970). *J. Org. Chem.* **35**, 4096. b. Bergamasco, R., and Porter, Q. N. (1977). *Aust. J. Chem.* **30**, 1061.
6. Fukuyama, T., and Yung, Y. M. (1981). *Tetrahedron Lett.* **22**, 3759. Other uses of this Diels–Alder adduct: Allan, R. D., and Twitchin, B. (1980). *Aust. J. Chem.* **33**, 599; Daluge, S., and Vince, R. (1978). *J. Org. Chem.* **43**, 2311.
7. Gillard, M., T'kint, C., Sonveaux, E., and Ghosez, L. (1979). *J. Am. Chem. Soc.* **101**, 5837.
8. Seitz, G., and Overheu, W. (1979). *Chem. Zeit.* **103**, 230.
9. For several examples of cobalt-mediated intramolecular cyclizations of nitriles to pyri-

dines which resemble Diels–Alder reactions, see Vollhardt, K. P. C. (1984). *Angew. Chem. Int. Ed. Engl.* **23**, 539.

10. Oppolzer, W. (1971). *Angew. Chem. Int. Ed. Engl.* **11**, 1031.
11. Butsugan, Y., Yoshida, S., Muto, M., Bito, T. Matsuura, T., and Nakashima, R. (1971). *Tetrahedron Lett.*, 1129.
12. a. Tagmazyan, K. T., Mkrtchyan, R. S., and Babayan, A. T. (1974). *J. Org. Chem. USSR (Engl. Transl.)* **10**, 1657. b. Taylor, E. C., and French, L. G. (1986). *Tetrahedron Lett.* **27**, 1967.
13. Schoeller, W. W. (1985). *J. Chem. Soc., Chem. Commun.*, 334.
14. a. Appel, R., and Zimmermann, R. (1983). *Tetrahedron Lett.* **24**, 3591. b. Appel, R., Knoch, F., and Zimmermann, R. (1985). *Chem. Ber.* **118**, 814.
15. a. Ko, Y. Y. C. Y. L., and Carrie, R. (1984). *J. Chem. Soc., Chem. Commun.*, 1640. b. Markl, G., Yu Jin, G., and Silbereisen, E. (1982). *Angew. Chem. Int. Ed. Engl.* **21**, 370. c. Pellon, P., and Hamelin, J. (1986). *Tetrahedron Lett.* **27**, 5611. d. Pellon, P., Lo, Y. Y. C. Y. L., Cosquer, P., Hamelin, J., and Carrie, R. (1986). *Tetrahedron Lett.* **27**, 4299. e. Rösch, W., and Regitz, M. (1986). *Z. Naturforsch.* **41b**, 931. f. Markovskii, L. N., Romanenko, V. D., and Kachkovskaya, L. S. (1985). *J. Gen. Chem. USSR* **55**, 2488.
16. Grobe, J., and LeVan, D. (1985). *Z. Naturforsch.* **40b**, 467.
17. a. Grobe, J., and LeVan, D. (1985). *Tetrahedron Lett.* **26**, 3681. b. Grobe, J., LeVan, D., Schulze, J., and Szameitat, J. (1986). *Phos. Sulf.* **28**, 239.
18. Navech, J., Majoral, J. P., Meriem, A., and Kraemer, R. (1983). *Phos. Sulf.* **18**, 27; Meriem, A., Majoral, J. P., Revel, M., and Navech, J. (1983). *Tetrahedron Lett.* **24**, 1975.
19. a. Deschamps, E., and Mathey, F. (1985). *J. Chem. Soc., Chem. Commun.*, 1010. b. Alcarez, J. M. A., and Mathey, F. (1984). *J. Chem. Soc., Chem. Commun.*, 508. c. Huy, N. H. T., and Mathey, F. (1987). *Organometallics* **6**, 207.
20. Heinicke, J., and Tzschach, A. (1983). *Tetrahedron Lett.* **24**, 5481.
21. Appel, R., Casser, C., and Knoch, F. (1984). *Chem. Ber.* **117**, 2693.
22. a. Mathey, F., Mercier, F., and Charrier, C. (1981). *J. Am. Chem. Soc.* **103**, 4595. b. De Lauzon, G., Charrier, C., Bonnard, H., and Mathey, F. (1982). *Tetrahedron Lett.* **23**, 511. c. Charrier, C., Bonnard, H., De Lauzon, G., Holand, S., and Mathey, F. (1983). *Phos. Sulf.* **18**, 51. d. Ko, Y. Y. C. Y. L., Carrie, R., Toupet, L., and DeSarlo, F. (1986). *Bull. Soc. Chim. Fr.*, 115. e. Deschamps, E., and Mathey, F. (1984). *J. Chem. Soc., Chem. Commun.*, 1214.
23. Sennyey, G., and Mathey, F. (1981). *Tetrahedron Lett.* **22**, 4713.
24. a. Ecker, A., Boie, I., and Schmidt, U. (1971). *Angew. Chem. Int. Ed. Engl.* **10**, 191. b. Schmidt, U. (1975). *Angew. Chem. Int. Ed. Engl.* **14**, 523. c. Ecker, A., Boie, I., and Schmidt, U. (1973). *Monatsh. Chem.* **104**, 503; see also Scheibye, S., Shabana, R., Lawesson, S.-O., and Romming, C. (1982) *Tetrahedron* **38**, 993.
25. Nakayama, S., Yoshifuji, M., Okazaki, R., and Inamoto, N. (1975). *Bull. Chem. Soc. Jpn.* **48**, 546.
26. Niecke, E., and Ruger, R. (1983). *Angew. Chem. Int. Ed. Engl.* **22**, 155; Niecke, E., Ruger, R., Lysek, M., and Schoeller, W. W. (1983). *Phos. Sulf.* **18**, 35.
27. Schmidt, U., Boie, I., Osterroht, C., Schroer, R., and Grutzmacher, H. F. (1968). *Chem. Ber.* **101**, 1381.
28. Diels, O., Blom, J. H., and Koll, W. (1925). *Liebigs Ann. Chem.* **443**, 242.
29. Reviews: Fahr, E., and Lind, H. (1966). *Angew. Chem. Int. Ed. Engl.* **5**, 372; Moody, C. J. (1982). *Adv. Heterocycl. Chem.* **30**, 1.
30. a. Daniels, R., and Roseman, K. A. (1966). *Tetrahedron Lett.* 1335. b. Burrage, M. E., Cookson, R. C., Gupte, S. S., and Stevens, I. D. R. (1975). *J. Chem. Soc., Perkin*

Trans. 2, 1325. c. Konovalov, A. I., Breus, I. P., Shosagin, I. A., and Kiselev, V. D. (1979). *J. Org. Chem. USSR (Engl. Transl.)* **15**, 315.

31. Sauer, J., and Schroder, B. (1967). *Chem. Ber.* **100**, 678.
32. Hoffmann, H. M. R. (1969). *Angew. Chem. Int. Ed. Engl.* **8**, 556.
33. Forrest, A. K., Schmidt, R. R., Huttner, G., and Jibril, I. (1984). *J. Chem. Soc., Perkin Trans. 1*, 1981.
34. Tokita, S., Hiruta, K., Kitahara, K., and Nishi, H. (1982). *Synthesis*, 229.
35. Coppola, G. M., and Gimelli, S. P. (1973). *J. Heterocycl. Chem.* **10**, 323.
36. Kitahara, Y., Murata, I., and Nitta, T. (1967). *Tetrahedron Lett.*, 3003.
37. Kuo, C. H., and Wendler, N. L. (1984). *Tetrahedron Lett.* **25**, 2291.
38. Dao, L. H., and MacKay, D. (1978). *Can. J. Chem.* **56**, 1724.
39. Rastetter, W. H., and Richard, T. J. (1978). *Tetrahedron Lett.*, 2995.
40. Bandlish, B. K., Brown, J. N., Timberlake, J. W., and Trefonas, L. M. (1973). *J. Org. Chem.* **38**, 1102.
41. Minami, T., Chikugo, T., and Hanamoto, T. (1986). *J. Org. Chem.* **51**, 2210.
42. a. Rodgman, A., and Wright, G. F. (1953). *J. Org. Chem.* **18**, 465. b. Carpino, L. A., Terry, P. H., and Crowley, P. J. (1961). *J. Org. Chem.* **26**, 4336.
43. Kresze, G., Morper, M., and Bijev, A. (1977). *Tetrahedron Lett.*, 2259.
44. Engel, P. S. (1980). *Chem. Rev.* **80**, 99.
45. For an exception, see Baranger, P., Levisalles, J., and Vuidart, M. (1953). *Comptes Rendus Acad. Sci. C* **236**, 1365.
46. Ahern, M. F., and Gokel, G. W. (1979). *J. Chem. Soc., Chem. Commun.*, 1019; Ahern, M. F., Leopold, A., Beadle, J. R., and Gokel, G. W. (1982). *J. Am. Chem. Soc.* **104**, 548.
47. Gapinski, D. P., and Ahern, M. F. (1982). *Tetrahedron Lett.* **23**, 3875.
48. Cookson, R. C., Gilani, S. S. H., and Stevens, I. D. R. (1962). *Tetrahedron Lett.*, 615.
49. For extensive lead references, see Paquette, L. A., and Doehner, R. F., Jr. (1980). *J. Org. Chem.* **45**, 5105.
50. Christl, M., Lang, R., Reimann, W., and Irngartinger, H. (1984). *Chem. Ber.* **117**, 959.
51. Danishefsky, S., Yan, C.-F., Singh, R. K., Gammill, R. B., McCurry, P. M., Fritsch, N., and Clardy, J. (1979). *J. Am. Chem. Soc.* **101**, 7001.
52. Johnson, M. P., and Moody, C. J. (1985). *J. Chem. Soc., Perkin Trans. 1*, 71.
53. Kobal, V. M., Gibson, D. T., Davis, R. E., and Garza, A. (1973). *J. Am. Chem. Soc.* **95**, 4420.
54. Adam, W., Erden, I., and Cox, O. (1979). *J. Org. Chem.* **44**, 860.
55. Korat, M., Tatarsky, D., and Ginsburg, D. (1972). *Tetrahedron* **28**, 2315.
56. Paquette, L. A., Carr, R. V. C., Charumilind, P., and Blount, J. F. (1980). *J. Org. Chem.* **45**, 4922.
57. Kjell, D. P., and Sheridan, R. S. (1985). *J. Photochem.* **28**, 205.
58. Wald, K., and Wamhoff, H. (1978). *Chem. Ber.* **111**, 3519.
59. Barton, D. H. R., Lusinchi, X., and Ramirez, J. S. (1983). *Tetrahedron Lett.* **24**, 2995; Tada, M., and Oikawa, A. (1978). *J. Chem. Soc., Chem. Commun.*, 727, and references cited therein.
60. Vanmaele, L., DeClercq, P. J., and Vandewalle, M. (1985). *Tetrahedron* **41**, 141. *Idem* (1982). *Tetrahedron Lett.* **23**, 995.
61. a. Corey, E. J., and Snider, B. B. (1973). *J. Org. Chem.* **38**, 3632. b. Moje, S. W., and Beak, P. (1974). *J. Org. Chem.* **39**, 2951.
62. a. Keana, J. F. W., Guzikowski, A. P., Ward, D. D., Morat, C., and Van Nice, F. L. (1983). *J. Org. Chem.* **48**, 2654. b. Keana, J. F. W., Guzikowski, A. P., Morat, C., and Volwerk, J. J. (1983). *J. Org. Chem.* **48**, 2661. c. Keana, J. F. W., and Ward, D. D. (1983). *Synth. Commun.* **13**, 729.

63. a. Gardlik, J. M., and Paquette, L. A. (1979). *Tetrahedron Lett.*, 3597. b. Paquette, L. A., and Trova, M. P. (1986). *Tetrahedron Lett.* **27,** 1895.
64. Nelsen, S. F., Hollinsed, W. C., Grezzo, L. A., and Parmelee, W. P. (1979). *J. Am. Chem. Soc.* **101,** 7347.
65. Stetter, H., and Woernle, P. (1969). *Liebigs Ann. Chem.* **724,** 150.
66. Hassall, C. H., and Ramachandran, K. L. (1977). *Heterocycles* **7,** 119. Bevan, K., Davies, J. S., Hassall, C. H., Morton, R. B., and Phillips, D. A. S. (1971). *J. Chem. Soc. (C),* 514.
67. Sheradsky, T., and Moshenberg, R. (1985). *J. Org. Chem.* **50,** 5604.
68. Lopez, B., Lora-Tomayo, M., Navarro, P., and Soto, J. L. (1974). *Heterocycles* **2,** 649.
69. Sheppard, W. A., and Webster, O. W. (1973). *J. Am. Chem. Soc.* **95,** 2695.
70. Carlson, B. A., Sheppard, W. A., and Webster, O. W. (1975). *J. Am. Chem. Soc.* **97,** 5291.
71. Meyer, K. H. (1919). *Chem. Ber.* **52,** 1468.
72. Bronberger, F., and Huisgen, R. (1984). *Tetrahedron Lett.* **25,** 57.
73. Nelsen, S. F., Blackstock, S. C., and Frigo, T. B. (1984). *J. Am. Chem. Soc.* **106,** 3366; Nelsen, S. F., Blackstock, S. C., and Frigo, T. B. (1986). *Tetrahedron* **42,** 1769.
74. Steliou, K., Gareau, Y., and Harpp, D. N. (1984). *J. Am. Chem. Soc.* **106,** 799.
75. Paetzold, P., Richter, A., Thijssen, T., and Wurtenberg, S. (1979). *Chem. Ber.* **112,** 3811.

Chapter 7

Oxabutadienes

INTRODUCTION

Since the thermal dimerizations of acrolein and methyl vinyl ketone were shown to provide the 3,4-dihydro-2H-pyrans **1**,[1,2] an extensive range of related observations have been disclosed. This work has been the subject of several reviews.[3-14] Only the work reported since the extensive Desimoni and Tacconi account[5] of the Diels-Alder reaction of α,β-unsaturated carbonyl compounds, 1-oxabutadienes bearing an oxygen atom at the diene terminus, has been detailed herein. The prior reviews should be consulted for an excellent discussion of the mechanism, scope, and application of the [4 + 2] cycloaddition reactions of α,β-unsaturated carbonyl compounds as well as for extensive tabular compilations of the work through 1974.[3-6]

The [4 + 2] cycloaddition reactions of 1-oxabutadienes generally exhibit excellent, predictable regioselectivity, and 2-substituted 3,4-dihydro-2H-pyrans are formed nearly exclusively at the expense of the 3-substituted isomers [Eq. (1)].[1]

$$R \overset{}{\underset{O}{\bigwedge}} \quad + \quad \underset{O}{\overset{}{\bigsqcup}}_R \quad \longrightarrow \quad \underset{1}{\overset{4}{\underset{R}{\bigcirc}}}\overset{3}{\underset{2}{}}_R \tag{1}$$

Predictions based on rigorous or simplified theoretical calculations support the formation of the predominant 2-substituted 3,4-dihydro-2H-pyran regioisomer and accommodate a preferred endo approach of the reactants in which the carbon–carbon bond formation is more advanced than carbon–oxygen bond formation, i.e., a concerted but nonsynchronous [4 + 2] cycloaddition.[5,15–20] Notable exceptions to the predicted regioselectivity of the Diels–Alder reactions of oxabutadienes have been observed, and without exception the examples have involved the poorly matched reaction of electron-deficient α,β-unsaturated carbonyl compounds (4π component) with electron-deficient dienophiles (2π component), e.g., methyl crotonate or methacrolein.[5,21,22]

Although the first recognized Diels–Alder reactions of oxabutadienes detailed their 4π participation in [4 + 2] cycloadditions with electron-deficient olefins, e.g., the thermal dimerization of α,β-unsaturated carbonyl compounds,[1,2,5,23–25] 1-oxabutadiene systems are electron-deficient and consequently participate preferentially in inverse electron demand (LUMO$_{diene}$ controlled) Diels–Alder reactions with electron-rich, strained, or simple olefinic and acetylenic dienophiles.[5,6,20] The additional, complementary substitution of the 1-oxabutadiene system with an electron-withdrawing group lowers the LUMO$_{oxabutadiene}$ energy level, accelerates the rate of 1-oxabutadiene 4π participation in inverse electron demand (LUMO$_{diene}$ controlled) Diels–Alder reactions, and generally enhances the observed regioselectivity of the cycloaddition reaction. In selected instances, the C-4 addition of strong, electron-donating substituents is sufficient to promote the nucleophilic participation of the oxabutadiene system in [4 + 2] cycloadditions with reactive, electrophilic olefins, e.g., ketenes, and represent examples of cycloaddition reactions proceeding through polar intermediates.[26]

The entropic assistance provided in the intramolecular Diels–Alder reaction is sufficient to promote reluctant oxabutadiene participation in [4 + 2] cycloadditions with electron-rich, unactivated, and electron-deficient dienophiles.[27] In addition, Lewis acid catalysis and pressure-promoted reaction conditions[28-30] have been successfully utilized to accelerate the rate of 4π participation of thermally sensitive oxabutadiene systems in thermally slow or sluggish Diels–Alder reactions. These latter techniques have proved exceptionally useful for promoting the typically poor reactions of simple α,β-unsaturated carbonyl compounds.

1. α,β-UNSATURATED CARBONYL COMPOUNDS (1-OXABUTADIENES)

Simple 1-Oxabutadiene Systems

Simple α,β-unsaturated aldehydes, ketones, and esters participate preferentially in inverse electron demand (LUMO$_{diene}$ controlled) Diels–Alder reactions with electron-rich, strained, or simple olefinic and acetylenic dienophiles.[3,5] The thermal reaction conditions for promoting the [4 + 2] cycloadditions of simple 1-oxabutadienes (R = H > alkyl, aryl > OR), cf. Eq. (1), are relatively harsh (150–250°C), and the reactions are characterized by competitive α,β-unsaturated carbonyl compound dimerization or polymerization. Usual experimental techniques employed to compensate for poor conversions include the addition of radical inhibitors to the reaction mixture and the use of excess 1-oxabutadiene for promoting the [4 + 2] cycloaddition. Recent efforts have demonstrated that Lewis acid catalysis and pressure-promoted reaction conditions[28-30] may be used successfully to conduct the [4 + 2] cycloaddition under mild thermal conditions (25–100°C).

Early extensive accounts of the 4π participation of α,β-unsaturated carbonyl compounds in [4 + 2] cycloadditions detailed their reactions with electron-deficient dienophiles including α,β-unsaturated nitriles, aldehydes, and ketones; simple unactivated olefins including allylic alcohols; and electron-rich dienophiles including enol ethers, enamines, vinyl carbamates, and vinyl ureas.[23-25,31-33] Subsequent efforts have recognized the preferential participation of simple α,β-unsaturated carbonyl compounds (α,β-unsaturated aldehydes > ketones > esters) in inverse electron demand [4 + 2] cycloadditions and have further explored their [4 + 2]-cycloaddition reactions with enol ethers,[34-48] acetylenic ethers,[48,49] ketene acetals,[36,50] enamines,[41,51-60,66] ynamines,[61-63] ketene aminals,[66] and selected simple olefins[64,65] (Scheme 7-I). Additional examples may be found in Table 7-I.

The most extensively studied Diels–Alder reaction of α,β-unsaturated carbonyl compounds is their [4 + 2] cycloaddition with enol ethers. Desimoni, Tacconi, and co-workers have conducted careful, detailed investi-

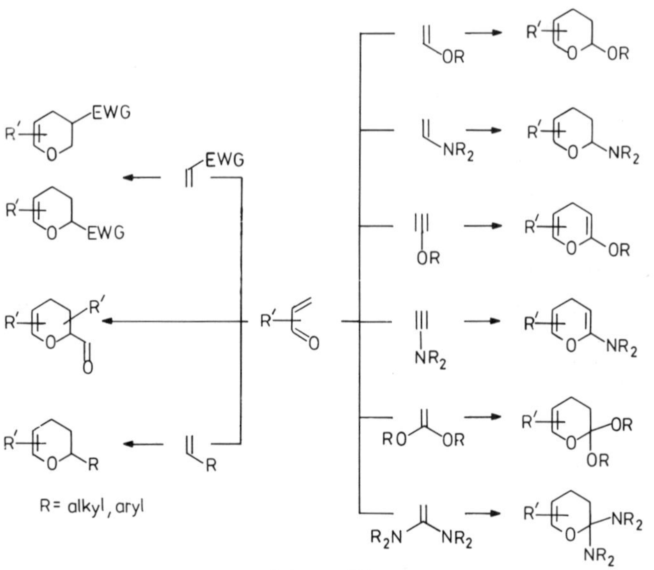

Scheme 7-I

gations of the [4 + 2] cycloadditions of 1-aryl-4-arylidene-5-pyrazolones (**2**) with electron-rich olefins including enol ethers as a representative example of the 4π participation of α,β-unsaturated carbonyl compounds in Diels–Alder reactions. Pyrazolone C-3 substituents (R′) have no effect on the rate of reaction; increasing the electron-withdrawing character of substituent X′ increases the rate of reaction, and a good correlation between $\log k$ and $\sigma(X')$ was obtained; increasing the electron-withdrawing character of the substituent X increases the rate of reaction, and a good correlation of $\log k$ and $\sigma^+(X)$ was obtained; and increasing the electron-donating character of R increases the rate of [4 + 2] cycloaddition, and a good correlation of $\log k$ with $\sigma^*(R)$ was observed [Eq. (2)].[48a–e]

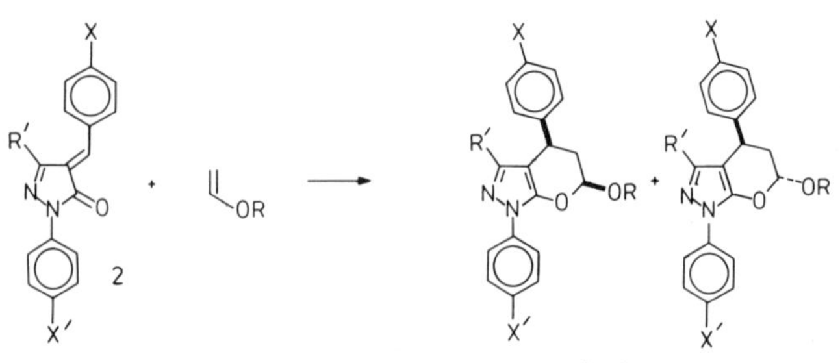

$$\underline{ca}. \ 4:1$$

(2)

TABLE 7-I.

Intermolecular Diels–Alder Reactions of Simple α,β-Unsaturated Carbonyl Compounds: Representative Applications to the Total Synthesis of Natural Products

Diels–Alder reaction	Conditions	Natural product	Ref.
(R = CH₃; R = H)	—	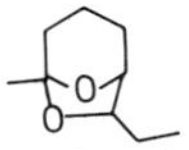 Brevicomin	76 77
	160°C, 1 hr 73% yield		78, 79
	—		80
(R = CH₃; R = H)	—	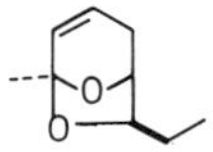	81, 78 77
	C₆H₆, 200°C, 2 hr	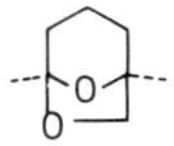 Frontalin	76
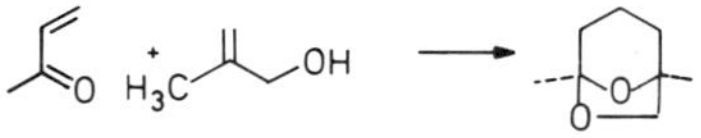	6–7% yield 35–40% yield		82 83

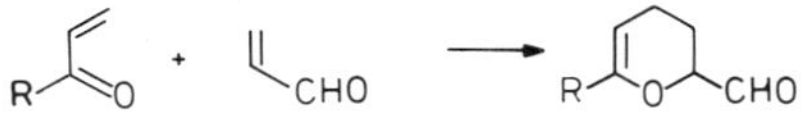

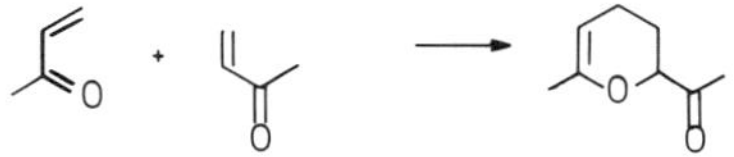

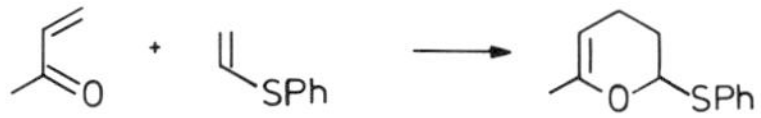

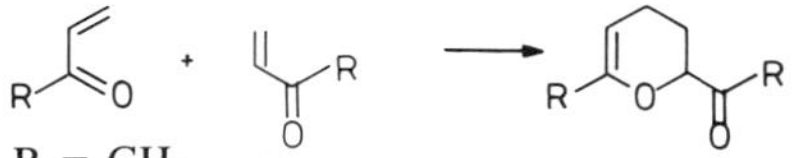

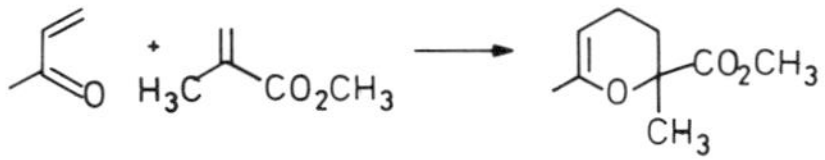

(continued)

TABLE 7-I (Continued)

Diels–Alder reaction	Conditions	Natural product	Ref.
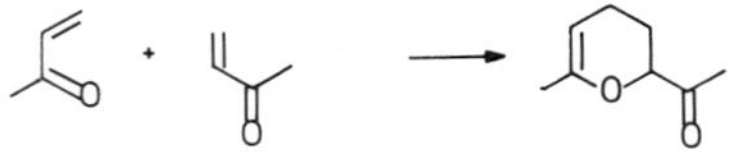	185°C, 2 hr	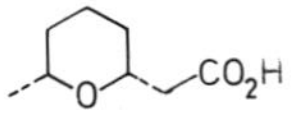 CO_2H	84
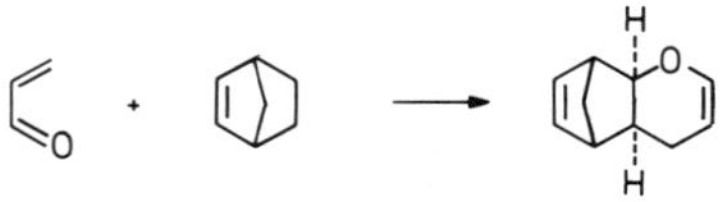	190°C, 44% yield	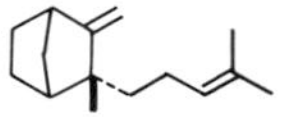 β-Santalene	85
		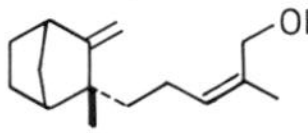 OH Epi-β-santalol	85
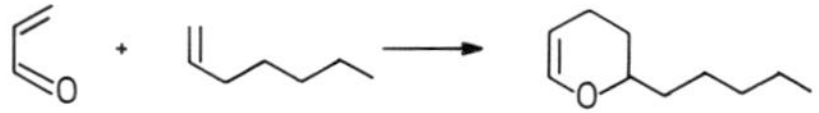	190°C, 25 hr, 50% yield	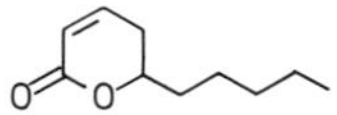 Massoia lactone	86
	—	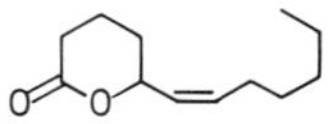	86
		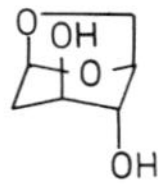	87
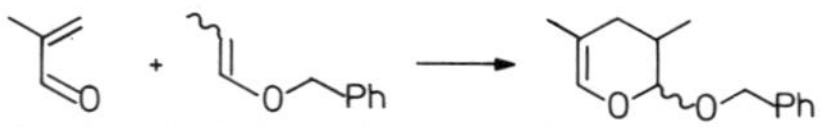	MoO₂(acac)₂ cat., 100°C, 5 hr, 40–67% yield	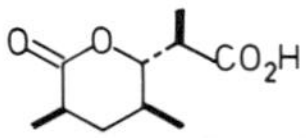 CO_2H	70

172

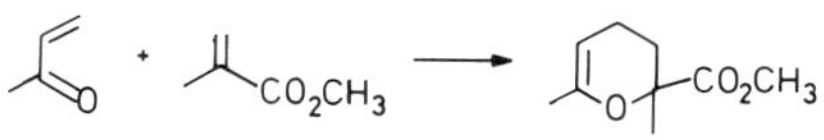

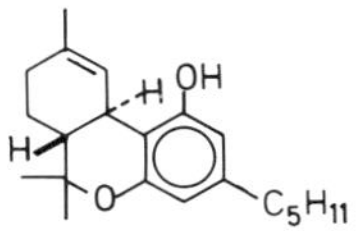

88

89

89

90

91

trans-Δ^1-Tetrahydrocannabinol

Iridoids

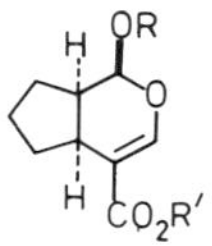

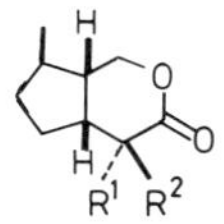

R^1 = H, R^2 = CH$_3$ Isoiridomymecin
R^1 = CH$_3$, R^2 = H Iridomymecin

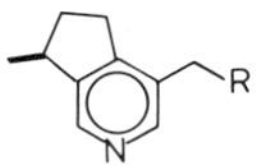

R = OCH$_3$ (−)-Valerianine
R = OH (−)-Tecostidine
R = H (−)-Actinidine

—

185°C, 95% yield

200°C, 64 hr, <5% yield

130°C, 48 hr, 42% yield

200°C, 47% yield

(continued)

173

TABLE 7-I (Continued)

Diels–Alder reaction	Conditions	Natural product	Ref.
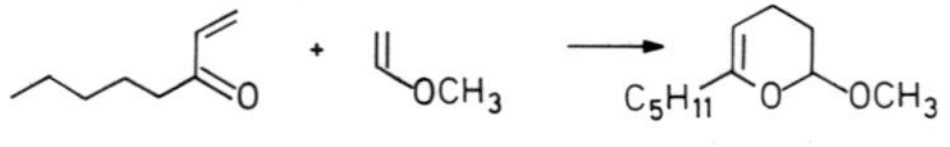	180°C, 40 hr, 30–60% yield	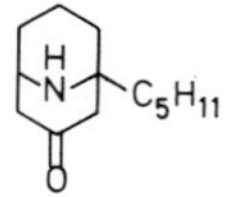 Adaline	92
	25°C, 4–6 days, 82% yield	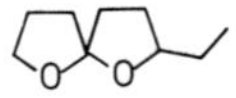 Chalcogran	93
	—	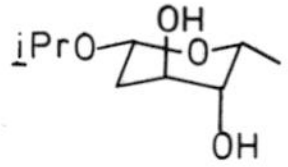 β-L-Boivinoside	47

Various experimental techniques for estimating the relative energy of the $LUMO_{oxabutadiene}$, including one-electron half-wave reduction potentials ($E_{1/2}^{red}$), were employed, and good correlations of $LUMO_{pyrazolone}$ with kinetic cycloaddition results were obtained. The results confirmed the 4π participation of the pyrazolones in inverse electron demand ($LUMO_{diene}$ controlled) Diels–Alder reactions under FMO control although Sustmann's approximations[67] of the general perturbation theory could not be used to adequately explain the experimental results. The Diels–Alder reactions of the 4-arylidene-5-pyrazolones, characteristic of the [4 + 2] cycloadditions of enol ethers with α,β-unsaturated carbonyl compounds, proceed with maintenance of dienophile olefin geometry preferentially, though not exclusively, through an endo transition state.[5,48]

The Diels–Alder reaction of simple α,β-unsaturated carbonyl compounds generally proceed only under vigorous thermal reaction conditions (150–250°C),[3,5,23-25,50] even when electron-rich dienophiles are employed. Under vigorous thermal conditions, dimerization and polymerization of the α,β-unsaturated carbonyl compounds or thermally sensitive electron-rich dienophiles will compete effectively with the desired [4 + 2] cycloaddition.[93] Two experimental methods have been employed to facilitate the 4π participation of simple α,β-unsaturated carbonyl compounds in Diels–Alder reactions. Lewis acid catalysis has proved effective in accelerating the rate of [4 + 2] cycloadditions of simple α,β-unsaturated carbonyl compounds with electron-rich olefins [Eqs. (3)–(5)],[68-71] and it is surprising that this technique has not been

R^1	R^2	R^3	R^4	R^5	Conditions($ZnCl_2$ cat)	
CH_3	H	H	H	H	60°C, 0.5h	75-85%[68a]
CH_3	H	H	H	CH_3	80°C, 1h	70-80%
CH_3	H	H	CH_3	H	80°C, 1h	70-75%
CH_3	H	CH_3	H	H	80°C, 0.5h	85-90%
CH_3	H	CH_3	H	OCH_3	80°C, 5h	40-50%
CH_3	H	Ph	H	H	80°C 1h	70-80%
CH_3	CH_3	H	H	H	60°C, 2h	75-85%
OCH_3	H	H	H	H	60°C, 2h	70-75%
OCH_3	OCH_3	H	H	H	80°C, 2h	70-75%

(3)

R	Conditions (cat.)	
H	140°C, 12h (---)	44%[68b]
H	90°C, 0.5h ($ZnCl_2$)	35%
CH_3	60°C, 1.5h ($ZnCl_2$)	61%

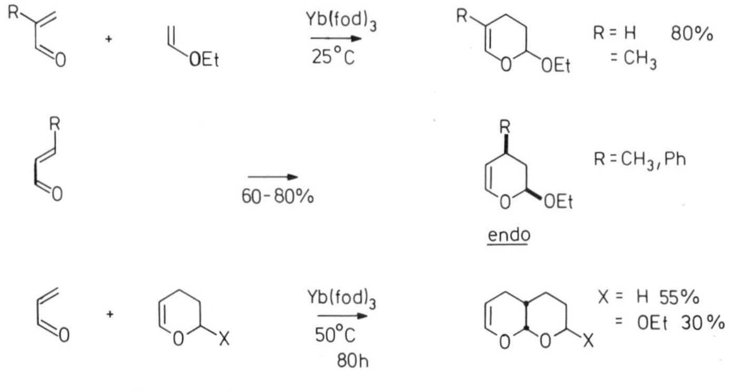

Yb(fod)3
25°C
R = H 80%
= CH3
60-80%
R = CH3, Ph
endo
Yb(fod)3
50°C
80h
X = H 55%
= OEt 30%

(4)

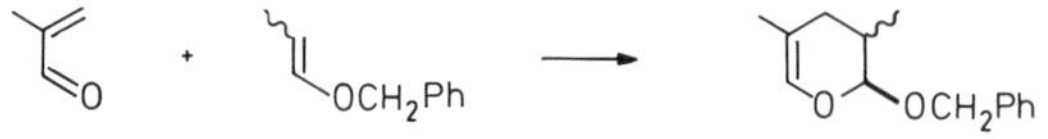

Catalyst	Temp.	Time	
none	250°C	10h	13%
$AlCl_3$ (1 equiv)	25°C	3h	63%
$Eu(fod)_3$ (0.05 equiv)	30°C	720h	0%
$VO(C_2O_4)\cdot H_2O$ (0.02 equiv)	100°C	5h	16%
$MoO_2(acac)_2$ (0.02 equiv)	100°C	5h	58%

(5)

more extensively employed. The zinc chloride-catalyzed reaction of ketene acetals with α,β-unsaturated ketones and aldehydes at low temperatures was shown to provide kinetic [2 + 2] oxetane cycloadducts, which were subsequently converted to the observed [4 + 2] cycloadducts at higher temperatures [Eq. (3)].[68a] A common problem accompanying the use of Lewis acid catalysts to promote such reactions has been the expected and documented instability of the 3,4-dihydro-2*H*-pyran [4 + 2] cycloadducts to the reaction conditions.[35,36,71]

An additional and exceptionally useful technique for promoting the Diels–Alder reaction of simple α,β-unsaturated carbonyl compounds with electron-rich or simple olefinic dienophiles employs the use of liquid high pressure techniques [Eqs. (6)–(8)].[23,28–30,69,72–73]

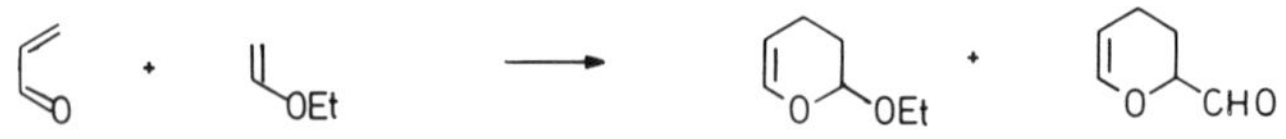

ether,	15kbar,	25°C,	20h,	69%	- [30]
heptane,	2kbar,	120°C,	24h,	3%	84%[73]
heptane,	1 atm,	120°C,	24h,	0.6%	17%[73]
none,	1 atm,	140°C,	12h,	84%	- [23]
none, cat. $Yb(fod)_3$	1 atm,	25°C,	24h,	80%	- [69]

(6)

(7)

acetone,	14kbar,	80°C,	5h,	35%	60%[72]
none,	1 atm,	230°C,	5h,	8%	nd [72]

(8)

R = H	ether,	15kbar,	24h,	75°C,	89%	(100% endo)[30]
R = CH$_3$	ether,	15kbar,	20h,	110°C,	68%	(10:1 endo:exo) [30]

Examples of the 4π participation of simple α,β-unsaturated carbonyl compounds in Diels–Alder reactions and their application to the total syntheses of natural products are summarized in Table 7-I. Reports of simple α,β-unsaturated carbonyl compounds participating in [4 + 2] cycloadditions with heterodienophiles have been detailed [Eq. (9)].[73b–75]

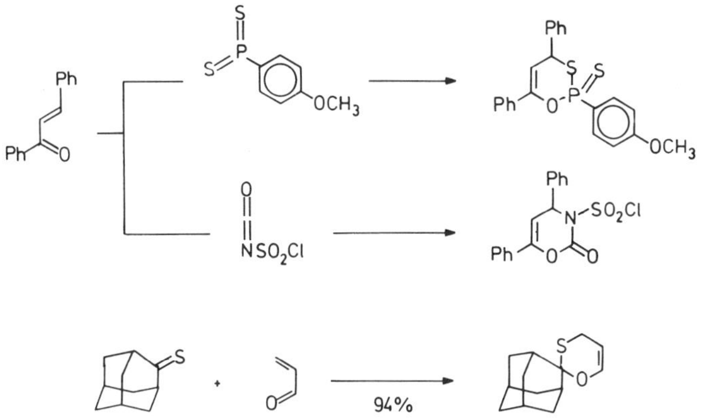

(9)

1-Oxabutadiene Systems Bearing C-2, C-3, or C-4 Electron-Withdrawing Substituents

Perhaps the most successful approach for promoting the 4π participation of 1-oxabutadiene systems in intermolecular Diels–Alder reactions employs α,β-unsaturated carbonyl compounds substituted with an additional C-3 (α) electron-withdrawing group. The addition of the C-3 electron-withdrawing substituent increases the electron-deficient character of the oxabutadiene system, decreases the LUMO$_{oxabutadiene}$, and as expected, enhances the observed [4 + 2] cycloaddition rate and regioselectivity with polarized, electron-rich dienophiles in inverse electron demand Diels–Alder reactions. These observations, first pursued by Tietze and co-workers[94–97,99,134–137,145] and subsequently investigated by a number of groups,[98,100–112] have been found to possess considerable potential in natural product total syntheses.[145]

Both the endo stereoselectivity and the potential enantioselectivity [Eq. (10)][94]

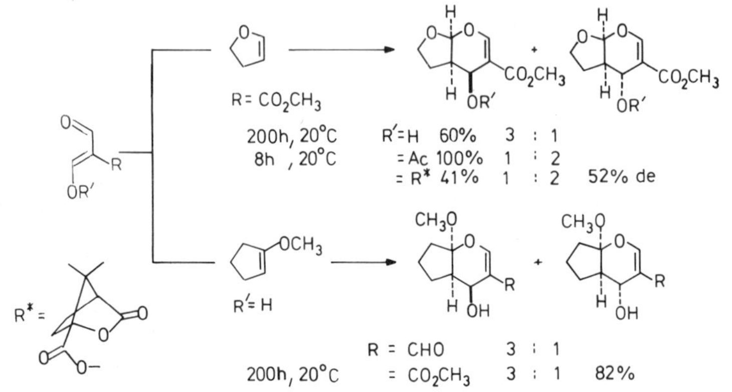

(10)

of the room temperature intermolecular regiospecific, inverse electron demand Diels–Alder reactions of triacyl compounds including 2-formylmalondialdehyde [Eq. (11)][95]

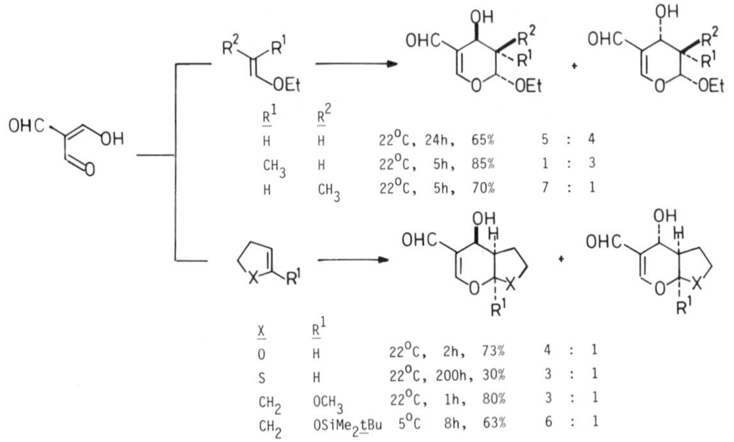

(11)

with electron-rich olefins have been investigated by Tietze and co-workers and have been shown to proceed with good to modest selectivity. Olefin geometry of the electron-rich dienophile is maintained in the course of the [4 + 2] cycloaddition and this observation is consistent with a concerted cycloaddition proceeding predominantly through an endo transition state.

A simple, *one-flask* stereoselective approach to the preparation of secoiridoids based on the [4 + 2] cycloaddition of alkylidene-1,3-dicarbonyl compounds has been described by Tietze and co-workers [Eq. (12)].[96]

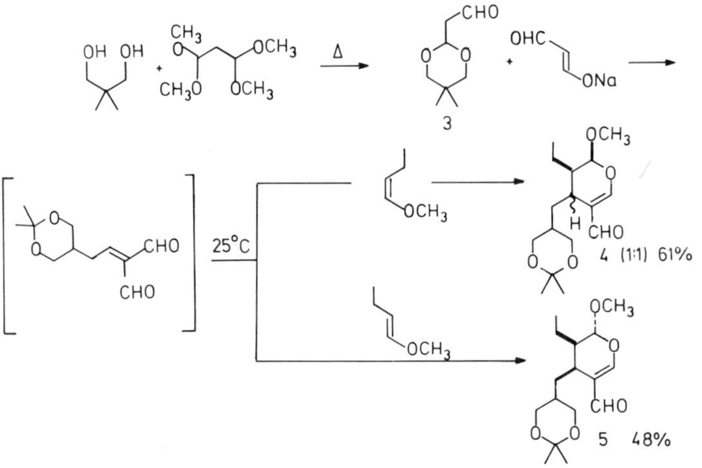

$$(12)$$

Treatment of the sodium salt of malondialdehyde with the monoacetal **3** (5°C, 48 hr), followed by the addition of (Z)- or (E)-1-methoxy-1-butene provides **4** and **5**, respectively. The secologanin derivative **5** possesses the relative configuration typical of the secoiridoids.[97]

Schreiber and co-workers have detailed the utility of the Tietze reagent, 2-formylmalondialdehyde, in stereoselective [4 + 2] cycloadditions with the Paterno–Buchi photoadducts derived from furan and aldehydes for use in the stereoselective assemblage of highly oxygenated aliphatic substrates [Eq. (13)].[98]

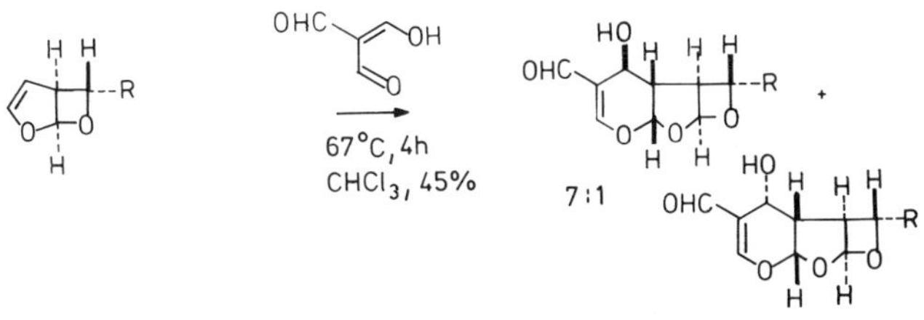

$$(13)$$

Additional, related efforts have illustrated and defined the [4 + 2] cycloadditions of N-acyl-2-methoxycarbonyl enaminecarboxaldehydes [Eq. (14)]

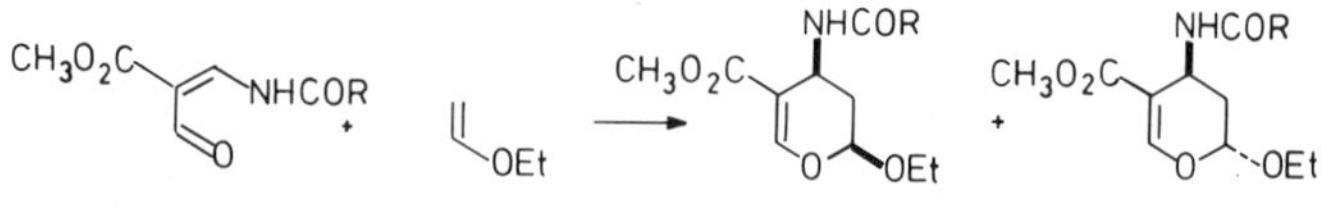

$$\text{(14)}$$

R					
CH$_3$	90°C,	12h.	93%,	1 :	1.9
$\underline{t}$Bu	120°C,	20h,	72%	1 :	1.2
CCl$_3$	22°C,	20h,	37%	1 :	1.3
Ph	70°C,	25h,	86%	1 :	1.7
$\underline{p}$NO$_2$Ph	22°C,	72h,	63%	1 :	2.0
CO$_2$CH$_3$	90°C,	12h,	59%	1 :	1.1
NHPh	95°C,	96h	71%	1 :	1.1

in the development of an approach to the branched amino sugars.[99]

The reported thermal [4 + 2] cycloaddition of arylmethylenemalondialdehydes with electron-rich olefins including enol ethers, ketene acetals, enamines, and cyclopentadiene (benzene, 25°C) or their Lewis acid-catalyzed reaction with simple olefins including isobutylene and 1,1-diphenylethylene (ZnI$_2$ catalyst, benzene, 25°C) further illustrates the Diels–Alder rate acceleration accompanying the substitution of an α,β-unsaturated aldehyde with a C-3 electron-withdrawing group [Eq. (15)].

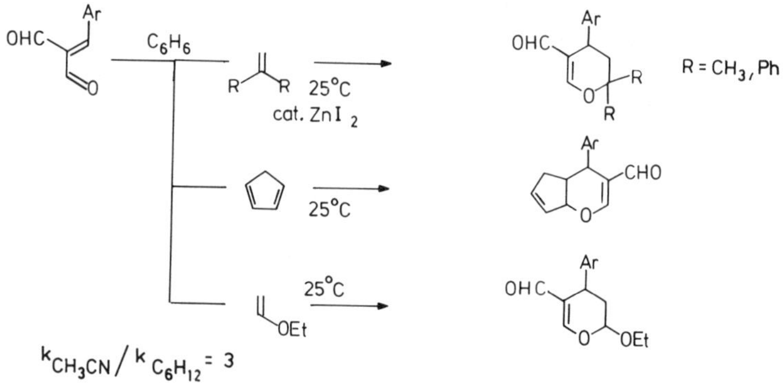

$$\text{(15)}$$

The small solvent rate dependence observed for the [4 + 2] cycloaddition ($k_{\text{acetonitrile}}/k_{\text{cyclohexane}} = 3$) is consistent with a concerted cycloaddition reaction.[100]

3-Acylchromones behave as 4π components of stereoselective [4 + 2] cycloadditions with enol ethers which proceed predominantly through an endo transition state.[101] The reactivity of the diene component was shown to be strongly dependent on the acyl substitution (R = H > CH$_3$ > Ph)[101a] [Eq. (16)].

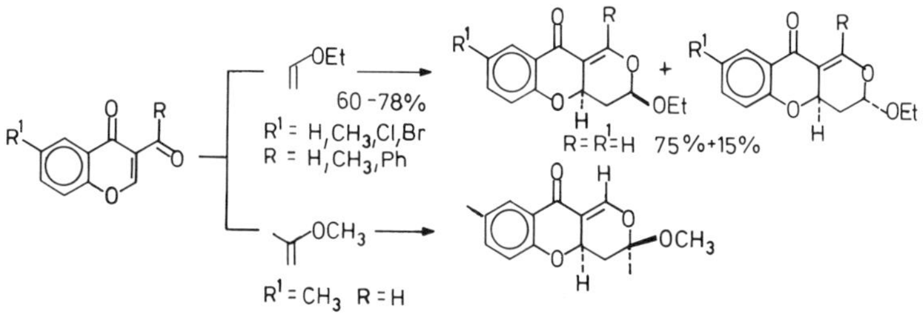

$$(16)$$

Benzylidenemalonaldehyde has been shown to exist in equilibrium with its dimer at room temperature [Eq. (17)].[102]

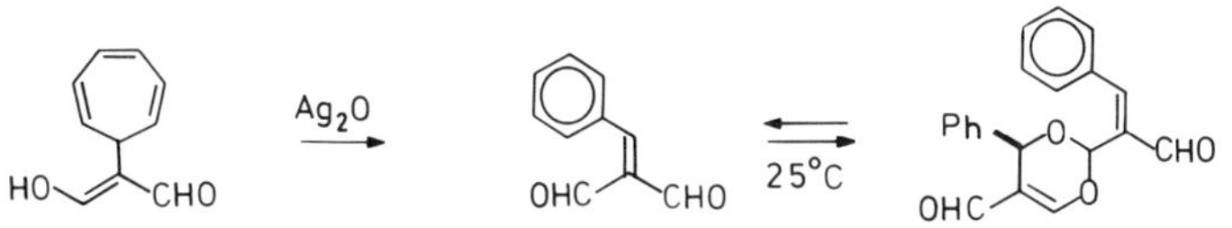

$$(17)$$

The reaction of substituted benzlidenemalondialdehydes with dimethylcyanamide provides the corresponding 4-aryl-2-dimethylamino-4H-1,3-oxazine-5-carboxaldehydes [Eq. (18)].[103]

$$(18)$$

In addition to studies detailing the effective use of electron-deficient α,β-unsaturated aldehydes, a number of related investigations have described the inverse electron demand Diels–Alder reactions of additional α,β-unsaturated carbonyl compounds bearing a C-3 (α) electron-withdrawing substituent and include 2-acetylcyclohex-2-enone [Eq. (19)],[104]

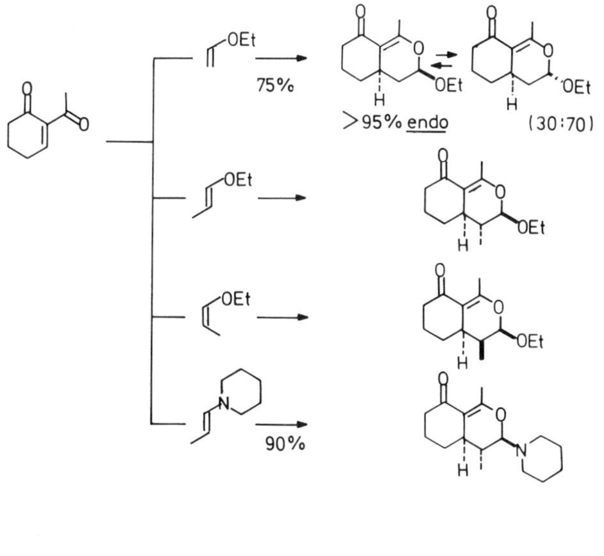

$$(19)$$

alkylidene derivatives of Meldrum's acid and 1,3-indandione [Eq. (20)],[105]

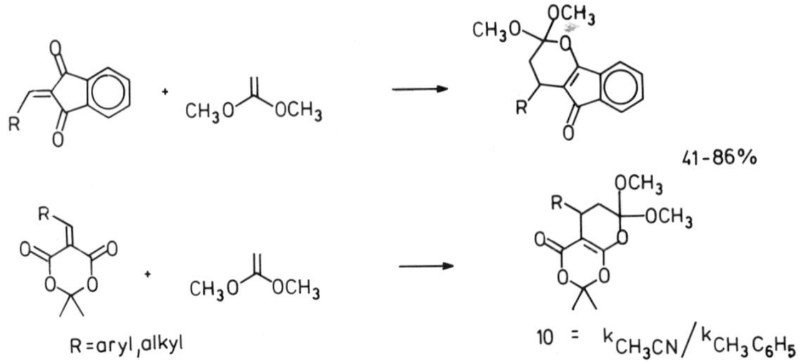

$$41-86\%$$

$$(20)$$

$$10 = k_{CH_3CN}/k_{CH_3C_6H_5}$$

R = aryl, alkyl

methyl methoxymethyleneacetoacetate and methoxymethyleneacetoacetone [Eq. (21)],[106]

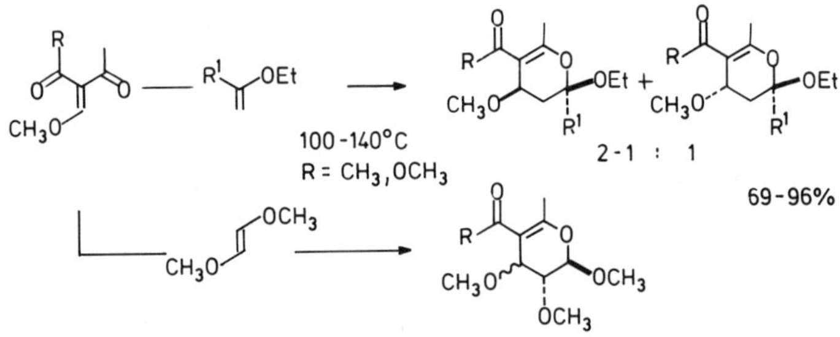

$$100-140°C$$
$$R = CH_3, OCH_3$$

$$2-1 : 1$$

$$69-96\%$$

$$(21)$$

methylene-1,3-dicarbonyl compounds [Eq. (22)],[107]

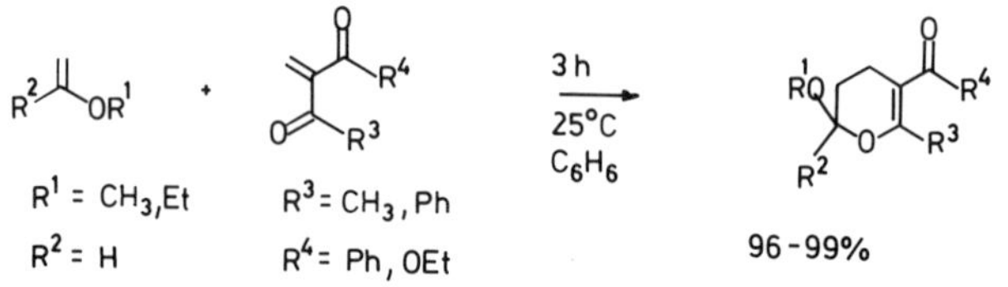

$$R^1 = CH_3, Et \qquad R^3 = CH_3, Ph$$
$$R^2 = H \qquad R^4 = Ph, OEt$$

$$3 h$$
$$25°C$$
$$C_6H_6$$

$$96-99\%$$

$$(22)$$

as well as a range of α-cyano- and α-methoxycarbonyl α,β-unsaturated esters [Eqs. (23)–(26)].[109–112]

$$(23)$$

R^1	R^2	R^3	R^4		
CN	H	H	iPr	25°C, C_6H_6	62%
CN	CN	H	Ph	25°C, $CDCl_3$	56%
CN	CO_2CH_3	H	iBu	25°C, $CDCl_3$	75%
CO_2CH_3	CO_2CH_3	H	iBu	25°C, $CDCl_3$	70%
CN	CO_2CH_3	CN	iBu	70°C, CH_3CN	34%

$$R = CO_2CH_3 \quad 0 - 30\%$$
$$= CN \quad 0 - 50\%$$

(24)

(25)

other products

(26)

The endo stereoselectivity (> 95%), the preservation of dienophile olefin geometry, and the enhanced reactivity of (E)-1-ethoxypropene versus (Z)-1-ethoxypropene [$k(E)/k(Z) = 35$] in the [4 + 2] cycloadditions of 2-acetylcyclohex-2-enone with electron-rich olefins have been interpreted as being consistent and characteristic of a concerted inverse electron demand Diels–Alder reaction [Eq. (19)].[104]

Until recently, the reaction of α,β-unsaturated esters with electron-rich olefins has been reported to afford cyclobutane [2 + 2] cycloaddition products. Amice and Conia first proposed the intermediacy of [4 + 2] cycloadducts in the reaction of ketene acetals with methyl acrylate,[108] and the first documented example of the 4π participation of an α,β-unsaturated ester in a Diels–Alder reaction appears to be the report of Snider and co-workers of the reversible, intramolecular cycloaddition of 1-allylic-2,2-dimethyl ethylenetricarboxylates.[142] Subsequent efforts have recognized that substitution of the α,β-unsaturated ester with a C-3 electron withdrawing substituent permits the 4π participation of such oxabutadiene systems in inverse electron demand Diels–Alder reactions with electron-rich olefins. In the instances studied, the rate of the [4 + 2] cycloaddition showed little dependence on solvent polarity [$k_{acetonitrile}/k_{cyclohexane} = 3$, Eq. (15); $k_{acetonitrile}/k_{toluene} = 10$, Eq. (20)], and reactions generally proceed with a maintenance of the dienophile stereochemistry consistent with a concerted [4 + 2] cycloaddition.

In the selected instances of the observation of ring-opened products, copolymerization reactions, and the loss of dienophile stereochemistry in the [4 + 2] reactions of α,β-unsaturated esters bearing an additional C-3 electron-withdrawing group as well as the lack of an observed rate dependency on the solvent polarity have led Hall and co-workers to conclude that such cycloadditions may proceed with the generation of biradical intermediates. However, such conclusions have been further cautioned by the detailed investigations of Hall and his co-workers in which they demonstrated that the primary [4 + 2] cycloadducts may not be entirely stable under the reaction conditions and may revert to biradical and dipolar intermediates at higher temperatures. This temperature dependency on the extent of products derived from biradical or dipolar intermediates and the demonstrated thermal instability of the primary [4 + 2] cycloadducts suggest that care should be taken in drawing conclusions on the concerted versus stepwise addition–cyclization [4 + 2] mechanisms based on such observations.

α,β-Unsaturated acyl cyanides, an oxabutadiene system possessing a C-2 electron-withdrawing group, have been shown to participate as 4π components in thermal and Lewis acid-catalyzed [4 + 2] cycloadditions with electron-rich and simple olefinic dienophiles [Eq. (27)].[113]

R	R^1		
OiPr	H	160°C, 10h, neat	76%
CH$_2$SiMe$_3$	H	25°C, 24h, 0.5 equiv AlCl$_3$, C$_6$H$_6$	48%
Ph	H	25°C, 18h, 0.2 equiv AlCl$_3$, C$_6$H$_6$	41%
CH=CHCH$_3$	H	25°C, 3h, 0.2 equiv AlCl$_3$, C$_6$H$_6$	52%
CH$_3$	CH$_3$	5°C, 3h, 0.2 equiv AlCl$_3$, C$_6$H$_6$	35%

(27)

In selected instances, their 4π participation in Lewis acid-catalyzed Diels–Alder reactions with dienes have been observed [Eq. (28)].[113]

(28)

In studies extending the thermal and Lewis acid-catalyzed [2 + 2] cycloadditions of electron-rich olefins with carbonyl groups and electron-deficient olefins, the reactions of ketene acetals with extended π systems including β,β-dicyano α,β-unsaturated carbonyl compounds have been investigated.[114] The reaction of 6 with ketene acetals and tetramethoxyethylene provided the regiospecific [4 + 2] cycloadducts with no apparent competing [2 + 2] cycloaddition [Eq. (29)].[114]

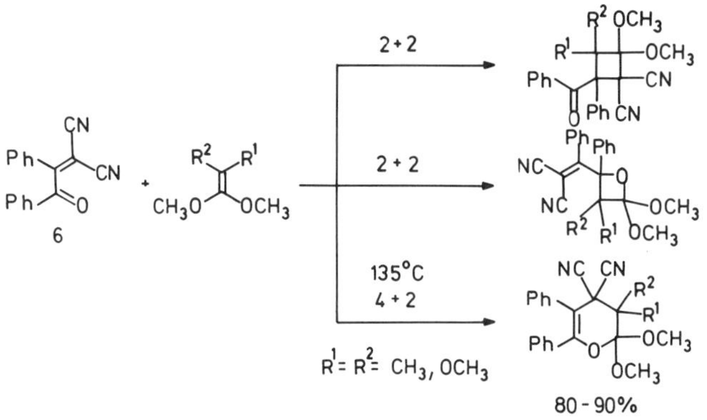

(29)

The reaction proceeds under mild conditions and zinc chloride could be used to accelerate the rate of reaction. The observed regiospecificity of the [4 + 2] cycloaddition is not obvious, and the potential of an undetected reversible, kinetic [2 + 2] cycloaddition preceding the formation of the thermodynamically stable [4 + 2] cycloaddition products was not ruled out.

1-Oxabutadiene Systems Bearing C-3 or C-4 Electron-Donating Substituents

The addition of electron-donating substituents to C-3 or C-4 of a 1-oxabutadiene system would be expected to decrease the facility with which the system participates as the 4π component of Diels–Alder reactions. As a consequence of this expected behavior only a select set of such systems have been investigated and shown to successfully participate as 4π components of Diels–Alder reactions.

3-Trimethylsilyloxybut-3-en-2-one [Eq. (30)][115]

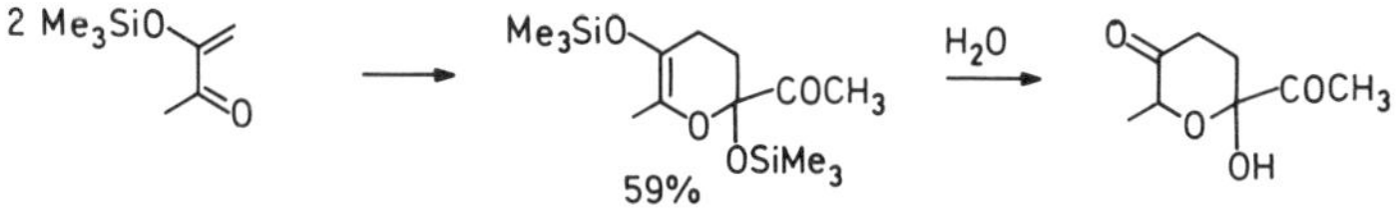

(30)

and 3-phenylthiobut-3-en-2-one [Eq. (31)][116,118]

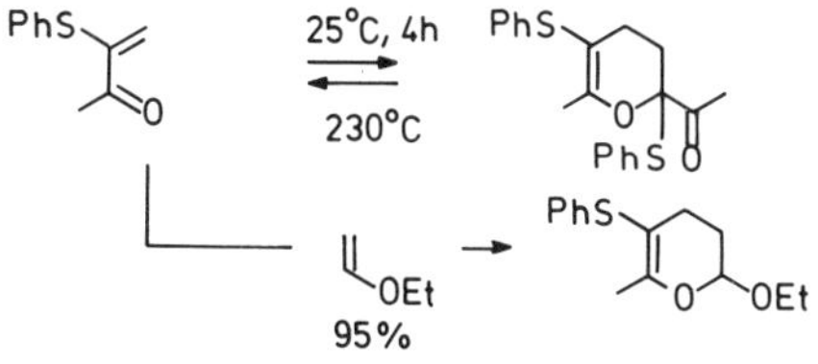

(31)

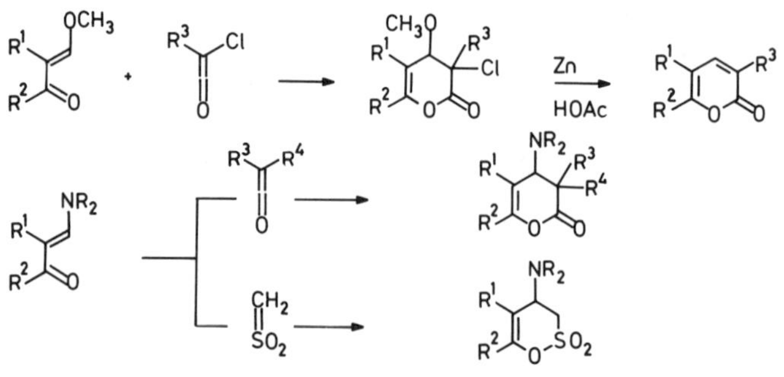

Scheme 7-II

readily dimerize with 2π and 4π Diels–Alder participation of the substituted α,β-unsaturated ketones. Subsequent studies have detailed effective intermolecular Diels–Alder reactions of 3-phenylthiobut-3-en-2-one with electron-rich olefins [Eq. (31)].[116] β-Acyloxy-α-phenylthio α,β-unsaturated ketones react with ethyl vinyl ether under surprisingly mild conditions to provide endo-specific 3,4-dihydro-2H-pyran [4 + 2] cycloadducts, which have been employed as precursors for deoxysugars with excellent diastereocontrol and modest enantiocontrol of four chiral centers [Eq. (32)].[117]

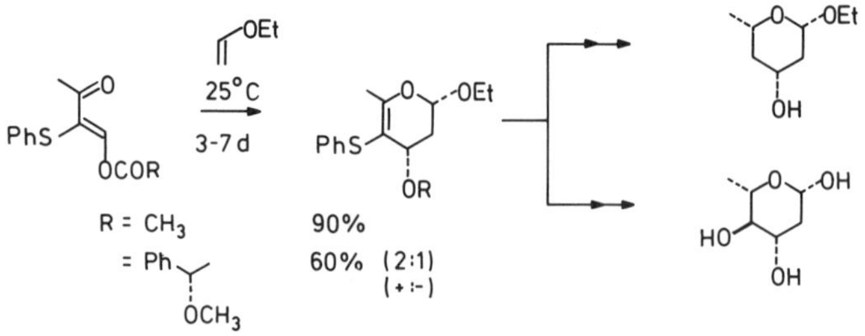

(32)

Clear demonstrations of the expected rate deceleration of the 4π participation of α,β-unsaturated carbonyl compounds possessing a C-4 electron-donating group in inverse electron demand Diels–Alder reactions with enol ethers have been detailed [Eq. (33)].[119,120]

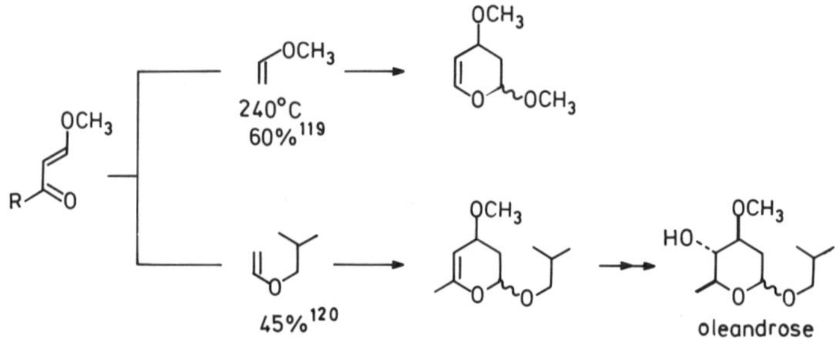

(33)

However, the C-4 substitution of the 1-oxabutadiene system with strong electron-donating substituents (—OR, —NR₂) does serve to increase the nucleophilic character of the α,β-unsaturated carbonyl compound to the extent that stepwise, addition–cyclization [4 + 2] cycloadditions, often in competition with [2 + 2] cycloadditions, with reactive electrophilic ole-fins, e.g., ketenes[121–127] or sulfenes,[127–130] have been observed and exten-sively investigated (Scheme 7-II). For early investigations, prior reviews should be consulted.[5]

2. INTRAMOLECULAR 1-OXABUTADIENE DIELS–ALDER REACTIONS

An effective approach to promoting the 4π participation of oxabuta-diene systems in Diels–Alder reactions employs the intramolecular cy-cloaddition reactions of α,β-unsaturated carbonyl compounds.[27] If this intramolecular variant of the oxabutadiene [4 + 2] cycloaddition is com-bined with the use or generation of α,β-unsaturated carbonyl compounds bearing an additional C-3 (α) electron-withdrawing group, exceptionally effective room temperature, regio-, diastereo-, and enantioselective [4 + 2] cycloadditions are observed.[134–138] Table 7-II details the intramolecular [4 + 2] cycloadditions of α,β-unsaturated carbonyl compounds.[27,131–145]

The reports of simple α,β-unsaturated aldehydes, ketones, or esters participating as 4π oxabutadiene components of Diels–Alder reactions are limited. Cyclization of aldehyde **7** in the presence of Lewis acid cata-lysts provides the bridged [4 + 2] adduct **8** exclusively and the *cis*-enal **9** is considered to be the reactive intermediate. The exclusive formation of the bridged product has been attributed to electronic effects.[131] In contrast, mild thermolysis of aldehyde **7** in the gas phase provided the fused [4 + 2] cycloadduct **10**, albeit in low yield, and subjecting **7** to more vigorous thermolysis conditions provided mixtures of **8** and **10** [Eq. (34)][131] and intramolecular ene reaction products.

(34)

TABLE 7-II

Intramolecular Diels–Alder Reactions of α, β-Unsaturated Carbonyl Compounds

Entry	Substrate	Conditions	Product(s)	Yield	Ref.
1	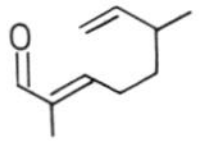	BF$_3$·OEt$_2$ cat.		49%	131
2		FVP, 350°C	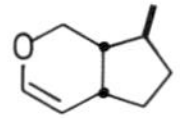	3%	131
3	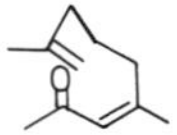	AlCl$_3$, CH$_2$Cl$_2$, 25°C, 4hr	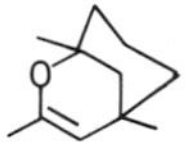	33%	132
4	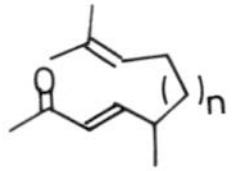	Me$_2$AlCl, CH$_2$Cl$_2$, 25°C, 8 hr	$n = 1$,	10%	133
			$n = 2$,	58%	
5	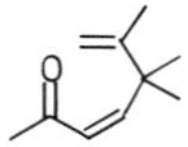	$h\nu$, CH$_3$OH, Pyrex	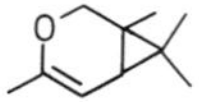	45%	139
6		C$_6$H$_{12}$, 140°C		—	140
				R = CF$_3$	
7	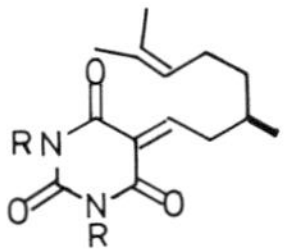	THF, C$_5$H$_5$N, H$_2$O, reflux	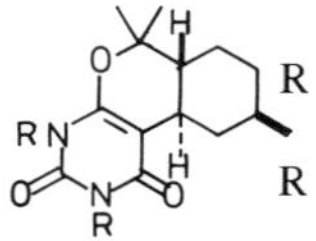 R = H,	34%	134
		H$_2$NCH$_2$CH$_2$NH$_3^+$CH$_3$CO$_2^-$; 30°C, 5 min	R = CH$_3$,	97%	134

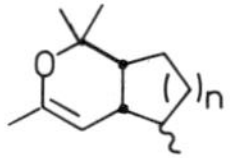

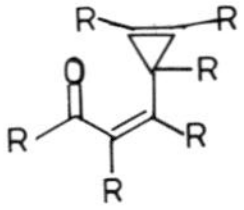

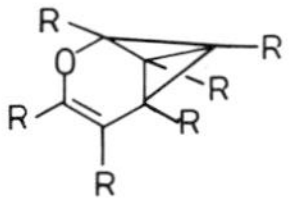

188

Entry	Conditions	Product	Yield	Ref.
8	DMF, 100°C, DMF, 100°C CH₃OH, 20°C, 1 hr	R = C₅H₁₁ R = H R = H	65% 64% 33%	134, 135 135 134, 136
9	H₂NCH₂CH₂NH₃⁺CH₃CO₂⁻, 30°C, 5 min H₂NCH₂CH₂NH₃⁺CH₃CO₂⁻, 20°C, 30 min	R = CH₃, R = H,	95% 61%	134 145
10	Base, iPrOH, 25°C, 24 hr	R = CH₃,	50%	146
10	CH₂Cl₂ or CH₃CN, 20°C, 5 hr	R = H, X = O R = CH₃, X = O R = CH₃, X = CH₂O R = CH₃, X = S R = CH₃, X = CH₂ R = CH₃, X = MsN	76% 91% 33% 67% 69% 53%	137
11	CH₂Cl₂ or CH₃CN, 20°C, 5 hr	X = O X = CH₂	73% 65%	137

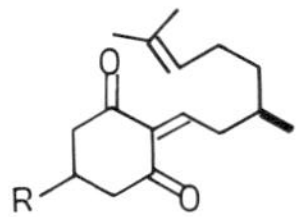

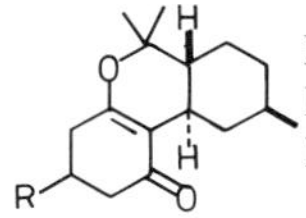

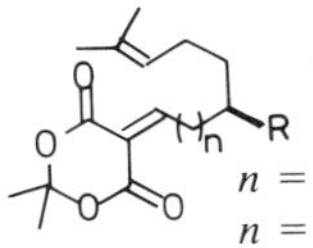

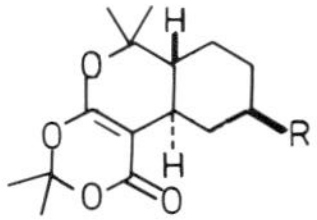

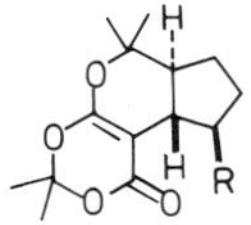

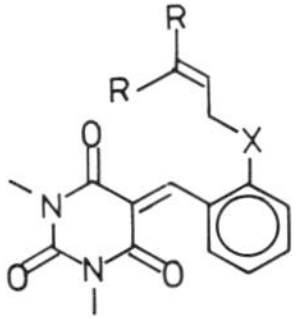

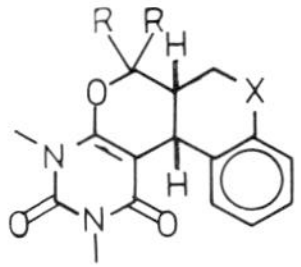

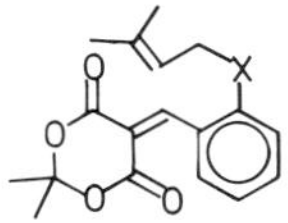

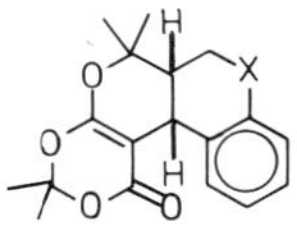

Entry 9: $n = 1$, $n = 1$

Entry 10 (substrate): $n = 0$

(continued)

189

TABLE 7-II (Continued)

Entry	Substrate	Conditions	Product(s)	Yield	Ref.
12		CH₂Cl₂, 20°C, 5 hr		81%	137
13	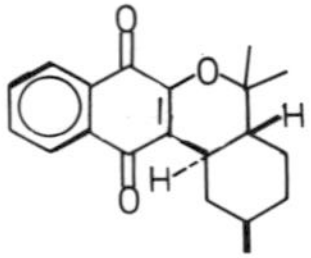	90°C, 3 hr	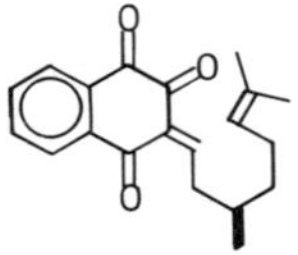	70%	138
14	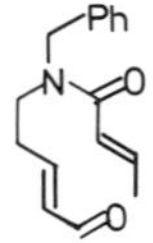	Xylene, 190°C, 14 hr	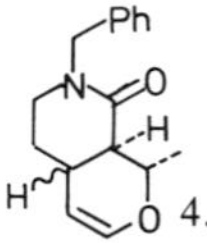 4.5 : 1 cis : trans	73%	141
15	 R¹ = CH₃, R² = H R¹ = H, R² = CH₃ R¹ = CH₃, R² = CH₃ R¹ = Ph, R² = H	C₆H₆, 135°C, 200 hr C₆H₆, 137°C, 200 hr C₆H₆, 85°C, 112 hr C₆H₆, 85°C, 112 hr	 1 : 1 Diels–Alder:ene products 1 : 9 Diels–Alder:ene products 3 : 7 Diels–Alder:ene products —	 22% Trace Trace 33%	142
16	 R = CN R = COCH₃ R = COCH₃	ZnBr₂ cat., oCl₂C₆H₄, 25°C ZnBr₂ cat., oCl₂C₆H₄, 25°C oCl₂C₆H₄, 180°C		 68% 88% 81%	143

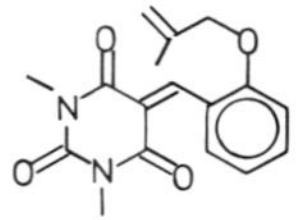

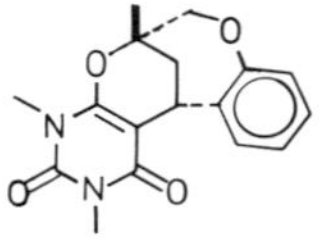

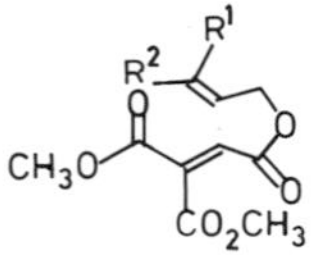

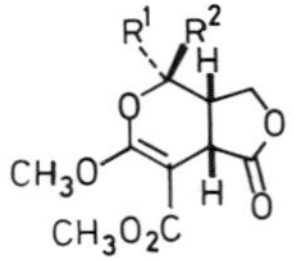

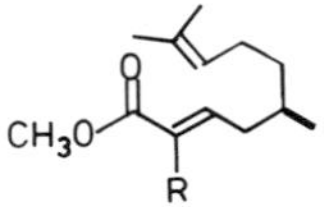

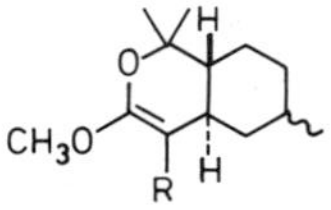

A related Lewis acid-catalyzed intramolecular formation of a bridged [4 + 2] cycloadduct derived from an *in situ* generated alkene α,β-unsaturated ketone had been previously observed (Table 7-II, entry 3).[132] The reversible, Lewis acid-catalyzed formation of intramolecular Diels–Alder [4 + 2] cycloadducts have been observed as competing intermediates generated in the Lewis acid-catalyzed irreversible ene reactions of α,β-unsaturated ketones[133] (Table 7-II, entry 4).

In what may prove to be a general and useful observation, the Lewis acid-catalyzed intramolecular [4 + 2] cyclization of (Z)-enal **11** has been shown to afford the cis-fused cyclopentapyrans stereoselectively while the isomeric (E)-enal **11** provided all four stereoisomeric products under identical conditions [Eq. (35)].[144]

$$\text{(35)}$$

The formal total syntheses of the heteroyohimbine alkaloids tetrahydroalstonine (**12**) and akuammigine (**13**) have been described in which the D/E ring system of the naturally occurring materials was constructed through the 4π participation of an α,β-unsaturated aldehyde in [4 + 2] cycloaddition with an electron-deficient α,β-unsaturated amide [Eq. (36)].[141]

$$\text{(36)}$$

α,β-Unsaturated carbonyl compounds bearing an additional electron-withdrawing substituent at the α position do participate readily in intramolecular Diels–Alder reactions even at room temperature with electron-rich *or* simple alkenes. The required alkyliden- or aryliden-1,3-dicarbonyl systems may be obtained by the *in situ* condensation of 1,3-dicarbonyl compounds with aldehydes. This simple, *in situ* generation of the heterodiene has proved to be a technically effective and efficient approach to implementing the use of [4 + 2] cycloadditions of 4π oxabutadiene systems. Much of this work has been reviewed.[145]

The condensation of *N,N*-dimethylbarbituric acid, barbituric acid, Meldrum's acid, and substituted cyclohexa-1,3-diones with (*R*)-citronellal each provided a single enantiomerically and diastereomerically pure tricy-

clic dihydropyran possessing a trans ring junction. The enantio- and diastereoselectivity of the intramolecular cycloaddition was shown to arise
from a preferred exo [4 + 2] transition state possessing a chair conformation and an equatorial methyl group. The E/Z stereochemistry of the
reacting oxabutadiene system, the type and size of the oxabutadiene–
alkene linking unit, and the alkene substitution each may have a pronounced effect on the rate, observed stereochemical course, and regioselectivity (fused versus bridged cycloadduct) of the intramolecular [4 + 2]
cycloaddition.[145] Entries 7–13 of Table 7-II detail the results of much of
this work.[145] The application of these observations in the enantio- and
diastereospecific total syntheses of either enantiomer of hexahydrocannabinol (**14**) [Eq. (37)][134,135]

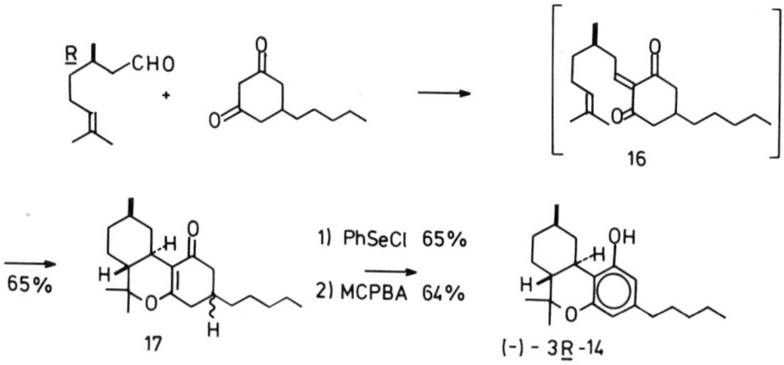

(37)

and 3-hydroxyhexahydrocannabinol (**15**) [Eq. (38)][145] have been detailed.

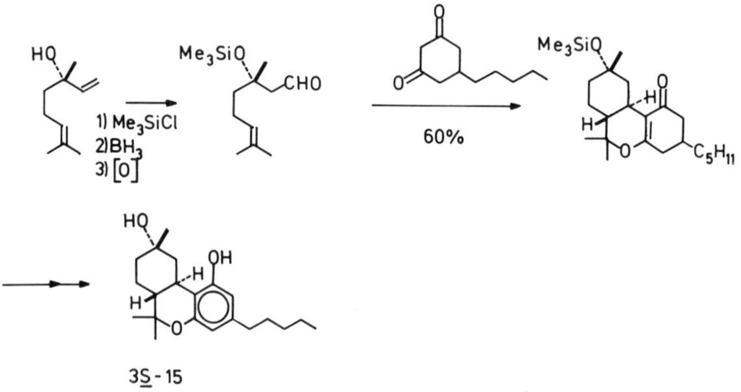

(38)

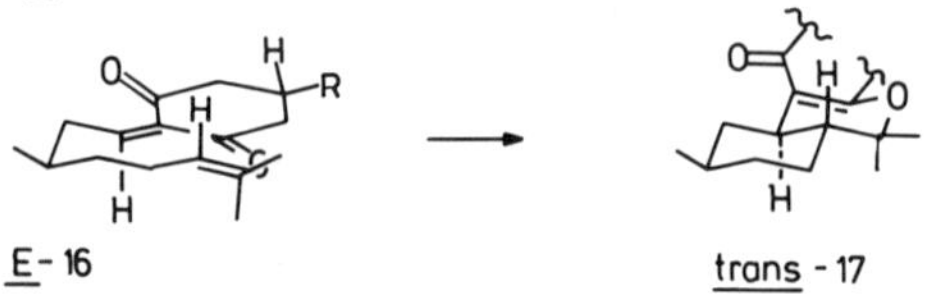

In studies which represent the first carefully documented example of the 4π participation of an α,β-unsaturated ester in a [4 + 2] cycloaddition,[142] Snider and co-workers systematically investigated the competing thermal, intramolecular Diels–Alder and ene reactions of 1-allylic-2,2-dimethyl ethylenetricarboxylates [Eq. (39)].[142]

$$\Delta H = -14.3 \text{ kcal/m}$$
$$\Delta S = -41 \text{ eu}$$

$$85°C \quad 2:1$$
$$115°C \quad 8:1$$

(39)

Lower reaction temperatures favor the reversible formation of the Diels–Alder products, and higher reaction temperatures promote the irreversible ene reactions. Related observations of the competing or predominating intramolecular Diels–Alder reactions of α-cyano or α-acetyl α,β-unsaturated esters have been reported in studies of the thermal and Lewis acid-promoted ene reaction (Table 7-II, entry 16).[143] The complementary use of o-quinone methide intermediates in intra- as well as intermolecular [4 + 2] cycloaddition reactions is discussed in the following section.

3. o-QUINONE METHIDES

The various methods of generating o-quinone methides,[4,5] including the thermal or (Lewis) acid-catalyzed elimination of a phenol Mannich base,[149,150,160,161,163] the thermal or (Lewis) acid-catalyzed dehydration of an o-hydroxybenzyl alcohol (ether),[147–149,151,153,156,157,162,163,165,168,171–175,178–183] the thermal 1,5-hydride shift of an o-hydroxystyrene,[171–173,175,178–183] the thermal dissociation of the corresponding spirochromane dimer,[158,163–164,166] the oxidation of substituted o-alkylphenols,[152,170] and the thermal or photochemical-promoted cheletropic extrusion[154,155,159] of carbon monoxide, carbon dioxide, or sulfur dioxide (Scheme 7-III), as well as their subsequent in situ participation in regiospecific, intermolecular [4 + 2] cycloadditions with simple olefins and acetylenes,[147,149,151,152,153,159,162–164]

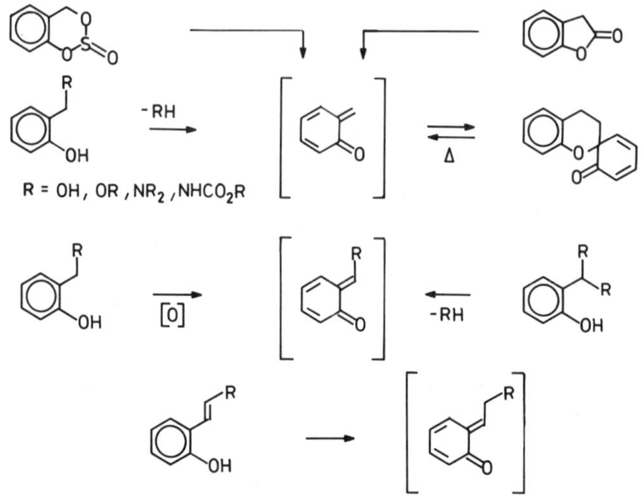

Scheme 7-III

dienes,[149,151,152] enol ethers,[151,152,153,163,164] ketene acetals,[152,154] enamines,[150] electron-deficient olefins or alkynes,[147,152,160,161,163–166] and in dimerization–trimerization reactions[147–149,152,156,157,158,163,164,166,167] (Scheme 7-IV) have been extensively reviewed.[4,5] Only the more recent observations on the generation and [4 + 2] cycloadditions of *o*-quinone methides are detailed herein, and prior reviews should be consulted for a full discussion and citations.

The general approach employed for the *in situ* generation of *o*-quinone methides relies on the elimination of water [Eq. (41)], alcohol, or second-

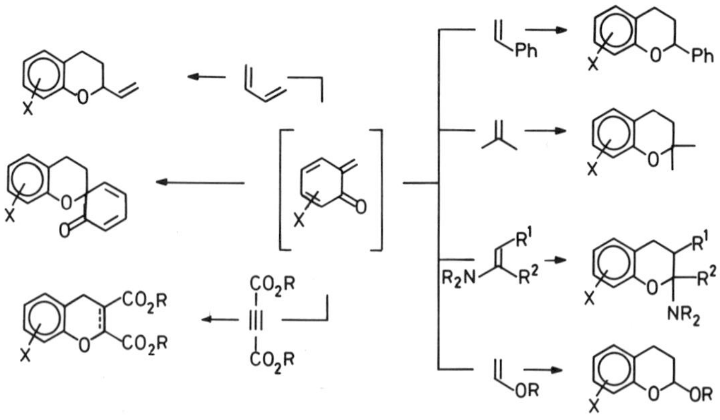

Scheme 7-IV

ary amine from a phenol Mannich base or o-hydroxybenzyl alcohol (ether). In many instances these methods are not suitable, and undesired side reactions are observed.[163] In the case of 1,2-naphthoquinone 1-methide (**18**), the thermal dissociation of the spirochromane dimer **19** provides a more suitable source of o-quinone methide than these methods[163,164,166] and permitted a careful study of its [4 + 2] cycloadditions with simple, electron-rich, and electron-deficient olefins.[163,164] This approach has been successfully employed to demonstrate that dienophile olefin geometry is often, though not always, maintained in the course of the [4 + 2] cycloaddition with an o-quinone methide [Eq. (40)] and to promote the [4 + 2] cycloadditions of **18** with sensitive electron-deficient heterocycles.[166]

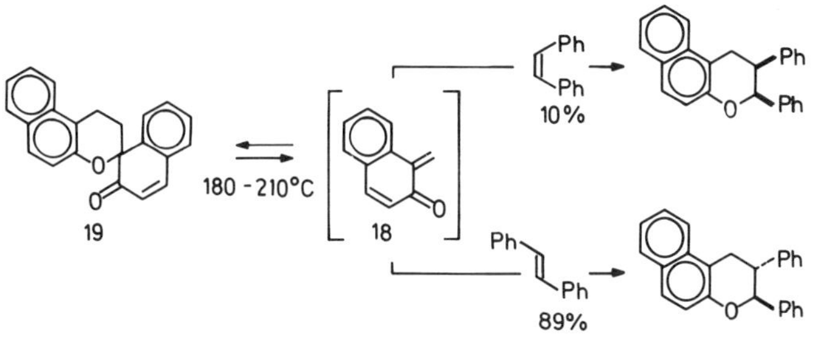

(40)

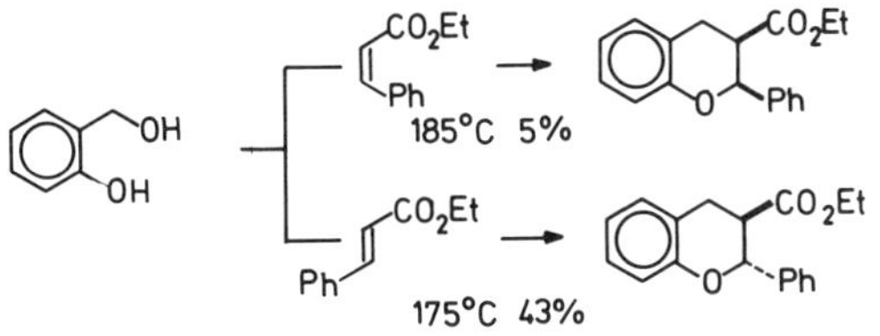

(41)

The reaction of chlorotrimethylsilane with hydroxymethylspiroepoxycyclohexadienone (**20**) affords the spiroannelated chroman **21**. The reaction presumably proceeds through an o-quinone methide intermediate which dimerizes to the chroman [Eq. (42)].[167]

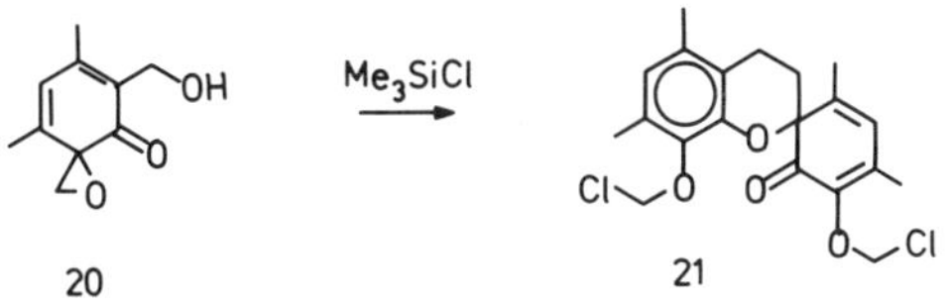

(42)

Two recent reports have detailed the [4 + 2] cycloadditions of o-quinone methides with heterodienophiles, thiones [Eq. (43)][168]

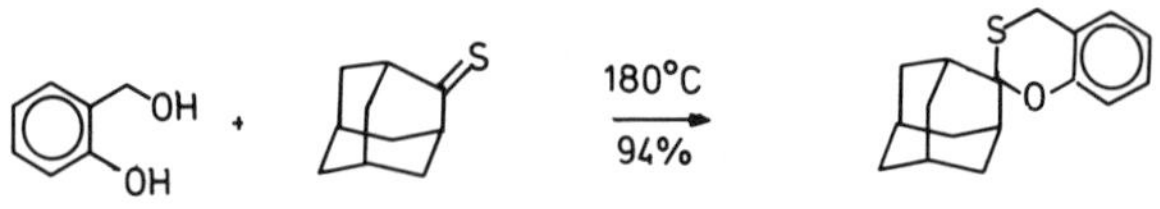

$$\text{(43)}$$

and nitriles.[169] These studies complement the earlier reports of the [4 + 2] cycloadditions of o-quinone methides with carbonyl compounds.[4,5]

Since the remarkable demonstration of the facility of an intramolecular Diels–Alder reaction (dimerization) of an o-quinone methide, generated by the oxidation of a substituted o-alkylphenol in the development of a biomimetic synthesis of carpanone [Eq. (44)],[170]

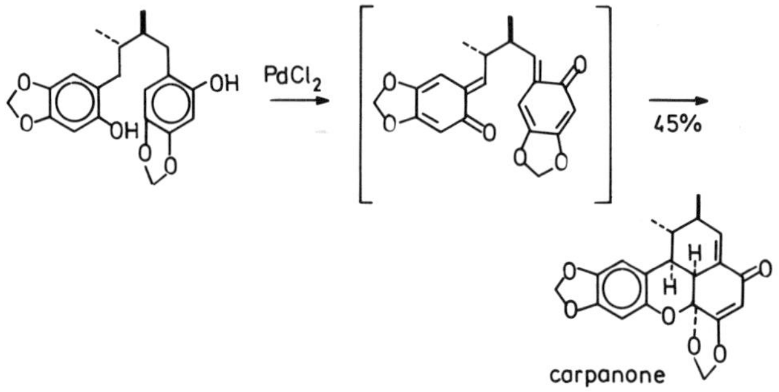

$$\text{(44)}$$

carpanone

a number of additional investigations have served to further develop and extend the scope of the intramolecular [4 + 2] cycloaddition reactions of o-quinone methides. Thermolysis of **22** at 147°C resulted in dehydration to the styrene derivative, and prolonged warming of **23** at 270°C provided the [4 + 2] cycloadducts **24** and **25** in a combined yield of 69%. Presumably a 1,5-hydrogen shift provides the o-quinone methide which participates in a Diels–Alder cycloaddition to provide the fused and bridged [4 + 2] cycloadducts [Eq. (45)].[171–173]

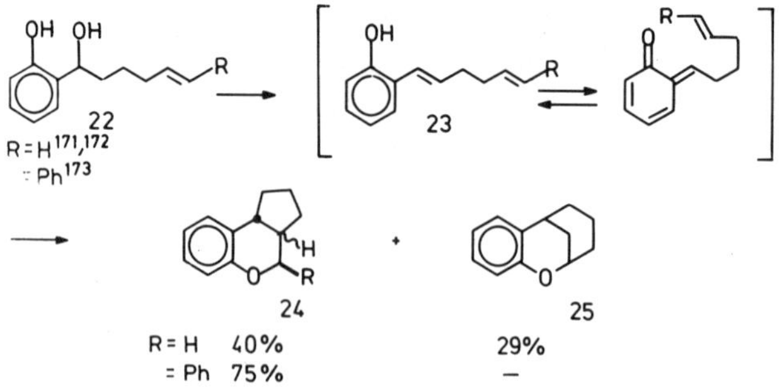

$$\text{(45)}$$

22		24		25
R=H[171,172]		R = H 40%		29%
= Ph[173]		= Ph 75%		—

The pyrolysis of **23** at 600°C has been reported to provide **24** in 12% yield.[172] In one instance, the 2π participation of a carbonyl group in an intramolecular Diels–Alder reaction with an o-quinone methide has been reported to provide a bridged [4 + 2] cycloadduct [Eq. (46)].[174]

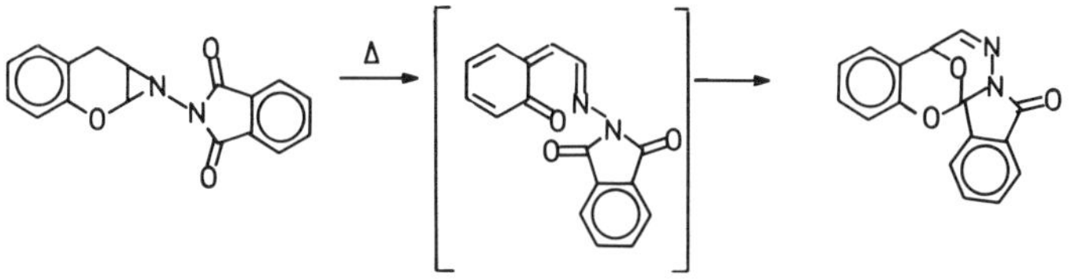

$$(46)$$

Full details of a careful study of the regio- and stereospecific intramolecular [4 + 2] cycloadditions of o-quinone methides generated by the thermal or acid-catalyzed (CF_3CO_2H) dehydration of o-hydroxybenzyl alcohols have been described[175] and have found application in the total synthesis of enantiomerically pure (3R)-26 and (3R)-27 possessing the ring system and correct absolute configuration of the cannabinol family [Eq. (47)].

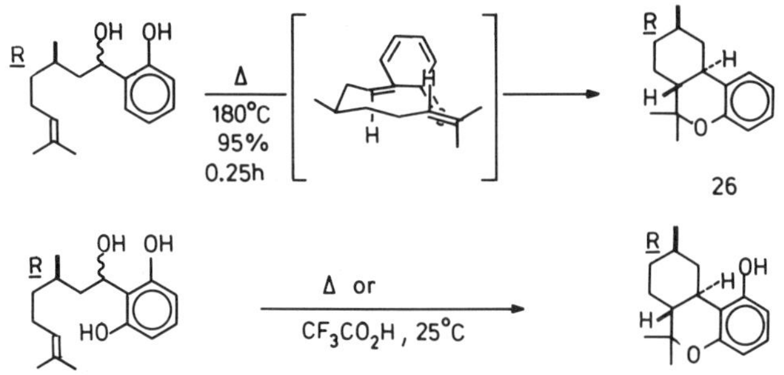

$$(47)$$

The exclusive formation of a single diastereomer with a trans ring fusion in the intramolecular Diels–Alder reaction of the o-quinone methide is consistent with the cycloaddition reaction proceeding through a preferred exo transition state from a chair conformation possessing an equatorial methyl substituent. The similar preference for a chair-exo transition state in the intramolecular [4 + 2] cycloadditions of substituted α,β-unsaturated carbonyl compounds[145] and substituted o-quinodimethanes[176] has been observed. A variant of these observations utilizes the room temperature fluoride-induced 1,4-elimination of trimethylsilynol for the in situ generation of an o-quinone methide enroute to the total synthesis of (+)- or (−)-hexahydrocannabinols [Eq. (48)].[177]

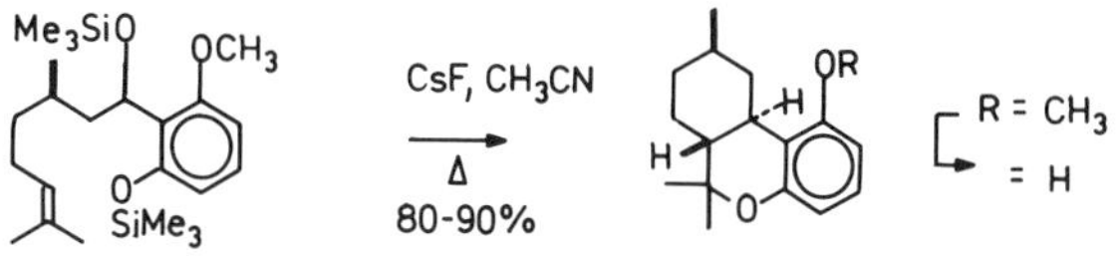

$$(48)$$

The condensation of polyhydroxylated benzenes with unsaturated aldehydes including citral or farnesal to provide citrans [Eq. (49)]

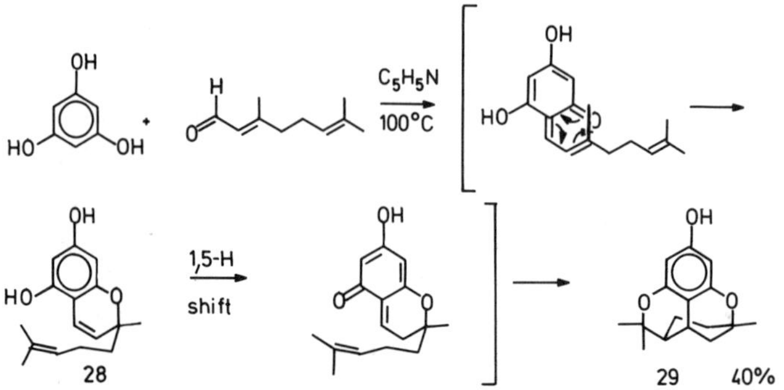

(49)

has been interpreted as proceeding by the intramolecular [4 + 2] cycloaddition of intermediate o-quinone methides because of the pronounced
stereoselectivity of the observed reactions.[178-184] Both the acid- and base-
catalyzed reactions are stereospecific, and consequently both are presumed to proceed with the *in situ* generation and subsequent [4 + 2]
cycloaddition of an o-quinone methide. The two processes differ only in
the mode of catalysis of o-hydroxystyrene to o-quinone methide tauto-
merism.[183] The intermediate chromenes **28** have been isolated and thermally converted to the observed citrans **29**.[184] The application of these
observations in the development of a two-step, biomimetic total synthesis
of ($\pm$)-deoxybruceol (**30**) [Eq. (50)][178]

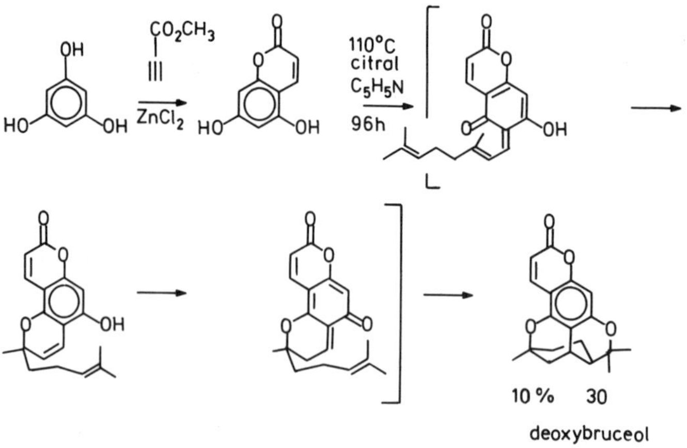

(50)

and in the preparation of cannabinoids[179-182] has been detailed.

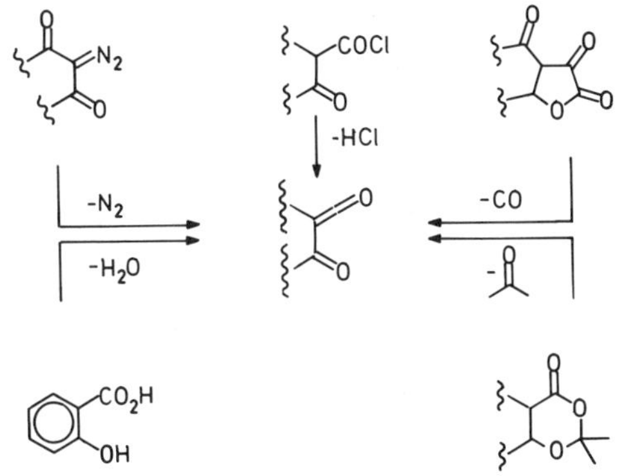

Scheme 7-V

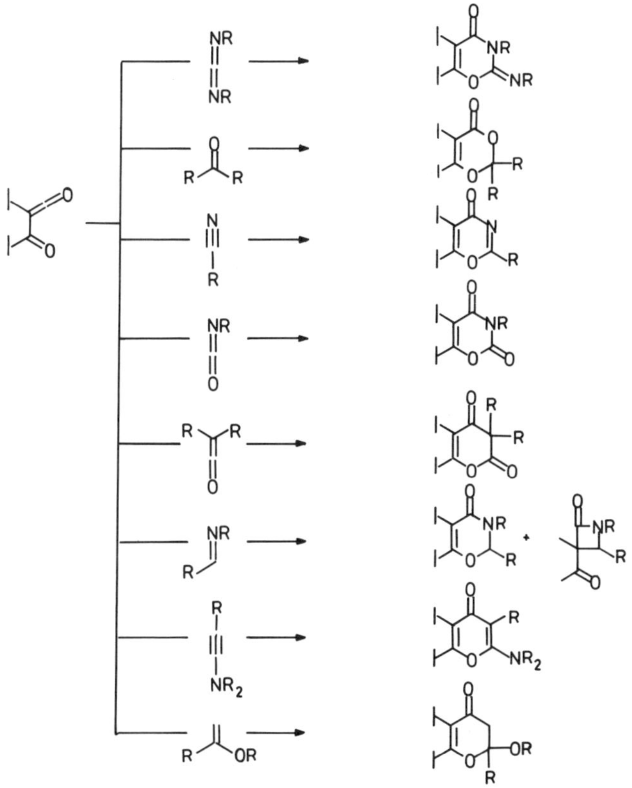

Scheme 7-VI

4. ACYL KETENES AND ALLENES

The methods for generating acyl ketenes (Scheme 7-V) and their subsequent *in situ* participation in [4 + 2] cycloadditions with a wide range of hetero- or olefinic and acetylenic dienophiles (Scheme 7-VI), including acyl ketenes,[185,186,197] carbonyl compounds,[186–188] nitriles,[187a,189,191] isocyanates and isothiocyanates,[186a,190,191] ketenes,[191] imines,[186a,187a,191,192] carbodiimides,[187c,190,191,193] ynamines,[194] ketene acetals,[186a,195] enol ethers,[186a,191,196] and N-sulfinylamines[197] have been extensively reviewed.[5,9,12] Two reports have detailed the 4π participation of allenic ketones in [4 + 2] cycloaddition reactions [Eq. (51)].[198,199]

$$\tag{51}$$

5. o-QUINONES AND 1,2-DICARBONYL COMPOUNDS (1,4-DIOXABUTADIENES)

The thermal and photochemical [4 + 2] cycloadditions of o-quinones with olefinic and acetylenic dienophiles have been extensively reviewed[4,5,200] and include their 4π heterodiene Diels–Alder reactions with olefins,[201–204] vinyl ethers,[205] enamines,[206] selected dienes,[207–209] diphenylketenimines,[210] ketenes,[209,210] fulvenes,[211] and selected heterocycles including furan,[207–209,212] benzofuran,[209,212,215] indoles,[213] azepines,[214] and 1,2-diazepines.[214] The tetrahalo-substituted o-quinones, tetrachloro- and tetrabromo-o-quinone, generally participate in heterodiene [4 + 2] cycloadditions at an increased rate over the unsubstituted systems and generally provide higher overall yields of the Diels–Alder products.[4,5] With simple olefins, the dienophile geometry is maintained in the course of the thermal [4 + 2] cycloadditions [Eq. (52)],[203,204]

$$\tag{52}$$

and this observation is consistent with the 4π heterodiene participation of the o-quinone in a concerted Diels-Alder reaction. With olefins and dienes, the competitive and often times predominate o-quinone all-carbon 4π and o-quinone olefin or carbonyl 2π participation in [4 + 2], [4 + 6], and [2 + 2] cycloadditions has been observed.[201,211,212,214,216-219]

Ketene acetals, e.g., 1,1-dimethoxypropene and tetramethoxythylene, generally afford [2 + 2] oxetane cycloadducts with carbonyl compounds including α-diketones. However, the dihydrodioxin [4 + 2] cycloadducts of α-diketones with 1,1-dimethoxypropene and tetramethoxyethylene have been observed in thermal and Lewis acid-catalyzed reactions when the [4 + 2] cycloaddition is accompanied by an increase in the stabilization energy of the diketone moiety, e.g., phenanthrene generation [Eq. (53)].[220]

$$(53)$$

6. HETERO-1-OXABUTADIENES

An extensive range of hetero-1-oxabutadiene systems containing nitrogen (Chapter 9) and sulfur (Chapter 8) have been investigated and have been found to participate as 4π components of Diels–Alder reactions. Several reviews have treated aspects of this work.[5,9,10,11,221] In summary, the incorporation of an additional heteroatom into the oxabutadiene system generally increases its electrophilic character and facilitates the 4π participation of the diene system in inverse electron demand Diels–Alder reactions with electron-rich dienophiles. This is especially evident in the studies of azaoxabutadiene systems (Chapter 9).

Hetero-1-oxabutadiene systems that have been shown to participate as 4π components of Diels–Alder reactions include vinylnitroso compounds (**31**, 2-aza-1-oxabutadiene, Chapter 9, Section 3), N-acylimines (**32**) as well as aromatic and aliphatic acylisocyanates (**33**, X = O) and isothiocyanates (**33**, X = S, 3-aza-1-oxabutadiene, Chapter 9, Section 3), acylnitroso compounds (**34**, 2-aza-1,4-dioxabutadiene, Chapter 9, Section 3), selected 4-aza-1-oxabutadienes, e.g., **35**, including o-quinone monoimines (**36**,

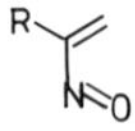

31 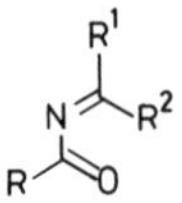   32 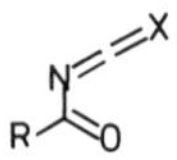   33   34 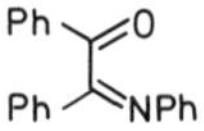   35 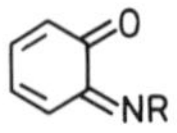   36

X= O,S

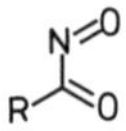

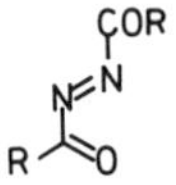

37 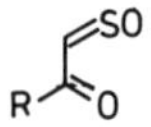   38 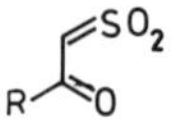   39 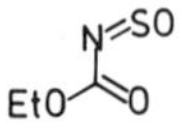   40 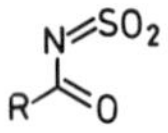   41

R = Ph
  = OEt

Chapter 9, Section 1), azodicarboxylates and related acyl and aroyl azo compounds (**37**, 2,3-diaza-1-oxabutadiene, Chapter 9, Section 5), acylsulfines (**38**) and -sulfenes (**39**, 4-oxa-1-thiabutadiene, Chapter 8, Section 5), N-sulfinylurethanes (**40**), and N-sulfonylamides and -urethanes (**41**, 2-aza-4-oxa-1-thiabutadiene, Chapter 9, Section 3). Chapters 8 and 9 detail the investigations on each of the systems indicated and should be consulted for citations and full discussions of their 4π participation in Diels–Alder reactions.

7. CATIONIC OXABUTADIENES, $[4^+ + 2]$ CYCLOADDITIONS

The generation of a limited number of oxabutadiene systems bearing a formal cationic charge and their subsequent *in situ* participation in $[4^+ + 2]$ polar cycloadditions have been detailed. The classification of cationic oxabutadiene $[4^+ + 2]$ cycloadditions as "polar" cycloadditions is not derived from an implied stepwise addition–cyclization reaction mechanism but was terminology introduced to distinguish cycloadditions employing cationic or anionic components from those employing dipolar or uncharged components.[222,223]

The o-hydroxybenzyl cation, generated *in situ* by the action of protic acid on an o-hydroxybenzyl alcohol or by the Lewis acid treatment of an o-hydroxybenzyl chloride, has been shown to participate in a range of regio- and stereospecific $[4^+ + 2]$ cycloadditions [Eq. (54)].[224,225]

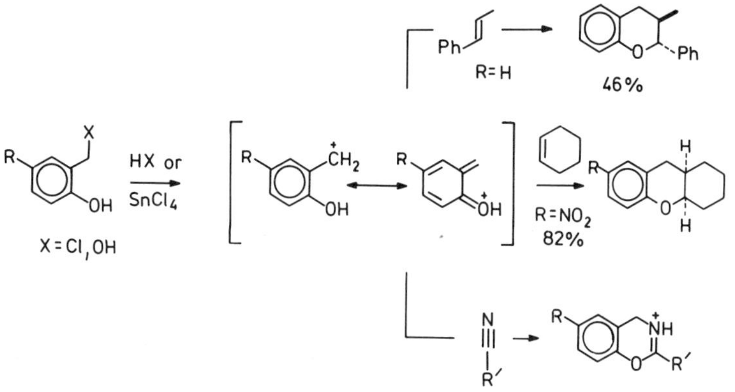

$$(54)$$

The dienophile olefin geometry is maintained in the course of the cycloaddition reaction.

Schmidt has demonstrated that selected β-acylvinyl carbocations, generated from β-chlorovinyl ketones[226] or from the *in situ* acylation of acetylenes, react with acetylenes to provide pyrylium salts presumably with 4π participation of the cationic oxabutadienes in [4+ + 2] cycloadditions [Eq. (55)].[226]

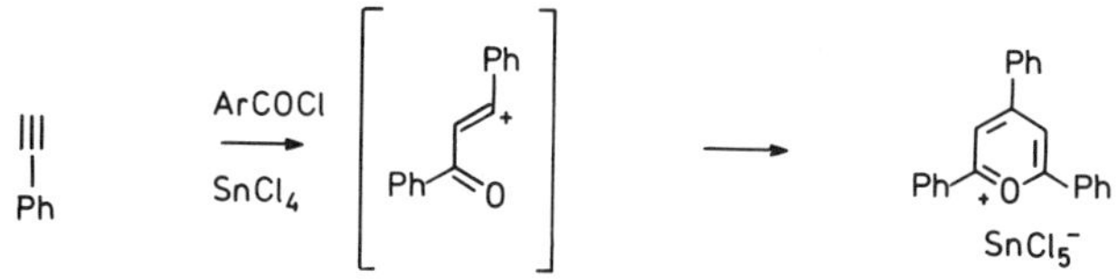

$$(55)$$

The generation and subsequent *in situ* 4π participation of the related *N*-methyl *N*-methylenium amides (**42**), simple *N*-methylenium amides (**43**), and related cationic 3-aza-1-oxabutadiene systems in [4+ + 2] cycloadditions are detailed in Chapter 9, Section 10.[222,223] These simple aliphatic cationic heterodiene systems have been more thoroughly investigated than any other aliphatic system. The maintenance of dienophile geometry in the [4+ + 2] cycloaddition,[227] the lack of observed rearrangement products potentially derived from dienophiles susceptible to carbocation rearrangement,[228] and the clean regioselectivity observed in the [4+ + 2] cycloadditions of *N*-methylenium amides[227,229] are consistent with the 4π

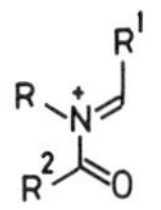

42 R = CH$_3$
43 R = H

participation of **42** and **43** in concerted but nonsynchronous [4$^+$ + 2] cycloaddition reactions (Chapter 9, Section 10).

REFERENCES

1. Sherlin, S. M., Berlin, A. Y., Serebrennikova, T. A., and Rabinovitch, R. F. (1938). *J. Gen. Chem. USSR* **8**, 22.
2. Alder, K., Offermanns, H., and Ruden, E. (1941). *Chem. Ber.* **74**, 905.
3. Colonge, J., and Descotes, G. (1967). *In* "1,4-Cycloaddition Reactions, The Diels–Alder Reaction in Heterocyclic Syntheses" (J. Hamer, ed.), Chapter 9, Academic Press, New York.
4. a. *o*-Quinones and 1,2-dicarbonyl compounds: Pfundt, G., and Schenck, G. O. (1967). *In* "1,4-Cycloaddition Reactions, The Diels–Alder Reaction in Heterocyclic Syntheses" (J. Hamer, ed.), Chapter 11. Academic Press, New York. b. *o*-Quinone methides: Wagner, H. U., and Gompper, R. (1974). *In* "The Chemistry of Quinonoid Compounds" (S. Patai, ed.), Vol. 2, p. 1145. Wiley, New York.
5. Desimoni, G., and Tacconi, G. (1975). *Chem. Rev.* **75**, 651.
6. Needleman, S. B., and Chang Kuo, M. C. (1962). *Chem. Rev.* **62**, 405; Wollweber, H. (1970). *In* "Methoden der Organische Chemie (Houben-Weyl)" (E. Muller, ed.), Teil 3, Vol. V/1c, pp. 981–1139. Thieme, Stuttgart; Jager, V., and Viehe, H. G. (1977). *In* "Methoden der Organische Chemie (Houben-Weyl)" (E. Muller, ed.), Vol./2a, pp. 809–877. Thieme, Stuttgart.
7. Sauer, J. (1967). *Angew. Chem. Int. Ed. Engl.* **6**, 16; Sauer, J., and Sustmann, R. (1980). *Angew. Chem. Int. Ed. Engl.* **19**, 779; Povarov, L. S. (1967). *Russ. Chem. Rev.* **36**, 656.
8. Onho, M., and Sasaki, T. (1984). *J. Syn. Org. Chem. Jpn.* **42**, 126.
9. Ulrich. H. (1967). "Cycloaddition Reactions of Heterocumulenes." Academic Press, New York.
10. Aliphatic and aromatic acyl isocyanates: Arbuzov, B. A., and Zobova, N. N. (1974). *Synthesis,* 461; Arbuzov, B. A., and Zobova, N. N. (1982). *Synthesis,* 433.
11. Thioacyl isocyanates and acyl isocyanates: Tsuge, O. (1979). *Heterocycles* **12**, 1067.
12. a. Kato, T., Katagiri, N., and Yamamoto, Y. (1980). *Heterocycles* **14**, 1333. b. Clemens, R. J. (1986). *Chem. Rev.* **86**, 241.
13. Seoane, C., Soto, J. L., and Quinteiro, M. (1980). *Heterocycles* **14**, 337.
14. Mundy, B. P., Lipkowitz, K. B., and Dirks, G. W. (1977). *Heterocycles* **6**, 51.
15. Woodward, R. B., and Katz, T. J. (1959). *Tetrahedron* **5**, 70.
16. McIver, J. W., Jr. (1972). *J. Am. Chem. Soc.* **94**, 4782.
17. Salem, L. (1968). *J. Am. Chem. Soc.* **90**, 553.
18. Alston, P. V., and Shillady, D. D. (1974). *J. Org. Chem.* **39**, 3402.
19. Eisenstein, O., Lefour, J. M., Anh, N. T., and Hudson, R. F. (1977). *Tetrahedron* **33**, 523; Minot, C., and Anh, N. T. (1977). *Tetrahedron* **33**, 533.
20. Burnier, J. S., and Jorgensen, W. L. (1983). *J. Org. Chem.* **48**, 3923.
21. Gore, W. E., Pearce, G. T., and Silverstein, R. M. (1976). *J. Org. Chem.* **41**, 603.
22. Lipowitz, K. B., Scarpone, S., Mundy, B. P., and Bornmann, W. G. (1979). *J. Org. Chem.* **44**, 486.
23. Longley, R. I., Jr., and Emerson, W. S. (1950). *J. Am. Chem. Soc.* **72**, 3079; Emerson, W. S., Birum, G. H., and Longley, R. I., Jr. (1953). *J. Am. Chem. Soc.* **75**, 1312;

Longley, R. I., Jr., Emerson, W. S., and Blardinelli, A. J. (1963). *Org. Syn.,* Coll. Vol. 4, 311.
24. Smith, C. W., Norton, D. G., and Ballard, S. A. (1951). *J. Am. Chem. Soc.* **73,** 5267, 5270.
25. Parham, W. E., and Holmquist, H. E. (1951). *J. Am. Chem. Soc.* **73,** 913; Rane, D. F., Fishman, A. G., and Pike, R. E. (1984). *Synthesis,* 694.
26. Gompper, R. (1969). *Angew. Chem. Int. Ed. Engl.* **8,** 312.
27. Ciganek, E. (1984). *Org. React.* **32,** 1; Fallis, A. G. (1984). *Can. J. Chem.* **62,** 183; Brieger, G., and Bennett, J. N. (1980). *Chem. Rev.* **80,** 63; Oppolzer, W. (1977). *Angew. Chem. Int. Ed. Engl.* **16,** 10.
28. Matsumoto, K., and Sera, A. (1986). *Synthesis,* 999.
29. Dauben, W. G., and Kozikowski, A. P. (1974). *J. Am. Chem. Soc.* **96,** 3664.
30. Dauben, W. G., and Krabbenhoft, H. D. (1977). *J. Org. Chem.* **42,** 282; Snider, B. B., and Phillips, G. B. (1983). *J. Org. Chem.* **48,** 2790.
31. Schultz, R. C., and Hartmann, H. (1962). *Chem. Ber.* **95,** 2735.
32. Quagliaro, R., Moreau, M., and Dieux, J. (1963). *Comptes Rendus Acad. Sci. Paris C* **257,** 2843; Schultz, H., and Wagner, H. (1950). *Angew. Chem.* **62,** 105.
33. Hall, R. H., and Howe, B. K. (1951). *J. Chem. Soc.,* 2480.
34. Mizuta, M., and Ishii, Y. (1963). *Kogyo Kagaku Zasshi* **66,** 1442.
35. Broquet, C., d'Angelo, J., and Thuy, V. M. (1968). *Bull. Soc. Chim. Fr.,* 341.
36. Thuy, V. M. (1970). *Bull. Soc. Chim. Fr.,* 4429.
37. Hoffmann, H., Schmidt, E., Jeschek, G., Kuerzinger, A., Schoenleben, W., Winderl, S., and Voges, D. (1971). German Patent 2,008,131; *Chem. Abstr.* **75,** 151674 (1971).
38. Sadikh-zade, S. I., Kasumov, F. Y., Kyazimov, S. K., and Sultanov, R. A. (1972). *Zh. Org. Khim.* **8,** 1788.
39. Mochalin, V. B., Smolina, Z. I., and Unkovskii, B. V. (1972). *Khim. Geterotsikl. Soedin,* 452, and references cited therein.
40. Skvortsova, G. G., Andriyankov, M. A., and Tyrina, S. M. (1972). *Khim. Geterotsikl. Soedin,* 1155, and references cited therein.
41. Belanger, A., and Brassard, P. (1972). *J. Chem. Soc., Chem. Commun.,* 863; Belanger, A., and Brassard, P. (1975). *Can. J. Chem.* **53,** 195, 201.
42. Pedersen, E. B., and Lawesson, S.-O. (1970). *Tetrahedron* **26,** 2959; Couturier, D., Fargeau, M.-C., and Maitte, P. (1972). *Bull. Soc. Chim. Fr.,* 4777.
43. Brannock, K. C. (1960). *J. Org. Chem.* **25,** 258.
44. Baganz, H., and Brinckmann, E. (1956). *Chem. Ber.* **89,** 1565.
45. Pike, J. E., Rebenstorf, M. A., Slomp, G., and MacKellar, F. A. (1963). *J. Org. Chem.* **28,** 2499.
46. Paul, R., and Tchelitcheff, S. (1954). *Bull. Soc. Chim. Fr.* **21,** 672.
47. Barili, P., Berti, G., Catelani, G., Colonna, F., and Mastrorilli, E. (1986). *J. Chem. Soc., Chem. Commun.,* 7.
48. a. Coda, A. C., Desmoni, G., Righetti, P. P., Tacconi, G., Buttafava, A., and Faucitano, F. M. (1983). *Tetrahedron* **39,** 331. b. Desimoni, G., Righetti, P. P., Gamba, A., and Tacconi, G. (1981). *Tetrahedron* **37,** 1779. c. Desimoni, G., Righetti, P. P., Selva, E., Tacconi, G., Riganti, V., and Specchiarello, M. (1977). *Tetrahedron* **33,** 2829. d. Desimoni, G., Gamba, A., Monticelli, M., Nicola, M., and Tacconi, G. (1976). *J. Am. Chem. Soc.* **98,** 2947. e. Desimoni, G., Righetti, P. P., Tacconi, G., and Vigliani, A. (1977). *Gazz. Chim. Ital.* **107,** 91. f. Desimoni, G., Tacconi, G., and Marinone, F. (1968). *Gass. Chim. Ital.* **98,** 1301; Desimoni, G., Gamba, A., Righetti, P. P., and Tacconi, G. (1971). *Gazz. Chim. Ital.* **101,** 899; Desimoni, G., Cellerino, G., Minoli, G., and Tacconi, G. (1972). *Tetrahedron* **28,** 4003; Tacconi, G., Marinone, F., and

Desimoni, G. (1972). *Gass. Chim. Ital.* **101**, 173; Tacconi, G., Iadarola, P., Marinone, F., Righetti, P. P., and Desimoni, G. (1975). *Tetrahedron* **31**, 1179; Desimoni, G., and Tacconi, G. (1968). *Gazz. Chim. Ital.* **98**, 1329; Desimoni, G., Astolfi, L., Cambierri, M., Gamba, A., and Tacconi, G. (1973). *Tetrahedron* **29**, 2627; Desimoni, G., Colombo, G., Righetti, P. P., and Tacconi, G. (1973). *Tetrahedron* **29**, 2635; Desimoni, G., Gamba, A., Righetti, P. P., and Tacconi, G. (1972). *Gazz. Chim. Ital.* **102**, 491; Tacconi, G., Gamba, A., Righetti, P. P., and Desimoni, G. (1980). *J. Prakt. Chem.* **322**, 711.

49. Desimoni, G., Righetti, P. P., Tacconi, G., Piccolini, A. C. C., Pesenti, M. T., and Oberti, R. (1979). *J. Chem. Soc., Perkin Trans. 1*, 863; see also Desimoni, G., Gamba, A., Righetti, P. P., and Tacconi, G. (1971). *Gazz. Chim. Ital.* **101**, 899; Desimoni, G., and Tacconi, G. (1968). *Gazz. Chim. Ital.* **98**, 1329.

50. McElvain, S. M., Degginger, E. R., and Behun, J. D. (1954). *J. Am. Chem. Soc.* **76**, 5736.

51. Schut, R. N., and Liu, T. M. H. (1965). *J. Org. Chem.* **30**, 2845.

52. Opitz, G., and Loschmann, I. (1960). *Angew. Chem.* **72**, 523; Opitz, G., and Holtmann, H. (1965). *Liebigs Ann. Chem.* **684**, 79.

53. Fleming, I., and Karger, M. H. (1967). *J. Chem. Soc. C*, 226.

54. Colonna, F. P., Fatutta, S., Risaliti, A., and Russo, C. (1970). *J. Chem. Soc. C*, 2377.

55. Lewis, J. W., Myers, P. L., Ormerod, J. A., and Selby, I. A. (1972). *J. Chem. Soc., Perkin Trans. 1*, 1549; Lewis, J. W., Myers, P. L., and Readhead, M. J. (1970). *J. Chem. Soc. C*, 771; Lewis, J. W., and Myers, P. L. (1970). *Chem. Ind. (London)*, 1625.

56. Rao, R. B., and Bhide, G. V. (1969). *Chem. Ind. (London)*, 1095.

57. Prasad, K. K., and Girijavallabhan, V. M. (1971). *Chem. Ind. (London)*, 426; Prasad, K. K. (1971). *Indian J. Chem.* **9**, 1239.

58. Penades, S., Kisch, H., Tortschanoff, K., Margaretha, P., and Polansky, O. E. (1973). *Monatsh. Chem.* **104**, 447.

59. Tacconi, G., Gamba, A., Marinone, F., and Desimoni, G. (1971). *Tetrahedron* **27**, 561.

60. Forchiassin, M., Risaliti, A., Russo, C., Calligaris, M., and Pitacco, G. (1974). *J. Chem. Soc., Perkin Trans. 1*, 660.

61. Ficini, J., and Krief, A. (1969). *Tetrahedron Lett.*, 1427; Ficini, J., and Krief, A. (1970). *Tetrahedron Lett.*, 885.

62. Ficini, J., Besseyre, J., D'Angelo, J., and Barbara, C. (1970). *Comptes Rendus Acad. Sci. Paris C* **271**, 468; Myers, P. L., and Lewis, J. W. (1973). *J. Heterocycl. Chem.* **10**, 165.

63. Desimoni, G., Righetti, P., Tacconi, G., and Oberti, R. (1979). *J. Chem. Soc., Perkin Trans. 1*, 856.

64. Mizuta, M., Kato, T., and Ishii, Y. (1964). *Kogyo Kagaku Zasshi* **67**, 1382; Mizuta, M., Haraki, H., and Ishii, Y. (1966). *Kogyo Kagaku Zasshi* **69**, 79.

65. Descotes, G., and Jullien, A. (1969). *Tetrahedron Lett.*, 3395.

66. Tacconi, G., Leoni, M., Righetti, P. P., Desimoni, G., Oberti, R., and Comin, F. (1979). *J. Chem. Soc., Perkin Trans. 1*, 2687.

67. Sustmann, R. (1974). *Pure Appl. Chem.* **40**, 569.

68. a. Bakker, C. G., Scheeren, J. W., and Nivard, R. J. F. (1981). *Recl. Trav. Chim. Pays-Bas* **100**, 13. b. Hall, S. S., Weber, G. F., and Duggan, A. J. (1978). *J. Org. Chem.*, 667.

69. Danishefsky, S., and Bednarski, M. (1984). *Tetrahedron Lett.* **25**, 721.

70. Yamamoto, Y., Suzuki, H., and Mora-Oka, Y. (1986). *Chem. Lett.*, 73.

71. Snider, B. B., and Phillips, G. B. (1983). *J. Org. Chem.* **48**, 2789; see also Menicagli, R., Malanga, C., and Lardicci, L. (1982). *J. Org. Chem.* **47**, 2288; Takahashi, M.,

Suzuki, H., Mora-Oka, Y., and Ikawa, T. (1982). *Tetrahedron Lett.*, 1097; Takahashi, M., Suzuki, H., Mora-Oka, Y., and Ikawa, T. (1981). *Chem. Lett.*, 1435.

72. Makin, S. M., Raifel'd, Yu. E., and El'yanov, B. S. (1976). *Izv. Akad. Nauk SSSR Ser. Khim.*, 1094.

73. a. Jenner, G., Abdi-Oskoui, H., and Rimmelin, J. (1977). *Bull. Soc. Chim. Fr.*, 983, 341. Jenner, G., Abdi-Oskoui, H., Rimmelin, J., and Libs, S. (1979). *Bull. Soc. Chim. Fr. 2*, 33. b. Katada, T., Eguchi, S., Esaki, T., and Sasaki, T. (1984). *J. Chem. Soc., Perkin Trans. 1*, 1869.

74. Dhar, D. N., Mehta, G., and Suri, S. C. (1976). *Indian J. Chem. (B)*, 477.

75. Scheibye, S., Shabana, R., and Lawesson, S.-O. (1982). *Tetrahedron* **38**, 993.

76. Mundy, B. P., Otzenberger, R. D., and DeBernardis, A. R. (1971). *J. Org. Chem.* **36**, 2390, 3830; Lipkowitz, K. B., Mundy, B. P., and Geeseman, D. (1973). *Synth. Commun.* **3**, 453.

77. Bhupathy, M., and Cohen, T. (1985). *Tetrahedron Lett.* **26**, 2619.

78. Lipkowitz, K. B., Scarpone, S., Mundy, B. P., and Bornmann, W. G. (1979). *J. Org. Chem.* **44**, 486.

79. Chaquin, P., Morizur, J.-P., and Kossanyi, J. (1977). *J. Am. Chem. Soc.* **99**, 903.

80. Cohen, T., and Matz, J. R. (1980). *J. Am. Chem. Soc.* **102**, 6900; Cohen, T., and Bhupathy, M. (1983). *Tetrahedron Lett.* **24**, 4163.

81. Mundy, B. P., and Bornmann, W. G. (1984). *J. Org. Chem.* **49**, 5264.

82. Kinzer, G. W., Fentiman, A. F., Page, T. F., Foltz, R. L., Vite, J. P., and Pitman, G. B. (1969). *Nature (London)* **221**, 477.

83. D'Silva, T. D. J., and Peck, D. W. (1972). *J. Org. Chem.* **37**, 1828.

84. Kim, Y., and Mundy, B. P. (1982). *J. Org. Chem.* **47**, 3556; Mundy, B. P., Lipkowitz, K. B., and Dirks, G. W. (1975). *Synth. Commun.* **5**, 7.

85. Snowden, R. L., Sonnay, P., and Ohloff, G. (1981). *Helv. Chim. Acta* **64**, 25.

86. Fehr, C., Galindo, J., and Ohloff, G. (1981). *Helv. Chim. Acta* **64**, 1247; see also Woods, G. F., and Sanders, H. (1946). *J. Am. Chem. Soc.* **68**, 2483.

87. Sweet, F., and Brown, R. K. (1968). *Can. J. Chem.* **46**, 2289.

88. Childers, W. E., Jr., and Pinnick, H. W. (1984). *J. Org. Chem.* **49**, 5276.

89. Tietze, L.-F. (1974). *Chem. Ber.* **107**, 2491; see also Ref. 97.

90. Korte, F., Buchel, K. H., and Zschocke, A. (1961). *Chem. Ber.* **94**, 1952.

91. Franck, B., Petersen, U., and Huper, F. (1970). *Angew. Chem. Int. Ed. Engl.* **9**, 891.

92. Tursch, B., Chome, C., Braekman, J. C., and Daloze, D. (1973). *Bull. Soc. Chim. Belg.* **82**, 699.

93. Ireland, R. E., and Habich, D. (1980). *Tetrahedron Lett.* **21**, 1389.

94. Tietze, L.-F., and Glusenkamp, K.-H. (1983). *Angew. Chem. Int. Ed. Engl.* **22**, 887.

95. Tietze, L.-F., Glusenkamp, K.-H., Harms, K., Remberg, G., and Sheldrick, G. M. (1982). *Tetrahedron Lett.* **23**, 1147.

96. Tietze, L.-F., Glusenkamp, K.-H., and Holla, W. (1982). *Angew. Chem. Int. Ed. Engl.* **21**, 793.

97. Tietze, L.-F. (1983). *Angew. Chem. Int. Ed. Engl.* **22**, 828; Tietze, L.-F., and Glusenkamp, K.-H. (1983). *Angew. Chem. Int. Ed. Engl.* **22**, 887; Tietze, L.-F. (1974). *Chem. Ber.* **107**, 2491. Related work: see Halpern, O., and Schmid, H. (1958). *Helv. Chim. Acta* **41**, 1109; Tietze, L.-F., and Niemeyer, U. (1978). *Chem. Ber.* **111**, 2423; Tietze, L.-F., Niemeyer, U., Marx, P., and Glusenkamp, K.-H. (1980). *Tetrahedron* **36**, 1231; Tietze, L.-F., Niemeyer, U., Marx, P., Glusenkamp, K.-H., and Schwenen, L. (1980). *Tetrahedron* **36**, 735.

98. Schreiber, S. L., Hoveyda, A. H., and Wu, H.-J. (1983). *J. Am. Chem. Soc.* **105**, 660.

99. Tietze, L.-F., Harms, K., and Sheldrick, G. M. (1985). *Tetrahedron Lett.* **26**, 5273.

100. Dvorak, D., and Arnold, Z. (1982). *Tetrahedron Lett.* **23**, 4401.
101. a. Wallace, T. W. (1983). *J. Chem. Soc., Chem. Commun.*, 228. b. Ghosh, C. K., Tewari, N., and Bhattacharyya, A. (1984). *Synthesis* **7**, 614. c. Dean, F. M., Al-Sattar, M., and Smith, D. A. (1983). *J. Chem. Soc., Chem. Commun.*, 535.
102. Reichardt, C., and Yun, K.-Y. (1982). *Tetrahedron Lett.* **23**, 3163.
103. Arnold, Z., and Dvorak, D. (1985). *Coll. Czech. Chem. Commun.* **50**, 2265.
104. Snider, B. B. (1980). *Tetrahedron Lett.* **21**, 1133.
105. Bitter, J., Leitich, J., Partale, H., Polansky, O. E., Reimer, W., Ritter-Thomas, U., Schlamann, B., and Stilkerieg, B. (1980). *Chem. Ber.* **113**, 1020. Competitive dimerization: Eaton, P. E., and Bunnelle, W. H. (1984). *Tetrahedron Lett.* **25**, 23.
106. Schmidt, R. R., and Maier, M. (1982). *Tetrahedron Lett.* **23**, 1789.
107. Yamauchi, M., Katayama, S., Baba, O., and Watanabe, T. (1983). *J. Chem. Soc., Chem. Commun.*, 281.
108. Conia, J. M., and Amice, Ph. (1974). *Bull. Soc. Chim. Fr.* **5/6**, 1015; Conia, J. M., and Amice, Ph. (1974). *Tetrahedron Lett.*, 479.
109. Hall, H. K., Jr., Rasoul, H. A. A., Gillard, M., Abdelkader, M., Nogues, P., and Sentman, R. C. (1982). *Tetrahedron Lett.* **23**, 603.
110. Hall, H. K., Jr., and Abdelkader, M. (1981). *J. Org. Chem.* **46**, 2948; Hall, H. K. Jr., Padias, A. B., and Hedrick, S. T. (1983). *J. Org. Chem.* **48**, 3787.
111. Rasoul, H. A. A., and Hall, H. K., Jr. (1982). *J. Org. Chem.* **47**, 2080.
112. Hall, H. K., Jr., Nogues, P., Rhoades, J. W., Sentman, R. C., and Detar, M. (1982). *J. Org. Chem.* **47**, 1451.
113. a. Ismail, Z. M., and Hoffmann, H. M. R. (1982). *Angew. Chem. Int. Ed. Engl.* **21**, 859; El-Abed, D., Jellal, A., and Santelli, M. (1984). *Tetrahedron Lett.* **25**, 4503; Santelli, M., Douniazad, E. A., and Jellal, A. (1986). *J. Org. Chem.* **51**, 1199. b. Romann, E., Frey, A. J., Stadler, P. A., and Eschenmoser, A. (1957). *Helv. Chim. Acta* **40**, 1900.
114. Bakker, C. G., Ooms, P. H. M., Scheeren, J. W., and Nivard, R. J. F. (1983). *Recl. Trav. Chim. Pays-Bas* **102**, 130.
115. Murai, S., Ryu, I., Kadono, Y., Katayama, H., Kondo, K., and Sonoda, N. (1977). *Chem. Lett.*, 1219.
116. a. Takaki, K., Okada, M., Yamada, M., and Negoro, K. (1982). *J. Org. Chem.* **47**, 1200. b. Takaki, K., Yamada, M., and Negoro, K. (1982). *J. Org. Chem.* **47**, 5246.
117. Schmidt, R. R., and Maier, M. (1985). *Tetrahedron Lett.* **26**, 2065.
118. Leyendecker, F., and Comte, M.-T. (1982). *Tetrahedron Lett.* **23**, 5031.
119. Eskenazi, C., and Maitte, P. (1974). *Comptes Rendus Acad. Sci. Paris C* **279**, 233; for comparison see Bertri, G., Catelani, G., Magi, S., and Monti, L. (1980). *Gazz. Chim. Ital.* **110**, 173.
120. Berti, G., Catelani, G., Colonna, F., and Monti, L. (1982). *Tetrahedron* **38**, 3067; Anselmi, C., Catelani, G., and Monti, L. (1983). *Gazz. Chim. Ital.* **113**, 167.
121. Scarpati, R., Sica, D., and Santacroce, C. (1964). *Tetrahedron* **20**, 2735; Bargagna, A., Evangelisti, F., and Schenone, P. (1979). *J. Heterocycl. Chem.* **16**, 93; Mosti, L., Schenone, P., and Menozzi, G. (1978). *J. Heterocycl. Chem.* **15**, 181.
122. Opitz, G., and Zimmerman, F. (1964). *Chem. Ber.* **97**, 1266.
123. Opitz, G., and Tempel, E. (1966). *Liebigs Ann. Chem.* **699**, 74.
124. Martin, J. C., Barton, K. R., Gott, P. G., and Meen, R. H. (1966). *J. Org. Chem.* **31**, 943.
125. Brady, W. T., and Shieh, C. H. (1984). *J. Heterocycl. Chem.* **21**, 1337; Brady, W. T., and Agho, M. O. (1983). *J. Org. Chem.* **48**, 5337; Brady, W. T., and Agho, M. O. (1983). *J. Heterocycl. Chem.* **20**, 501; Brady, W. T., and Watts, R. D. (1983). *J. Org. Chem.* **46**, 4047.

126. Bignardi, G., Evangelisti, F., Schenone, P., and Bargagna, A. (1972). *J. Heterocycl. Chem.* **9**, 1071; Gandini, A., Schenone, P., and Bignardi, G. (1967). *Monatsh. Chem.* **98**, 1518; Bignardi, G., Schenone, P., and Evangelisti, F. (1971). *Ann. Chim. (Rome)* **61**, 326; Evangelisti, F., Bignardi, G., Bargagna, A., and Schenone, P. (1978). *J. Heterocycl. Chem.* **15**, 511; Schenone, P., Evangelisti, F., Bignardi, G., and Bargagna, A. (1974). *Ann. Chim. (Rome)* **64**, 613; Mosti, L., Bignardi, G., Evangelisti, F., and Schenone, P. (1976). *J. Heterocycl. Chem.* **13**, 1201; Bargagna, A., Cafaggi, S., and Schenone, P. (1980). *J. Heterocycl. Chem.* **17**, 507; Bargagna, A., Evangelisti, F., and Schenone, P. (1981). *J. Heterocycl. Chem.* **18**, 111; Schenone, P., Bargagna, A., Bondavalli, F., and Longobardi, M. (1982). *J. Heterocycl. Chem.* **19**, 257; Schenone, P., Menozzi, G., Mosti, L., and Cafaggi, S. (1982). *J. Heterocycl. Chem.* **19**, 937; Schenone, P., Mosti, L., Menozzi, G., and Cafaggi, S. (1982). *J. Heterocycl. Chem.* **19**, 1031; Schenone, P., Mosti, L., Menozzi, G., and Romussi, G. (1982). *J. Heterocycl. Chem.* **19**, 1227; Schenone, P., Mosti, L., and Menozzi, G. (1982). *J. Heterocycl. Chem.* **19**, 1355; Menozzi, G., Mosti, L., and Schenone, P. (1983). *J. Heterocycl. Chem.* **20**, 539; Bargagna, A., Bignardi, G., Schenone, P., and Longobardi, M. (1983). *J. Heterocycl. Chem.* **20**, 839; Bargagna, A., Schenone, P., Bignardi, G., and Longobardi, M. (1983). *J. Heterocycl. Chem.* **20**, 1549; Mosti, L., Menozzi, G., and Schenone, P. (1984). *J. Heterocycl. Chem.* **21**, 361; Menozzi, G., Mosti, L., and Schenone, P. (1984). *J. Heterocycl. Chem.* **21**, 1441; Bargagna, A., Schenone, P., and Longobardi, M. (1985). *J. Heterocycl. Chem.* **22**, 1471.
127. Bargagna, A., Schenone, P., Bondavalli, F., and Longobardi, M. (1980). *J. Heterocycl. Chem.* **17**, 1201; Bargagna, A., Schenone, P., Bondavalli, F., and Longobardi, M. (1980). *J. Heterocycl. Chem.* **17**, 33.
128. Truce, W. E., Abraham, D. J., and Son, P. (1967). *J. Org. Chem.* **32**, 990.
129. Schenone, P., Mosti, L., and Bignardi, G. (1976). *J. Heterocycl. Chem.* **13**, 225; Schenone, P., Bignardi, G., and Morasso, S. (1972). *J. Heterocycl. Chem.* **9**, 1341.
130. Opitz, G., and Tempel, E. (1964). *Angew. Chem. Int. Ed. Engl.* **3**, 754; Opitz, G., and Tempel, E. (1966). *Liebigs Ann. Chem.* **699**, 68, 74.
131. Snider, B. B., and Duncia, J. V. (1980). *J. Org. Chem.* **45**, 3461.
132. Cookson, R. C., and Tuddenham, R. M. (1978). *J. Chem. Soc., Perkin Trans 1*, 678.
133. Snider, B. B., Karras, M., Price, R. T., and Rodini, D. J. (1982). *J. Org. Chem.* **47**, 4538; see also Naves, Y.-R., and Ardizio, P. (1953). *Bull. Soc. Chim. Fr.*, 494; Naves, Y.-R., Wahl, R., Ardizio, P., and Favre, C. (1953). *Bull. Soc. Chim. Fr.*, 873; Berkoff, C. E., and Chrombie, L. (1960). *J. Chem. Soc.*, 3734.
134. Tietze, L.-F., and von Kiedrowski, G. (1981). *Tetrahedron Lett.* **22**, 219.
135. Tietze, L.-F., von Kiedrowski, G., and Berger, B. (1982). *Angew. Chem. Int. Ed. Engl.* **21**, 221; Tietze, L.-F., von Kiedrowski, G., and Berger, B. (1982). *Tetrahedron Lett.* **23**, 51.
136. Tietze, L.-F., von Kiedrowski, G., Harms, K., Clegg, W., and Sheldrick, G. (1980). *Angew. Chem. Int. Ed. Engl.* **19**, 134.
137. Tietze, L.-F., Stegelmeier, H., Harms, K., and Brumby, T. (1982). *Angew. Chem. Int. Ed. Engl.* **21**, 863.
138. Ferreira, V. F., Coutada, L. C., Pinto, M. C. F. R., and Pinto, A. V. (1982). *Synth. Commun.* **12**, 195.
139. Schneider, R. A., and Meinwald, J. (1967). *J. Am. Chem. Soc.* **89**, 2023.
140. Kobayashi, Y., Hanzawa, Y., Miyashita, W., Kashiwagi, T., Nakano T., and Kumadaki, I. (1979). *J. Am. Chem. Soc.* **101**, 6445.
141. Martin, S. F., and Benage, B. (1984). *Tetrahedron Lett.* **25**, 4863; Martin, S. F., Benage, B., Williamson, S. A., and Brown, S. P. (1986). *Tetrahedron* **42**, 2903.

142. Snider, B. B., Roush, D. M., and Killinger, T. A. (1979). *J. Am. Chem. Soc.* **101,** 6023; Snider, B. B., and Roush, D. M. (1979). *J. Org. Chem.* **44,** 4229.
143. Tietze, L.-F., and Beifuss, U. (1986). *Tetrahedron Lett.* **27,** 1767.
144. Denmark, S. E., unpublished observations.
145. Tietze, L.-F. (1984). *In* "Selectivity—A Goal for Synthetic Efficiency" (W. Bartmann and B. M. Trost, eds.), pp. 299–316. Verlag Chemie, Weinheim.
146. Takana, S., Shigeki, S., and Ogasawara, K. (1985). *Heterocycles* **23,** 41.
147. Hultzsch, K. (1941). *Chem. Ber.* **74,** 898, 1539; Hultzsch, K. (1948). *Angew. Chem.* **60,** 179; Wakschmann, M., and Vilkas, M. (1964). *Comptes Rendus Acad. Sci. Paris C* **258,** 1526.
148. Gardner, P. D., Rafsanjani, H. S., and Land, L. (1959). *J. Am. Chem. Soc.* **81,** 3364.
149. Brudigou, J., and Christol, H. (1963). *Comptes Rendus Acad. Sci. Paris C* **256,** 3149, 3326; Brudigou, J., and Christol, H. (1962). *Bull. Soc. Chim. Fr.,* 1284.
150. von Strandtmann, M., Cohen, M. P., and Shavel, J., Jr. (1970). *J. Heterocycl. Chem.* **7,** 1311; von Strandtmann, M., Cohen, M. P., and Shavel, J., Jr. (1965). *Tetrahedron Lett.,* 3103; von Strandtmann, M., Cohen, M. P., and Shavel, J., Jr. (1965). *J. Org. Chem.* **30,** 3240; Quagkaro, R., Moreau, M., and Dreux, J. (1963). *Comptes Rendus Acad. Sci. Paris C,* 2843.
151. Merten, R., and Muller, G. (1964). *Chem. Ber.* **97,** 682; Paquette, L. A., and Stucki, H. (1966). *J. Org. Chem.* **31,** 1232.
152. Bolon, D. A. (1970). *J. Org. Chem.* **35,** 715, 3666.
153. Moreau, M., Quagliaro, R., Longeray, R., and Dreux, J. (1968). *Bull. Soc. Chim. Fr.,* 4251.
154. Chapman, O. L., and McIntosh, C. L. (1971). *J. Chem. Soc., Chem. Commun.,* 383.
155. Gutsche, C. D., and Oude-Alink, B. A. M. (1968). *J. Am. Chem. Soc.* **90,** 5855.
156. Dimerization: Cavitt, S. B., Sarrafizadeh, H., Gardner, R., and Gardner, P. D. (1962). *J. Org. Chem.* **27,** 1211; Gardner, P. D., and Serrafizadeh, H. (1960). *J. Org. Chem.* **25,** 641; Schonberg, A., Schutz, G., and Latif, N. (1961). *Chem. Ber.* **94,** 2540; Merijan, A., Shoulders, B. A., and Gardner, P. D. (1963). *J. Org. Chem.* **28,** 2148; Brown, T. L., Curtin, D. Y., and Fraser, R. R. (1958). *J. Am. Chem. Soc.* **80,** 4339; Smith, L. I., Tess, R. W. H., and Ullyot, G. E. (1944). *J. Am. Chem. Soc.* **66,** 1320; see also Refs. 158, 163, 164, and 166.
157. Gardner, P. D., Sarrapizadeh, H., and Rand, R. (1959). *J. Am. Chem. Soc.* **81,** 3364.
158. Dean, F. M., and Matkin, D. A. (1977). *J. Chem. Soc., Perkin Trans. 1,* 2289; Chauhan, M. S., Dean, F. M., McDonald, S., and Robinson, M. S. (1973). *J. Chem. Soc., Perkin Trans. 1,* 359.
159. Sheppard, W. A. (1968). *J. Org. Chem.* **33,** 3297.
160. Balasubramanian, K. K., and Selvaraj, S. (1980). *J. Org. Chem.* **45,** 3726.
161. Lanteri, P., Accary, A., Liu, R. P. T., Mathieu, D., and Longeray, R. (1981). *Bull. Soc. Chim. Fr. 2,* 415.
162. Lanteri, P., Longeray, R., and Royer, J. (1981). *J. Chem. Res. (Suppl.),* 168.
163. Chauhan, M. S., Dean, F. M., Matkin, D., and Robinson, M. S. (1973). *J. Chem. Soc., Perkin Trans. 1,* 120.
164. Chauhan, M. S., Dean, F. M., McDonald, S., and Robinson, M. S. (1973). *J. Chem. Soc., Perkin Trans. 1,* 359.
165. Silwa, M., Silwa, H., and Maitte, P. (1974). *Bull. Soc. Chim. Fr.,* 958.
166. Chauhan, M. S., and McKinnon, D. M. (1981). *Cand. J. Chem.* **59,** 2223.
167. Cacioli, P., Mackay, M. F., and Reiss, J. A. (1980). *Tetrahedron Lett.,* 4973.
168. Katada, T., Eguchi, S., Esaki, T., and Sasaki, T. (1984). *J. Chem. Soc., Perkin Trans. 1,* 2649.

169. Zagorevskii, V. A., Lopatina, K. I., and Klyuev, S. M. (1969). *Chem. Abstr.* **70**, 28929.
170. Chapman, O. L., Engel, M. R., Springer, J. P., and Clardy, J. C. (1971). *J. Am. Chem. Soc.* **93**, 6696; Matsumoto, M., and Kuroda, K. (1981). *Tetrahedron Lett.* **22**, 4437; see also Brophy, G. C., Mohandas, J., Slaytor, M., Sternhell, S., Watson, T. R., and Wilson, L. A. (1969). *Tetrahedron Lett.*, 5159.
171. Hug, R., Hansen, H.-J., and Schmid, H. (1972). *Helv. Chim. Acta* **55**, 1675.
172. Boekelheide, V., and Mao, Y.-L. (1980). *Proc. Natl. Acad. Sci. U.S.A.* **77**, 1732.
173. Oude-Alink, B. A. M., Chan, A. W. K., and Gutsche, C. D. (1973). *J. Org. Chem.* **38**, 1993.
174. Jones, D. W. (1972). *J. Chem. Soc., Perkin Trans. 1*, 225.
175. Talley, J. J. (1985). *J. Org. Chem.* **50**, 1695.
176. Funk, R. L., and Vollhardt, K. C. P. (1977). *J. Am. Chem. Soc.* **99**, 5483.
177. Marino, J. P., and Dax, S. L. (1984). *J. Org. Chem.* **49**, 3671.
178. Crombie, L., and Ponsford, R. (1971). *J. Chem. Soc. C*, 788; Crombie, L., and Ponsford, R. (1968). *J. Chem. Soc., Chem. Commun.*, 368; Crombie, L., and Ponsford, R. (1968). *Tetrahedron Lett.*, 4557; Begley, M. J., Crombie, L., Slack, D. A., and Whiting, D. A. (1976). *J. Chem. Soc., Chem. Commun.*, 140; Begley, M. J., Crombie, L., Slack, D. A., and Whiting, D. A. (1977). *J. Chem. Soc., Perkin Trans. 1*, 2402.
179. Crombie, L., and Ponsford, R. (1971). *J. Chem. Soc. C*, 796; Crombie, L., and Ponsford, R. (1968). *J. Chem. Soc., Chem. Commun.*, 894.
180. Kane, V. V., and Razdan, R. K. (1968). *J. Am. Chem. Soc.* **90**, 6551; Kane, V. V., and Razdan, R. K. (1969). *Tetrahedron Lett.*, 591.
181. Montero, J. L., and Winternitz, F. (1973). *Tetrahedron* **29**, 1243.
182. Mechoulam, R., Yagnitinsky, Y. Y., and Gaoni, Y. (1968). *J. Am. Chem. Soc.* **90**, 2418.
183. Clarke, D. G., Crombie, L., and Whiting, D. A. (1973). *J. Chem. Soc., Chem. Commun.*, 582; Crombie, L., Redshaw, S. D., and Whiting, D. A. (1979). *J. Chem. Soc., Chem. Commun.*, 630; Begley, M. J., Crombie, L., King, R. W., Slack, D. A., and Whiting, D. A. (1977). *J. Chem. Soc., Perkin Trans. 1*, 2393; Begley, M. J., Crombie, L., King, R. W., Slack, D. A., and Whiting, D. A. (1976). *J. Chem. Soc., Chem. Commun.*, 138.
184. Bandaranayake, W. M., Crombie, L., and Whiting, D. A. (1971). *J. Chem. Soc. C*, 804.
185. Clemens, R. J., and Hyatt, J. A. (1985). *J. Org. Chem.* **50**, 2431.
186. a. Hyatt, J. A., Feldman, P. L., and Clemens, R. J. (1984). *J. Org. Chem.* **49**, 5105. b. Andreichikov, Y. S., Nalimova, U. A., Kozlov, A. P., and Rusakov, I. A. (1978). *J. Org. Chem. USSR (Engl. Transl.)* **14**, 2245.
187. a. Stetter, H., and Kiehs, K. (1965). *Chem. Ber.* **98**, 2099. b. Jager, G. (1972). *Chem. Ber.* **105**, 137. c. Huynh, C., and Julia, S. (1972). *Bull. Soc. Chim. Fr.*, 1794. d. Dehmlow, E. V., and Shamout, A. R. (1982). *Liebigs Ann. Chem.*, 1753; Sato, M., Sekiguchi, K., Ogasawara, H., and Kaneko, C. (1985). *Synthesis*, 224; see Hyatt, J. A. (1984). *J. Org. Chem.* **49**, 5102.
188. Maujean, A., and Chuche, J. (1976). *Tetrahedron Lett.*, 2905.
189. Kollenz, G., Ziegler, E., and Ott, W. (1973). *Org. Prep. Proc. Int.* **5**, 261; Hunig, S., Benzing, E., and Hubner, K. (1961). *Chem. Ber.* **94**, 486.
190. Kollenz, G., Igel, H., and Ziegler, E. (1972). *Monatsh. Chem.* **103**, 450.
191. Jager, G. (1972). *Chem. Ber.* **105**, 137; Jager, B., and Wenzelburger, J. (1976). *Liebigs Ann. Chem.*, 1689.
192. Ziegler, E., Kollenz, G., and Ott, W. (1973). *Synthesis*, 679.
193. Kurzer, F., and Douraghi-Zadeh, K. (1967). *Chem. Rev.* **67**, 125.

194. Ficini, J., and Pouliquen, J. (1972). *Tetrahedron Lett.*, 1131; Ficini, J., and Pouliquen, J. (1972). *Tetrahedron Lett.*, 1135.
195. Capuano, L., Tammer, T., and Zander, R. (1976). *Chem. Ber.* **109**, 3497.
196. Sato, M., Ogasawara, H., Oi, K., and Kato, T. (1983). *Chem. Pharm. Bull. Tokyo* **31**, 1896; Sato, M., Ogasawara, H., Yoshizumi, E., and Kato, T. (1983). *Chem. Pharm. Bull. Tokyo* **31**, 1902.
197. Minami, T., Yamauchi, Y., Ohshiro, Y., Agawa, T., Murai, S., and Sonoda, N. (1977). *J. Chem. Soc., Perkin Trans. 1*, 904; Capuano, L., Urhahn, G., and Willmes, A. (1979). *Chem. Ber.* **112**, 1012.
198. Bertrand, M., and Le Gras, J. (1967). *Bull. Chim. Soc. Fr.*, 4336.
199. Saalfrank, R. W., Paul, W., and Schierling, P. (1980). *Chem. Ber.* **113**, 3477.
200. Finley, K. T. (1974). *In* "The Chemistry of Quinoid Compounds" (S. Patai, ed.), Vol. 2, Chapter 17. Wiley, New York.
201. Buldt, E., Debaerdemaeker, T., and Friedrichsen, W. (1980). *Tetrahedron* **36**, 267; Nunn, E. E., Wilson, W. S., and Warrener, R. N. (1972). *Tetrahedron Lett.*, 175; see also Warrener, R. N., Nunn, E. E., and Paddon-Row, M. N. (1976). *Tetrahedron Lett.*, 2355; Pritschins, W., and Grimme, W. (1979). *Tetrahedron Lett.*, 4545; Friedrichsen, W., Buldt, E., and Schmidt, R. (1975). *Tetrahedron Lett.*, 1137; Friedrichsen, W., Buldt, E., Betz, M., and Schmidt, R. (1974). *Tetrahedron Lett.*, 2469.
202. Friedrichsen, W., and Epbinder, R. (1973). *Tetrahedron Lett.*, 2059.
203. Lown, J. W., and Aidoo, A. S. K. (1966). *Can. J. Chem.* **44**, 2507; Bryce-Smith, D., and Gilbert, A. (1968). *J. Chem. Soc., Chem. Commun.*, 1701, 1702.
204. Scott, W., Joseph, T. C., and Chow, Y. L. (1976). *J. Org. Chem.* **41**, 2223; Chow, Y. L., Joseph, T. C., Quon, H. H., and Tam, J. N. S. (1970). *Can. J. Chem.* **48**, 3045.
205. Bryce-Smith, D., and Gilbert, A. (1968). *J. Chem. Soc., Chem. Commun.*, 1702.
206. Ried, W., and Torok, E. (1964). *Naturwissenschaften* **51**, 265.
207. Friedrichsen, W. (1969). *Tetrahedron Lett.*, 4425.
208. Horspool, W. M., Tedder, J. M., and Din, Z. U. (1969). *J. Chem. Soc. C*, 1692, 1694.
209. Friedrichsen, W. (1969). *Tetrahedron Lett.*, 4425.
210. Ried, W., and Radt, W. (1965). *Liebigs Ann. Chem.* **688**, 170, 174.
211. Allmann, R., Debaerdemaeker, T., Friedrichsen, W., Jurgens, H. J., and Betz, M. (1976). *Tetrahedron* **32**, 147; Friedrichsen, W., Schroer, W.-D., and Schmidt, R. (1976). *Liebigs Ann. Chem.*, 793; Friedrichsen, W., Betz, M., Buldt, E., Jurgens, H.-J., Schmidt, R., Schwarz, I., and Visser, K. (1978). *Liebigs Ann. Chem.*, 440.
212. Friedrichsen, W., Kallweit, I., and Schmidt, R. (1977). *Liebigs Ann. Chem.*, 116.
213. Omote, Y., Tomotake, A., and Kashima, C. (1984). *Tetrahedron Lett.* **25**, 2993; Komatsu, T., Nishio, T., and Omote, Y. (1978). *Chem. Ind. (London)*, 95.
214. Saito, K., Iida, S., and Mukai, T. (1982). *Heterocycles* **19**, 1197.
215. Heinicke, J., and Tzschach, A. (1983). *Tetrahedron Lett.* **24**, 5481.
216. Ansell, M. F., and Leslie, V. J. (1971). *J. Chem. Soc. C*, 1423.
217. Ansell, M. F., Bignold, A. J., Gosden, A. F., Leslie, V. J., and Murray, R. A. (1971). *J. Chem. Soc. C*, 1414; Ansell, M. F., and Leslie, V. J. (1971). *J. Chem. Soc. C*, 1423.
218. Vogel, E., Ippen, J., and Buch, V. (1975). *Angew. Chem. Int. Ed. Engl.* **14**, 566.
219. Herndon, W. C., and Giles, W. B. (1969). *J. Chem. Soc., Chem. Commun.*, 497.
220. Bakker, C. G., Scheeren, J. W., and Nivard, R. J. F. (1983). *Recl. Trav. Chim. Pays-Bas* **102**, 97. Photochemical cycloadditions: Muruyama, K., Muraoka, M., and Naruta, Y. (1981). *J. Org. Chem.* **46**, 983; Barlow, M. G., Coles, B., and Haszeldine, R. N. (1980). *J. Chem. Soc., Perkin Trans. 1*, 2523.
221. Boger, D. L. (1983). *Tetrahedron* **39**, 2869.
222. Schmidt, R. R. (1973). *Angew. Chem. Int. Ed. Engl.* **12**, 212.

223. Bradsher, C. K. (1974). *Adv. Heterocycl. Chem.* **16**, 289.
224. Schmidt, R. R. (1969). *Tetrahedron Lett.,* 5279; Wakselman, M., and Vilkas, M. (1964). *Comptes Rendus Acad. Sci. Paris C* **258**, 1526.
225. Lopatina, K. J., Klynev, S. M., and Zagoreuskii, V. A. (1970). *Khim. Geterotsikl. Soedin,* 43.
226. Schmidt, R. R. (1965). *Chem. Ber.* **98**, 344; Schmidt, R. R. (1964). *Angew. Chem. Int. Ed. Engl.* **3**, 387; see also Schroth, W., Fischer, G. W., and Rottman, J. (1969). *Chem. Ber.* **102**, 1202. Comparable reactions with nitriles: Schmidt, R. R. (1965). *Chem. Ber.* **98**, 3892; Lora-Tamayo, M., Madronero, R., Munoz, G. G., and Leipprand, H. (1964). *Chem. Ber.* **97**, 2234. Related observations: Schmidt, R. R., Schwille, D., and Sommer, U. (1969). *Liebigs Ann. Chem.* **723**, 111.
227. Schmidt, R. R. (1969). *Angew. Chem.* **81**, 576.
228. Schmidt, R. R. (1970). *Chem. Ber.* **103**, 3242.
229. Seeliger, W., and Diepers, W. (1966). *Liebigs Ann. Chem.* **697**, 171.

Chapter **8**

Thiabutadienes

INTRODUCTION

In contrast to the thorough studies of the Diels–Alder reactions of other heterodienes, the [4 + 2] cycloaddition of 1-thia-1,3-butadienes bearing a sulfur atom at the diene terminus have not been extensively investigated.[1] In a large measure this may be due to the difficulty currently encountered in the preparation, stability, and handling of the thiocarbonyl derivatives. In general, thiabutadienes that participate as effective 4π components of Diels–Alder reactions are electron-deficient systems and consequently should be ideally suited for regiospecific [4 + 2] cycloaddition reactions with electron-rich or strained dienophiles in inverse electron demand (LUMO$_{diene}$ controlled) Diels–Alder reactions. While this general feature of the Diels–Alder reactions of thiabutadienes has been recognized and experimentally verified, most investigations have detailed the 4π Diels–Alder reactions of thiabutadiene derivatives in normal (HOMO$_{diene}$ controlled) Diels–Alder reactions with typical, electron-deficient dienophiles. In such instances, the complementary addition of electron-donating sub-

214

stituents to C-2/C-4 of the 1-thia-1,3-butadiene system increases both the rate and regioselectivity of the normal (HOMO$_{diene}$ controlled) Diels–Alder reaction. In contrast to the results obtained with many oxabutadiene systems, simple FMO theory has been used to accurately predict the regioselectivity of the [4 + 2] cycloadditions of thiabutadienes.[2]

1. α,β-UNSATURATED THIOCARBONYL COMPOUNDS (1-THIABUTADIENES)

Thiocarbonyl compounds have been employed in Diels–Alder reactions to prepare thiopyranyl systems,[1] and generally the carbon–sulfur double bond serves as the 2π dienophile component of the [4 + 2] cycloaddition (Chapter 5). However, there are several reported examples of α,β-unsaturated thioaldehydes, thioketones, and dithioesters participating as the diene partners of Diels–Alder reactions in which the thiocarbonyl group comprises a component of the 4π diene system.

The *in situ* generation of methyl vinyl thioketone in refluxing pyridine provided each of the four possible mixed Diels–Alder products derived from 4π and 2π methyl vinyl thioketone participation in [4 + 2] cycloadditions with methyl vinyl ketone [Eq. (1)].[3]

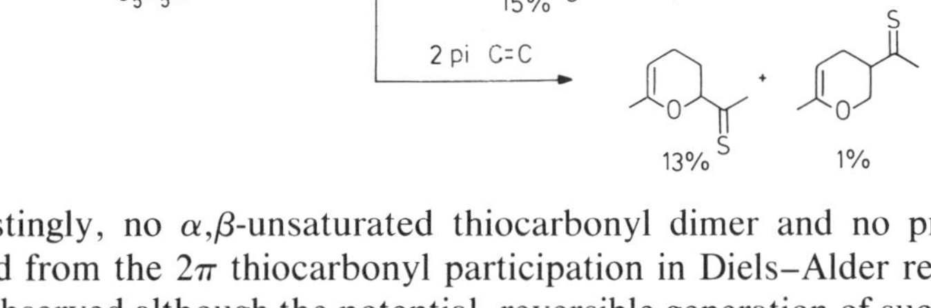

$$\tag{1}$$

Interestingly, no α,β-unsaturated thiocarbonyl dimer and no products derived from the 2π thiocarbonyl participation in Diels–Alder reactions were observed although the potential, reversible generation of such products could not be ruled out. In contrast, the low temperature dimerization of α,β-unsaturated thioketones including methyl vinyl thioketone, generated by flash vacuum pyrolysis (FVP), provided **1** derived from the clean, regiospecific Diels–Alder dimerization with 4π and 2π thiocarbonyl participation [Eq. (2)].[4]

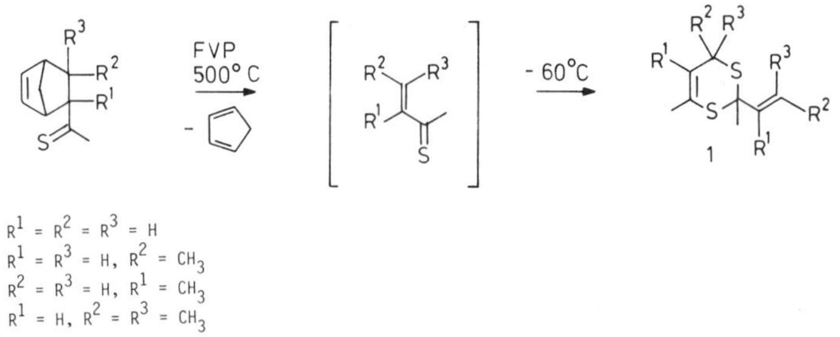

FVP
500° C
- 60°C
1
R1 = R2 = R3 = H
R1 = R3 = H, R2 = CH3
R2 = R3 = H, R1 = CH3
R1 = H, R2 = R3 = CH3

(2)

The *in situ* generation of the α,β-unsaturated thioaldehyde, thiomethacrolein, has been reported to provide a 2 : 1 ratio of the Diels–Alder dimerization products **2** and **3** derived from the 4π and carbon–carbon double bond 2π participation of thiomethacrolein in a [4 + 2] cycloaddition which proceeds with little regioselectivity [Eq. (3)].[5]

P4S10
C5H5N
Δ
[3,3]
210°C
10 min
4
5
6
CHO
2
2:1
3

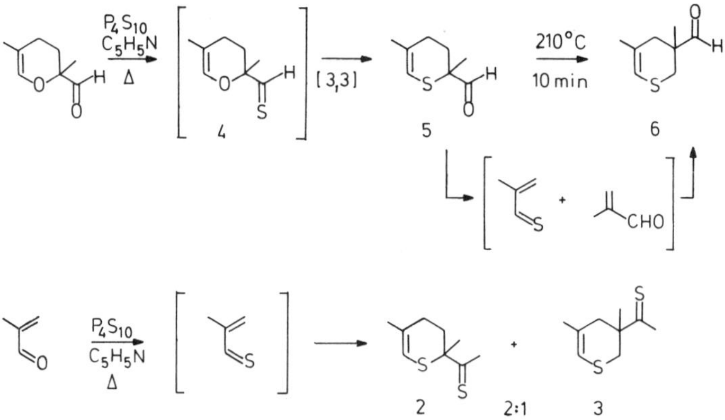

(3)

The *in situ* conversion of methacrolein dimer to the corresponding thioaldehyde is followed by a [3,3]-sigmatropic rearrangement, providing **5.** On prolonged warming (110°C) or on rapid thermolysis, **5** cleanly isomerizes to **6** via a retro Diels–Alder reaction and regioselective [4 + 2] recombination [Eq. (3)].[5] Consequently, the accurate interpretation of the preferred kinetic mode and regioselectivity of the Diels–Alder dimerization reactions of α,β-unsaturated thioketones and thioaldehydes presently is clouded by the reversible nature of many of the observed cycloadditions and the potential participation of the products in subsequent, low temperature rearrangements.

The C-2/C-4 addition of conjugating substituents (e.g., thiochalcone and related compounds)[6,7] or electron-donating substituents (e.g., β-enaminothioketones,[8–15] -thioaldehydes,[15] -dithioesters,[19,20] and α,β-unsaturated dithioesters[16–18] or thioamides[21]) to the α,β-unsaturated thiocarbonyl compounds stabilizes the monomeric forms. It also slows the [4 + 2] dimerization reaction, improves the 4π participation in Diels–Alder reactions with typical electron-deficient dienophiles, and predictably improves the observed regioselectivity of the intermolecular [4 + 2] cycloadditions.   The initial observations[6] on the reversible and apparently regioselective dimerization of thiochalcones and related α,β-unsaturated thioketones [Eq. (4)]

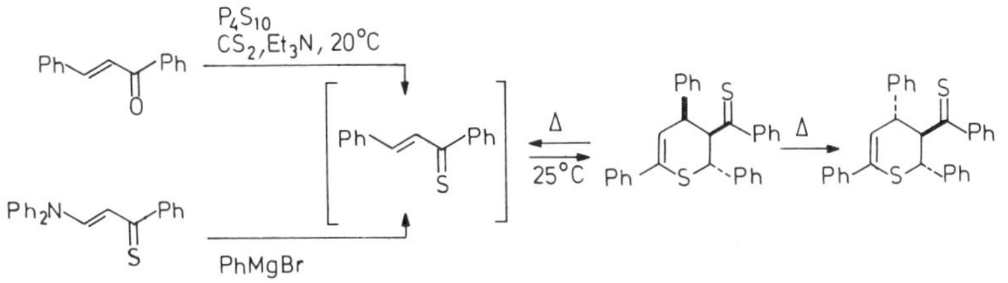

$$(4)$$

have been carefully investigated.[6,7] The kinetic [4 + 2] dimerization products are derived from 4π and 2π carbon–carbon double bond participation of the α,β-unsaturated thioketone in a Diels–Alder reaction which proceeds with complete regioselectivity exclusively through an endo transition state. Prolonged warming of the kinetic [4 + 2] dimer provides the thermodynamically more stable product derived from a regiospecific, exo Diels–Alder reaction [Eq. (4)]. Warming the dimers reversibly regenerate the monomeric α,β-unsaturated thioketones, which have been shown to participate in regioselective Diels–Alder reactions with a range of dienophiles, including electron-deficient,[7a,b] strained,[7b] and electron-rich olefins[7b] as well as cumulenes,[7c] to provide dihydrothiopyrans in good to excellent yield (Scheme 8-I).

The formal 4π participation of the monomeric 2-arylmethylene-1-tetralinthiones in Diels–Alder reactions with isoprene and (E,E)-1,4-diphenyl-1,3-butadiene provides 3,4-dihydro-2H-thiopyran derivatives [Eq. (5)].[7d]

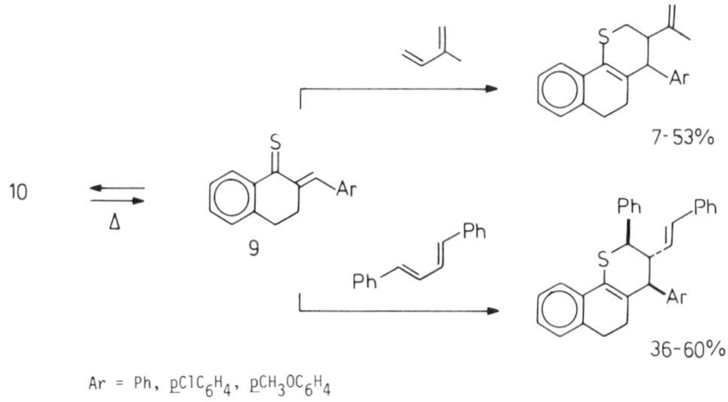

$$(5)$$

Ar = Ph, p-ClC$_6$H$_4$, p-CH$_3$OC$_6$H$_4$

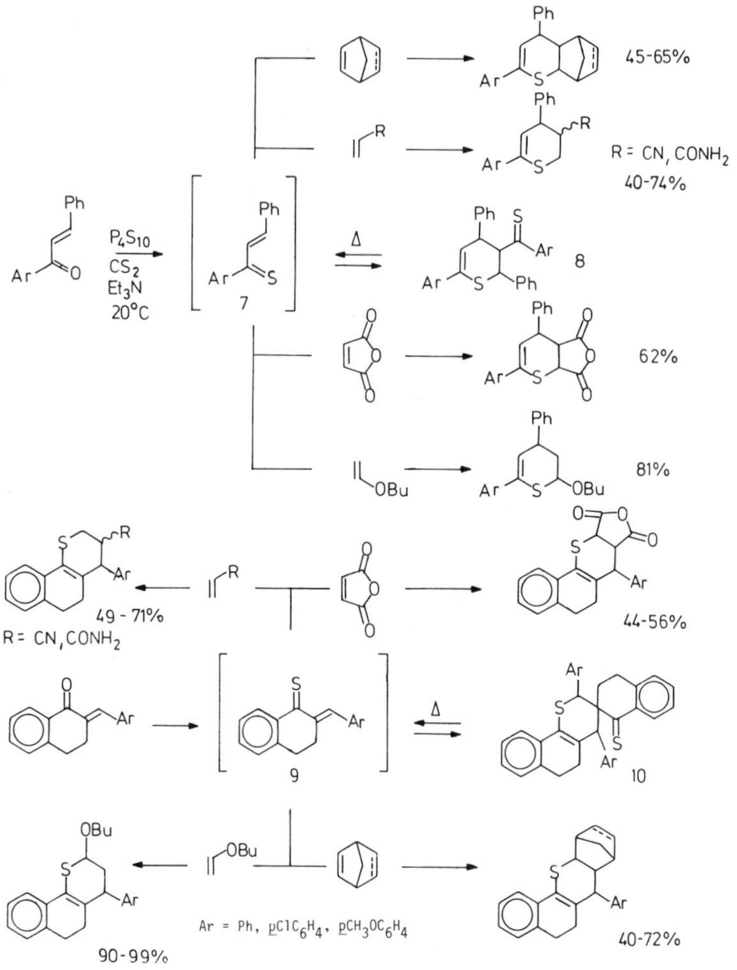

Scheme 8-I

At present it has not been determined unambiguously whether this reaction constitutes an example of direct α,β-unsaturated thioketone 4π participation in a [4 + 2] cycloaddition or 2π thiocarbonyl Diels–Alder reaction with the dienes followed by an undetected [3,3]-sigmatropic rearrangement.[7d]

β-Enaminothioketones and β-enaminothioaldehydes (α,β-unsaturated thiocarbonyl compounds bearing a C-4 electron-donating substituent have been the most thoroughly investigated thiabutadiene system capable of regiospecific 4π Diels–Alder reactions with typical, electron-deficient dienophiles[8–11] or reactive olefins including ketenes[12,14] and sulfenes.[12,13] The β-enaminothiocarbonyl compounds do exist as monomers at room temperature, and those bearing C-4 secondary amines exist predominantly in the S-cis diene conformation, thus facilitating their 4π participation in Diels–Alder reactions.[8] Proper choice of the reaction conditions often permits the clean isolation of the primary Diels–Alder products, 4-aminodihydrothiopyrans, or secondary products derived from a subsequent elimination of the C-4 amino group (Scheme 8-II). In the carefully examined cases, the regiospecific Diels–Alder cycloadditions have been shown to proceed preferentially through an endo [4 + 2] transition state. Unambiguous labeling studies have determined that the [4 + 2] Diels–Alder reaction mechanism is operative for the reaction of the β-enaminothioketones with reactive, electron-deficient alkynes. The plausible and precedented alternative mechanism of [2 + 2] enamine–alkyne cycloaddition followed by subsequent electrocyclic cylobutene ring opening and an electrocyclic thiahexatriene ring closure is not observed[9b] (cf. **11**). The extension of these observations to the generation and subsequent Diels–Alder reactions of β-enaminothioaldehydes has been reported [Eq. (6)].[15]

$$(6)$$

The Diels–Alder dimerization of α,β-unsaturated dithioesters has been reported to proceed with complete regioselectivity and exclusively through an endo [4 + 2] transition state to provide the cycloadducts, e.g., **12**, derived from 4π and 2π carbon–carbon double bond participation of E-α,β-unsaturated dithioesters [Eq. (7)].[16–18]

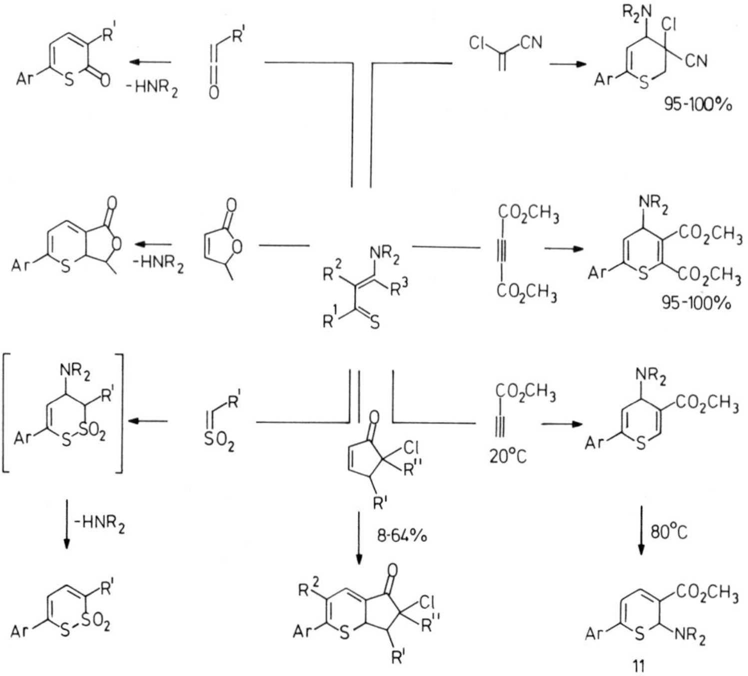

$NR_2 = N(CH_2)_4$, $N(CH_2)_3$, $NHPh$, NMe_2
$Ar = Ph$, $pClC_6H_4$, $pCH_3OC_6H_4$, $pNO_2C_6H_4$, $pBrC_6H_4$
$R' = Ph$, H

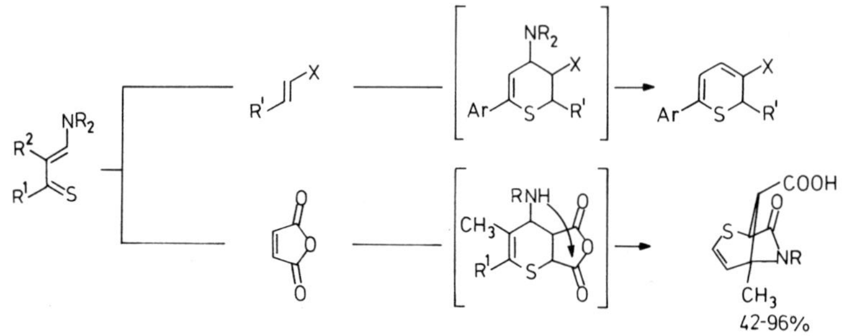

$NR_2 = N(CH_2)_4$, NEt_2; $R^1 = Ar$; $R^2 = H$
$Ar = Ph$, $pClC_6H_4$, $pCH_3OC_6H_4$
$X = CN$, $CONH_2$, CO_2CH_3, CHO, $COCH_3$, $CONR_2$; $R' = H$, OEt, SEt

$NR_2 = NHPh$, $NHCH_2Ph$, NH_2, $NHCH_3$
$R^1 = CH_3$, Ph; $R^2 = CH_3$

Scheme 8-II

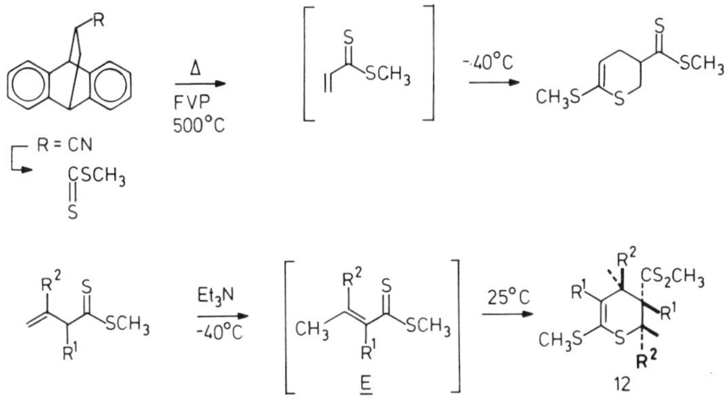

$$(7)$$

With the development of dependable approaches to the generation of α,β-unsaturated dithioesters[16-18] and with the recognition that additional alkyl substitution stabilizes the monomeric form,[18] a study of their intermolecular Diels–Alder reactions has been described (Scheme 8-III).[18] Representative electron-deficient, strained, and electron-rich olefins participate in [4 + 2] cycloadditions with the *in situ* generated α,β-unsaturated dithioesters. The HOMO$_{diene}$ controlled Diels–Alder reactions proceed preferentially through an endo [4 + 2] transition state, and the LUMO$_{diene}$ con-

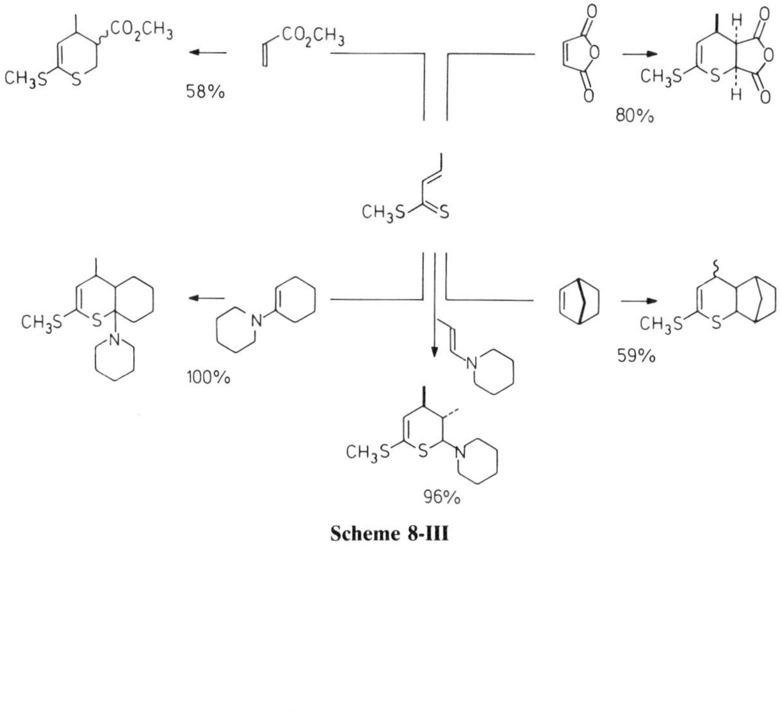

Scheme 8-III

trolled Diels–Alder reactions with electron-rich olefins proceed with the greatest facility.

The reaction of methyl dithiocrotonate with cyclopentadiene was shown to provide the cycloadduct **13**. The initial assumption that the product was derived from dithioester 2π participation in an all-carbon Diels–Alder reaction with cyclopentadiene followed by a [3,3]-sigmatropic rearrangement was shown not to be operative. Consequently, the generation of **13** resulted from the direct 4π participation of the α,β-unsaturated dithioester in a Diels–Alder reaction with cyclopentadiene [Eq. (8)].[18]

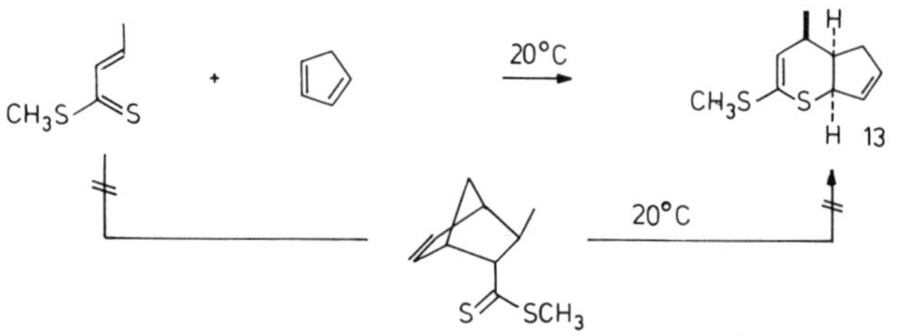

$$\text{(8)}$$

The stable, monomeric β-enaminodithioesters, like their thioketone counterparts, have been shown to participate in regiospecific, intermolecular Diels–Alder reactions with a range of representative electron-deficient and reactive dienophiles (Scheme 8-IV).[19,20] Cinnamic acid thioamides exist in equilibrium with their dimers at room temperature, although to date no reports of their cycloaddition with other dienophiles have been detailed [Eq. (9)].[21]

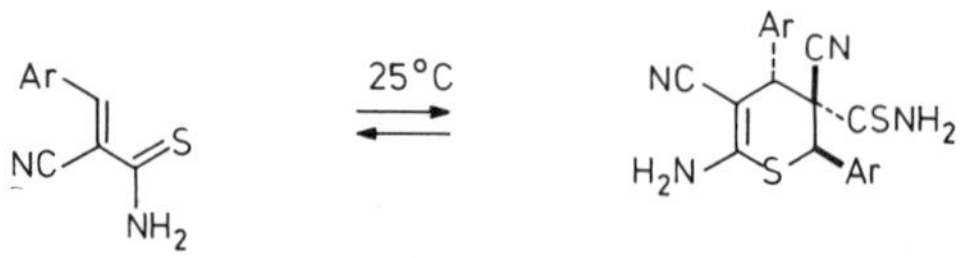

$$\text{(9)}$$

The [4 + 2] Diels–Alder dimerization is regiospecific and proceeds exclusively through an endo transition state.

An interesting and reactive class of 4π thiabutadienes capable of useful participation in normal ($\text{HOMO}_{\text{diene}}$ controlled) Diels–Alder reactions are the stable thioacylketene thioacetals,[22] generated by the photoaddition of olefins to 1,2-dithiole-3-thiones[23] [Eq. (10)].

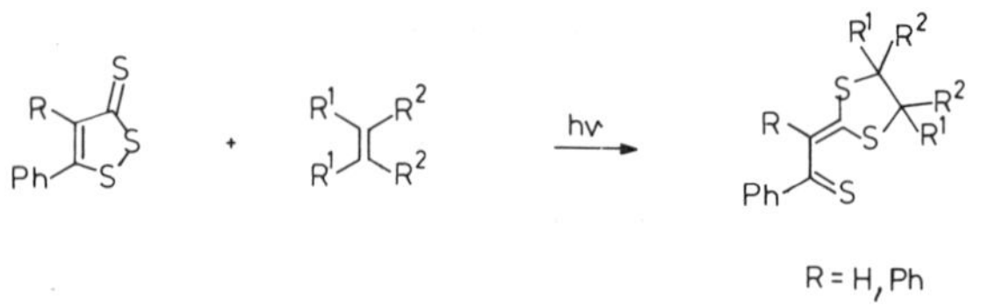

$$\text{(10)}$$

$$R = H, Ph$$

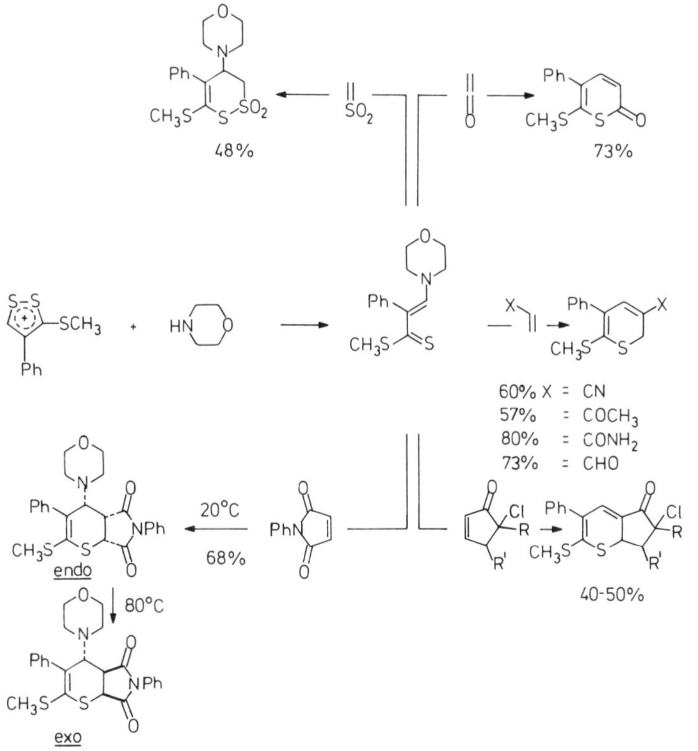

Scheme 8-IV

Olefinic, acetylenic, and typical electron-deficient dienophiles participate in apparent concerted [4 + 2] cycloadditions with the thioacylketene thioacetals (Scheme 8-V), while enamines provide modest yields of [4 + 2] cycloadducts via a polar, stepwise addition–cyclization.[22c]

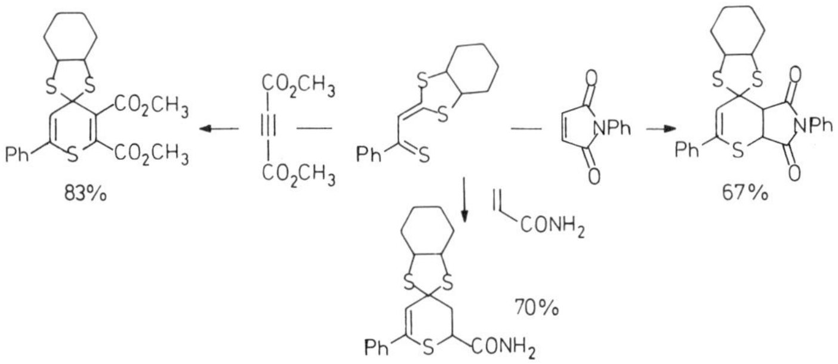

Scheme 8-V

The photochemical-promoted reaction of 1,3-dimethyl-2-thioparabanate with dimethyl acetylenedicarboxylate provided the 1:2 adduct **15** [Eq. (11)].[24]

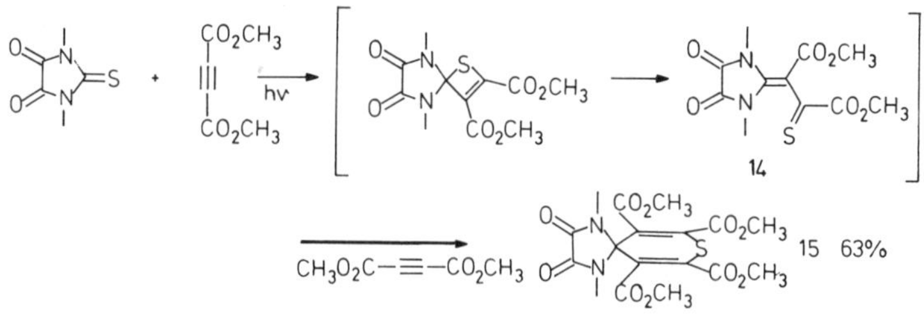

$$(11)$$

The final reaction of the sequence was the [4 + 2] cycloaddition of the thioacyl ketene aminal **14** with dimethyl acetylenedicarboxylate.

2. ARYL THIOKETONES

Since the initial report of the photochemical-promoted [4 + 2] cycloaddition of thiobenzophenone with acetylenic compounds [Eq. (12)],[25]

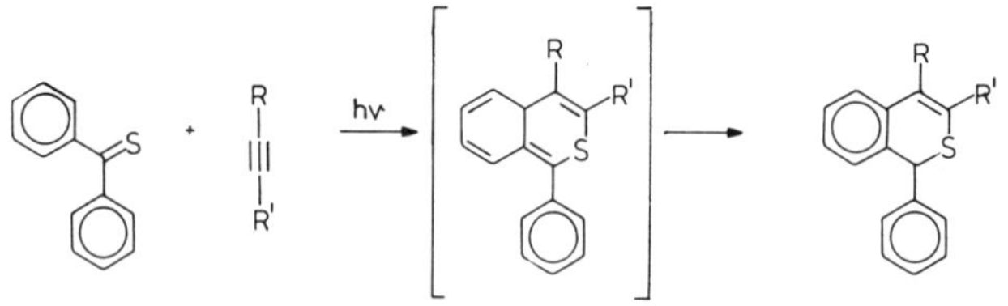

$$(12)$$

R	R'	
CO_2H	H	26%
Ph	H	56%
CH_2OH	H	28%
CN	H	20%
CO_2CH_3	CO_2CH_3	28%
OEt	H	0 %

a reaction initially investigated in anticipation of promoting a [2 + 2] acetylenic thioketone Paterno–Buchi reaction, a number of additional and selected examples of the photochemical and thermal 4π [4 + 2] cycloadditions of aryl thioketones have been described. The photochemical-promoted [4 + 2] cycloaddition does proceed by a stepwise, radical addition–cyclization,[25-27] and the thermal, dark Diels–Alder reactions of most aryl thioketones with reactive, electron-deficient alkynes including dimethyl acetylenedicarboxylate proceed by a stepwise, polar addition-cyclization.[26,27] Competitive [2 + 2] cycloaddition often is observed. Representa-

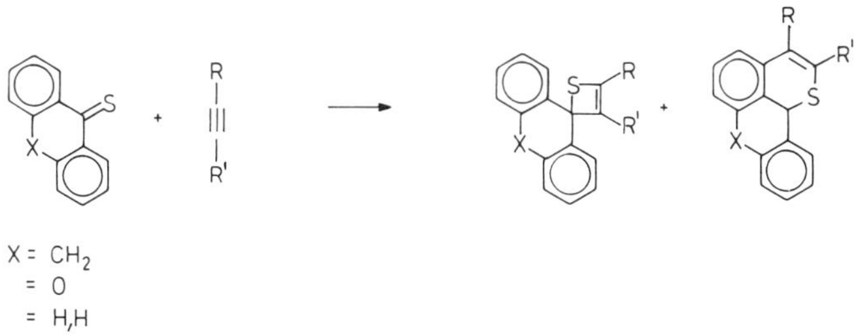

Scheme 8-VI

tive examples of the [4 + 2] cycloadditions of aryl thioketones are summarized in Scheme 8-VI.

Reports of the 4π participation of 2-thienyl and 2-furyl thioketones [Eq. (13)][28]

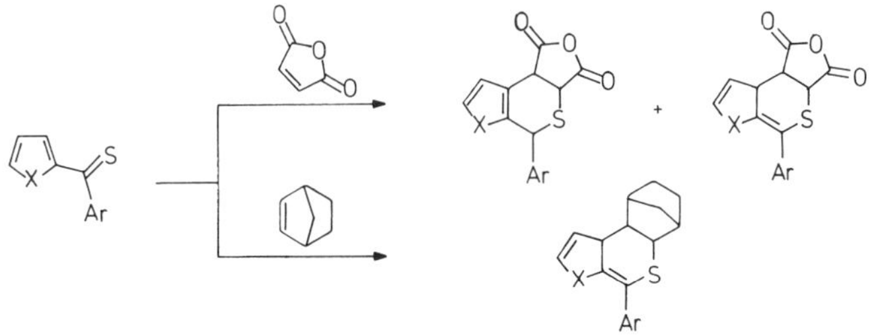

(13)

and of 1,8-naphthalic trithioanhydride [Eq. (14)][29]

(14)

in Diels–Alder reactions represent apparent concerted cycloaddition processes. The first two represent the 4π participation of an aryl thioketone in a normal (HOMO$_{diene}$ controlled) Diels–Alder reaction while the last represents the 4π participation of an aryl thiocarbonyl compound in an inverse electron demand (LUMO$_{diene}$ controlled) Diels–Alder reaction.

3. o-THIOBENZOQUINONE METHIDES

The efforts on the generation and subsequent *in situ* Diels–Alder trap of o-thiobenzoquinone methide (**16**) are limited.[30–32] The successful methods employed in the preparation of **16** are summarized in Scheme 8-VII, and

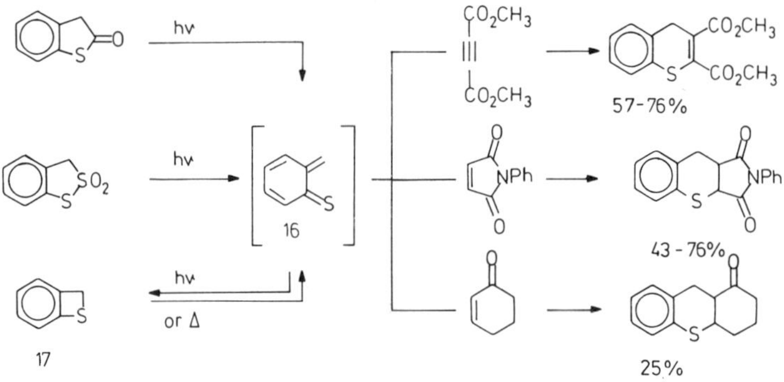

Scheme 8-VII

initial proof of its generation rested on a single successful Diels–Alder reaction with *N*-phenylmaleimide under selected conditions (benzene solvent).[30,31] Efforts to trap *o*-thiobenzoquinone methide (**16**) with other representative dienophiles or the use of alternative reaction solvents initially proved unsuccessful. The more recent reversible, thermal or photolytic generation of **16** from benzothiete (**17**) has proved to be the most dependable and convenient source of *o*-thiobenzoquinone methide, and its participation in a number of representative [4 + 2] cycloadditions with electron-deficient dienophiles including regioselective Diels–Alder reactions with unsymmetrical dienophiles has been described.[32]

A series of stabilized and isolable *o*-thiobenzoquinone methides and related *o*-thioquinone methides, initially prepared by the photochemical addition of olefins to 1,2-benzodithiole-3-thione [Eq. (15)],[33,34]

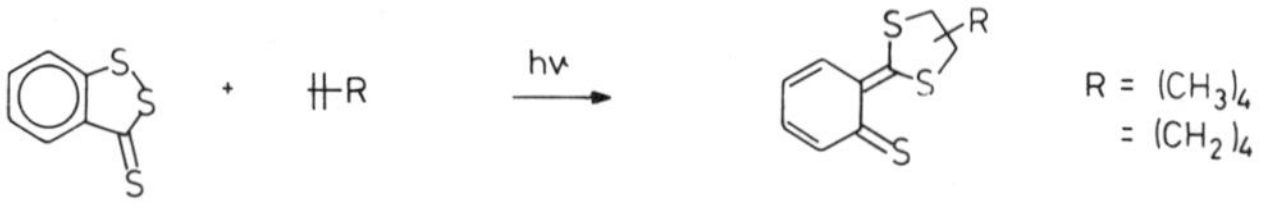

$$R = (CH_3)_4 = (CH_2)_4 \qquad (15)$$

have been shown to participate in a range of Diels–Alder reactions with representative olefinic and acetylenic dienophiles (Scheme 8-VIII).[33] The *o*-thiobenzoquinone methide ketene thioacetals, ketene aminals, and ketene acetals have been prepared and the scope of their Diels–Alder reactions examined. Electron-rich and reactive olefins including enamine and ketene derivatives provide Diels–Alder products derived from stepwise, polar addition–cyclization [4 + 2] reactions. Neutral and typical electron-deficient dienophiles appear to participate in concerted Diels–Alder reactions with the *o*-thiobenzoquinone methide ketene thioacetals and related *o*-thioquinone methide ketene acetal derivatives. Interestingly, no reports of the generation and subsequent *in situ* 4π heterodiene participation of *o*-

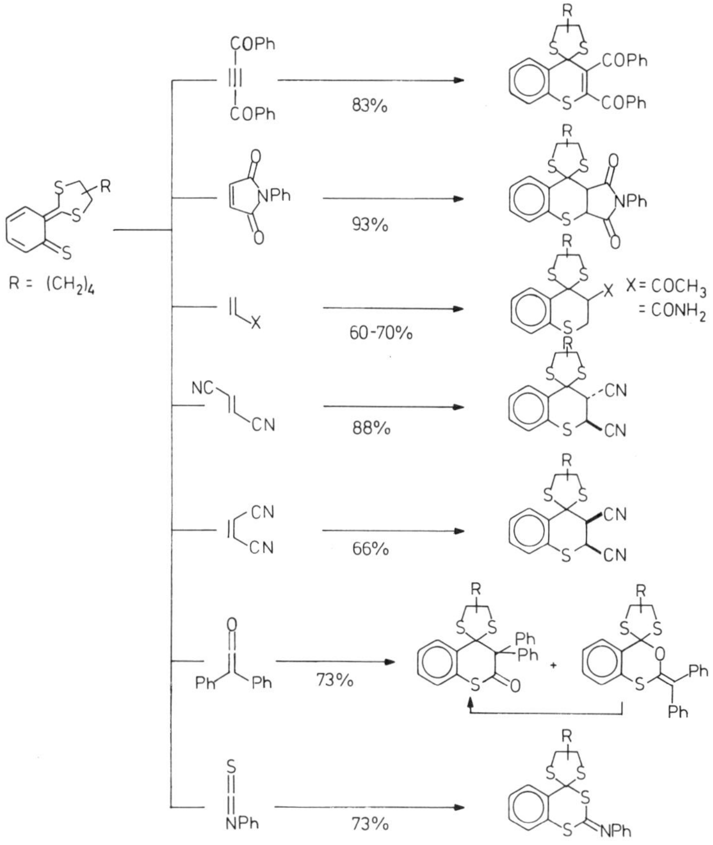

Scheme 8-VIII

monothiobenzoquinone or o-dithiobenzoquinone and related thio-o-quinones have been detailed.[35]

4. 1,2-DITHIOCARBONYL COMPOUNDS (1,4-DITHIABUTADIENES)

The reactivity of the thiocarbonyl group of 1,2-dithiocarbonyl compounds is well suited for 2π participation in Diels–Alder reactions with typical dienes (Chapter 5). In addition, frequent and well-defined Diels–Alder reactions with 4π participation of symmetrical 1,2-dithio-carbonyl compounds with olefinic and acetylenic dienophiles have been detailed.[36–39]

The stability of the α-dithiocarbonyl compounds follows closely their potential utility for 4π participation in Diels–Alder reactions with typical olefinic and acetylenic dienophiles [α-dithioamide $>$ α-dithionoester $>$ α-dithionothioester (dimer equilibrium at 25°C) $>$ α-dithione $>$ α-dithioaldehyde]. The α-dithiocarbonyl compounds are electron-deficient and consequently react rapidly with electron-rich and strained olefins in inverse electron demand (LUMO$_{diene}$ controlled) Diels–Alder reactions and more slowly with unactivated or electron-deficient dienophiles in apparent normal (HOMO$_{diene}$ controlled) Diels–Alder reactions (Scheme 8-IX).[36] Subtle differences in the reactivity and observed course of reactions of the various α-dithiocarbonyl compounds have been detailed.[36]

The *in situ* photochemical generation of 1,2-dithioketones and -aldehydes including dithioglyoxal, 3,3-dimethyl-2-thioxobutanethial, and camphor dithioquinone was confirmed by the subsequent trap of the 1,2-dithiocarbonyl compounds in [4 + 2] cycloadditions. The 1,2-dithioketones and 1,2-dithioaldehydes exhibited a preference for 4π participation in inverse electron demand (LUMO$_{diene}$ controlled) Diels–Alder reactions with electron-rich dienophiles (Scheme 8-X).[37]

The stable of 3,4-bis(trifluoromethyl)-1,2-dithietene (**18**)[38] [Eq. (16)]

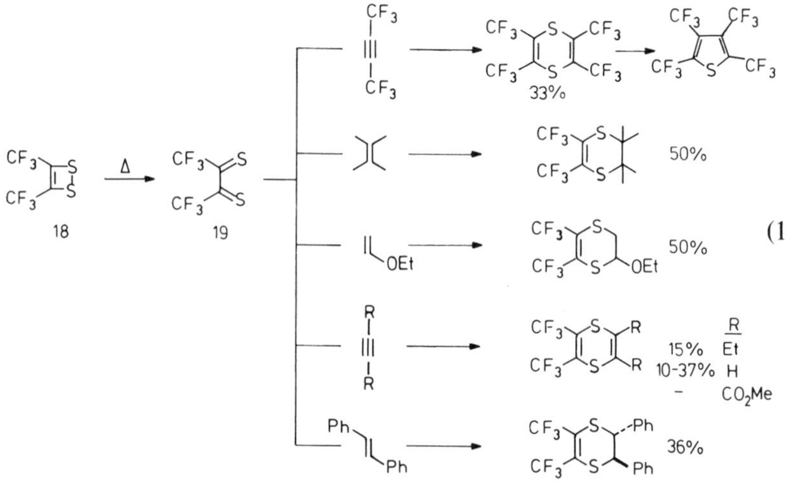

$$(16)$$

and the *in situ* generated 3,4-dicyano-1,2-dithietene (**20**)[39] [Eq. (17)]

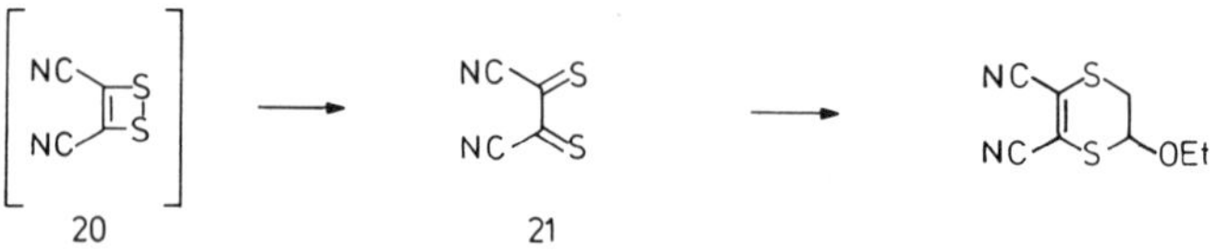

$$(17)$$

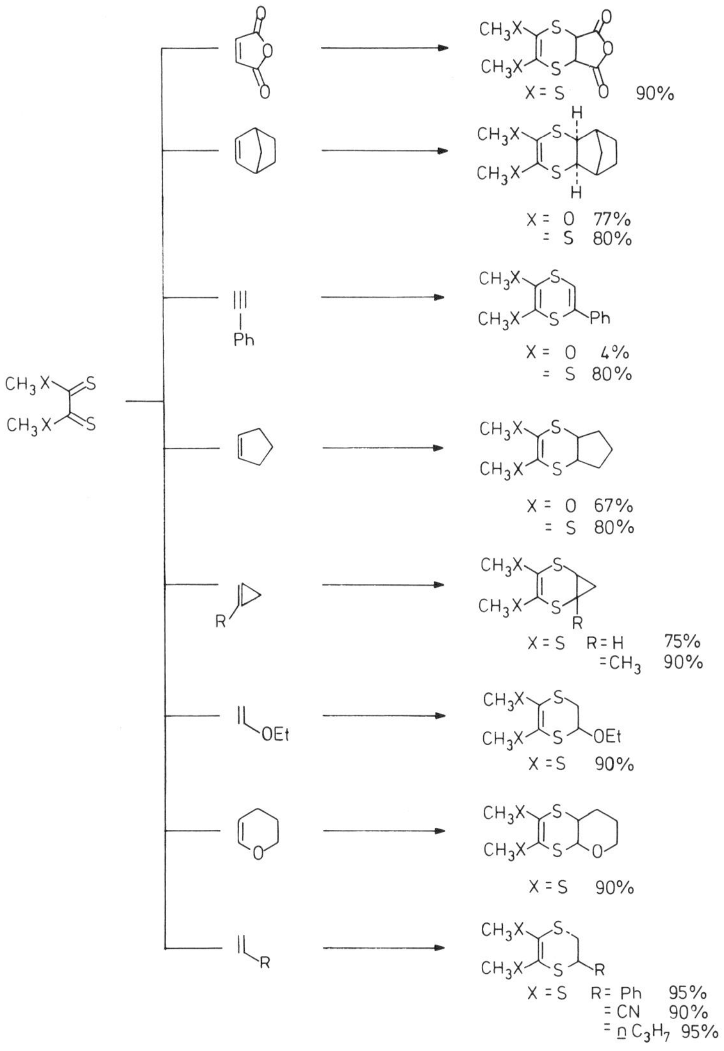

Scheme 8-IX

suffer thermal cyclobutene electrocyclic ring opening to provide the corresponding electron-deficient 1,2-dithiocarbonyl compounds **19** and **21**, which participate as 4π components in a range of [4 + 2] cycloadditions. Reactions with electron-rich dienophiles proceed with the greatest facility.

Apparently, the instability of simple 1,2-dithiocarbonyl compounds and

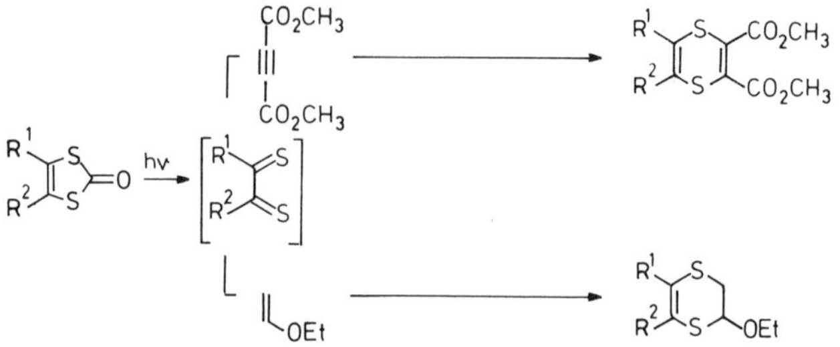

Scheme 8-X

the lack of dependable or general synthetic entries into this class of compounds have precluded additional studies including those that would provide regiochemical information derived from the reactions of unsymmetrical 1,2-dithiocarbonyl compounds with unsymmetrical dienophiles.

5. VINYL AND ACYL SULFINES, VINYL AND ACYL SULFENES

Electrophilic sulfenes, $R_2C=SO_2$, are recognized for their ability to react with nucleophiles and for their 2π participation in [2 + 2] and [4 + 2] cycloadditions (Chapter 5). Less well recognized is the demonstrated 4π participation of vinyl sulfenes in Diels–Alder reactions with a select set of dienophiles. Thermolysis of thiete 1,1-dioxides including the parent thiete 1,1-dioxide (**22**) in the presence of norbornenes provided the Diels–Alder products derived from 4π participation of the vinyl sulfenes [Eq. (18)].[40]

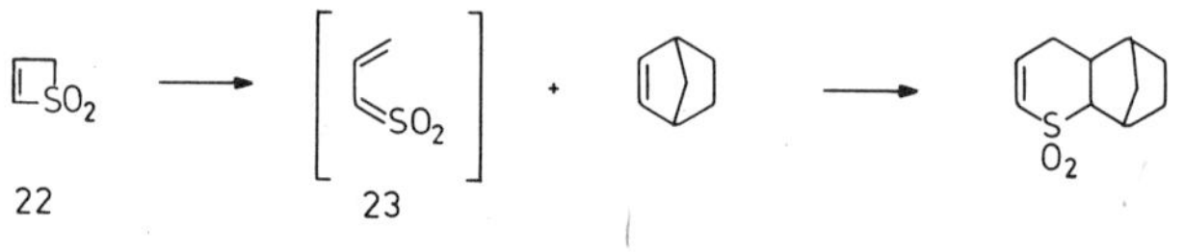

$$(18)$$

Neutral, unstrained olefins were essentially unreactive toward **23**,[40] and typical electron-rich olefins apparently provide low yields of products derived from [4 + 2] cycloaddition.[41]

The *in situ* generation of the α,β-unsaturated sulfine **24** via thermolytic retro Diels–Alder reaction of its dimer and subsequent reactions with the representative dienophiles, acrylonitrile and styrene, provided a regioisomeric mixture of Diels–Alder products in modest yields [Eq. (19)].

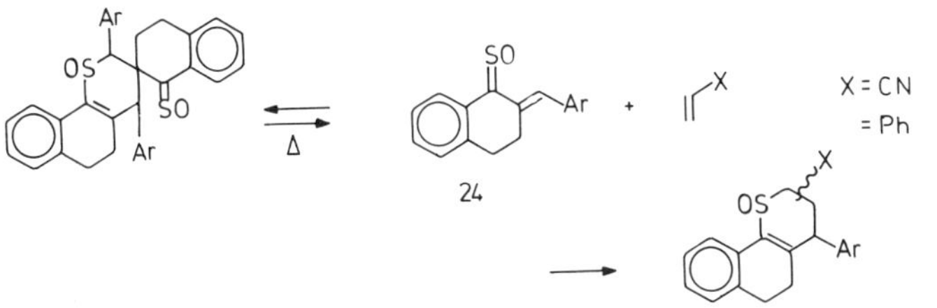

(19)

Maleic anhydride and electron-rich olefins failed to undergo [4 + 2] cycloaddition with **24**.[42]

Acyl sulfines, α-oxo sulfines, participate readily as 2π components of Diels–Alder reactions with a range of dienes (Chapter 5).[43–45] In addition, stable acyl sulfines (Scheme 8-XI) have been shown to participate as 4π components in Diels–Alder reactions with ethyl vinyl ether[43] and isobutylene.[44]

Acyl sulfenes, like all sulfenes, prefer to participate as 2π components of [2 + 2] or [4 + 2] cycloadditions (Chapter 5). Nonetheless, a range of [4 + 2] cycloaddition reactions of acyl sulfenes have been described[46,47] (Scheme 8-XII), including their 4π participation in dimerization reactions[46–48] and reactions with imines,[49] carbodiimides,[50] ketenimines,[51] 1-azirines,[52] vinyl ethers,[53] and ketenes.[47] The reactions often provide mixtures of [4 + 2] and [2 + 2] cycloadducts, and the observed course of the reaction usually depends on the reaction conditions. Consequently, many of the observed [4 + 2] cycloadditions of acyl sulfenes proceed by a stepwise, polar addition–cyclization reaction.

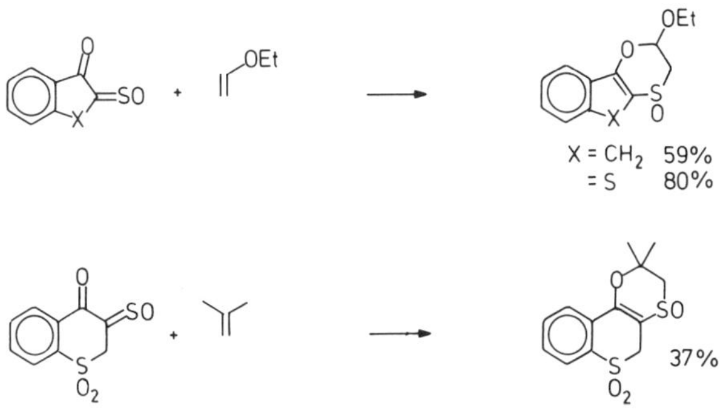

Scheme 8-XI

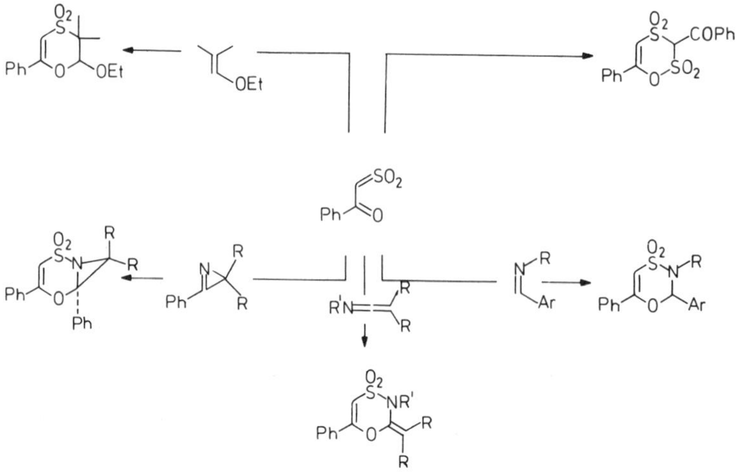

Scheme 8-XII

6. HETERO-1-THIABUTADIENES

The 4-oxa-1-thia-1,3-butadienes, the acyl sulfines **25** and acyl sulfenes **26,** which have been shown to be capable of 4π participation in selected Diels–Alder reactions, have been summarized in the preceding section. It

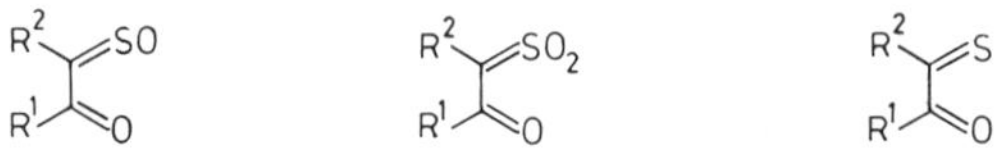

does not appear as if there has been a report of 4π participation of a monothio-1,2-dicarbonyl compound, e.g., an acyl thione **27,** in Diels–Alder reactions, and this may reflect the difficulty in securing such compounds for further study. Their 2π thione participation in Diels–Alder reactions (Chapter 5) and their [2 + 2] dimerization have been detailed.[54]

The aza-1-thia-1,3-butadienes are the most widely recognized and thoroughly investigated hetero-1-thiabutadienes capable of 4π Diels–Alder participation. A single report has detailed the [4 + 2] cycloaddition of the selected 4-aza-1-thia-1,3-butadiene **28** as the 4π component of a Diels–Alder reaction [Eq. (20)].[55]

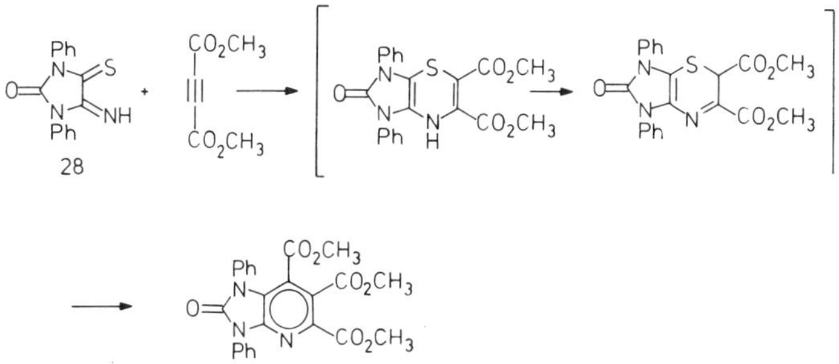

(20)

A wide range of 3-aza-1-thia-1,3-butadiene systems has been extensively investigated as 4π components of Diels–Alder reactions. The work with each of these systems is summarized in Chapter 9 and includes the 4π Diels–Alder reactions of appropriately substituted *N*-thioacylimines (e.g., **29,** Chapter 9, Section 3),[47,56] thioacyl isocyanates (e.g., **30,** Chapter 9, Section 3),[57,58] thioacyl isothiocyanates (e.g., **31,** Chapter 9, Section 3),[57,59] and *N*-thioacyl dithioimidocarbonates (e.g., **32**).[22,23] Examples of

29; 30 X = O; 31 = S; 32; 33; 34; 35 R = Ph; 36 = OEt; 37 X = CR_2, = O,S

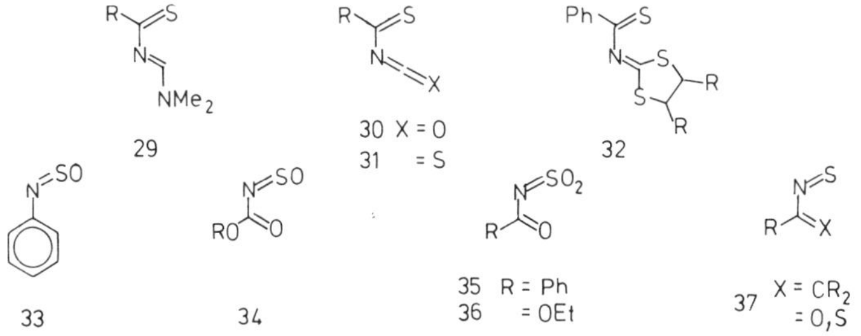

the 4π Diels–Alder participation of 2-aza-1-thia-1,3-butadiene systems include *N*-sulfinylaniline (**33**) and related *N*-arylsulfinylimines (Chapter 9, Section 3),[60] *N*-sulfinylurethanes (e.g., **34,** Chapter 9, Section 3),[61] *N*-sulfenylamides and -urethanes (e.g., **35** and **36,** respectively, Chapter 9, Section 3).[61] No reports of the 4π participation of vinyl- or acylthionitroso compounds (e.g., **37**) have been detailed.[62]

7. CATIONIC THIABUTADIENES, [4⁺ + 2] CYCLOADDITIONS

A growing class of potentially useful polar [4⁺ + 2] cycloadditions employing *in situ* generated 2-thieniumbutadienes have been reported.[63–65] The apparent 4π participation of arylthienium salts in polar [4⁺ + 2] cyclizations with alkenes,[63,64] alkynes,[63] and nitriles[65] has been detailed [Eqs. (21)–(23)],

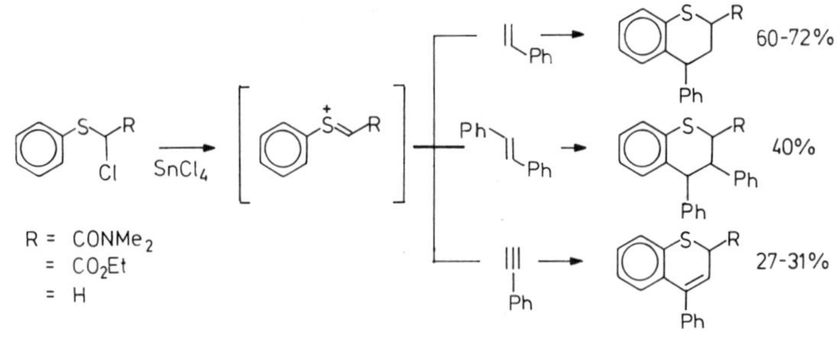

(21)

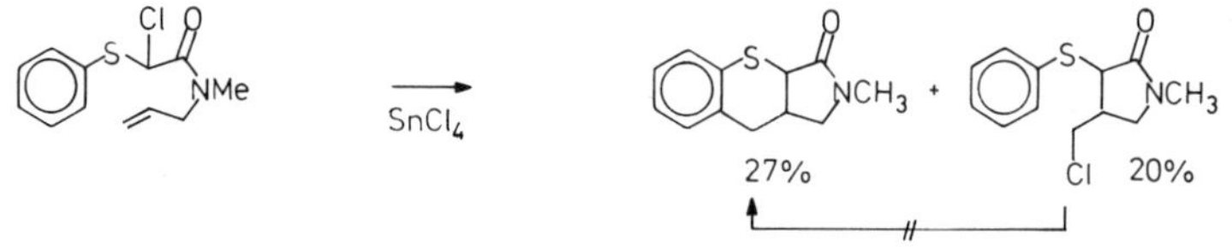

R = CONMe$_2$
  = CO$_2$Et
  = H

(22)

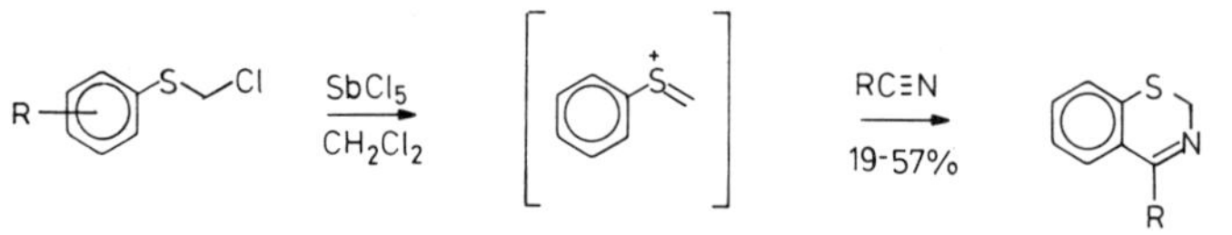

27%   20%

(23)

19-57%

and it complements the 2π participation of simple thienium salts in [4 + 2$^+$] cycloadditions (Chapter 5).[66] Thioamidomethylium cations generated *in situ* from the corresponding hydroxymethylthioamides by acid catalysis participate in regio- and stereospecific [4$^+$ + 2] cycloadditions with olefins to provide 5,6-dihydro-4H-1,3-thiazinium salts [Eq. (24)][67]

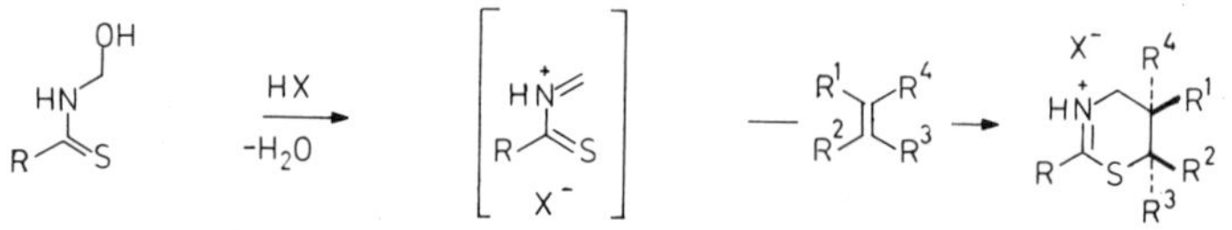

(24)

although the isolated yields of product are lower than that derived from the reaction of amidomethylium cations (Chapter 7, Section 7).[68,69]

REFERENCES

1. Ohno, M., and Sasaki, T. (1984). *J. Syn. Org. Chem. Jpn.* **42**, 125; Ingall, A. H. (1984). "Comprehensive Heterocyclic Chemistry," (A. J. Boulton and A. McKillop, eds.) Vol. 3, pp. 885–942. Pergamon, Oxford.
2. Eisenstein, O., Lefour, J. M., Ahn, N. T., and Hudson, R. F. (1977). *Tetrahedron* **33**, 523; Minot, C., and Ahn, N. T. (1977). *Tetrahedron* **33**, 533; Fleming, I. (1976). "Frontier and Orbitals and Organic Chemical Reactions." Wiley, New York.

3. Lipkowitz, K. B., and Mundy, B. P. (1977). *Tetrahedron Lett.*, 3417.
4. Beslin, P., Lagain, D., Vialle, J., and Minot, C. (1981). *Tetrahedron* **37**, 3839; Beslin, P., Lagain, D., and Vialle, J. (1979). *Tetrahedron Lett.*, 2677; see also Giles, H. G., Marty, R. A., and de Mayo, P. (1974). *J. Chem. Soc., Chem. Commun.*, 409; Giles, H. G., Marty, R. A., and de Mayo, P. (1976). *Can. J. Chem.* **54**, 537; Beslin, P., Lagain, D., and Vialle, J. (1980). *J. Org. Chem.* **45**, 2517.
5. Lipkowitz, K. B., Scarpone, S., Mundy, B. P., and Bornmann, W. G. (1979). *J. Org. Chem.* **44**, 486.
6. Pradere, J.-P., Bouet, G., and Quiniou, H. (1972). *Tetrahedron Lett.*, 3471. Guemas, J.-P., Reliquet, A., Reliquet, F., and Quiniou, H. (1979). *Comptes Rendus Acad. Sci. Paris C* **288**, 89.
7. a. Karakasa, T., and Motoki, S. (1978). *J. Org. Chem.* **43**, 4147. b. Karakasa, T., and Motoki, S. (1979). *J. Org. Chem.* **44**, 4151. c. Karakasa, T., Yamaguchi, H., and Motoki, S. (1980). *J. Org. Chem.* **45**, 927. d. Karakasa, T., and Motoki, S. (1980). *Chem. Lett.*, 879.
8. Quiniou, H. (1981). *Phos. Sulf.* **10**, 1.
9. a. Rasmussen, J. B., Shabana, R., and Lawesson, S.-O. (1982). *Tetrahedron* **38**, 1705; Pradere, J.-P., Quiniou, H., Rabiller, C., and Martin, G. J. (1976). *Bull. Soc. Chim. Fr.*, 991. b. Rasmussen, J. P., Shabana, R., and Lawesson, S.-O. (1981). *Tetrahedron* **37**, 3693. c. Rasmussen, J. P., Shabana, R., and Lawesson, S.-O. (1981). *Tetrahedron* **37**, 197, 1819. d. Nishio, T., Nakajima, N., and Omote, Y. (1980). *J. Heterocycl. Chem.* **17**, 405.
10. Pradere, J.-P., N'Guessan, Y. T., Quiniou, H., and Tonnard, F. (1975). *Tetrahedron* **31**, 3059; Pradere, J.-P., and Quiniou, H. (1972). *Comptes Rendus Acad. Sci. Paris C* **275**, 677; Pradere, J.-P., and Quiniou, H. (1973). *Ann. Chim. Ital.* **63**, 563.
11. Adiwidjaja, G., Proll, T., and Walter, W. (1981). *Tetrahedron Lett.* **22**, 3175.
12. Meslin, J.-C., N'Guessan, Y. T., Quiniou, H., and Tonnard, F. (1975). *Tetrahedron* **31**, 2679.
13. Bard, M., Meslin, J.-C., and Quiniou, H. (1973). *J. Chem. Soc., Chem. Commun.*, 672.
14. Meslin, J.-C., and Quiniou, H. (1972). *Bull. Soc. Chim. Fr.*, 2517; Meslin, J.-C., and Quiniou, H. (1971). *Comptes Rendus Acad. Sci. Paris C* **273**, 148; Meslin, J.-C. (1973). *Comptes Rendus Acad. Sci. Paris C* **277**, 1391.
15. Gokou, C. T., Pradere, J.-P., and Quiniou, H. (1985). *J. Org. Chem.* **50**, 1545.
16. Gosselin, P., Masson, S., and Thuillier, A. (1980). *Tetrahedron Lett.* **21**, 2421; Gosselin, P., Masson, S., and Thuillier, A. (1978). *Tetrahedron Lett.*, 2175, 2717; see also Westmijze, H., Kleijn, H., Meijer, J., and Vermeer, P. (1979). *Synthesis*, 432.
17. Hoffmann, R., and Hartke, K. (1980). *Chem. Ber.* **113**, 919. The α,β-unsaturated thionoesters behave similarly.
18. Lawson, K. R., Singleton, A., and Whitham, G. H. (1984). *J. Chem. Soc., Perkin 1*, 859,865.
19. a. Meslin, J.-C., Pradere, J.-P., and Quiniou, H. (1976). *Bull. Soc. Chim. Fr.*, 1195. b. Pradere, J.-P., and Hadjukovic, G. (1978). *Comptes Rendus Acad. Sci. Paris C* **286**, 553. c. Pradere, J.-P., Quiniou, H., Rabiller, C., and Martin, G. J. (1976). *Bull. Soc. Chim. Fr.*, 991.
20. Kalish, R., Smith, A. E., and Smutny, E. J. (1971). *Tetrahedron Lett.*, 2241.
21. Brunskill, J. S. A., De, A., and Ewing, D. F. (1978). *J. Chem. Soc., Perkin 1*, 629; Brunskill, J. S. A., De, A., and Ewing, D. F. (1980). *J. Chem. Soc., Perkin 2*, 4.
22. a. O-oka, M., Kitamura, A., Okazaki, R., and Inamoto, N. (1978). *Bull. Chem. Soc. Jpn.* **51**, 301. b. Okazaki, R., Kitamura, A., and Inamoto, N. (1975). *J. Chem. Soc., Chem. Commun.*, 257; Okazaki, R., O-oka, M., and Inamoto, N. (1976). *J. Chem. Soc.*,

Chem. Commun., 562. c. Okazaki, R., Ishii, F., and Inamoto, N. (1978). *Bull. Chem. Soc. Jpn.* **51**, 309.

23. Okazaki, R., Ishii, R., Ozawa, Kazumi, Ozawa, Kenji, and Inamoto, N. (1975). *J. Chem. Soc., Perkin Trans. 1,* 270.

24. Gotthardt, H., Nieberl, S., and Donecke, J. (1980). *Liebigs Ann. Chem.,* 873; Gotthardt, H., and Nieberl, S. (1976). *Tetrahedron Lett.,* 3563.

25. Ohno, A., Koizumi, T., and Ohnishi, Y. (1971). *Bull. Chem. Soc. Jpn.* **44**, 2511; Ohno, A., Koizumi, T., Ohnishi, Y., and Tsuchihashi, G. (1970). *Tetrahedron Lett.,* 2025.

26. Gotthardt, H., and Nieberl, S. (1980). *Liebigs Ann. Chem.,* 867; Gotthardt, H., Nieberl, S., and Donecke, J. (1980). *Liebigs Ann. Chem.,* 873; Gotthardt, H., and Nieberl, S. (1976). *Tetrahedron Lett.,* 3563; see also Gotthardt, H. (1972). *Chem. Ber.* **105**, 2004.

27. Brouwer, A. C., George, A. V. E., and Bos, H. J. T. (1983). *Recl. Trav. Chim. Pays-Bas* **102**, 83; Brouwer, A. C., and Bos, H. J. T. (1983). *Recl. Trav. Chim. Pays-Bas* **102**, 91; Kamphius, J., Visser, R. G., and Bos, H. J. T. (1982). *Tetrahedron Lett.* **23**, 3603; see also Brouwer, A. C., and Bos, H. J. T. (1976). *Tetrahedron Lett.,* 209; Coyle, J. D., Rapley, P. A., Kamphuis, J., and Bos, H. J. T. (1985). *Tetrahedron Lett.* **26**, 2249.

28. Ohmura, H., and Motoki, S. (1984). *Bull. Chem. Soc. Jpn.* **57**, 1131; Ohmura, H., and Motoki, S. (1981). *Chem. Lett.,* 235.

29. Lakshmikantham, M. V., Carroll, P., Furst, G., Levinson, M. I., and Cava, M. P. (1984). *J. Am. Chem. Soc.* **106**, 6084.

30. Jacqmin, G., Nasielski, J., Billy, G., and Remy, M. (1973). *Tetrahedron Lett.,* 3655.

31. Hortmann, A. G., Aron, A. J., and Bhattacharya, A. K. (1978). *J. Org. Chem.* **43**, 3374.

32. Kanakarajan, K., and Meier, H. (1983). *J. Org. Chem.* **48**, 881. Preparations of benzothiete: see Voigt, E., and Meier, H. (1977). *Chem. Ber.* **110**, 2242; Voigt, E., and Meier, H. (1976). *Angew. Chem. Int. Ed. Engl.* **15**, 117. See also van Tilborg, W. J. M., and Plomp, R. (1977). *Recl. Trav. Chim. Pays-Bas* **96**, 282; van Tilborg, W. J. M., and Plomp, R. (1977). *J. Chem. Soc., Chem. Commun.,* 130; Schulz, R., and Schweig, A. (1980). *Tetrahedron Lett.* **21**, 343.

33. Okazaki, R., Sunagawa, K., Kang, K.-T., and Inamoto, N. (1979). *Bull. Chem. Soc. Jpn.* **52**, 496; Kang, K.-T., Okazaki, R., and Inamoto, N. (1979). *Bull. Chem. Soc. Jpn.* **52**, 3640; Okazaki, R., Sunagawa, K., Kotera, M., Kang, K.-T., and Inamoto, N. (1982). *Bull. Chem. Soc. Jpn.* **55**, 243; see also Ref. 22.

34. de Mayo, P., and Ng, H. Y. (1977). *Can. J. Chem.* **55**, 3763; Giles, H. G., Marty, R. A., and de Mayo, P. (1974). *J. Chem. Soc., Chem. Commun.,* 409; Giles, H. G., Marty, R. A., and de Mayo, P. (1976). *Can. J. Chem.* **54**, 537.

35. Schulz, R., and Schweig, A. (1981). *Angew. Chem. Int. Ed. Engl.* **20**, 570; see also de Mayo, P., Weedon, A. C., and Wong, G. S. K. (1979). *J. Org. Chem.* **44**, 1977; Chapman, O. L., and McIntosh, C. L. (1971). *J. Chem. Soc., Chem. Commun.,* 383.

36. Hartke, K., Henssen, G., and Kissel, T. (1980). *Liebigs Ann. Chem.,* 1665; Hartke, K., Quante, J., and Kampchen, T. (1980). *Liebigs Ann. Chem.,* 1482; Hartke, K., Kissel, T., Quante, J., and Henssen, G. (1978). *Angew. Chem. Int. Ed. Engl.* **17**, 953; Hartke, K., Kissel, T., Quante, J., and Matusch, R. (1980). *Chem. Ber.* **113**, 1898; Hartke, K., and Hoffman, R. (1980). *Chem. Ber.* **113**, 919; Gillmann, T., and Hartke, K. (1986). *Chem. Ber.* **119**, 2859.

37. Kusters, W., and de Mayo, P. (1974). *J. Am. Chem. Soc.* **96**, 3502.

38. Krespan, C. G., McKusick, B. C., and Cairns, T. L. (1960). *J. Am. Chem. Soc.* **82**, 1515; Krespan, C. G. (1961). *J. Am. Chem. Soc.* **83**, 3434; Krespan, C. G., and McKusick, B. C. (1961). *J. Am. Chem. Soc.* **83**, 3438.

39. Simmons, H. E., Blomstrom, D. C., and Vest, R. D. (1962). *J. Am. Chem. Soc.* **84**, 4756, 4772, 4782; Simmons, H. E., Vest, R. D., Blomstrom, D. C., Roland, J. R., and Cairns, T. L. (1962). *J. Am. Chem. Soc.* **84**, 4746.

40. Dittmer, D. C., McCaskie, J. E., Babiarz, J. E., and Ruggeri, M. V. (1977). *J. Org. Chem.* **42**, 1910; see also Sedergran, T. C., Yokoyama, M., and Dittmer, D. C. (1984). *J. Org. Chem.* **49**, 2408.

41. Truce, W. E., and Norell, J. R. (1963). *J. Am. Chem. Soc.* **85**, 3231; see also Truce, W. E., and Liu, L. K. (1969). *Mech. React. Sulf. Compd.* **4**, 145.

42. Ohmura, H., Karakasa, T., Satsumabayashi, S., and Motoki, S. (1982). *Bull. Chem. Soc. Jpn.* **55**, 333; Karakasa, T., and Motoki, S. (1979). *Tetrahedron Lett.*, 3961; Karakasa, T., Ohmura, H., and Motoki, S. (1980). *Chem. Lett.*, 825.

43. Lenz, B. G., Regeling, H., van Rozendaal, H. L. M., and Zwanenburg, B. (1985). *J. Org. Chem.* **50**, 2930; Lenz, B. G., Regeling, H., and Zwanenburg, B. (1984). *Tetrahedron Lett.* **25**, 5947; Lenz, B. G., Haltiwanger, R. C., Zwanenburg, B. (1984). *J. Chem. Soc., Chem. Commun.*, 502.

44. Still, I. W. J., and Ablenas, F. J. (1985). *J. Chem. Soc., Chem. Commun.*, 524.

45. Review: Zwanenburg, B. (1982). *Recl. Trav. Chim. Pays-Bas* **101**, 1.

46. Desimoni, G., and Tacconi, G. (1975). *Chem. Rev.* **75**, 651.

47. Tsuge, O. (1979). *Heterocycles* **12**, 1067.

48. Fusco, R., Rossi, S., Maiorana, S., and Pagani, G. (1965). *Gazz. Chim. Ital.* **95**, 774.

49. a. Tsuge, O., and Twanami, S. (1970). *Bull. Chem. Soc. Jpn.* **43**, 3543. b. Tsuge, O., and Noguchi, M. (1974). *Chem. Lett.*, 113.

50. Tsuge, O., and Iwanami, S. (1971). *Nippon Kagaku Zasshi* **92**, 448.

51. Tsuge, O., and Iwanami, S. (1971). *Org. Prep. Proc. Int.* **3**, 283.

52. Tsuge, O., and Noguchi, M. (1978). *Heterocycles* **9**, 423.

53. Opitz, G. (1967). *Angew. Chem. Int. Ed. Engl* **6**, 107; cf. p. 121.

54. Hartnedy, R. C., and Dittmer, D. C. (1984). *J. Org. Chem.* **49**, 4752, and references cited therein.

55. Khattak, I., and Ketcham, R. (1983). *J. Chem. Soc., Chem. Commun.*, 260.

56. Gokou, C. T., Pradere, J.-P., and Quiniou, H. (1985). *J. Org. Chem.* **50**, 1545; Pradere, J.-P., Roze, J. C., Duguay, G., Guevel, A., Tea Gokou, C., and Quiniou, H. (1983). *Sulf. Lett.* **1**, 115; Meslin, J.-C., and Quiniou, H. (1979). *Bull. Soc. Chim. Fr.* **2**, 347; Meslin, J.-C., and Quiniou, H. (1975). *Tetrahedron* **31**, 3055; Meslin, J.-C., and Quiniou, H. (1974). *Synthesis*, 298; Meslin, J.-C., Reliquet, A., Reliquet, F., and Quiniou, H. (1980). *Synthesis*, 453; Roze, J.-C., Pradere, J.-P., Duguay, G., Guevel, A., Quiniou, H., and Poignant, S. (1983). *Can. J. Chem.* **61**, 1169; Giordano, C., Belli, A., and Abis, L. (1979). *Tetrahedron Lett.*, 1537; Giordano, C., Belli, A., Erbea, R., and Panossian, S. (1979). *Synthesis*, 801; Giordano, C. (1975). *Gazz. Chim. Ital.* **105**, 1265; Burger, K., Huber, E., Schontag, W., and Ottlinger, R. (1983). *J. Chem. Soc. Chem. Commun.*, 945; Burger, K., and Goth, H. (1980). *Angew. Chem. Int. Ed. Engl.* **19**, 810; Burger, K., and Ottlinger, R. (1978). *J. Fluorine Chem.* **11**, 29; Burger, K., Ottlinger, R., and Albanbauer, J. (1977). *Chem. Ber.* **110**, 2114; Burger, K., Gott, H., Schoentag, W., and Firl, J. (1982). *Tetrahedron* **38**, 287; Burger, V. K., Partscht, H., Huber, E., Gieren, A., Hubner, T., Kaerlein, C.-P. (1984). *Chem. Zeit.* **108**, 209; Burger, K., Albanbauer, J., and Foag, W. (1975). *Angew. Chem. Int. Ed. Engl.* **14**, 767; Giordano, C., Belli, A., and Bellotti, V. (1978). *Synthesis*, 443.

57. Ulrich, H. (ed.) (1967). "Cycloaddition Reactions of Heterocumulenes." Academic Press, New York.

58. Weiss, R. (1967). *Chem. Ber.* **100**, 685; Goerdeler, J., and Schenk, H. (1965). *Chem. Ber.* **98**, 3831; Goerdeler, J., and Schulze, A. (1982). *Chem. Ber.* **115**, 1259; Goerdeler, J., Tiedt, M.-L., and Nandi, K. (1981). *Chem. Ber.* **114**, 2713; Goerdeler, J., and Schenk, H. (1965). *Chem. Ber.* **98**, 2954; Goerdeler, J., and Nandi, K. (1975). *Chem. Ber.* **108**, 3066; Goerdeler, J., and Nandi, K. (1981). *Chem. Ber.* **114**, 549; Goerdeler, J., and Schenk, H. (1963). *Angew. Chem. Int. Ed. Engl.* **2**, 552; Goerdeler, J. and Schulze,

A. (1982). *Chem. Ber.* **115**, 1252; Goerdeler, J., and Jonas, K. (1966). *Chem. Ber.* **99**, 3572; Goerdeler, J., and Schenk, H. (1965). *Chem. Ber.* **98**, 2954; Tsuge, O., Tashiro, M., Mizuguchi, R., and Kanemasa, S. (1966). *Chem. Pharm. Bull. Tokyo* **14**, 1055; Tsuge, O., and Iwanami, S. (1971). *Bull. Chem. Soc. Jpn.* **44**, 2750; Tsuge, O., and Kanemasa, S. (1972). *Tetrahedron* **28**, 4737; Wedekind, E., and Schenk, D. (1911). *Chem. Ber.* **44**, 198; Tsuge, O., and Kanemasa, S. (1972). *Bull. Chem. Soc. Jpn.* **45**, 2877; Niess, R., and Robins, R. K. (1970). *J. Heterocycl. Chem.* **7**, 243; Tsuge, O., and Sakai, K. (1972). *Bull. Chem. Soc. Jpn.* **45**, 1534; Nair, V., and Kim, K. H. (1974). *Tetrahedron Lett.*, 1487; Nair, V., and Kim, K. H. (1974). *J. Org. Chem.* **39**, 3763; Tsuge, O., and Inaba, A. (1976). *Bull. Chem. Soc. Jpn.* **49**, 2828; Tsuge, O., and Kanemasa, S. (1972). *Bull. Chem. Soc. Jpn.* **45**, 3591; Tsuge, O., and Kanemasa, S. (1974). *Bull. Chem. Soc. Jpn.* **47**, 2676; Goerdeler, J., and Sappelt, R. (1967). *Chem. Ber.* **100**, 2064; Goerdeler, J., and Weiss, R. (1967). *Chem. Ber.* **100**, 1627; Schulze, A., and Goerdeler, J. (1974). *Tetrahedron Lett.* 221; Schulze, A., and Goerdeler, J. (1982). *Chem. Ber.* **115**, 3063; Goerdeler, J., Schimpf, R., and Tiedt, M.-L. (1972). *Chem. Ber.* **105**, 3322; Milzner, K., and Seckinger, K. (1974). *Helv. Chim. Acta* **57**, 1614; Goerdeler, J., and Ludke, H. (1970). *Chem. Ber.* **103**, 3393.

59. Goerdeler, J., and Ludke, H. (1968). *Tetrahedron Lett.*, 2455; Goerdeler, J., and Ludke, H. (1970). *Chem. Ber.* **103**, 3393.

60. a. Review: Kresze, G., and Wucherpfennig, W. (1967). *Angew. Chem. Int. Ed. Engl.* **6**, 149. b. Kataev, E. G., and Plemenkov, V. V. (1968). *Z. Org. Chim.* **4**, 1094. c. Collins, G. R. (1964). *J. Org. Chem.* **29**, 1688. d. Macaluso, A., and Hamer, J. (1967). *J. Org. Chem.* **32**, 506. e. Kresze, G., Maschke, A., Albrecht, R., Bederke, K., Patzchke, H. P., Smalla, H., Trede, A., and Munchen, T. H. (1960). *Angew. Chem.* **74**, 135; Hanson, P., and Stone, T. W. (1984). *J. Chem. Soc., Perkin Trans. 1*, 2429.

61. a. Beecken, H. (1967). *Chem. Ber.* **100**, 2159, 2167. b. Horhold, H.-H., and Eibisch, H. (1968). *Chem. Ber.* **101**, 3567; Atkins, G. M., Jr., and Burgess, E. M. (1972). *J. Am. Chem. Soc.* **94**, 6135; Burgess, E. M., and Williams, W. M. (1973). *J. Org. Chem.* **38**, 1249; Kloek, J. A., and Leschinsky, K. L. (1980). *J. Org. Chem.* **45**, 721; Kobelt, D., Paulus, E. F., and Kampe, K.-D. (1970). *Tetrahedron Lett.*, 123; Kobelt, D., Paulus, E. F., and Kampe, K.-D. (1971). *Tetrahedron Lett.*, 1211.

62. Mayer, R., Bleisch, S., and Domschke, G. (1978). *Z. Chem.* **18**, 323.

63. Tamura, Y., Ishiyama, K., Mizuki, Y., Maeda, H., and Ishibashi, H. (1981). *Tetrahedron Lett.* **22**, 3773.

64. Wada, M., Shigehisa, T., Kitani, H., and Akiba, K. (1983). *Tetrahedron Lett.* **24**, 1715.

65. Thakur, D. K., and Vankar, Y. D. (1983). *Synthesis,* 223.

66. Related work on the [4 + 2] addition–cyclization reactions of unsaturated sulfur yields: Hori, M., Kataoka, T., Shimizu, H., Narita, K., Ohno, S., and Aoki, H. (1974). *Chem. Lett.*, 1101; see also Hori, M., Kataoka, T., Shimizu, H., Ohno, S., Narita, K., Takayanagi, H., Ogura, H., and Iitaka, Y. (1979). *Tetrahedron Lett.*, 4315; Hori, M., Kataoka, T., Shimizu, H., and Ohno, S. (1980). *J. Org. Chem.* **45**, 2468, and references cited therein.

67. Schmidt, R. R. (1965). *Angew. Chem. Int. Ed. Engl.* **4**, 241; Giordano, C. (1974). *Gazz. Chim. Ital.* **104**, 849.

68. Reviews of polar cycloadditions: Schmidt, R. R. (1973). *Angew. Chem. Int. Ed. Engl.* **12**, 212; Bradsher, C. K. (1974). *Adv. Heterocycl. Chem.* **16**, 289.

69. Gompper, R. (1969). *Angew. Chem. Int. Ed. Engl.* **8**, 312.

Chapter **9**

Azabutadienes

INTRODUCTION

The observation that conjugated systems containing nitrogen, typified by the 1- and 2-aza-1,3-butadiene systems,[1-17] show diminished reactivity toward representative electron-deficient dienophiles has focused attention on the fact that the introduction of a nitrogen atom into conjugated systems will confer electrophilic character to the system. These observations and the recognized shortcomings of attempting Diels–Alder reactions with 2π and 4π components of a similar electrophilic nature have led to development of several general approaches to the implementation of useful azadiene Diels–Alder reactions.

Recognition of the electrophilic character of azadienes led to the investigation, demonstration, and subsequent development of the inverse electron demand (LUMO$_{\text{diene}}$ controlled) Diels–Alder reaction. Substitution

239

of an azadiene system with complementary electron-withdrawing groups accentuates the electron-deficient nature of the azadiene and permits the use of electron-rich, strained, or even simple olefins as dienophiles. It is the magnitude of the $LUMO_{diene}$–$HOMO_{dienophile}$ energy separation that has been related to the rate of inverse electron demand [4 + 2] cycloaddition. Consequently, the addition of features to the azadiene system that lower the $LUMO_{azadiene}$ further accelerates the rate of azadiene participation in an inverse electron demand ($LUMO_{diene}$ controlled) Diels–Alder reaction.[16b]

Alternatively, the complementary addition of strong electron-donating substituents to the azadiene system increases the nucleophilic character of the azadiene system and permits the use of conventional electron-deficient dienophiles in Diels–Alder reactions. In such instances, the azadiene systems are participating in normal ($HOMO_{diene}$ controlled) Diels–Alder reactions. The appropriate introduction of electron-donating substituents to the azadiene system raises the $HOMO_{azadiene}$, reduces the magnitude of the $HOMO_{azadiene}$–$LUMO_{dienophile}$ energy separation, and accounts for the accelerated participation of nucleophilic azadienes in normal Diels–Alder cycloadditions.[16b]

In many instances, the entropic assistance provided in the intramolecular Diels–Alder reaction is sufficient to promote azadiene participation in Diels–Alder reactions.[12] The incorporation of the azadiene system, or dienophile, into a reactive or sensitive system, e.g., heterocumulene or strained olefin, allows a number of specialized azadiene Diels–Alder reactions. Many such examples may represent stepwise, polar [4 + 2] cycloaddition reactions.

1. 1-AZA-1,3-BUTADIENES

Early reports of the successful participation of 1-aza-1,3-butadienes in Diels–Alder reactions include only the reactions of benzisoxazole[18] and unsaturated 3,4-dihydroisoquinolines[19] [Eqs. (1)–(3)][18b,19]

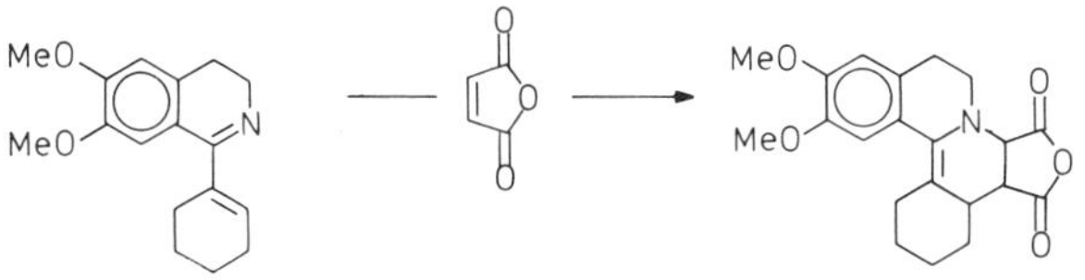

$$(3)$$

and represent the early exceptions to the general observations that such systems fail to undergo [4 + 2] cycloaddition. Attempts to repeat and confirm the initial work with benzisoxazole has led to the recognition of the sensitivity of the attainable Diels–Alder products to further transformations.[18b,c] More recent work utilizing selected 1-aza-1,3-butadienes incorporated into unsaturated heterocyclic systems is summarized in Table 9-I, and in each instance required the use of reactive dienophiles for observable reaction.[20–23] 2H-1,3-Oxazin-2-ones (Table 9-I entries 13 and 14) failed to react with dimethyl acetylenedicarboxylate, diethyl azodicarboxylate, or enamines, and their reactions with ynamines took a different course.[23]

Unsaturated imines, simple 1-aza-1,3-butadienes, participate in Diels–Alder reactions with typical electron-deficient dienophiles preferentially through their enamine tautomer [Eq. (4)],[24]

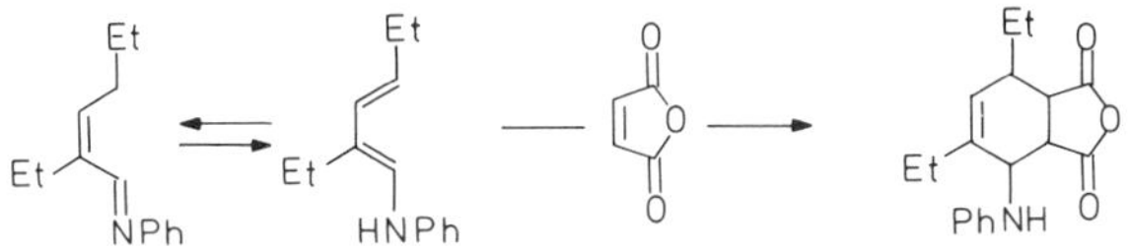

$$(4)$$

and in instances where isomerization is prevented [2 + 2] cycloaddition usually intervenes [Eqs. (5) and (6)].[25a,b,26]

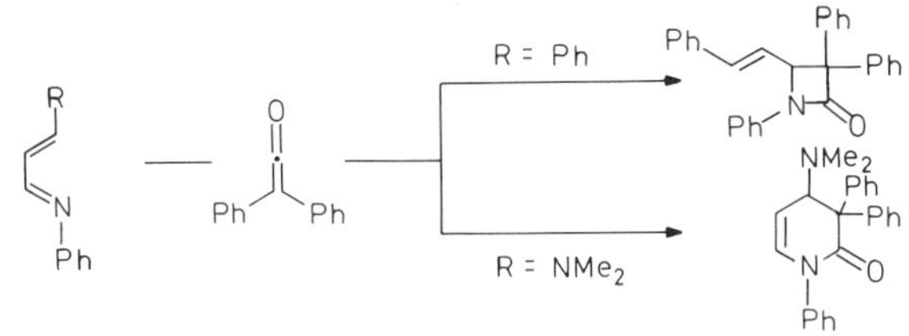

$$(5)$$

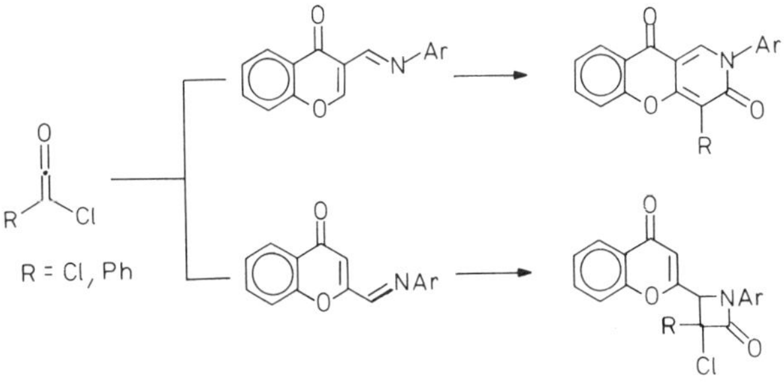

$$(6)$$

TABLE 9-I

Diels–Alder Reactions of Unsaturated Heterocyclic 1-Azadienes

Entry	1-Azadiene	Dienophile	Conditions	Product	Yield	Ref.
1	(benzo-fused isoxazole structure)	(N-phenylmaleimide) NPh	Neat, 125–130°C, 1 hr	(cycloadduct) NPh	17%	18a–c
			Xylene, reflux, 20 hr	(quinoline imide) NPh	—	18b
2		CO_2CH_3 / $\equiv$ / CO_2CH_3 (DMAD)	—	(structure) CO_2CH_3, CO_2CH_3, N+, O–	—	18b
3	R— (benzisoxazole)	(maleimide) N—CN	58°C	R— (cycloadduct) N—CN R = H		18c
			Xylene, 138°C	R— (CHO, NH) N—CN 37% R = H	37%	
			$o\text{Cl}_2\text{C}_6\text{H}_4$, 180°C, 41 hr	R— (product) N—CN 75% R = H	75%	
			$o\text{Cl}_2\text{C}_6\text{H}_4$, 180°C, 24 hr	R = 5-Cl	29%	
			$o\text{Cl}_2\text{C}_6\text{H}_4$, 180°C, 24 hr	R = 6-Cl	54%	
			$o\text{Cl}_2\text{C}_6\text{H}_4$, 180°C, 18 hr	R = 8-Cl	20%	
			$o\text{Cl}_2\text{C}_6\text{H}_4$, 180°C, 192 hr	R = 6-NO$_6$	29%	
			$o\text{Cl}_2\text{C}_6\text{H}_4$, 180°C, 18 hr	R = 7-NO$_2$	49%	
			$o\text{Cl}_2\text{C}_6\text{H}_4$, 180°C, 18 hr	R = 6-OCH$_3$	26%	
4		HO—$(CH_2)_n$	Xylene, 140–150°C	(product) $(CH_2)_n$		18c
				$n = 3$	23%	
				$n = 4$	11%	
				$n = 5$	39%	
				$n = 6$	25%	
5		$(CH_2)_n$	Xylene, 150°C	$n = 3$	19%	18e
				$n = 4$	6%	

TABLE 9-I (*Continued*)

Entry	1-Azadiene	Dienophile	Conditions	Product	Yield	Ref.
6		 (alkyne with Ph)	Xylene, 140°C	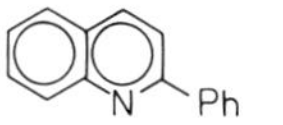 (quinoline with Ph, N)	11%	18d
7		 (benzene)	—	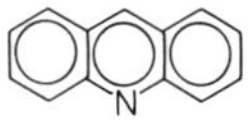 (acridine structure)	5%	18e
8	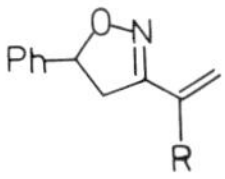 (Ph-isoxazoline with R)	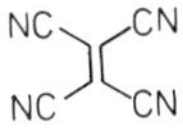 NC–C(CN)=C(CN)–CN	Neat, 140°C 4–15 min, reversible	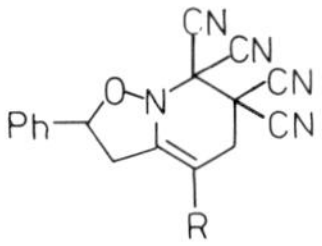 (Ph polycyclic with CN groups, R)		18f
				R = H	33%	
				R = Ph	52%	
9	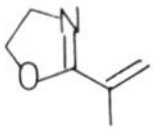 (oxazoline with methyl)	 O=N–Ph	—	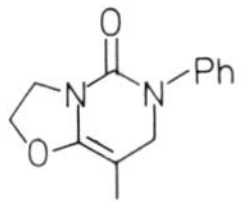 (bicyclic product, Ph)	—	20
10	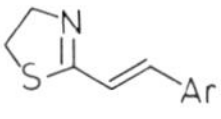 (thiazoline with Ar)	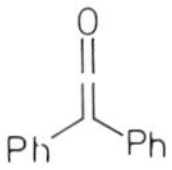 Ph–C(=O)–Ph	Xylene, reflux 10 hr	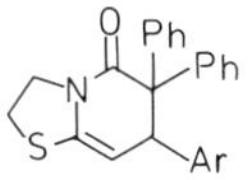 (bicyclic product, Ph, Ph, Ar)		22
				Ar = C$_6$H$_5$	39%	
				Ar = *p*ClC$_6$H$_4$	48%	
				Ar = *p*NO$_2$C$_6$H$_4$	78%	
11	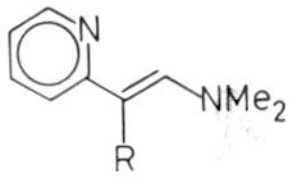 (pyridyl with NMe$_2$, R)	 O=C=O	—	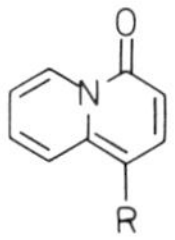 (quinolizinone with R)	—	21
				R = CN		
				R = CO$_2$Et		
12	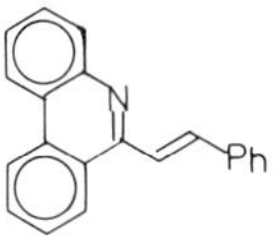 (phenanthridine with Ph)	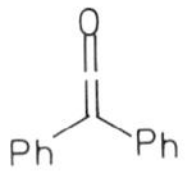 Ph–C(=O)–Ph	—	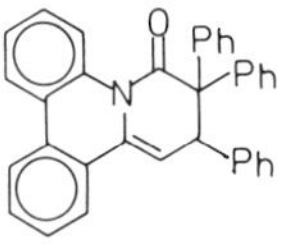 (polycyclic product, Ph, Ph, Ph)	28%	22
13	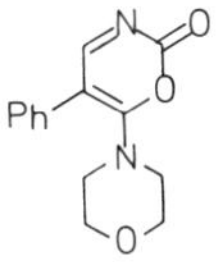 (Ph, morpholino oxazinone)	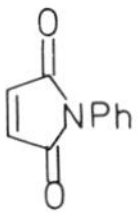 (maleimide, NPh)	—	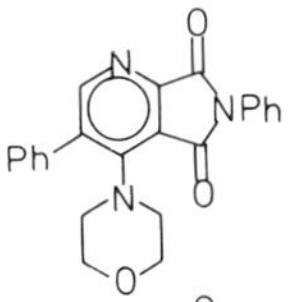 (Ph, morpholino bicyclic, NPh)	—	23
14	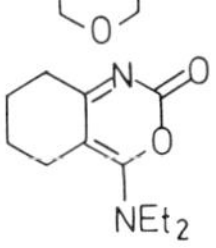 (cyclohexane-fused oxazinone, NEt$_2$)	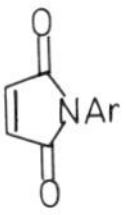 (maleimide, NAr)		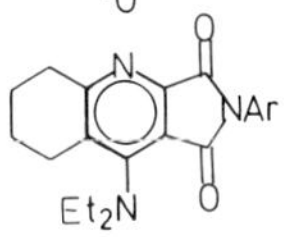 (cyclohexane-fused bicyclic, Et$_2$N, NAr)		23
				Ar = C$_6$H$_5$	49%	
				Ar = *p*ClC$_6$H$_4$	38%	

243

Reported exceptions to this observation include the [4 + 2] cycloadditions of selected imines with diphenylketene,[25,a,b] chlorocyanoketene,[25c] dichloroketene,[25b,d,26] chlorophenylketene,[25b,26] benzoylsulfene,[27] maleic acid,[28a] and (Z)-3-phenyl-3-trimethylsilyloxyacrylonitrile[28b] [Eqs. (5)–(9)].[25a,b,c,26–28]

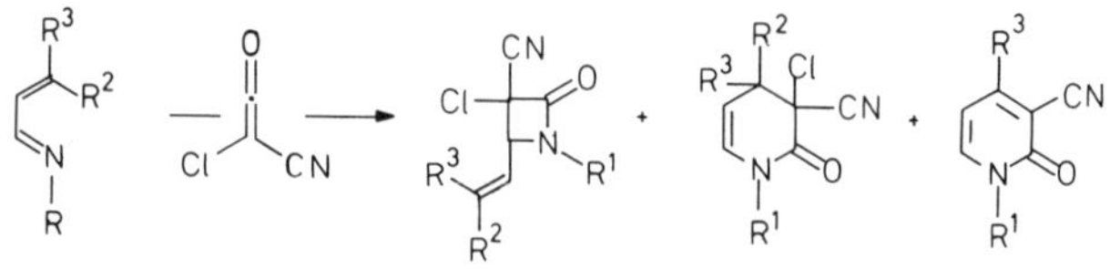

R^1	R^2	R^3				(7)
t-Bu	H	Ph	--	56%	19%	
C_6H_{11}	H	Ph	--	71%	21%	
p-CH$_3$OC$_6$H$_4$	H	Ph	17%	42%	22%	
t-Bu	Ph	Ph	86%	--	--	
C_6H_{11}	Ph	Ph	58%	36%	--	
p-CH$_3$OC$_6$H$_4$	Ph	Ph	90%	--	--	

10 – 43%

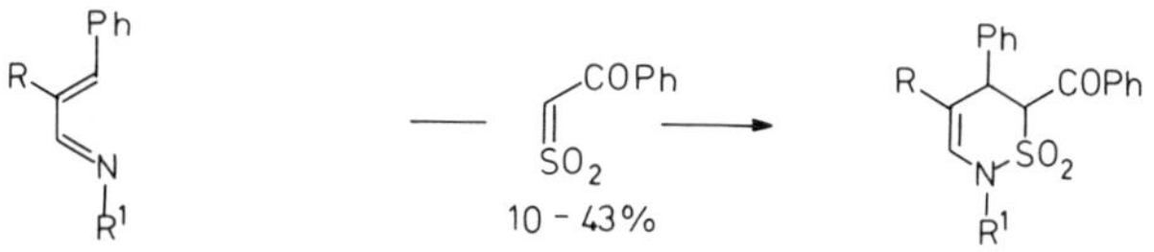

R = H, Alkyl
R'= Aryl, Alkyl

(8)

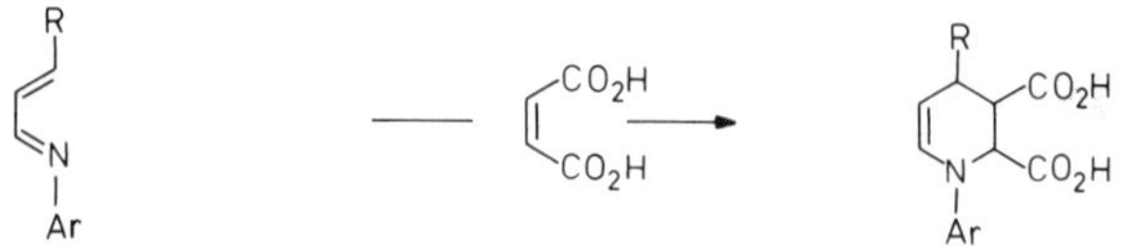

R = H, CH$_3$, Ph, furyl, thienyl

(9)

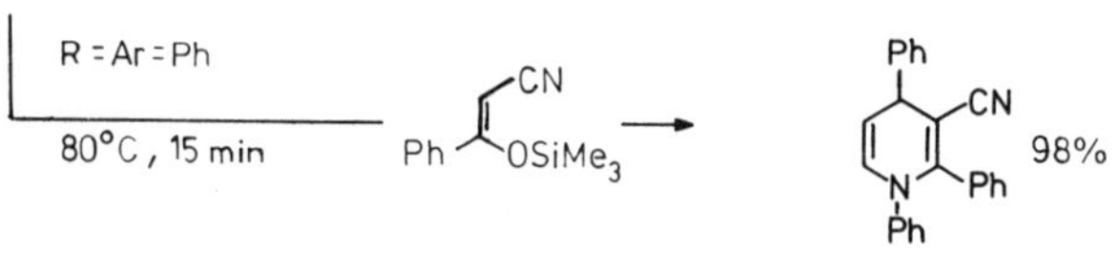

R = Ar = Ph

80°C , 15 min

98%

The acid-catalyzed reaction of methanimines with β-substituted enamines, utilized for the preparation of symmetrical 3,5-disubstituted pyridines[29a] [Eq. (10)], has been carefully investigated and shown to proceed by *in situ* generation and subsequent [4 + 2] cycloaddition of electrophilic 1-aza-1,3-butadienes with the electron-rich enamine.[29b] Additional studies have reduced this process to a controlled [4 + 2] reaction of the isolated *N-tert*-butyl-1-aza-1,3-butadienes with enamines or related electron-rich olefins including ketene *O,N*-acetals[29c] [Eq. (11)].

$$(10)$$

$$(11)$$

The full scope of the acid-catalyzed one-step conversion of N-*tert*-butylmethanimine, generated *in situ* from 1,3,5-tri-*tert*-butylhexahydro-1,3,5-triazine, to symmetrical 3,5-disubstituted pyridines employing enamines has been detailed.[29c] The generation of 2,3,5,6-tetrasubstituted pyridines from the thermal reaction of enolizable ketones with hexamethylphosphoric triamide has been proposed to proceed by *in situ* generation and subsequent [4 + 2] cycloaddition of N-methyl-1-aza-1,3-butadienes with intermediate dimethylamino enamines.[30]

In efforts to increase the reactivity of simple 1-aza-1,3-butadienes toward typical electron-deficient dienophiles, Ghosez and co-workers found that the α,β-unsaturated hydrazone 1 behaves as a well-defined electron-rich diene in regioselective, normal (HOMO$_{diene}$ controlled) Diels–Alder reactions with a number of representative dienophiles.[31] Reductive cleavage of the nitrogen–nitrogen bond concurrent with reduction of the carbon–carbon double bond has provided substituted piperidines, and aromatization with elimination of dimethylamine has provided substituted pyridines (Scheme 9-I).[31] Similar reactions of simple[31a] or functionalized[31c] α,β-unsaturated oximes failed to provide an observable [4 + 2] cycloaddition. It remains to be determined whether this process is general for α,β-unsaturated hydrazones capable of tautomerization.

Reports of the effective *in situ* generation of unstable 1-aza-1,3-butadienes and their subsequent intramolecular Diels–Alder reactions have been described[32–39] (Table 9–II). Fowler and co-workers have shown that the gas-phase pyrolysis of N-acyl-O-acetyl-N-allylhydroxylamines gener-

TABLE 9-II
Intramolecular Diels–Alder Reactions of 1-Aza-1,3-butadienes

Entry	Substrate	Conditions	Product	Yield	Ref.
1		FVP (650°C, 1 μsec)	X = CH₂, R = H X = CH₂, R = Ph X = (CH₂)₂, R = H X = O, R = H	75% 74% 69% 70%	32
2	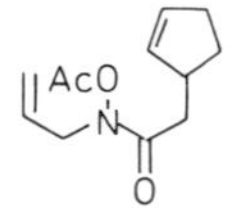	FVP (650°C, 1 μsec)	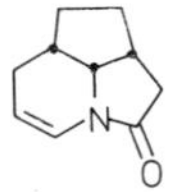	69%	32
3	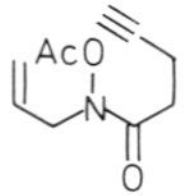	FVP (650°C, 1 μsec)	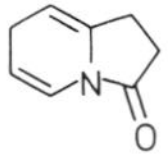	90%	32
4		FVP (650°C, 1 μsec)	R = H, 2:1 trans:cis R = CH₃, 4.5:1 trans:cis	60% 68%	32
5	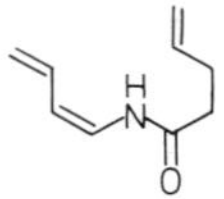	FVP (650°C, 1 μsec)	R = H	63%	32
6	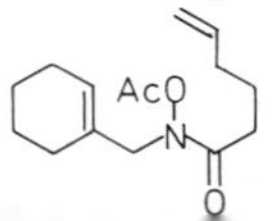	FVP (650°C, 1 μsec)	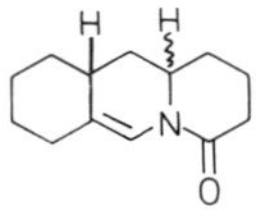	63%	32

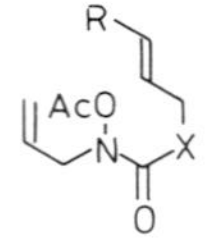

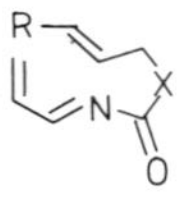

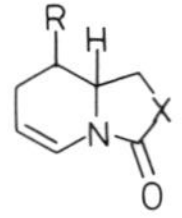

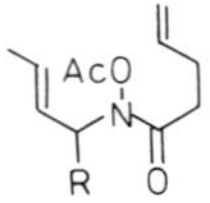

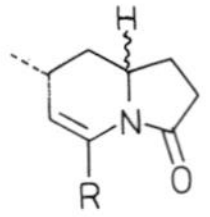

	Substrate	Conditions	Product	Yield	Ref.
7		FVP (650°C, 1 μsec)	R = CH₃, R¹ = H, 3:1 trans:cis; R = H, R¹ = CH₃, 1:1 trans:cis	69% / 70%	32c
8	(EtO)₂PO	Toluene, 110°C 12 hr	(EtO)₂PO; R = Ph, X = CH₂; R = Ph, X = (CH₂)₂	46% / 63%	33
9		195°C, 3 hr; 190°C, 10 hr	R = iPr; R = Me₂	42% / 35%	34
10	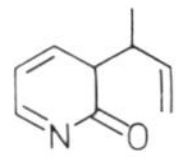	245°C	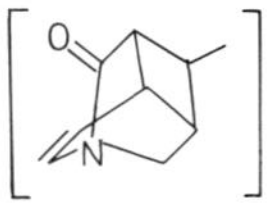	~27%	35
11	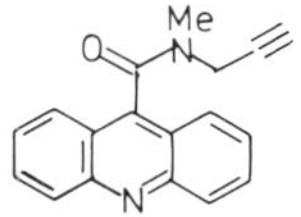	200°C, 15 hr	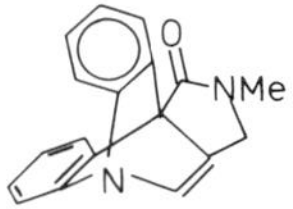	35%	36

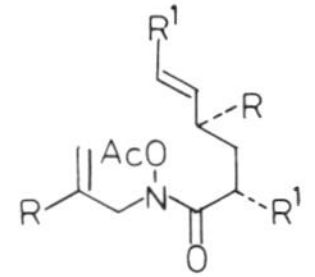

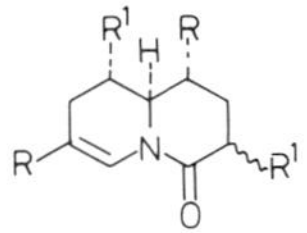

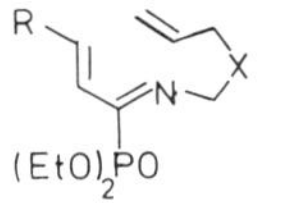

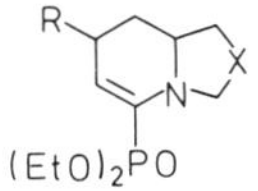

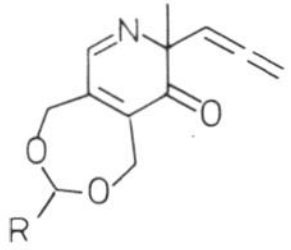

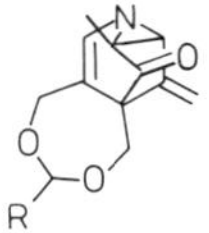

(continued)

247

TABLE 9-II (*Continued*)

Entry	Substrate	Conditions	Product	Yield	Ref.
12	(structure: EtO_2C, R, $(CH_2)_n$, NH, O)	TMSCl, Et_3N, $ZnCl_2$, toluene, 180–185°C, 7–10 hr	(bicyclic structure: CO_2Et, H, R, $(CH_2)_n$, N, =O)		
			$n = 1$, R = H	55%	37c
			$n = 1$, R = CH_3	72%	(+ 8% β-R)
			$n = 1$, R = Ph	75%	(+11% β-R)
			$n = 2$, R = H	56%	
			$n = 2$, R = CH_3	58%	(+17% β-R)
		t-$BuMe_2SiOTf$, Et_3N, CH_2Cl_2, 25°C, 1 hr	$n = 1$, R = CH_3	78%	37b
13	(structure: MeO, MeO, EtO_2C, NH, O, R^1, R^2)	TMSCl, Et_3N, $ZnCl_2$, 1,2-$(Cl)_2C_6H_3$, 16 hr	(tricyclic structure: MeO, MeO, H, EtO_2C, N, =O, R^1, R^2)		
			$R^1 = R^2 = H$	44%	37c
			$R^1 = H$, $R^2 = CH_3$	52%	
			$R^1 = H$, $R^2 = Ph$	47%	
		t-$BuMe_2SiOTf$, Et_3N, CH_2Cl_2, 25°C, 1 hr	$R^1 = R^2 = CH_3$	66%	37b
			$R^1 = H$, $R^2 = Ph$	83%	
14	(structure: EtO_2C, Ar, Ar, N, H, O)	t-$BuMe_2SiOTf$, Et_3N, CH_2Cl_2, 25°C, 5.5 hr	(bicyclic structure: CO_2Et, H, Ar, Ar, N, =O)	68%	37b

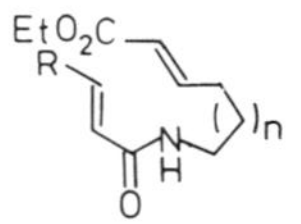

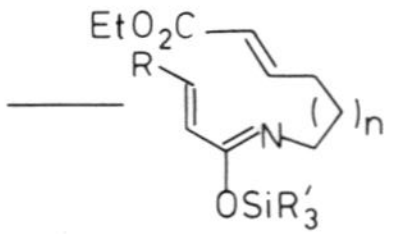

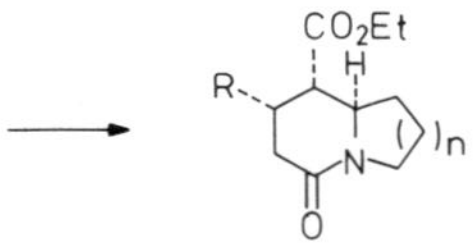

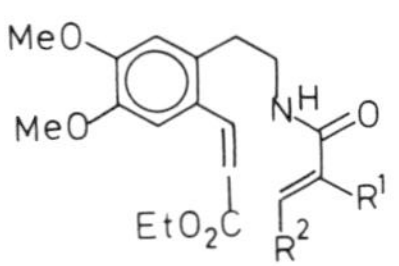

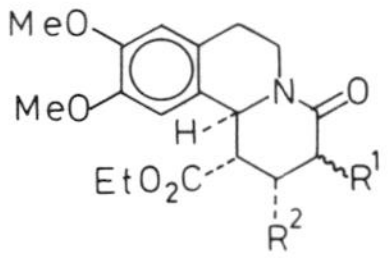

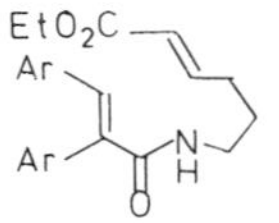

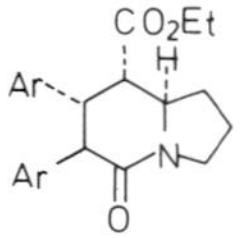

Ar = 3,4-$(OCH_3)_2C_6H_3$

248

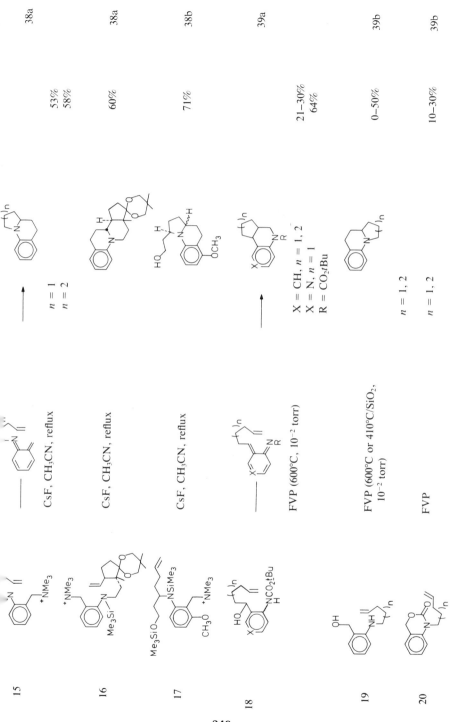

15	CsF, CH₃CN, reflux	$n = 1$ 53% $n = 2$ 58%	38a
16	CsF, CH₃CN, reflux	60%	38a
17	CsF, CH₃CN, reflux	71%	38b
18	FVP (600°C, 10⁻² torr)	X = CH, n = 1, 2 21–30% X = N, n = 1 64%	39a
19	FVP (600°C or 410°C/SiO₂, 10⁻² torr)	n = 1, 2 0–50%	39b
20	FVP	n = 1, 2 10–30%	39b

249

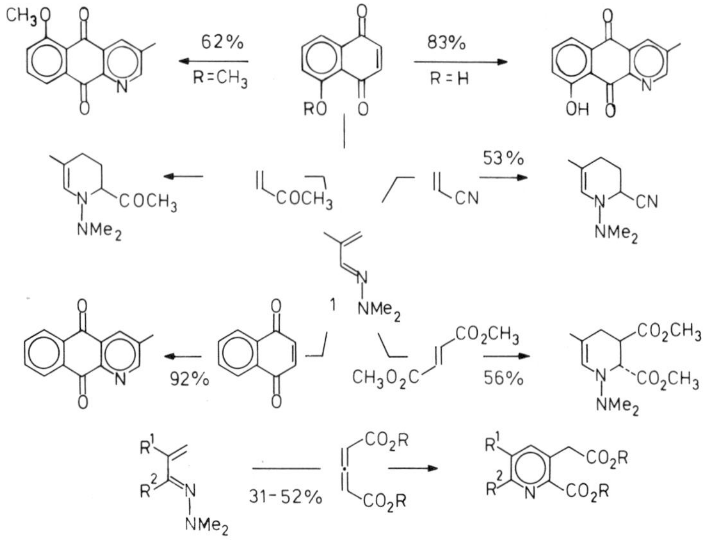

Scheme 9-I[31]

ates *N*-acyl-1-aza-1,3-butadienes capable of participation in intramolecular Diels–Alder reactions [Eq. (12)].[32]

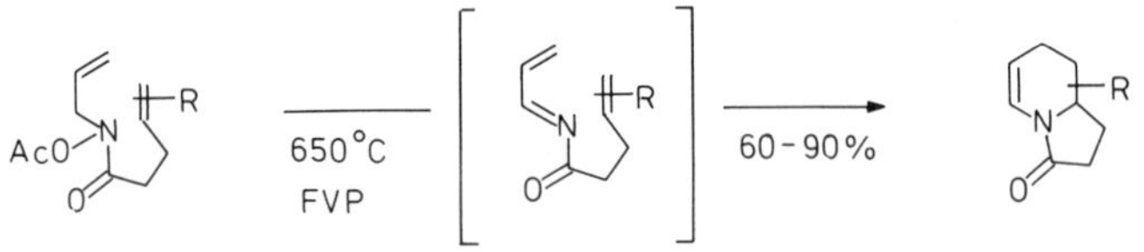

$$(12)$$

The entropic assistance provided by the intramolecular cycloaddition, the inability of the dienes to tautomerize, and the relative stability of the product (acyl versus alkyl enamine) contribute to the success of this process. A total synthesis of (−)-deoxynupharidine based on the implementation of the *in situ* generation and subsequent intramolecular Diels–Alder reaction of an *N*-acyl-1-aza-1,3-butadiene has been detailed [Eq. (13)].[32c]

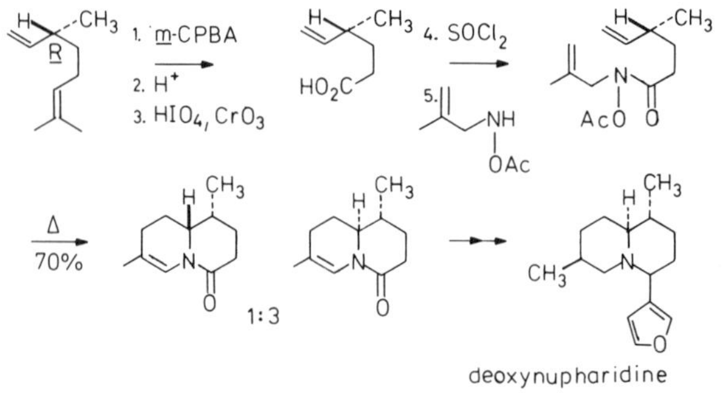

$$(13)$$

deoxynupharidine

The intramolecular Diels–Alder reactions of *N*-acyl-1-aza-1,3-butadienes proceed preferentially through an exo transition state.

Kametani and co-workers have demonstrated what may prove to be the most convenient and effective approach to the utilization of 1-aza-1,3-butadienes for intramolecular [4 + 2] cycloadditions.[37] Treatment of α,β-unsaturated enamides with trimethylsilylchloride–zinc chloride provides *in situ* generation of 2-trimethylsilyloxy-1-aza-1,3-butadienes which are capable of effective intramolecular [4 + 2] cycloaddition (Table 9-II, entries 12–14). Selected conditions have permitted this *in situ* generation and subsequent intramolecular Diels–Alder cycloaddition of 2-trimethylsilyloxy-1-aza-1,3-butadienes to be conducted at 25°C,[37b] and these observations have found application in the total syntheses of (±)-epilupinine and (±)-tylophorine [Eqs. (14) and (15)].[37c,b]

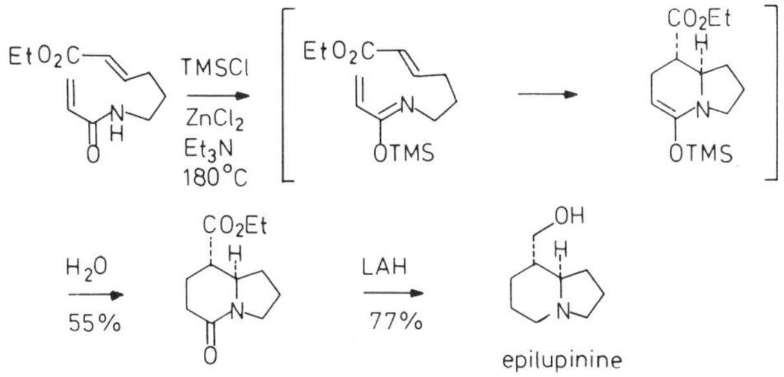

(14)

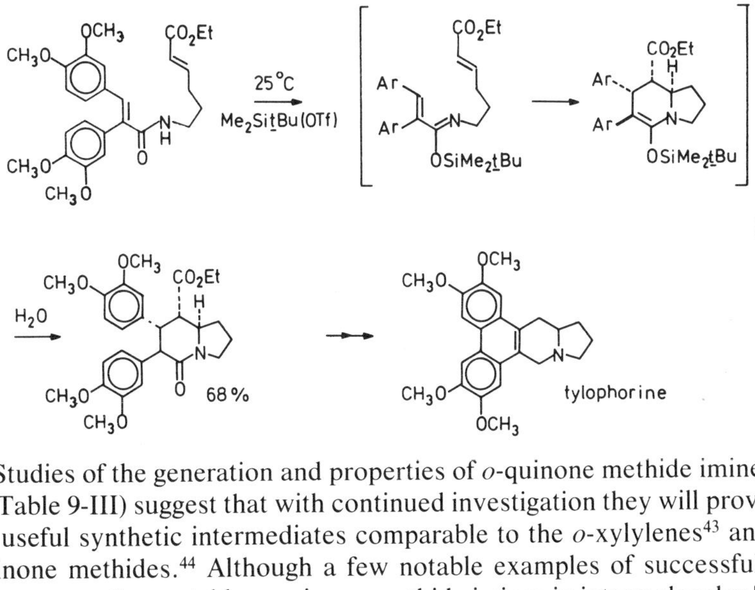

epilupinine

(15)

tylophorine

Studies of the generation and properties of *o*-quinone methide imines[38–42] (Table 9-III) suggest that with continued investigation they will prove to be useful synthetic intermediates comparable to the *o*-xylylenes[43] and *o*-quinone methides.[44] Although a few notable examples of successful efforts to trap the unstable *o*-quinone methide imines in intermolecular [4 +

TABLE 9-III
Intermolecular Diels–Alder Reactions of *o*-Quinone Methide Imines

Substrate	Dienophile	Conditions	Product	Yield	Ref.
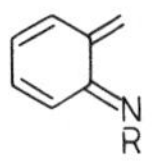 [structure, *o*-quinone methide imine with =N–R]	*o*-Quinone methide imines				38–40
[benzazetidine with –NPh]	[maleimide, NPh]	200°C	[tricyclic product with NPh, N–Ph]	—	40a
[benzothiazine SO₂, CH₃]	[Cl–CH=CH–CO₂H]	*hν*	[product with CO₂H, N–CH₃, Cl]	—	40b
	[alkyne CO_2CH_3 / CO_2CH_3]	*hν*	[CH_3O_2C, CH_3, H_3C, CO_2CH_3 product]	88%	40c
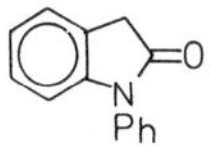 [indolinone, =O, N–Ph]	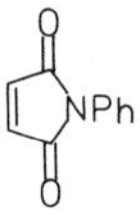 [maleimide, NPh]	*hν*, Et₂O, 6 hr	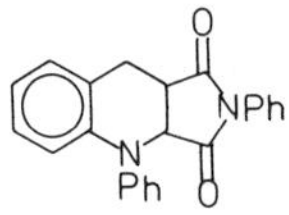 [product with NPh, N–Ph]	14%	40c
		hν, Et₂O, 8 hr	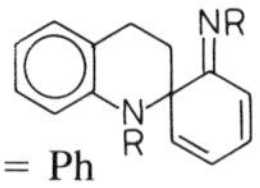 [product NR, R = Ph]	98%	40c
[⁺NMe₃ / N(SiMe₃)Me benzyl]		CsF, CH₃CN, reflux	R = Ph R = CH₃	77%	38a
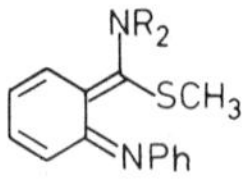 [NR₂, SCH₃, =NPh]	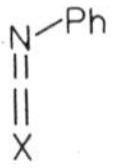 [N–Ph, N, X imine]	—	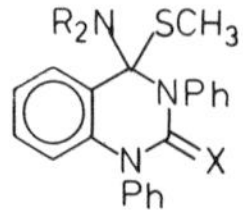 [R_2N, SCH_3, NPh, X product] X = O, S	—	40d
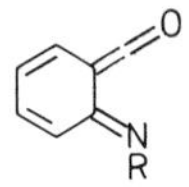 [*o*-quinone, =O, =N–R]	Ketene *o*-quinone methide imines	See Eq. (18)			41 Review, 41a
[*o*-quinone, =O, =N–R]	*o*-Quinone imines	See Scheme 9-II			42a,b
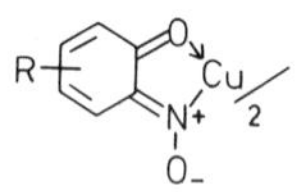 [R–, O, Cu, N⁺, O⁻ complex]	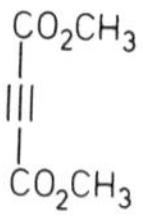 [alkyne CO_2CH_3 / CO_2CH_3]	DME, reflux	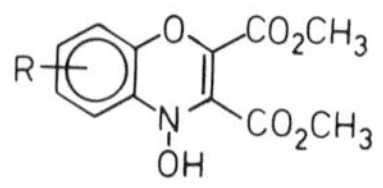 [R–, O, CO_2CH_3, N, OH, CO_2CH_3 product]	70–98%	42b
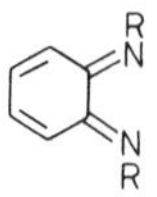 [R, =N, =N–R diimine]	*o*-Quinone diimines	See Section 7			42c

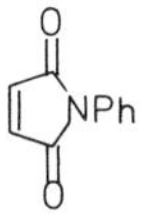

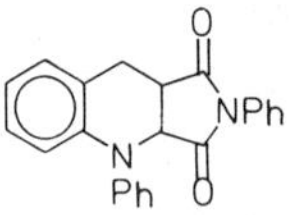

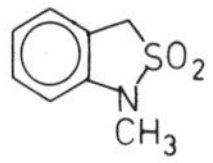

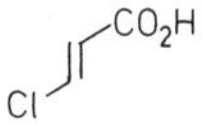

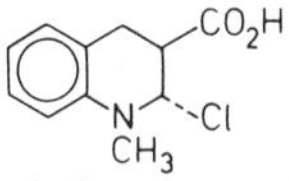

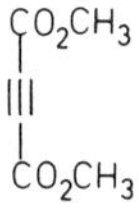

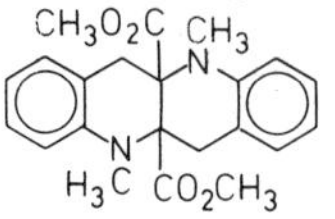

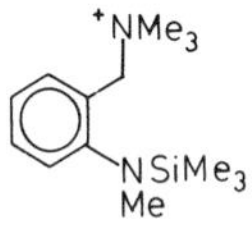

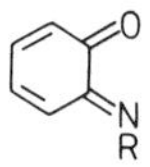

2] cycloadditions have been described[40a-d] (Table 9-III), it is their intramolecular Diels–Alder reactions[38-39] that have proved most useful (Table 9-II, entries 15–20). A mild, fluoride anion-induced 1,4-elimination of [o-[(trimethylsilyl)alkylamino]benzyl] trimethylammonium halides generates the o-quinone methide N-alkylimines[38a] (Table 9-II, entries 15–17) capable of effective intramolecular Diels–Alder reactions, and this methodology has been applied to the total synthesis of 9-azaestro-1,3,5(10)-trien-17-one [Eq. (16)][38a]

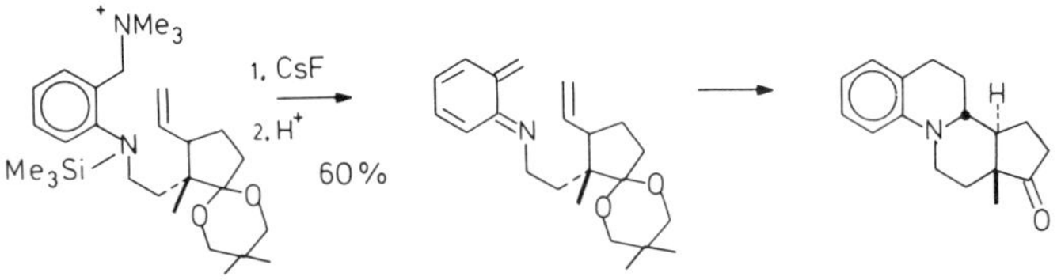

$$(16)$$

and gephyrotoxin [Eq. (17)].[38b]

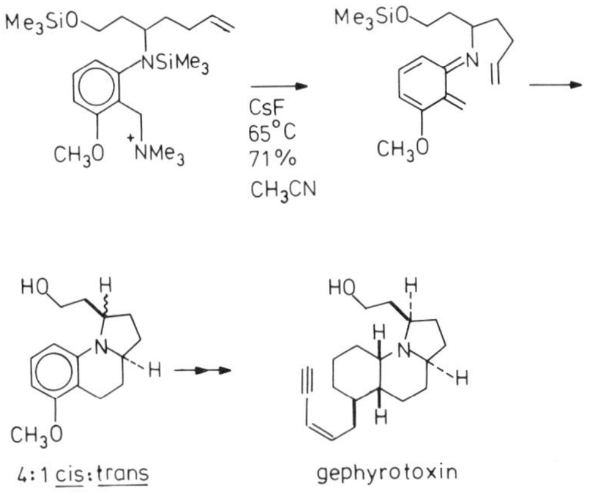

$$(17)$$

Until recently, attempts to trap the unstable o-quinone methide imines generated by this fluoride-induced 1,4-elimination in intermolecular Diels–Alder reactions with typical dienophiles were unsuccessful.[38a]

Successful and effective examples of the *in situ* generation and subsequent intermolecular [4 + 2] cycloaddition of apparent ketene o-quinone methide imines[40] have been detailed in the completion of total syntheses of evodiamine and rutecarpine [Eq. (18)].[41e]

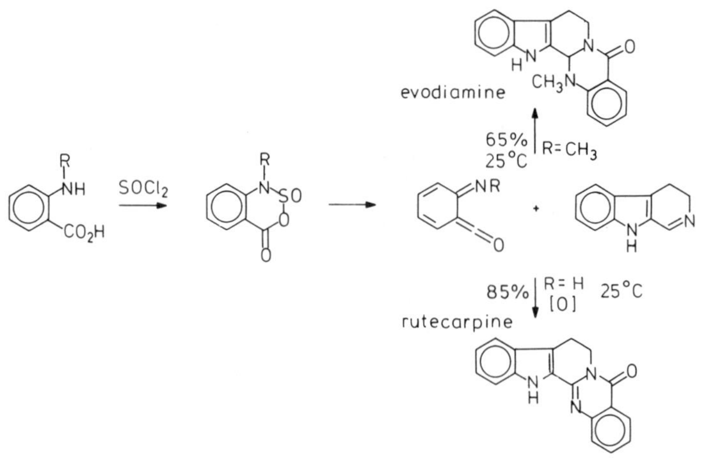
evodiamine
CH_3N
65%
25°C
R=CH₃
R
NH
CO_2H
$SOCl_2$
N
SO
O
NR
N
H
O
85%
R=H
[O]
25°C
rutecarpine

(18)

With the exception of studies with *o*-quinone mono- and diimines,[41–42] which exhibit accelerated participation in [4 + 2] cycloaddition reactions with electron-rich dienophiles (Table 9-III and Scheme 9-II[42a]), only two additional reports of [4 + 2] cycloaddition reactions of hetero-1-aza-1,3-butadienes have been detailed [Eqs. (19) and (20)].[45,46]

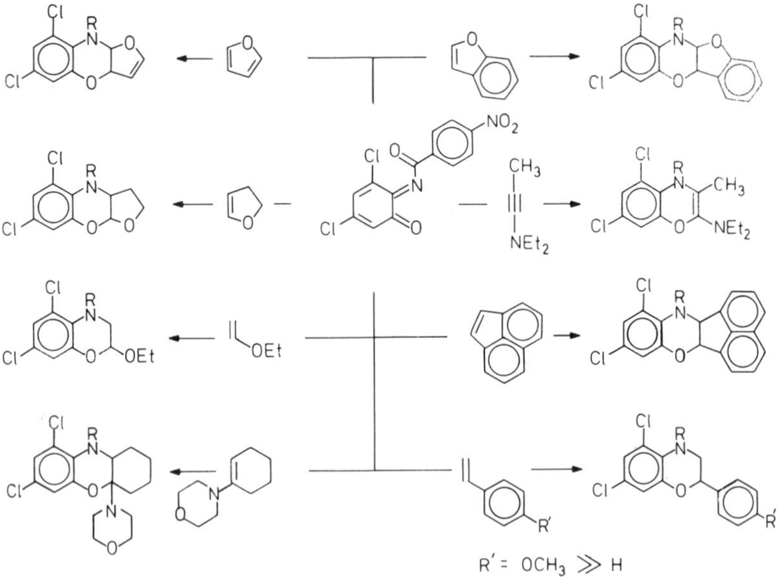
Cl
R
N
O
NO_2
CH_3
NEt_2
OEt
R′
R′ = OCH_3 ≫ H

Scheme 9-II[42a]

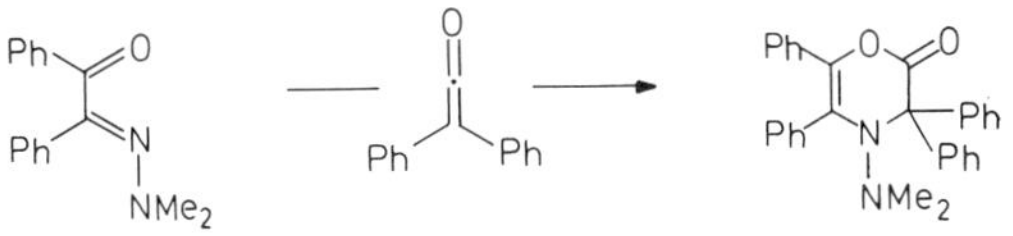

(19)

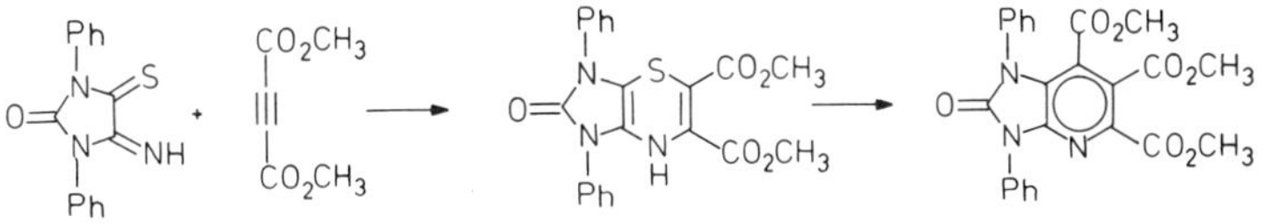

(20)

2. 2-AZA-1,3-BUTADIENES

A selected number of [4 + 2] cycloadditions of simple 2-aza-1,3-butadienes have been described [Eqs. (21)–(27)],[47–53]

(21)

(22)

R1	R2	R3	
H	H	CO_2CH_3	16%
Ph	H	CO_2CH_3	50%
Ph	H	H	29%
H	Ph	H	12%
H	Ph	CO_2CH_3	27%

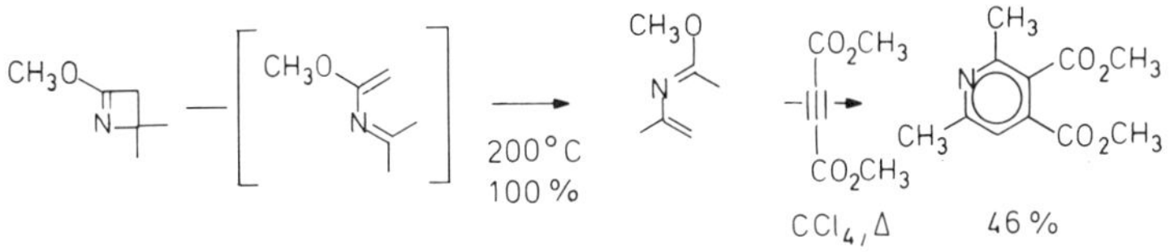

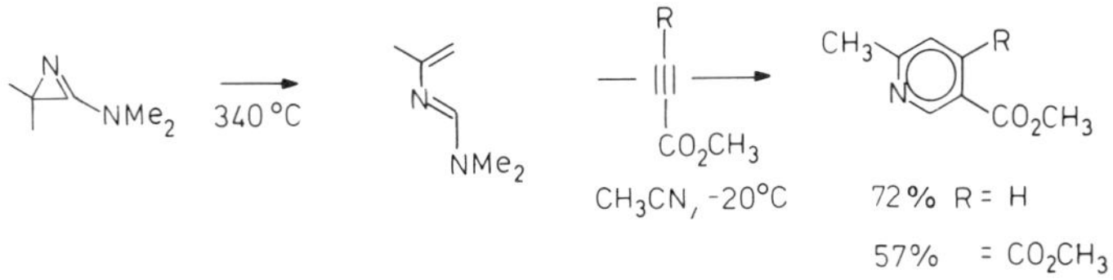

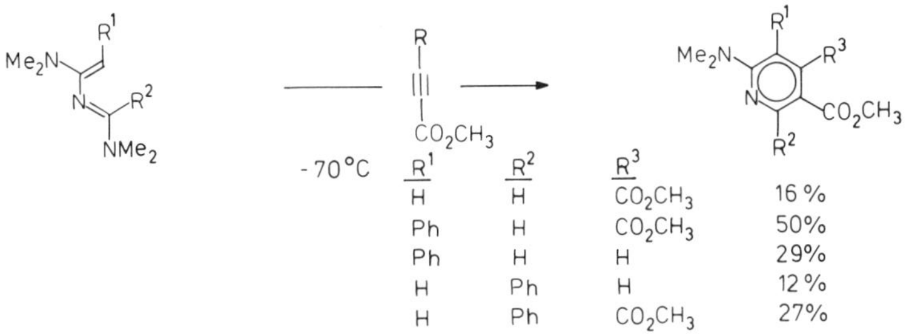

(23)

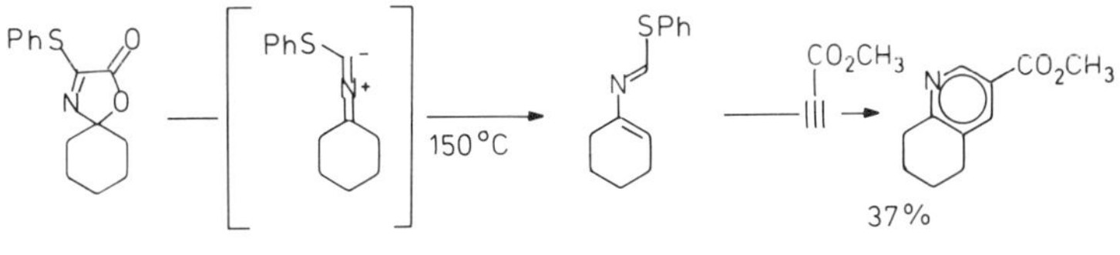

(24)

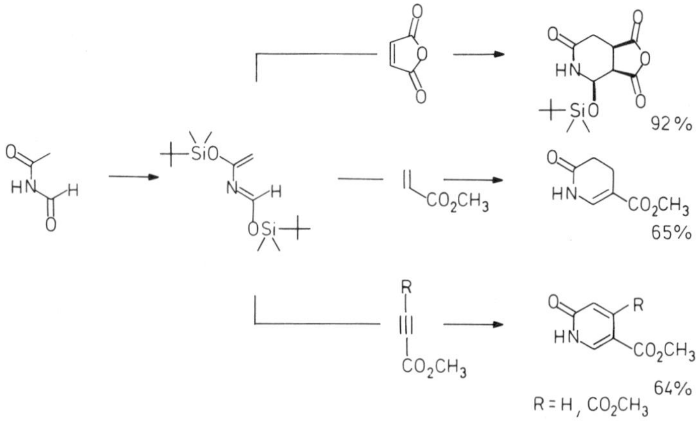

(25)

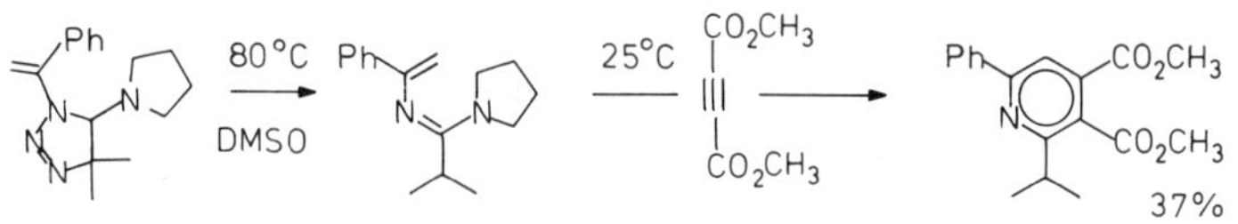

(26)

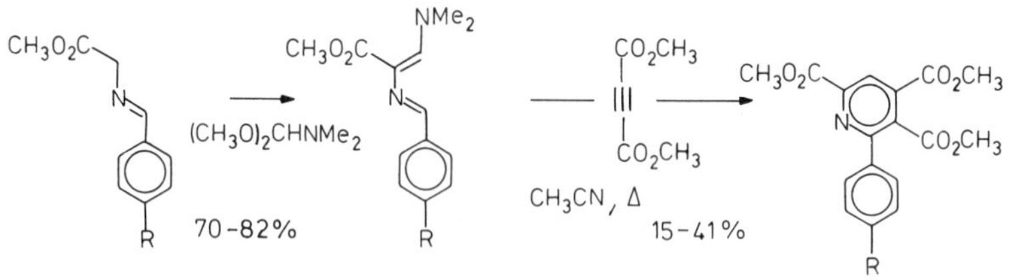

(27)

and in each instance the diene is substituted with strong electron-donating substituents capable of enhancing its reactivity toward representative electron-deficient dienophiles. The apparent ease of preparation of the 1,3-bis(*tert*-butyldimethylsilyloxy)-2-aza-1,3-butadienes from imides and the facility with which they participate in Diels–Alder reactions with a range of typical electron-deficient dienophiles [Eq. (25)] should prove exceptionally useful.[51] Products derived from the apparent [4 + 2] cycloaddition of simple, unactivated 2-aza-1,3-butadienes have been isolated from the flash vacuum pyrolytic (FVP) generation of imines and 2-aza-1,3-butadienes,[54a] and a [4 + 2] product of a simple, unsaturated imine with 4-phenyl-1,2,4-triazoline-3,5-dione has been characterized.[54b]

Substituted 1,3-oxazin-6-ones participate in exothermic inverse electron demand Diels–Alder reactions with electron-rich or reactive dienophiles[55] [Eq. (28)][55b]

$$(28)$$

including ynamines,[55a,b] enamines,[55a] ketene N,O-acetals,[55a,b] ketene acetals,[55a] enol ethers,[55a] benzynes,[55a] strained olefins,[55a] and reactive dienes.[55a] Selected electron-deficient[55a,c] and neutral alkynes[55a] and selected heterodienophiles including azirines[55a] have also been shown to undergo Diels–Alder reactions with 1,3-oxazin-6-ones. The potential utilization of such systems would seem to be limited only by the stability and ease of preparation of the parent ring system.[55a,56] $4H$-1,3-Benzoxazinones behave similarly with selected electron-rich dienophiles, but the solvent, diene substituent (R), and dienophile play an important role in determining the course of the reaction [Eq. (28)].[55d] Enamines, for instance, react with the $4H$-1,3-benzoxazinones in a different manner.[55e] Comparable, selected observations on the [4 + 2] cycloadditions of $2(1H)$-pyrazinones and $2(1H),5(4H)$-pyrazindiones have been detailed (see Chapter 10, Section 10, Table 10-VII).

$2H$-Pyrroles, cyclic 1-aza-1,3-butadienes, including pentachloro-$2H$-pyrrole (2) and the alkenyl $2H$-pyrrole 3, rearrange to the corresponding $3H$-pyrroles, cyclic 2-aza-1,3-butadienes, prior to participating in inter- or intramolecular Diels–Alder reactions[57,58] [Eq. (29)] (see top of p. 258). The initial reports of 2 participating directly in inverse electron demand Diels–Alder reactions as a cyclic 1-aza-1,3-diene have been shown to be incorrect.[57b] The *in situ* generation of 2,4,5-triphenyl-$3H$-pyrrol-3-one and its subsequent Diels–Alder reaction with dimethyl acetylenedicarboxylate represents an additional example of $3H$-pyrroles participating as 2-azadienes in [4 + 2] cycloadditions.[57c]

Several examples of the intramolecular Diels–Alder reactions of unactivated 2-aza-1,3-butadienes have been detailed[12,59–62] [e.g., Eq. (30)].[59,60]

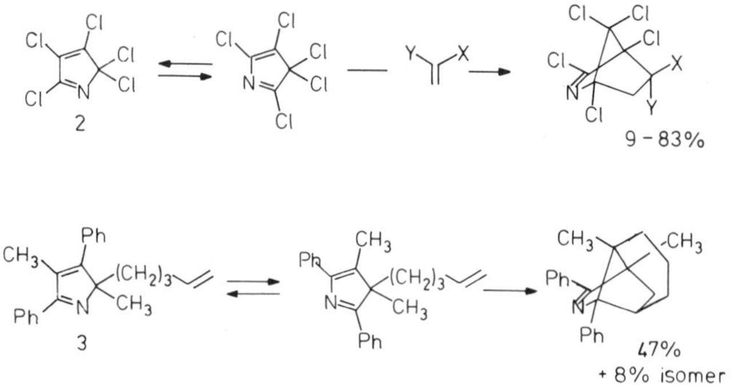

2

9 - 83%

3

47%
+ 8% isomer

(29)

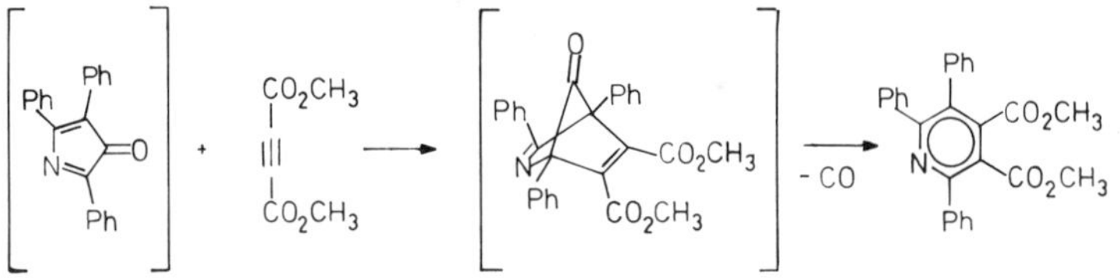

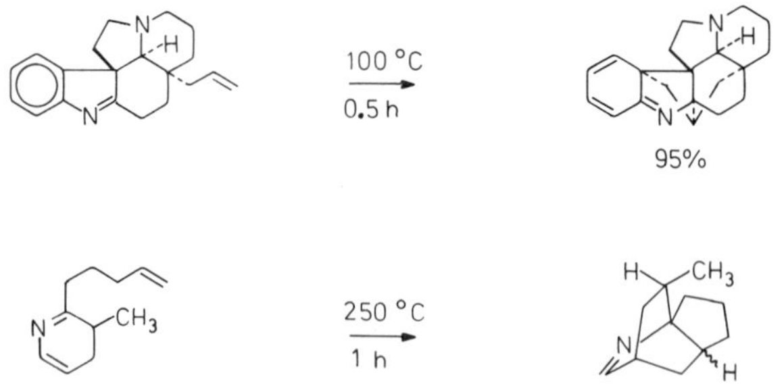

CH₃
CO₂CH₃

CO_2CH_3

$- CO$

100 °C
0.5 h

95%

(30)

250 °C
1 h

It is the entropic assistance provided by the intramolecular reaction that provides the driving force necessary to overcome the reluctant participation of the 2-azadiene in a Diels–Alder reaction. Prior reviews[12] should be consulted for compiled, tabulated examples.

Imines derived from anilines and arylaldehydes are a class of widely recognized and extensively studied 2-aza-1,3-diene systems that have been shown to participate in a range of [4 + 2] cycloadditions with electron-rich olefins in the presence of protic or Lewis acid catalysts [Eq. (31)].[4,9,63,64] These systems failed to participate in Diels-Alder reactions with typical electron-deficient dienophiles.[1,9]

In efforts to increase the rate of simple 2-aza-1,3-butadiene participation in [4 + 2] cycloadditions, Mariano and co-workers investigated the related boron trifluoride etherate-catalyzed intermolecular cycloaddition reactions of (1E,3E)- and (1E,3Z)-phenyl-2-aza-1,3-pentadiene with a range of representative electron-rich dienophiles including enol ethers and enamines. Reductive work-up (sodium borohydride) of the cycloaddition reactions (benzene, 25°C) provided piperidine products derived from regio- and stereoselective $[_\pi 2_s + _\pi 4_s]$ cycloadditions [Eq. (31)].[63b]

50-91%[64]

$BF_3 \cdot Et_2O$ [63]

33%

A

F_3B

A

27%

(31)

A = 25°C, C_6H_6;
NaBH$_4$

31%   +   8%

Only the 1*E*,3*E* stereoisomer of the 2-azadiene was found to be reactive under the reaction conditions. Consequently, each of the piperidine products was found to possess the cis 2-phenyl, 5-methyl relative stereochemistry. Moreover, the cycloaddition products derived from reactions of enol ethers possessed the all-cis 2-phenyl-4-alkoxy-5-methyl stereochemistry necessarily derived from exclusive endo cycloaddition. Dienophile geometry is maintained during the course of the [4 + 2] cycloaddition, and no products derived from a potential stepwise, addition–cyclization reaction were detected. Representative neutral and electron-deficient dienophiles failed to undergo cycloaddition. A related boron trifluoride etherate-catalyzed [4 + 2] cycloaddition of simple 2-aza-1,3-butadienes with carbonyl compounds provides 5,6-dihydro-2*H*-1,3-oxazines and appears to proceed preferentially through an endo [4 + 2] transition state although evidence supporting a stepwise, addition–cyclization was occasionally detected.[63c]

The incorporation of a 2-aza-1,3-diene into reactive systems including *N*-arylketenimines,[65a,66] *N*-arylvinylketenimines,[65b,66] aryl and vinyl isocyanates,[67,68] vinyl isothiocyanates,[68] and vinyldiimides[69] has been shown to provide 2-aza-1,3-diene systems capable of participating in [4 + 2] cycloaddition reactions with hetero- or electron-rich dienophiles [Eq. (32)].[7,65,67,68]

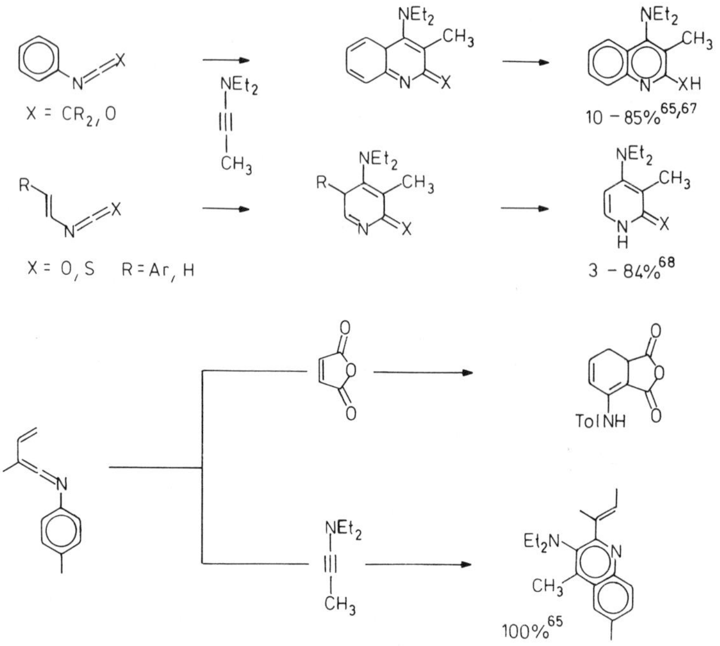

$$(32)$$

In the instances studied,[65,67] the effect of solvent polarity on the rate of [4 + 2] cycloaddition is consistent with the reactions proceeding by stepwise, addition–cyclization reactions.

6-Phenyl-5-azaazulene[70a] [Eq. (33)]

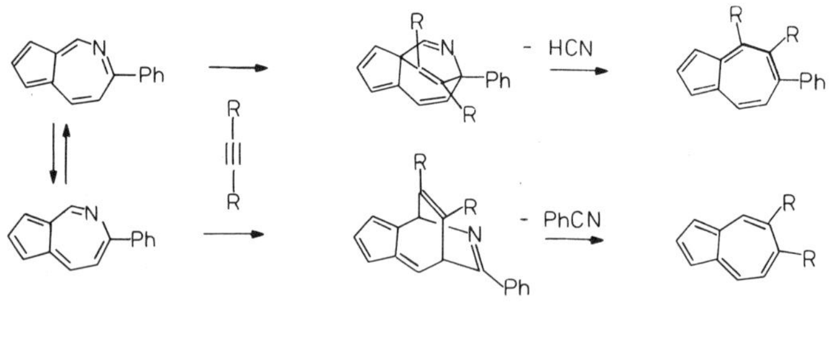

$$(33)$$

and 3,5-dimethylthio-1,2,4-triazepine[70b] have been shown to participate in Diels–Alder reactions. The [4 + 2] cycloaddition occurs across the 2-azadiene system.

3. HETERO-2-AZA-1,3-BUTADIENES

N-Acylimines, e.g., **4**, are the most widely recognized and the most extensively investigated hetero-2-azadiene system capable of participation in Diels–Alder reactions, and comprehensive reviews have been published.[7,71] In general, substituents X and Y are strongly electron-withdrawing groups and consequently the *N*-acylimines participate as electron-deficient partners in cycloaddition reactions with electron-rich dienophiles. Diels–Alder reactions of electron-deficient *N*-acylimines with vinyl ethers, enamines, olefins, sulfenes, acetylenes, and the carbon–carbon or carbon–oxygen double bond of ketenes have been detailed.[5,7,71] This 4π participation of electron-deficient *N*-acylimines does complement the ability of many simple *N*-acylimines to behave as 2π dienophile components in Diels–Alder reactions with typical electron-rich dienes.[6]

4   X = CCl₃, Y = H

X = Y = CF₃

Selected examples of *N*-thioacylimines, e.g., **5**, have been investigated and shown to behave in a similar manner.[71,72] *N*-Thioacyl dithioimidocarbonates (e.g., **6**), 4-thia-2-aza-1,3-butadienes possesing two electron-donating substituents, have been shown to undergo regiospecific, normal ($\text{HOMO}_{\text{diene}}$ controlled) Diels–Alder reactions with typical electron-deficient alkynes and olefins including dimethyl acetylenedicarboxylate, *N*-phenylmaleimide, methyl acrylate, and methyl vinyl ketone (22–84%).[72b]

The intramolecular 4π participation of an *N*-acylimine in a Diels–Alder reaction[73] and the use of *N*-acylimines bearing chiral auxiliaries in inverse electron demand Diels–Alder reactions with an observable high degree of asymmetric induction[74] illustrate additional capabilities for the applications of such systems which have not been fully explored [Eqs. (34) and (35)].[73,74]

(34)

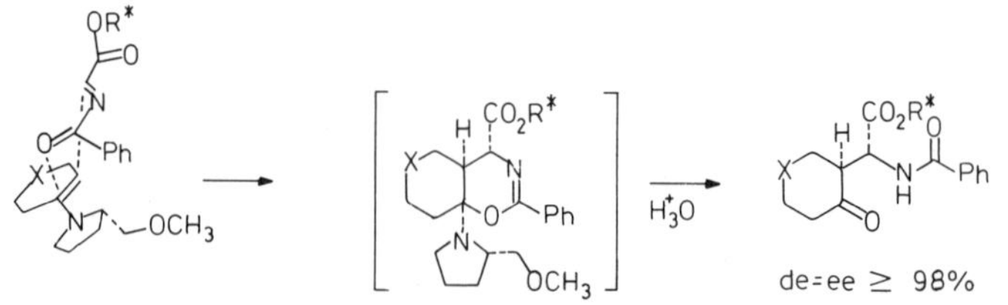

$$de \fallingdotseq ee \geq 98\% \tag{35}$$

An extensively investigated and useful hetero-2-azadiene system capable of 4π participation in Diels–Alder reactions is the vinylnitroso compounds.[75-78] The complementary addition of electron-withdrawing substituents to the 3 position of the vinylnitroso system enhances the rate of diene participation in inverse electron demand Diels–Alder reactions with electron-rich or neutral dienophiles (simple olefins).[75,76] Table 9-IV summarizes a series of representative examples of the 4π participation of vinylnitroso compounds in Diels–Alder reactions, and an extensive review has summarized much of this work.[5,75]

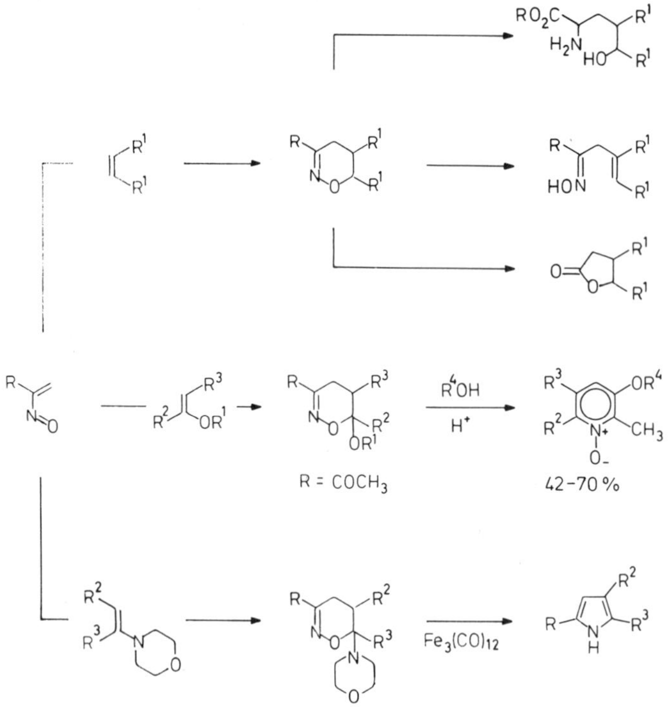

Scheme 9-III

TABLE 9-IV

**Diels–Alder Reactions of Vinylnitroso
Compounds (4π Component) with
Representative Olefins and Dienes**[a]

Vinylnitroso	Dienophile				Product

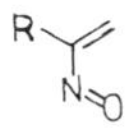

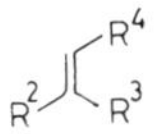

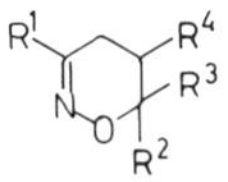

R[1]	R[2]	R[3]	R[4]	Yield
Ph	Morpholine	Ph	H	92%
Ph	Morpholine	$(CH_2)_4$		91%
CO_2Et	Morpholine	$(CH_2)_4$		100%
Ph	OEt	H	H	87%
CO_2Et	OEt	H	H	82%
CO_2Et	Ph	CH_3	H	43%
CHO	Ph	CH_3	H	43%
$COCH_3$	Ph	CH_3	H	44%
Ph	Ph	CH_3	H	26%
$4-NO_2C_6H_4$	Ph	CH_3	H	83%
$COCH_3$	Ph	H	Ph	46%
CHO	Ph	H	Ph	13%
Ph	Ph	H	Ph	5%
$4-NO_2C_6H_4$	Ph	H	Ph	57%
CHO	H	$(CH_2)_6$		42%
$COCH_3$	H	$(CH_2)_4$		22%
Ph	CH_3	$C(CH_3){=}CH_2$	H	41%

R[1]	 X	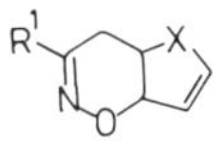 Yield
CO_2Et	CH_2	79%
Ph	CH_2	90%, 87%
$4-BrC_6H_4$	CH_2	79%
2-Furyl	CH_2	62%
H	CH_2	19%, 13%
CH_2Cl	CH_2	72%
CH_3	CH_2	23%
CN	CH_2	40%
Ph	CH_2CH_2	48%
CO_2Et	O	46%
$COCH_3$	O	75%
Ph	O	45%

[a] From Ref. 75.

The use of the 4π diene participation of vinylnitroso compounds in Diels–Alder reactions in the preparation of aryl pyruvate oximes,[76a,c,e,g,h] amino acids,[76a,c,g] γ-hydroxynitriles,[76a,d,g] γ-lactones,[76d,g] pyridine N-oxides,[76f] and pyrroles[77] is summarized in Scheme 9-III. The intramolecular Diels–Alder reactions of *in situ* generated and unactivated vinylnitroso compounds have been shown to proceed readily employing electron-rich olefins, and the course of the reaction has been shown to proceed preferentially through an endo transition state [Eq. (36)].[79]

$$\text{(36)}$$

R = OCH₃   X = H, CH₃   65-70%

One report has described the use of thiocarbonyl compounds as the heterodienophile 2π component of a Diels–Alder reaction with vinylnitroso compounds.[80]

Vinylnitroso compounds may act as 2π or 4π components of Diels–Alder reactions with typical dienes [Eq. (37)].

$$\text{(37)}$$

Of the four possible pathways, it has been empirically determined that vinylnitroso compounds bearing β substituents participate as 2π dienophile components in Diels–Alder reactions with dienes [Eq. (37), path b][6,78] whereas vinylnitroso compounds lacking β substitutents participate as 4π diene components in their Diels–Alder reactions with dienes [Eq. (37), path a].[6,76] The potential that the observed 4π participation of vinylnitroso compounds in Diels–Alder reactions with dienes [Eq. (37), path c] may arise from a sequence initiated by the all-carbon Diels–Alder reaction [Eq. (37), path d] followed by a [3,3]-sigmatropic rearrangement to the observed oxazine product [Eq. (38)]

$$\text{(38)}$$

has not received experimental verification. The related N-vinyl-N-alkylni-trosonium ions, participate in similar cycloaddition reactions (Section 10).

Acylnitroso compounds are best recognized for their ability to partici-pate as dependable 2π components of Diels–Alder reactions with dienes, and these observations have found substantial application in the total synthesis of natural products (Chapter 3).[81] However, Mackay and co-workers have detailed the isolation of 5,6-dihydro-1,4,2-dioxazine **9a** (10%) from the reaction of acylnitroso **7a** with cyclopentadiene [Eq. (39)].[82]

$$\text{(39)}$$

8a $R^1 = \underline{t}\text{-Bu}$

7a $R^1 = \underline{t}\text{-Bu}$
7b $= 4\text{-}C_6H_4NO_2$

9a $R^1 = \underline{t}\text{-Bu}$ $R^2 = R^3 = H$
9b $R^1 = 4\text{-}C_6H_4NO_2$ $R^2 = CH_3$ $R^3 = Ph$

Although this product may arise from the [3,3]-sigmatropic rearrangement of the oxazine **8a**, the acylnitroso **7b** provided 5,6-dihydro-1,4,2-dioxazine **9b** in excellent yield without the apparent intermediacy of the oxazine **8b**.[82] These observations suggest the potentially useful 4π participation of acylnitroso compounds in Diels–Alder reactions.

A number of heterocumulenes have been shown to function as hetero-2-azadienes in [4 + 2] cycloadditions with appropriate dienophiles. Alkyl- and arylacyl isocyanates (**10**, X = Y = O), isothiocyanates (**10**, X = O, Y

10

= S), thioacyl isocyanates (**10**, X = S, Y = O), or thioacyl isothiocyanates (**10**, X=Y=S) will participate in selected [4 + 2] cycloaddition reactions, and substantial differences in reactivity exist between these related he-tero-2-azadienes. Much of this work has been reviewed.[7,71]

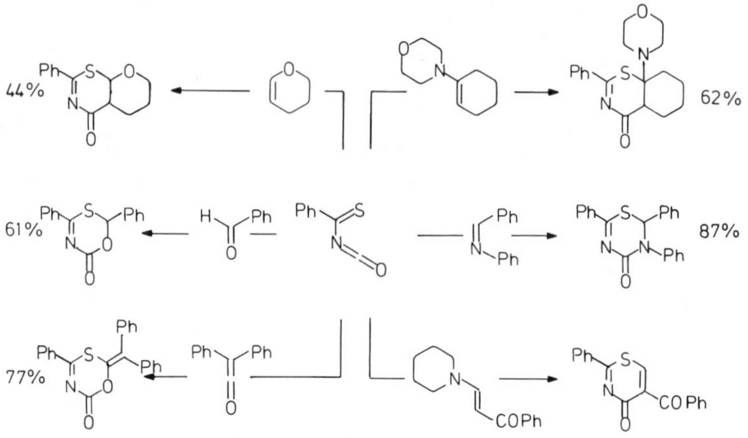

Scheme 9-IV

Because of their predictable behavior and reactivity, thioacyl isocyanates comprise the bulk of this work, and extensive studies of their [4 + 2] reactions with olefins,[83] enamines,[84] enol ethers,[84a] thioacyl isocyanates,[85] imines,[85d,86] carbodiimides,[84a,87] isocyanates,[84a] azirines,[88] β-enaminoketones,[89] dianils,[86d] azines,[90] hydrazones,[91] imidazoline-4,5-diones,[92] aryl cyanates,[93] disubstituted cyanamides,[93] aldehydes,[94] ketones,[94] ketenes,[94] alkyl or aryl iminodithiocarbonates,[95] and the carbon-carbon double bond of ketenimines[96] have been detailed. In an extensive comparative study of the [4 + 2] cycloaddition reactions of thioacyl isocyanates, the heterocumulenes bearing strong electron-withdrawing substituents were found to be more stable and less prone to participate in cycloaddition reactions.[84d] Representative examples are summarized in Scheme 9-IV.

Aromatic and aliphatic acyl isocyanates participate in a similar range of [4 + 2] cycloadditions although [2 + 2] and simple addition reactions often are observed. The acyl isocyanate substituents may determine or alter the observed course of the reaction, and the substituent effects have been detailed in extensive reviews.[7,71] Observed [4 + 2] cycloadditions of acyl isocyanates with selected olefins, p-quinones, allenes, the carbon–carbon double bond of ketenes, electron-rich acetylenes, imines, dianils, ethylenediimines, enamines, enol ethers, ketene acetals, carbodiimides, azirines, and vinyl sulfides have been described.[7b,c] The reaction of aromatic acyl isocyanates with carbodiimides is not a simple, direct [4 + 2] cycloaddition but proceeds by a kinetic [2 + 2] cycloaddition followed by a subsequent rearrangement to provide the observed [4 + 2] cycloadduct [Eq. (40)].[97]

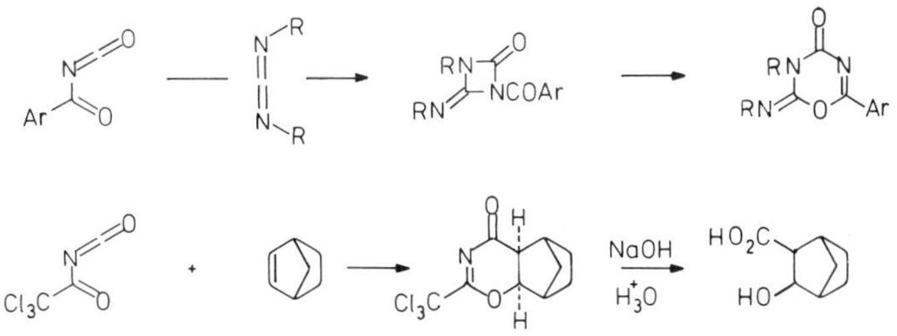

$$(40)$$

Acyl isothiocyanates have been shown to participate in [4 + 2] cycloadditions with enamines,[98] imines,[99] ketenimines,[100] and hydrazones.[101] Similar reports have detailed the [4 + 2] cycloadditions of thioacyl isothiocyanates with the carbon–carbon double bond of ketenes, ketenimines, and enamines as well as the carbon–nitrogen double bond of imines, carbodiimides, and isocyanates.[102]

N-Sulfinylaniline and related N-arylsulfinylimines participate as 2π components of [4 + 2],[6] [2 + 2], and [3 + 2] cycloadditions in addition to their observed 4π participation in Diels–Alder reactions with strained and reactive olefins [Eq. (41).][103]

$$(41)$$

N-Sulfinylurethanes and N-sulfonylamides and -urethanes similarly participate in Diels–Alder reactions with strained and electron-rich olefins [Eq. (42)].[104]

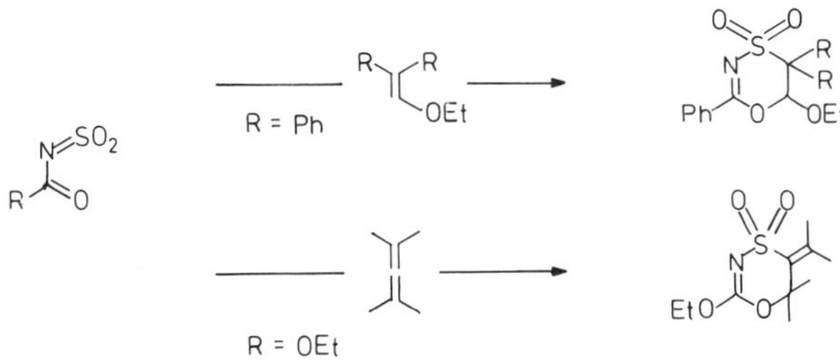

$$(42)$$

4. 1,2-DIAZA-1,3-BUTADIENES

Electron-deficient azoalkenes, 1,2-diaza-1,3-butadienes, have been shown to participate as 4π components in [4 + 2] cycloadditions with selected dienes (Table 9-V)[76a,b] and electron-rich or reactive olefins

TABLE 9-V

Diels–Alder Reactions of Azoalkenes (4π Component) with Dienes

	Azoalkene		Dienophile	Product

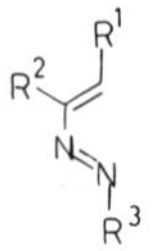

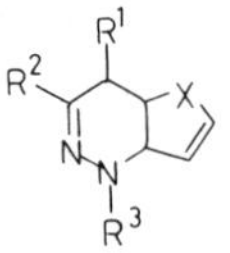

R^1	R^2	R^3	X	Yield
H	Ph	2,4-$(NO_2)_2C_6H_3$	CH_2	96%
H	Ph	$SO_2C_6H_4CH_3$	CH_2	72%
H	Ph	CO_2Et	CH_2	96%
H	CH_3	2,4-$(NO_2)_2C_6H_3$	CH_2	78%
CO_2Et	CH_3	2,4-$(NO_2)_2C_6H_3$	CH_2	83%
CO_2Et	CH_3	4-$NO_2C_6H_4$	CH_2	88%
CO_2Et	CH_3	Ph	CH_2	61%
H	H	Ph	CH_2	14%
H	Ph	2,4-$(NO_2)_2C_6H_3$	O	89%
H	CH_3	2,4-$(NO_2)_2C_6H_3$	O	22%
CO_2Et	CH_3	Ph	O	0%
H	CH_3	2,4-$(NO_2)_2C_6H_3$	$C=C(CH_3)_2$	68%

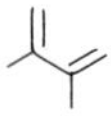

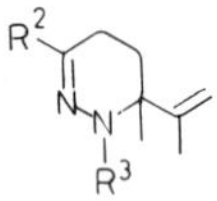

R^1	R^2	R^3	Yield
H	Ph	2,4-$(NO_2)_2C_6H_3$	67%

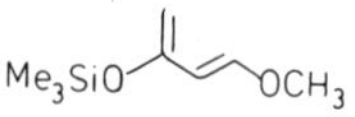

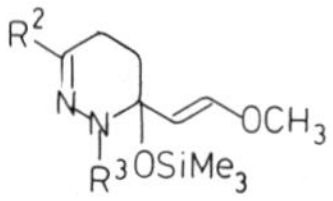

R^1	R^2	R^3	Yield
H	Ph	2,4-$(NO_2)_2C_6H_3$	32%

(Scheme 9-V)[105–107] in processes best characterized as concerted, inverse electron demand Diels–Alder reactions. Careful studies have demonstrated that the [4 + 2] cycloadditions show high endo selectivity and no rate dependency on the polarity of the reaction solvent.[106]

The feasibility of the 4π participation of azoalkenes in intramolecular Diels–Alder reactions with unactivated olefins has been detailed [Eq. (43)].[106b]

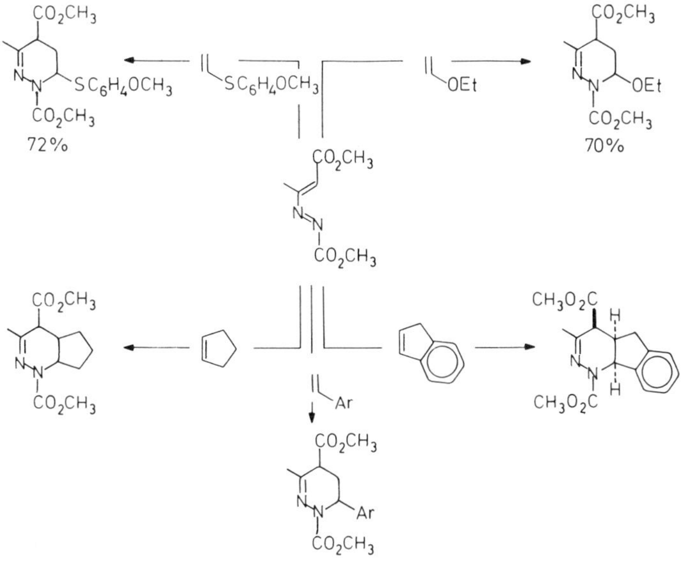

72%
70%
Ar

Scheme 9-V

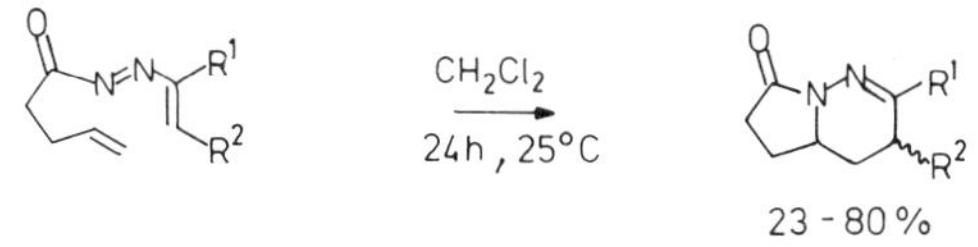

CH$_2$Cl$_2$
24h, 25°C
23 - 80%

(43)

In a complementary series of observations the simple azoalkenes, 1,2-diaza-1,3-butadienes, were shown to participate in apparent normal (HOMO$_{diene}$ controlled) Diels–Alder reactions with typical electron-deficient dienophiles [Eq. (44)][108]

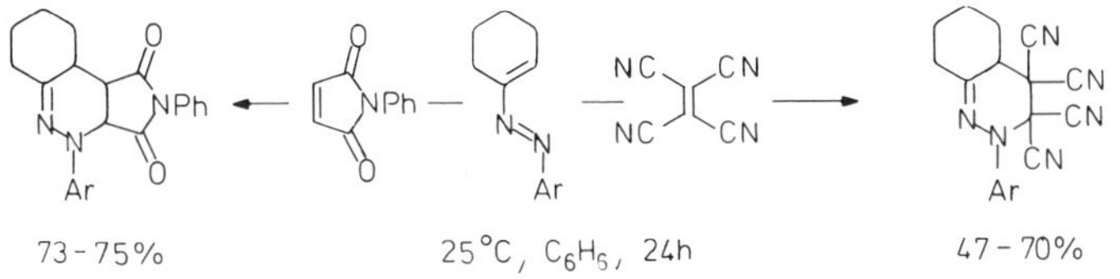

73-75%
25°C, C$_6$H$_6$, 24h
47 - 70%

(44)

as well as in [4 + 2] cycloadditions with electrophilic heterodienophiles including azodicarboxylates [Eq. (45)],[109]

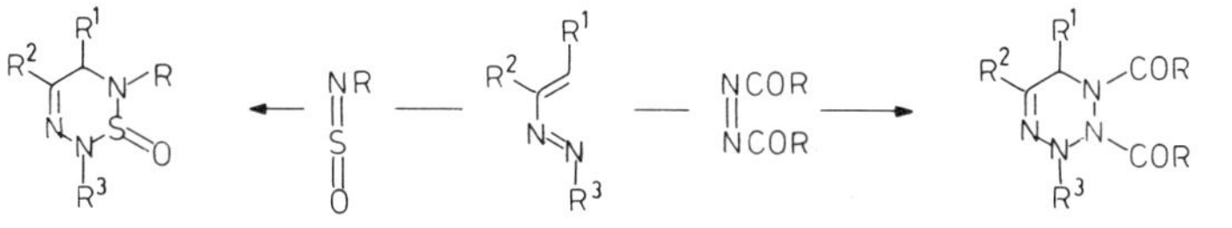
(45)

thioisocyanates [Eq. (45)],[110] thiones,[111] and sulfines.[111] Selected reactive olefins including ketenes provide mixtures of [2 + 2] and [4 + 2] cycloaddition products with the simple azoalkenes.[112]

5. HETERO-1,2-DIAZA-1,3-BUTADIENES

Azodicarboxylates are best recognized for their ability to participate as 2π components in normal (HOMO$_{diene}$ controlled) Diels–Alder reactions with dienes (Chapter 6)[5,71,113] and for their effective participation in ene reactions with reactive olefins.[113] In addition, electron-rich or reactive olefins that do not contain a reactive allylic hydrogen atom and consequently cannot enter into an ene reaction do possess the ability to participate in competing [2 + 2] and [4 + 2] cycloadditions with azodicarboxylates.[5,71,113] The Diels–Alder [4 + 2] cycloaddition of azodicarboxylates with olefins provides 1,3,4-oxadiazines (Scheme 9-VI) and is sensitive to solvent and substituent effects.[114–120] In general, the competing [2 + 2] cycloaddition to provide 1,2-diazetidines intervenes and predominates as the olefin nucleophilicity and the reaction solvent polarity are increased [Eqs. (46) and (47)].[113]

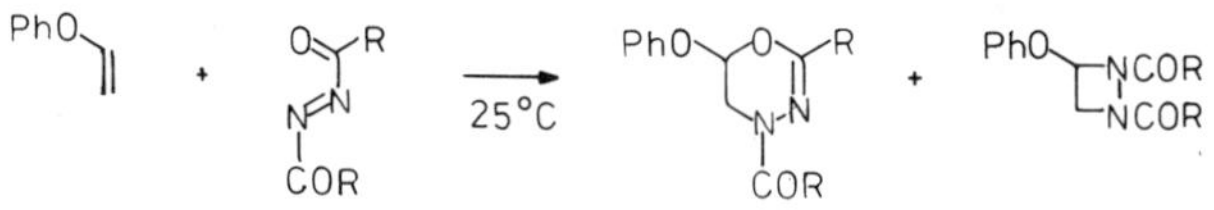

R = OCH$_2$CCl$_3$	85%	>95	<5
R = OCH$_3$	73%	77	23
R = OEt	84%	65	35
R = Ph	66%	<5	>95

	Solvent		
R = OCH$_3$	CH$_3$CN	80	20
	neat	65	35
	Et$_2$O	33	67
	C$_6$H$_6$	33	67

(46)

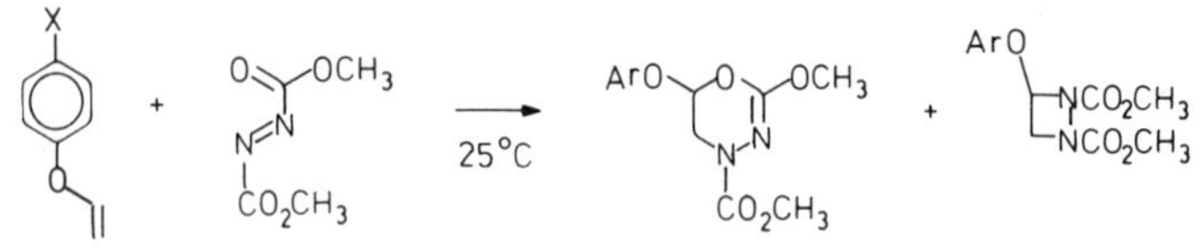

X = OCH$_3$	82%	13	87
X = CH$_3$	85%	16	84
X = H	73%	23	77
X = Cl	56%	33	67
X = NO$_2$	40%	95	5

(47)

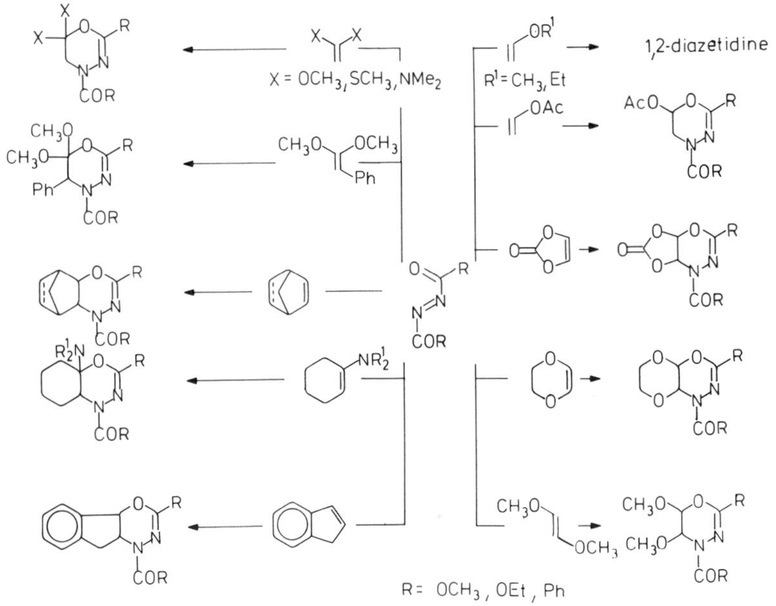

Scheme 9-VI

Alkoxycarbonylaroyldiimides (**11a**, $R^1 = OR$, $R^2 = Ar$),[116,120,122] diaroyl diimides (**11a**, $R^1 = R^2 = Ar$),[116–118, 121] and arylaroyldiimides (**11b**, $R^1 = R^2$

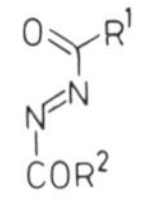

$= Ar$)[122] participate with increasing selectivity as 4π components in [4 + 2] cycloadditions, and much of this work has been reviewed.[5,71,113] In selected instances, the expected bicyclodiazine products of the normal Diels–Alder reaction of cyclopentadienes and stable cyclopentadienones with 2π participation of azodicarboxylates and diacyldiimides may rearrange to the corresponding 1,3,4-oxadiazines, the formal products of Diels–Alder 4π participation of the azodicarboxylate or diacyldiimides with the dienes [Eq. (48)].[123]

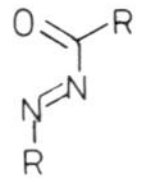

$$R = \underline{t}\text{-Bu} > Ph > CH_3 > OR$$

6. 1,3-DIAZA-1,3-BUTADIENES

A select group of 1,3-diaza-1,3-butadienes has been reported to undergo [4 + 2] cycloadditions, but the lack of extensive efforts with this system reflects the current difficulty encountered in the preparation of stable 1,3-diaza-1,3-butadienes and their reluctance to participate in the Diels–Alder reactions. Successful efforts which have been described include the thermal isomerization of an unsaturated N-silylurea with the *in situ* generation and subsequent Diels–Alder reaction of a 2-trimethylsilyloxy-1,3-diaza-1,3-butadiene [Eq. (49)],[124]

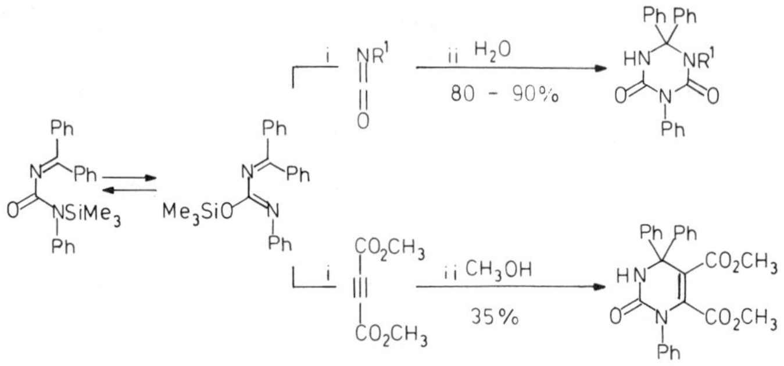

(49)

the [4 + 2] cycloadditions of 4,6-diaryl-1,2,3,5-oxathiodiazine 2-oxides with nucleophilic heterodienophiles[125] or electron-rich olefins[126] [Eq. (50)],

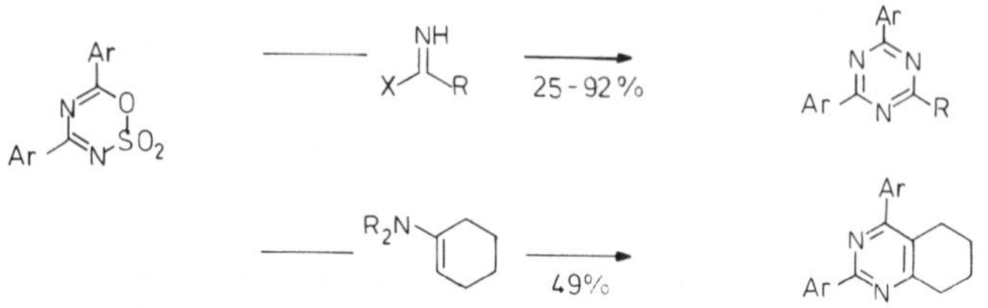

(50)

the *in situ* generation and subsequent intramolecular Diels–Alder reactions of 2-thiomethyl-1,3-diaza-1,3-butadienes [Eq. (51)],[127a]

(51)

selected examples of the Diels–Alder reactions of 1,2-dihydro-1,3,5-triazines,[127b] and two examples of stable, simple 1,3-diaza-1,3-butadienes

TABLE 9-VI

[4 + 2] Cycloadditions of Heterocyclic 1,3-Diaza-1,3-butadienes

1,3-Diaza-1,3-butadiene	Dienophile	Conditions	Product	Ref.
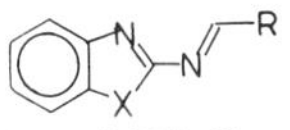 $X = O, S, NR'; R = 4-NO_2C_6H_4, 4-NMe_2C_6H_4, 4-CH_3OC_6H_4$	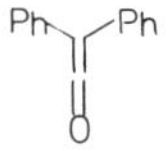	Xylene, reflux, 50–91% yield	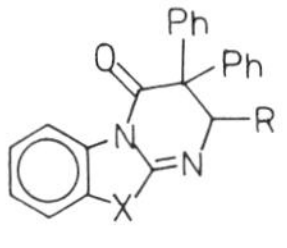	131
$X = S; R = Me_2N$		—	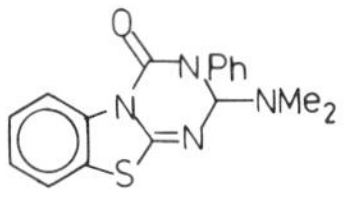	132
$X = O, NH; R = 4-NO_2C_6H_4$	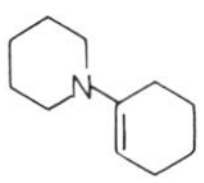	CHCl₃, reflux, 33–37% yield	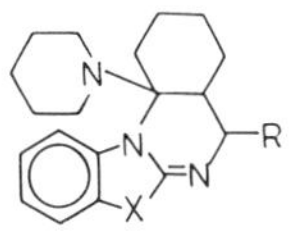	136
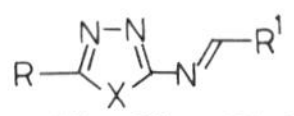 $R = Ph, Ch_3, H; X = O, S, Se; R^1 = Ar$	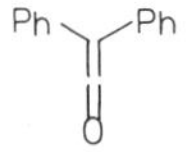	—, 33–87% yield	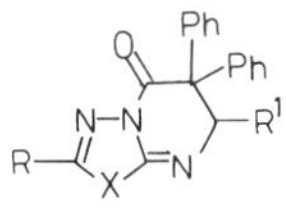	131
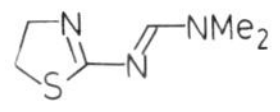		—	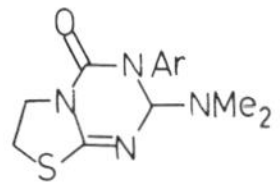	132
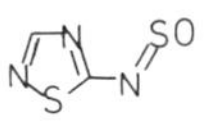		—	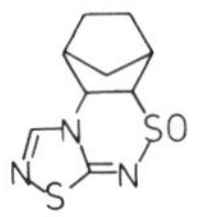	104a
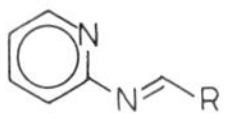 $R = OEt, Ph$	 $R^1 = Ph, SiMe_3$	—	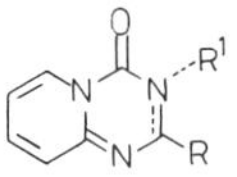	133
$R = =O, =S$	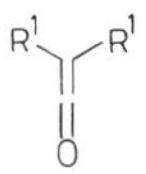 $R^1 = H$	—	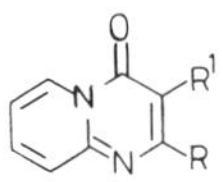	135
$R = Ph, OEt$	$R^1 = Cl$	—		134

which provide [4 + 2] cycloaddition products with selected dienophiles [Eqs. (52) and (53)].[128,129]

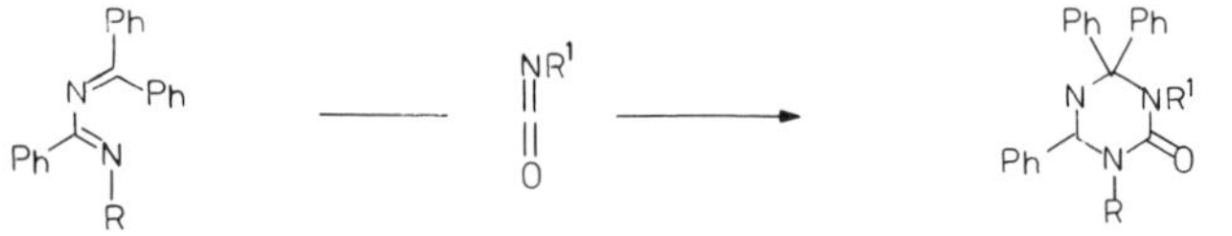

(52)

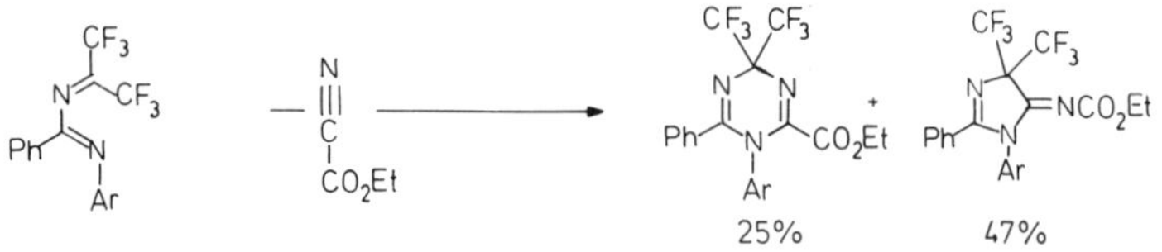

(53)

25%   47%

The incorporation of the 1,3-diaza-1,3-butadiene system into a reactive heterocumulene system, imidoyl isothiocyanates and imidoyl isocyanates [Eq. (54)],[130,137]

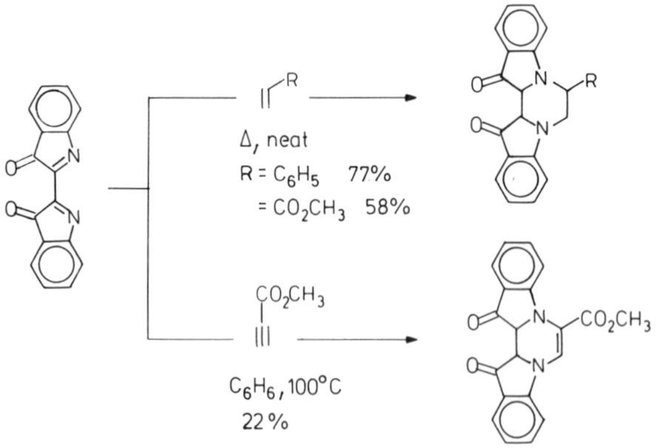

Δ, neat
R = C$_6$H$_5$ 77%
  = CO$_2$CH$_3$ 58%

CO$_2$CH$_3$

C$_6$H$_6$, 100°C
22%

(54)

or its incorporation into aromatic and heteroaromatic systems (Table 9-VI)[131-136] have permitted the observation of selected [4 + 2] cycloadditions with reactive olefins or heterodienophiles and represent processes proceeding through discrete dipolar intermediates.

7. 1,4-DIAZA-1,3-BUTADIENES

Only a select group of 1,4-diaza-1,3-butadienes has been shown to function as 4π components of [4 + 2] cycloadditions. Dehydroindigo affords Diels–Alder products on reaction with styrene, vinyl aryls, acrylonitrile, methyl acrylate, and methyl propiolate under forcing conditions [Eq.

(54)],[137] the cyclic thiocarbamate diimine **12** affords products derived from an initial [4 + 2] cycloaddition with dimethyl acetylenedicarboxylate [Eq. (55)],[46]

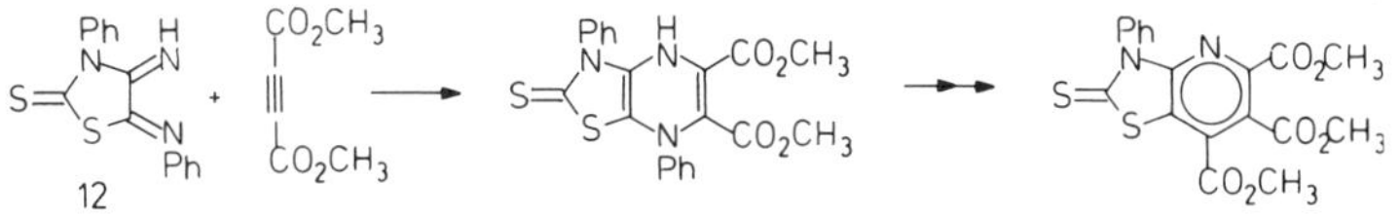

$$(55)$$

and N-aryl-1,4-diaza-1,3-butadienes have provided Diels–Alder products on reaction with reactive dienophiles [Eq. (56)].[138]

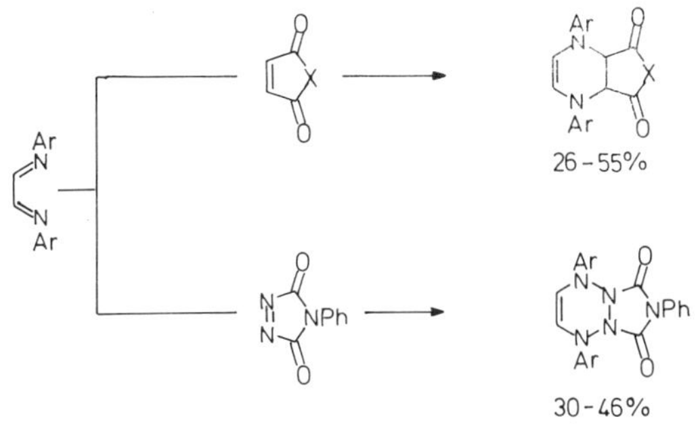

$$(56)$$

Early studies on the dimerization of substituted o-benzoquinone diimines[139a] and their reactions with diarylketenes[139c] [Eq. (57)][139]

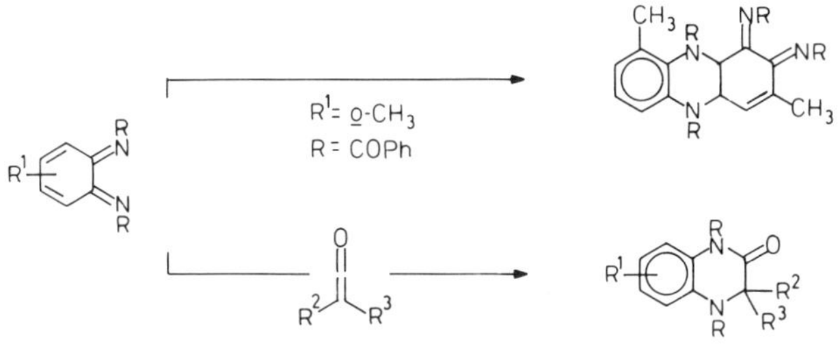

$$(57)$$

revealed that the reactions exhibit solvent effects, substituent effects, and activation parameters consistent with a moderately polar transition state with the diimines participating as electron-deficient heterodienes. Continued investigations have highlighted the heterodiene participation of o-benzoquinone diimines in inverse electron demand Diels–Alder reactions with electron-rich or strained, reactive olefins.[42a,139b] o-Benzoquinone diimines do participate as all-carbon 4π components of Diels–Alder reactions with typical electron-deficient dienophiles.[139b]

Diiminosuccinonitrile readily participates in [4 + 2] cycloadditions with electron-rich olefins [Eq. (58)].[140]

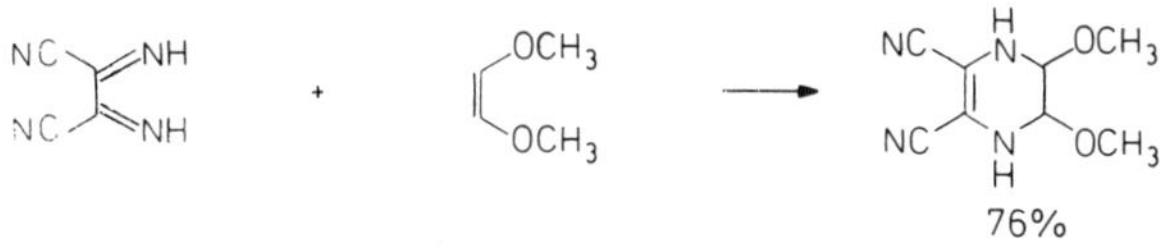

(58)

76%

A common and competing reaction of α-diimines, 1,4-diaza-1,3-butadienes, with electron-rich or reactive olefins is [2 + 2] cycloaddition to afford azetidine or β-lactam cycloaddition products [Eq. (59)],

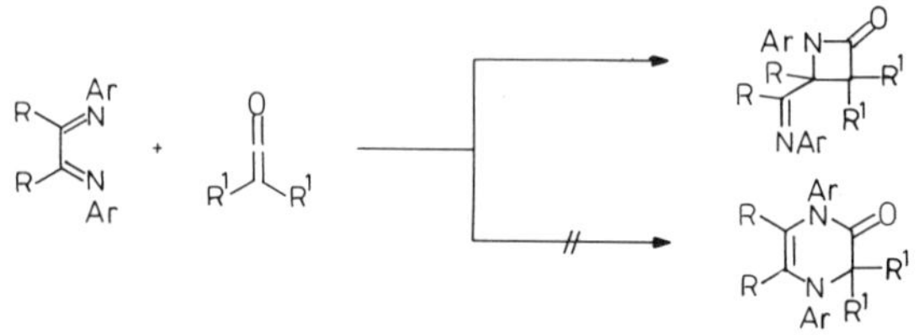

(59)

and early reports of apparent [4 + 2] cycloaddition reactions often are incorrect.[141]

8. 2,3-DIAZA-1,3-BUTADIENES

Conjugated dienes possessing the 2,3-diaza-1,3-butadiene system rarely participate in effective [4 + 2] cycloadditions. Typical efforts to promote the Diels–Alder cycloadditions of such systems with representative dienophiles afford 2:1 adducts or [3 + 2] criss–cross products.[11]

In selected instances 2,5-diphenyl-1-3,4-diazacyclopentadienone (3,6-diphenyl-4H-pyrazole-4-one) [Eq. (60)],[142]

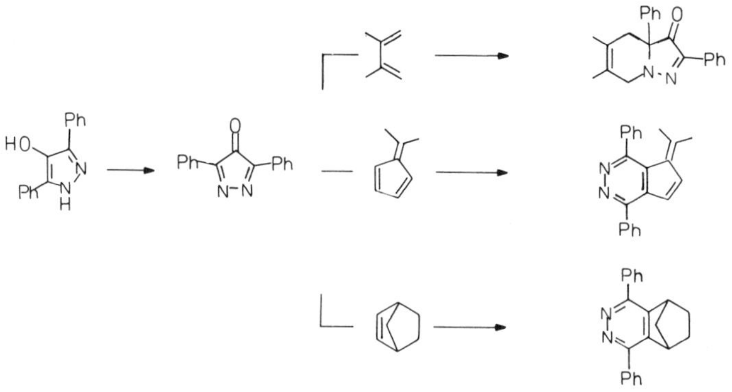

(60)

4,4-dimethyl-3,6-diphenyl-4H-pyrazole [Eq. (61)],[143]

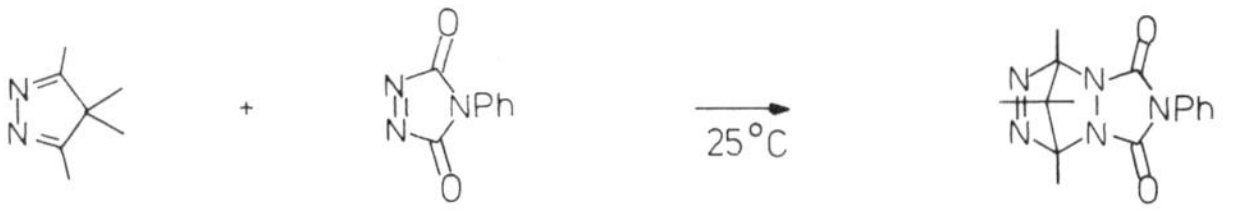

$$\text{(61)}$$

2,5-diphenyl-6-oxo-1,3,4-oxadiazine [Eq. (62)],[144]

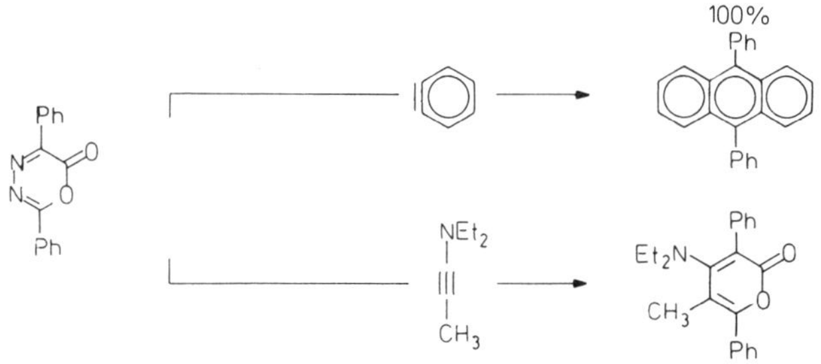

$$\text{(62)}$$

and the 2,3-diazadiene *13* [Eq. (63)][145]

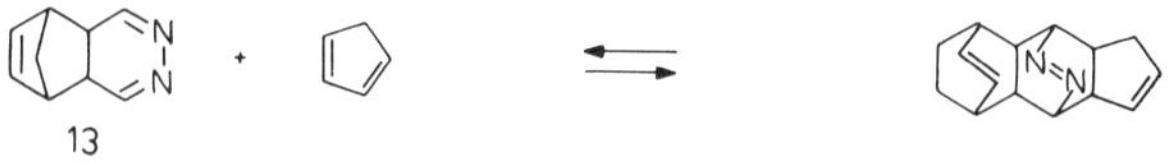

$$\text{(63)}$$

afford products derived from their [4 + 2] cycloaddition with dienophiles. 3,6-Diphenyl-4*H*-pyrazole-4-one [Eq. (60)][142] fails to react with conventional dienophiles including maleic anhydride, dimethyl acetylenedicarboxylate, diphenylacetylene, dimethyl fumarate, isobutyl vinyl ether, cyclopentene, and cyclohexene.[142b] The [4 + 2] cycloadditions of 2,5-diphenyl-6-oxo-1,3,4-oxadiazine [Eq. (62)] and related 6-oxo-1,3,4-oxadiazines with strained and reactive olefins or alkynes including cyclopropenes, cyclobutenes, norbornene, norbornadiene, styrenes, benzynes, ynamines, and cyclooctyne have been detailed, and their participation in intramolecular Diels–Alder reactions with unactivated alkenes and alkynes has been investigated.[144c] The reactions of 6-oxo-1,3,4-oxadiazines with simple and electron-rich dienophiles may often take an unanticipated course.[144,144b]

Additional examples of the Diels–Alder reactions of cyclic 2,3-diazabutadienes include the reactions of 4,5-dihydropyridazines, the products of the Diels–Alder reactions of 1,2,4,5-tetrazines with olefins. Specific examples of these reactions are summarized in Chapter 10, Section 14.

9. 1,2,4-TRIAZA-1,3-BUTADIENES

One selected heterocyclic 1,2,4-triazadiene system has been reported to participate in a range of [4 + 2] cycloadditions [Eq. (64)],[146a]

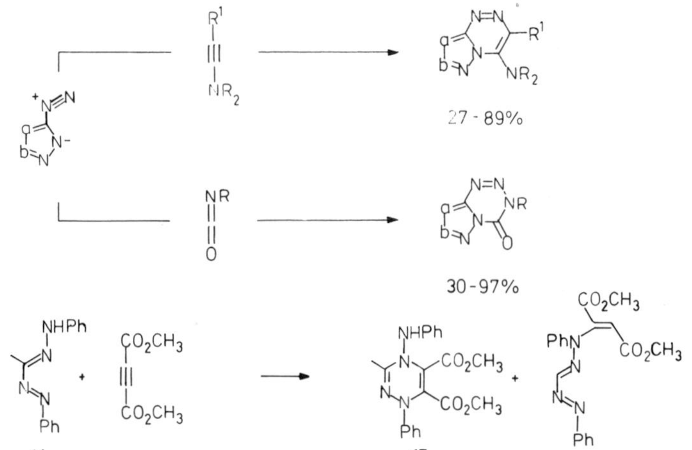

27 - 89%

(64)

30 - 97%

14   15

and the reaction of the formazan **14** with dimethyl acetylenedicarboxylate
provides a mixture of products which includes the [4 + 2] adduct **15** [Eq.
(64)].[146b] Intramolecular [4 + 2] cycloadditions of an electron-deficient
1,2,4-triaza-1,3-butadiene have been detailed [Eq. (65)].[147]

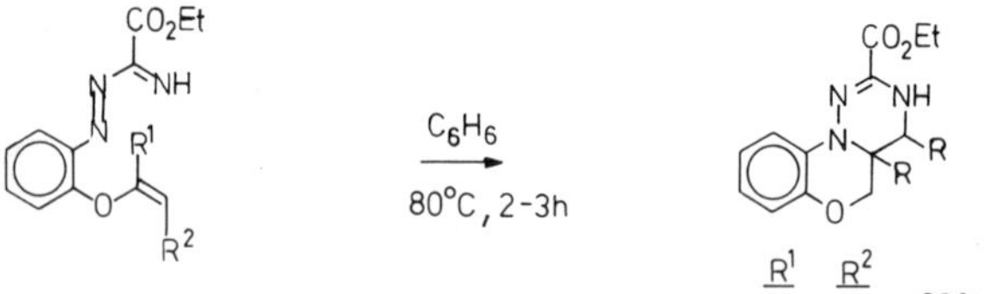

R¹	R²	
H	H	63%
CH₃	H	69%
H	Ph	79%

(65)

10. CATIONIC AZADIENES, [4⁺ + 2] CYCLOADDITIONS

The preparation or generation of azabutadiene systems bearing a formal
cationic charge, a quaternary iminium salt, provides an enhancement of
the electron-deficient nature of the azabutadiene systems and increases
their potential for participation in cycloaddition reactions with electron-
rich or neutral dienophiles. The classification of cationic azadiene [4⁺ + 2]
cycloadditions as "polar" cycloadditions is not derived from an implied
stepwise addition–cyclization mechanism for the reaction but was terminology introduced to distinguish cycloadditions employing cationic or
anionic components from those employing dipolar or uncharged components.[8b]

The most thoroughly investigated and widely recognized class of cationic azadiene systems capable of participation in [4+ + 2] cycloaddition reactions is the aromatic quaternary salts including the acridizinium and isoquinolinium cations. Extensive studies of Bradsher,[148] Fields,[149] and co-workers have outlined the scope, synthetic applications, and mechanistic details of cycloaddition reactions of the cationic aromatic azadienes typified by the acridizinium/isoquinolinium cations, and much of this work has been reviewed.[150] Nucleophilic alkenes including ketene acetals react in minutes at room temperature with acridizinium and isoquinolinium cations while typical electron-deficient dienophiles require more vigorous reaction conditions (100–160°C, 10–90 hr) (Scheme 9-VII).

The [4+ + 2] cycloadditions proceed in a regiospecific manner with the nucleophilic carbon of the electron-rich dienophile attaching to the expected electrophilic site of the aromatic quaternary salt (e.g., C-6 for acridizinium salts, C-1 for isoquinolinium salts) and the rate of reaction (ketene aminals, ketene acetals, enamines > cyclopentadiene > 2,3-dimethylbutadiene > styrene > dihydropyran > 2-butene > maleic anhydride) does increase as the nucleophilic character of the dienophile is enhanced [Eq. (66)].[148e,149]

$$X = CH_3O \ (4.3); \ CH_3 \ (1.7); \ H \ (1.0); \ NO_2 \ (0.34)$$

(66)

In the exploration of an effect comparable to the recognized Alder and Stein endo addition rule[151] for the normal (HOMO$_{diene}$ controlled) Diels–Alder reaction, Bradsher and co-workers have carefully defined and explored the stereoselectivity of the acridizinium/isoquinolinium salt [4+ + 2] cycloadditions.[148b,e] The observed stereoselectivity may be best rationalized by electrostatic interactions at or enroute to the transition state for the [4+ + 2] cycloaddition. Electron-withdrawing dienophile substituents often are found cis to the quaternary nitrogen in the cycloaddition products[148b,e] while electron-donating or alkyl/alkenyl dienophile substituents are found trans to the quaternary nitrogen[148b–e] (Scheme 9-VII). Although most investigators in this area are content to represent the [4+ + 2] cycloaddition of aromatic quaternary salts as stepwise, addition–cyclization reactions, the experimental results to date are consistent with the participation of the aromatic quaternary salts in concerted, but nonsynchronous [4+ + 2] cycloadditions in which polar influences dominate the regio- and stereospecificity of the cycloaddition processes.[148e,150]

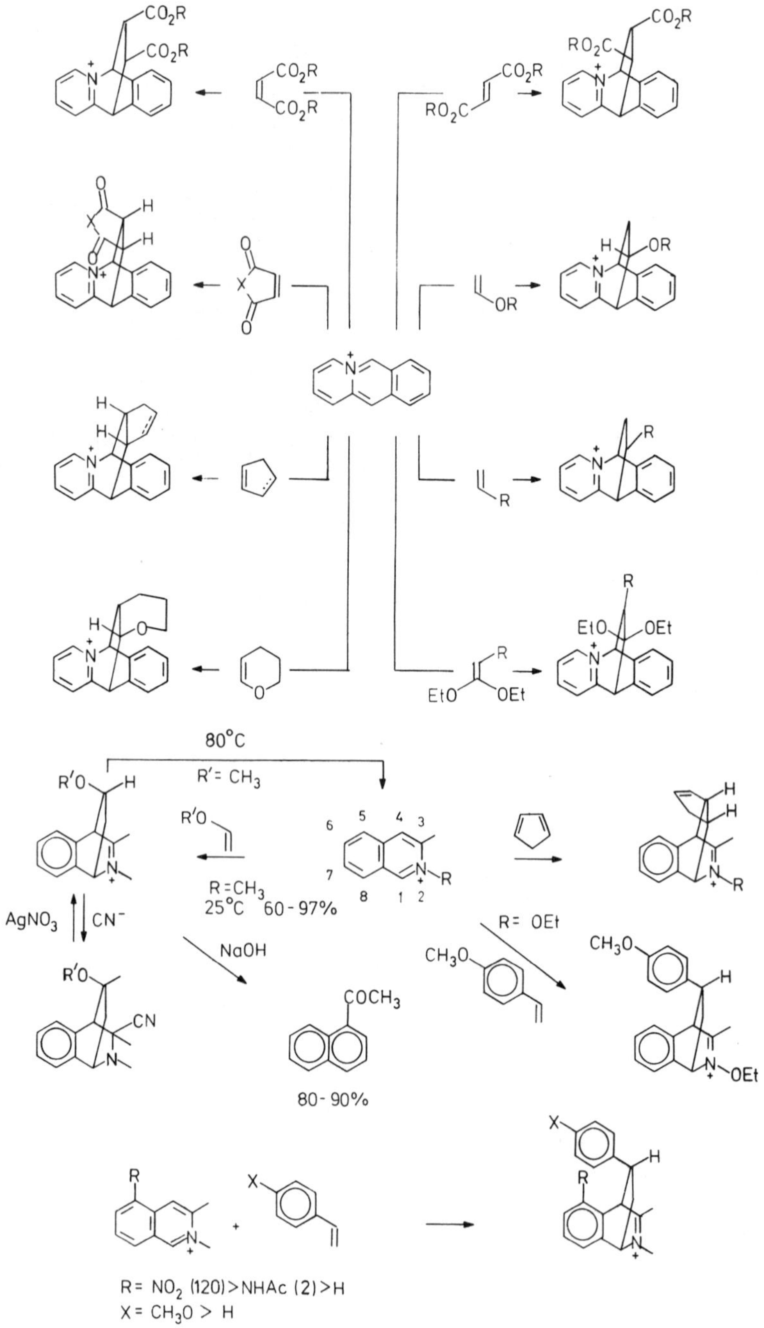
80°C
R′= CH3
R′O
R=CH3
25°C 60-97%
AgNO3
CN⁻
NaOH
COCH3
80-90%
R= OEt
CH3O
OEt
EtO
R= NO2 (120)>NHAc (2)>H
X= CH3O > H

Scheme 9-VII

Two experimental techniques have improved the observed participation of simple aromatic quaternary salts in [4+ + 2] cycloadditions. The addition of hydroquinone, which forms a stabilized 1:2 complex with the isoquinolinium salts, to the reaction mixture accelerates the rate of observed isoquinolinium salt [4+ + 2] cycloaddition.[148b] In addition, the use of 2,4-dinitrophenyl aromatic quaternary salts, prepared from the parent base and 2,4-dinitrophenyl bromide or chloride, accentuates the electron-deficient character of the quaternary salt and accelerates their participation in [4+ + 2] cycloadditions with electron-rich dienophiles.[152] These observations of Falck and co-workers have proved useful for promoting the [4+ + 2] cycloaddition of isoquinolinium systems previously regarded unmanageable. The application of these observations in the total syntheses of 14-epicorynoline (**16**) and methyl arrnothanamide (**17**) have been detailed [Eq. (67)],[152]

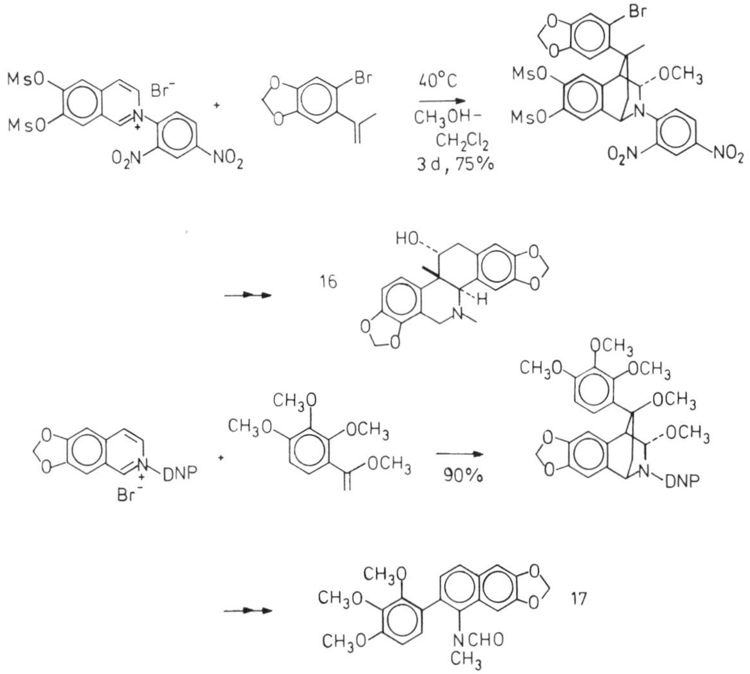

$$(67)$$

and the utility of the [4+ + 2] cycloadducts as precursors in the preparation of 1-naphthaldehydes [Eqs. (68) and (69)]

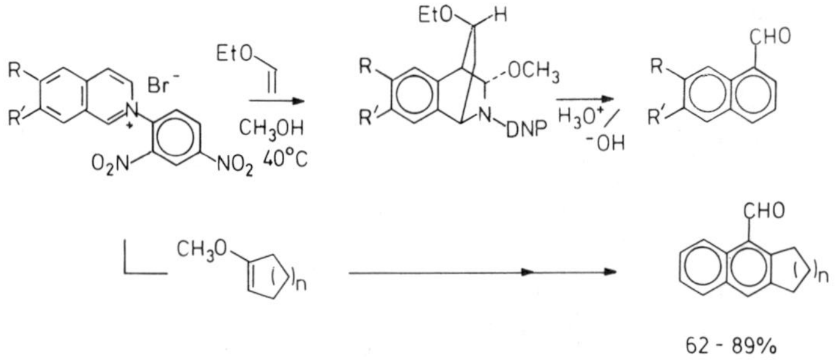

(68)

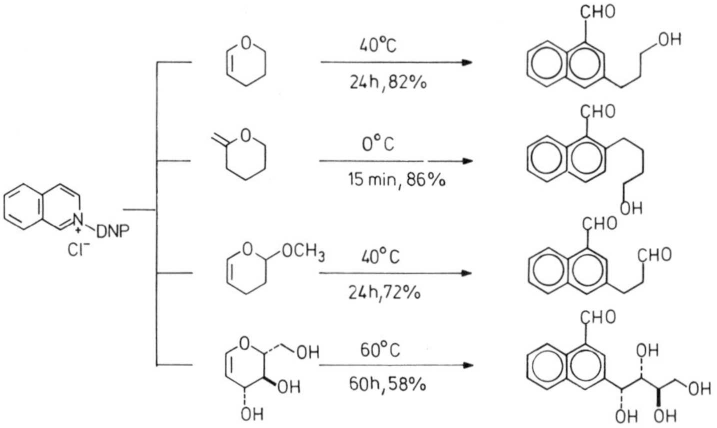

62 - 89%

(69)

have been reported.[152,153]

A one-step synthesis of the protoberberine alkaloid karachine (**18**), employing a presumed Mannich reaction of 4-methyl-2-trimethylsilyloxy-2,4-pentadiene with berberine, has been detailed by Stevens and Pruitt.[154] It is plausible that the initial step of the one-flask reaction sequence leading to karachine is, in fact, a $[4^+ + 2]$ cycloaddition of the quaternary aromatic salt, berberine chloride, with the terminal electron-rich olefin of the silyloxydiene [Eq. (70)].

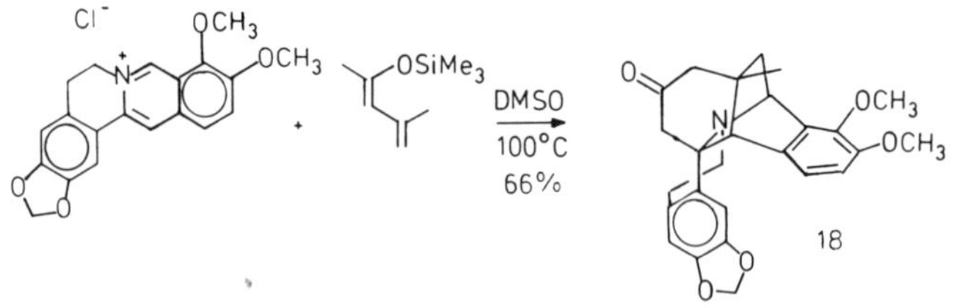

(70)

The thermally induced intramolecular $[4^+ + 2]$ cycloadditions of isoquinolinium salts bearing an unactivated alkenyl side chain have been investigated and shown to proceed with addition across the isolated double bond and the 1,4 positions of the isoquinolinium system to produce tetracyclic adducts [Eq. (71)].[155]

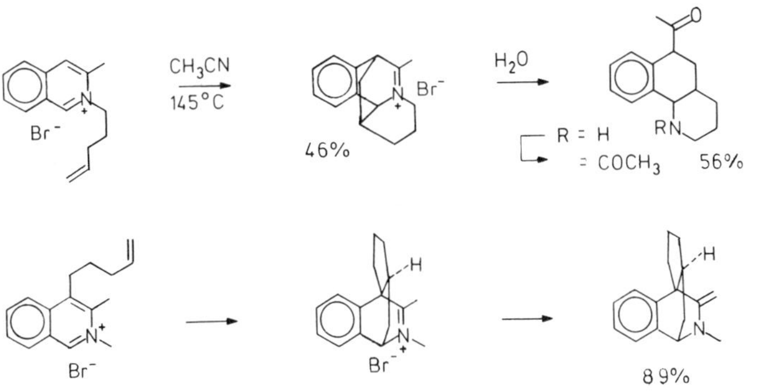

$$(71)$$

In related efforts, the reaction of N-methylquinolinium iodide with ynamines [Eq. (72)],[156]

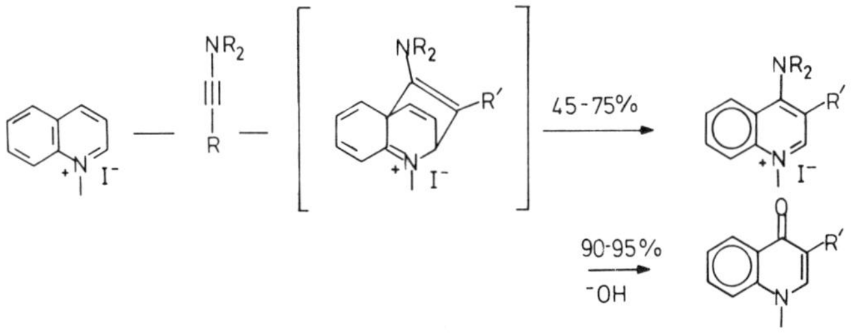

$$(72)$$

the reaction of isoxazolium salts with enamines [Eq. (73)],[157]

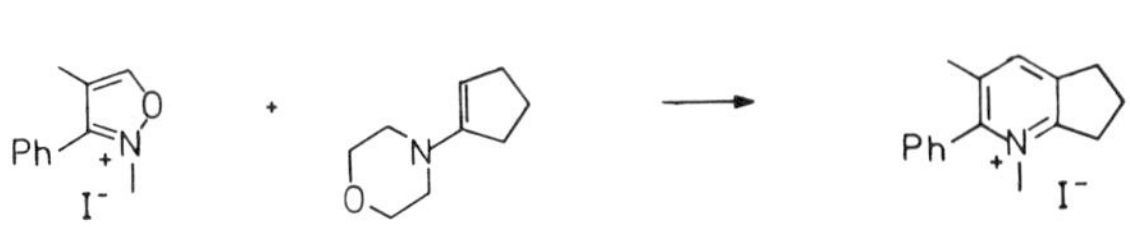

$$(73)$$

and the reaction of benzofuroxans with enamines[158] and ynamines[159] have been shown to afford products derived from $[4^+ + 2]$ cycloadditions.

The acid-catalyzed, *in situ* generation of N-alkylaryliminum salts, e.g., **19** [Eq. (74)],[160–162]

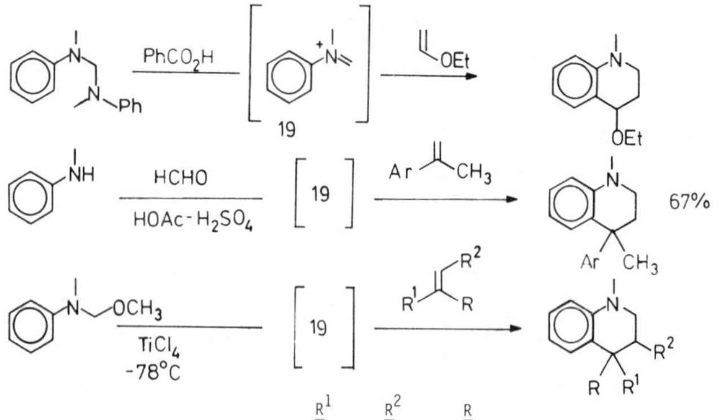

$$(74)$$

R^1	R^2	R	
H	H	Ph	84%
		$(CH_2)_2CH_3$	58%
		OEt	64%
CH_3	H	Et	89%
H	Et	$OSiMe_3$	61%
CH_3	iPr		42%
nBu	H		61%
H	Et	morpholine	81%

provides a cationic azadiene system which has been shown to be capable of participating in formal $[4^+ + 2]$ cycloadditions. An exceptionally versatile and useful procedure has been developed for the preparation of tetrahydroquinolines and julolidines utilizing the *in situ* generation of *N*-methylaryliminium salts and their subsequent participation in $[4^+ + 2]$ cycloadditions with electron-rich or simple olefins. For example, treatment of *N*-methyl-*N*-phenylmethoxymethylamine with Lewis acids (TiCl$_4$, BF$_3$·OEt$_2$, −78°C) in the presence of electron-rich or neutral olefins provides the *N*-methyltetrahydroquinolines in excellent yield [Eq. (74)].[162]

The early observation[163] that treatment of *N*-methylacetanilide with phosphoryl chloride provides **20** presumably by the $[4^+ + 2]$ reaction of an α-chloroenamine with its tautomeric iminium ion [Eq (75)],

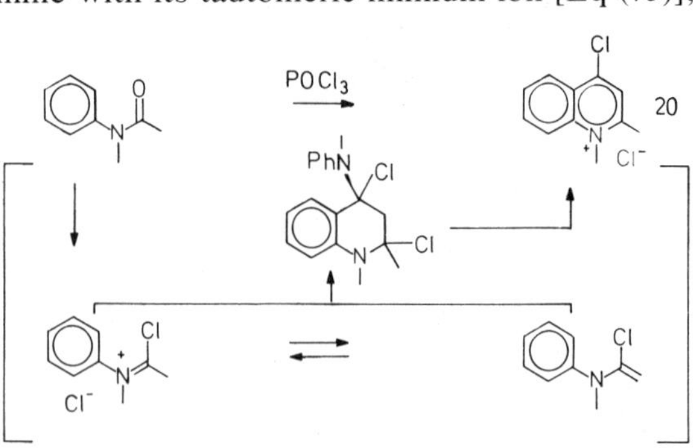

$$(75)$$

has provided the basis for the development of a controlled approach to the preparation of isoquinolines. Treatment of N-methylformanilide with phosphoryl chloride in the presence of electron-rich alkenes provides N-methylquinolinium salts, presumably by way of a [4+ + 2] reaction [Eq. (76)].[164]

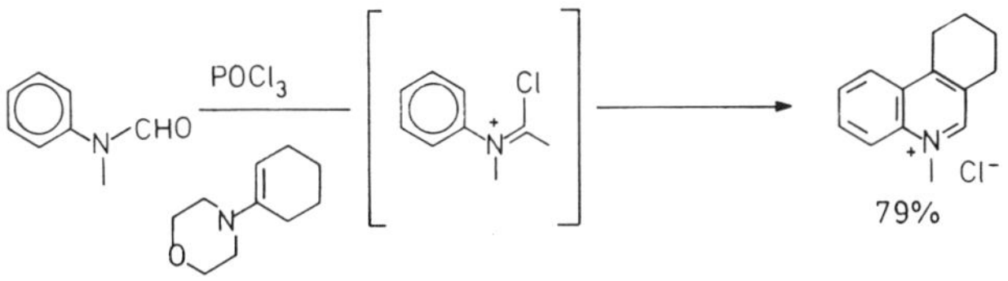

$$(76)$$

79%

Related studies employing N-protonated vinyl-[165] or aryliminium salts,[64,161] Lewis salts of aryl-[63a] and vinylimines,[63b] 1,3,4-oxadiazolium salts,[166] and nitrilium salts[167] constitute additional examples of reaction processes proceeding through cationic azadienes which formally participate in [4+ + 2] cycloaddition reactions.[168]

The preparation of N-vinyl-2-ethoxypyrrolidiniminium tetrafluoroborate (21) and its subsequent cycloaddition reactions with terminal akenes have been detailed [Eq. (77)].[169]

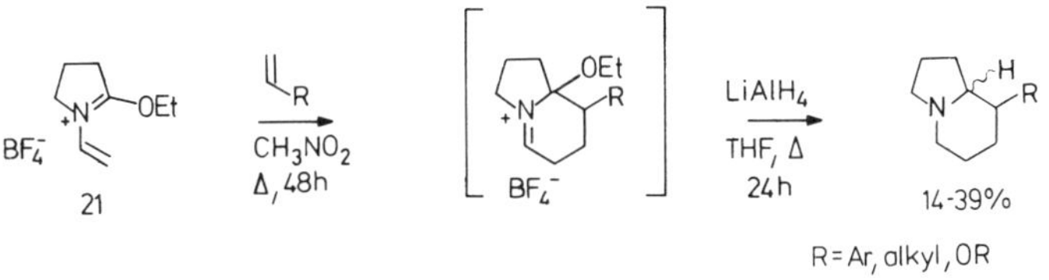

$$(77)$$

R = Ar, alkyl, OR

The unstable [4+ + 2] cycloadducts were reduced with lithium aluminum hydride to provide the cis/trans isomeric octahydroindolizines in low yields. At present, the reported regioselectivity of this [4+ + 2] cycloaddition is not easily rationalized. Typical, electron-deficient dienophiles were unreactive toward 21.

Two cationic heteroazadiene systems that participate in well-defined [4+ + 2] cycloadditions have been investigated extensively. The N-methyl-N-methylenium amide[170] or carbamate[172] cations as well as simple N-methylenium amides[170,171] and vinylnitrosonium cations[173–175] have been shown to participate as effective cationic azadienes in [4+ + 2] cycloadditions. N-Methyl-N-methylenium amide cations, e.g., 22, participate in [4+ + 2] cycloadditions with neutral or conjugated dienophiles [Eq. (78)],[170]

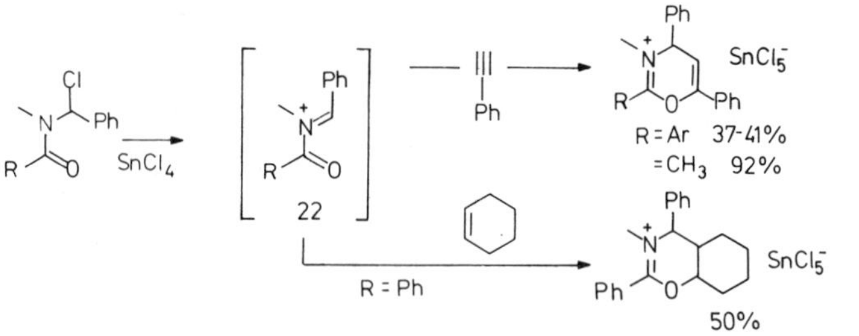

$$(78)$$

and the simple N-methylenium benzamide cation (23) participates in regio- *and* stereospecific[171] [4$^+$ + 2] cycloadditions with a range of representative electron-rich, neutral, or electron-deficient dienophiles (Scheme 9-VIII).[170,171] The demonstration that the stereochemistry of the olefinic dienophile is preserved in the course of the [4$^+$ + 2] cycloaddition satisfies an important requirement for concerted cycloaddition. While this observation does not exclude the possibility of a two-step addition–cyclization reaction, it does require that for such a mechanism to be operative the cyclization must be more rapid than carbon–carbon bond rotation.

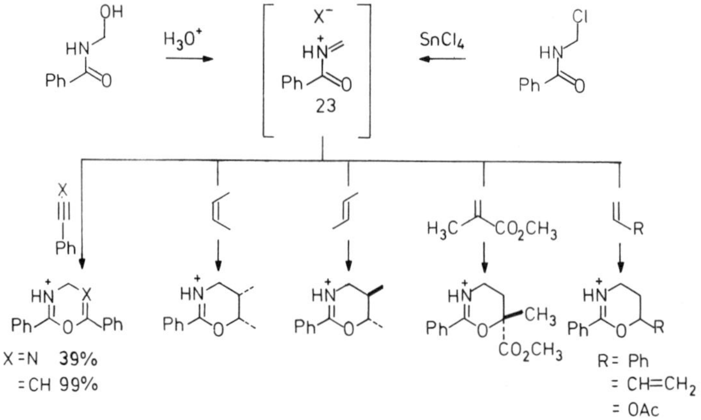

Scheme 9-VIII

In attempts to promote the acid-catalyzed intermolecular Diels–Alder reactions of the acylimine 24 with neutral dienes the predominate formation of α-acylamino-δ-alkenyl-δ-butyrolactones was observed. It has been suggested that the major products are derived from the 4π participation of the acyliminium salts in a [4$^+$ + 2] cycloaddition with the dienes followed by a subsequent acid-catalyzed rearrangement to the observed products [Eq. (79)].[172]

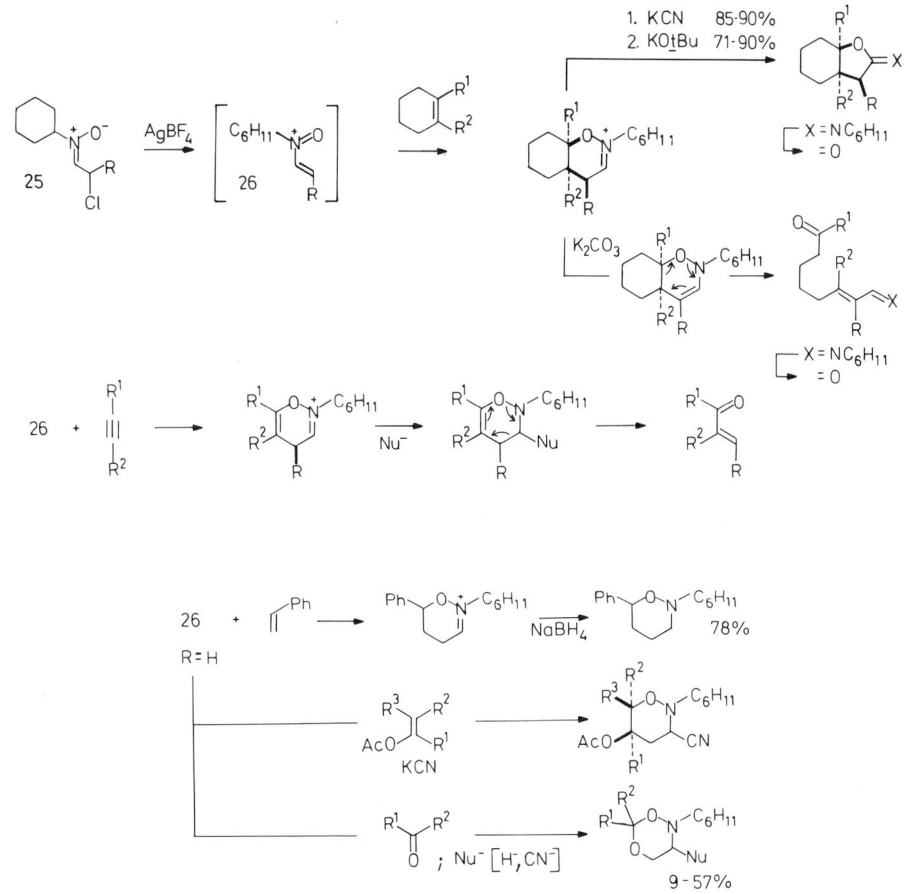

Scheme 9-IX

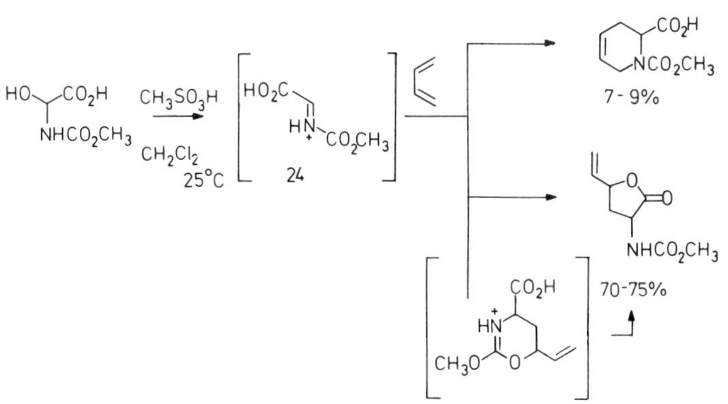

(79)

The *in situ* generation of *N*-vinyl-*N*-cyclohexylnitrosonium cations from the corresponding α-chloronitrone, e.g., *N*-(2-chloroethylidene)cyclohexylamine *N*-oxide (**25**), and their subsequent [4⁺ + 2] cycloaddition with electron-rich, neutral, or selected electron-deficient dienophiles have been extensively investigated (Scheme 9-IX).[173–175] The intermediate iminium salts which are derived from the regio- and stereoselective [4⁺ + 2] cycloadditions are not generally isolated but are subjected to the useful transformations illustrated in Scheme 9-IX. The intramolecular [4⁺ + 2] cycloadditions of vinylnitrosonium cations and nitroalkenes have been detailed.[176]

REFERENCES

1. Hamer, J. (ed.) (1967). "1,4-Cycloaddition Reactions, the Diels–Alder Reactions in Heterocyclic Syntheses." Academic Press, New York.
2. Diels–Alder syntheses with heteroatomic compounds: Needleman, S. B., and Chang-Kuo, M. C. (1962). *Chem. Rev.* **62**, 405.
3. Onishenko, A. S. (1964). "Diene Synthesis." Oldbourne Press, London.
4. a. Wollweber, H. (1970). *In* "Methoden der Organische Chemie (Houben-Weyl)" (E. Muller, ed.), Teil 3, Vol. V/1c, pp. 981–1139, cf. pp. 1128–1137. Thieme, Stuttgart. b. Jager, V., and Viehe, H. G. (1977). *In* "Methodender Organische Chemie (Houben-Weyl)" (E. Muller, ed.), Teil 4, Vol. V/2a, pp. 809–877, cf. pp. 858–861. Thieme, Stuttgart.
5. Heterodienes: a. Boger, D. L. (1983). *Tetrahedron* **39**, 2869. b. Ohno, M., and Sasaki, T. (1984). *J. Syn. Org. Chem. Jpn.* **42**, 125. c. Kita, Y., and Tamura, Y. (1984). *J. Syn. Org. Chem. Jpn.* **42**, 860.
6. Heterodienophiles: a. Weinreb, S. M., and Staib, R. R. (1982). *Tetrahedron* **38**, 3087. b. Weinreb, S. M., and Levin, J. I. (1979). *Heterocycles* **12**, 949.
7. a. Ulrich, H. (ed.) (1967). "Cycloaddition Reactions of Heterocumulenes." Academic Press, New York. b. Cycloaddition reactions of aliphatic and aromatic acyl isocyanates: Arbuzov, B. A., and Zobova, N. N. (1974). *Synthesis*, 461; Arbuzov, B. A., and Zobova, N. N. (1982). *Synthesis*, **433**. c. Thioacyl isocyanates and acyl isocyanates: Tsuge, O. (1979). *Heterocycles* **12**, 1067. d. Dondoni, A. (1980). *Heterocycles* **14**, 1547.
8. a. Cycloaddition with polar intermediates: Gompper, R. (1969). *Angew. Chem. Int. Ed. Engl.* **8**, 312. b. Polar cycloadditions: Schmidt, R. R. (1973). *Angew. Chem. Int. Ed. Engl.* **12**, 212.
9. Vinyl ethers as dienophiles: Povarov, L. S. (1967). *Russ. Chem. Rev.* **36**, 656.
10. Wagner-Jauregg, T. (1980). *Synthesis*, 165, 769.
11. Wagner-Jauregg, T. (1976). *Synthesis*, 349.
12. Intramolecular Diels–Alder reactions: a. Brieger, G., and Bennett, J. N. (1980). *Chem. Rev.* **80**, 63. b. Oppolzer, W. (1977). *Angew. Chem. Int. Ed. Engl.* **16**, 10. c. Carlson, R. G. (1974). *Ann. Rep. Med. Chem.* **9**, 270. d. Fallis, A. G. (1984). *Can. J. Chem.* **62**, 183. Ciganek, E. (1984). *Org. React.* **32**, 1.
13. Asymmetric Diels–Alder reactions: a. Oppolzer, W. (1984). *Angew. Chem. Int. Ed. Engl.* **23**, 876. b. Wurziger, H. (1984). *Kantakte* **2**, 3. c. Paquette, L. A. (1984). *In* "Asymmetric Synthesis" (J. D. Morrison, ed.), Vol. 3. p. 455. Academic Press, Orlando.

14. Retro-Diels–Alder reactions: a. Ripoll, J.-L., Rouessac, A., and Rouessac, F. (1978). *Tetrahedron* **34**, 19. b. Kwart, H., and King, K. (1968). *Chem. Rev.* **68**, 415. c. Lasne, M.-C., and Ripoll, J.-L. (1985). *Synthesis*, 121.
15. Woodward, R. B., and Hoffmann, R. (1970). "The Conservation of Orbital Symmetry." Academic Press, New York.
16. a. Houk, K. N. (1975). *Acc. Chem. Res.* **8**, 361. b. Namura, Y., Takeuchi, Y., and Tomeda, S. (1981). *Bull. Chem. Soc. Jpn.* **54**, 2779.
17. a. Martin, J. G., and Hill, R. K. (1961). *Chem. Rev.* **61**, 537. b. Huisgen, R. (1963). *Angew. Chem. Int. Ed. Engl.* **2**, 565. c. Sauer, J. (1966). *Angew. Chem. Int. Ed. Engl.* **5**, 211; Sauer, J. (1967). *Angew. Chem. Int. Ed. Engl.* **6**, 16. d. Sauer, J., and Sustmann, R. (1980). *Angew. Chem. Int. Ed. Engl.* **19**, 779. e. Goerdeler, J. (1970). *Q. Rep. Sulf. Chem.* **5**, 169. f. Fleischhauer, J., Asaad, A. N., Schleker, W., and Scharf, H.-D. (1981). *Liebigs Ann. Chem.*, 306.
18. a. Nenitzescu, C. D., Cioranescu, E., and Birladeanu, L. (1958). *Commun. Acad. Rep. Populare Romine* **8**, 775 [*Chem. Abstr.* **53**, 18003 (1958)]. b. Taylor, E. C., Eckroth, D. R., and Bartulin, J. (1967). *J. Org. Chem.* **32**, 1899; Eckroth, D. R. (1966). Ph.D. thesis, Princeton University [*Diss. Abstr. Int. B* **27**, 102 (1966)]. c. Johnson, J. L., Ladner, D. W., Cross, B., Doehner, R. F., Jr., and Wong, W. (1985). Personal communication with J. L. J. of American Cyanamide. d. Wilk, M., Schwab, H., and Rochlitz, J. (1966). *Liebigs Ann. Chem.* **698**, 149. e. Campbell, C. D., and Rees, C. W. (1969). *J. Chem. Soc. C*, 748. f. Wade, P. A., Amin, N. V., Yen, H.-K., Price, D. T., and Huhn, G. F. (1984). *J. Org. Chem.* **49**, 4595.
19. a. Tomimatsu, Y. (1957). *J. Pharm. Soc. Jpn.* **77**, 7. b. Tomimatsu, Y. (1957). *J. Pharm. Soc. Jpn.* **77**, 186.
20. Seeliger, W., Aufderhaar, E., Diepers, W., Feinauer, R., Nehring, R., Thier, W., and Hellmann, H. (1966). *Angew. Chem. Int. Ed. Engl.* **5**, 875.
21. Kato, T., Chiba, T., and Tanaka, S. (1974). *Chem. Pharm. Bull. Tokyo* **22**, 744; Kato, T., and Chiba, T. (1969). *J. Pharm. Soc. Jpn.* **89**, 1464.
22. Sakamoto, M., Miyazawa, K., Kuwabara, K., and Tomimatsu, Y. (1979). *Heterocycles* **12**, 231.
23. Baydar, A. E., Boyd, G. V., Lindley, P. F., and Watson, F. (1979). *J. Chem. Soc., Chem. Commun.*, 178.
24. a. Snyder, H. R., Cohen, H., and Tapp, W. J. (1939). *J. Am. Chem. Soc.* **61**, 3560. b. Snyder, H. R., and Robinson, J. C., Jr. (1941). *J. Am. Chem. Soc.* **63**, 3279.
25. a. Gompper, R., and Paul, K. P., unpublished observations, cf. Gompper, R. (1969). *Angew. Chem. Int. Ed. Engl.* **8**, 322. b. Brady, W. T., and Shieh, C. H. (1983). *J. Org. Chem.* **48**, 2499; Ohshiro, Y., Komatsu, M., Uesaka, M., and Agawa, T. (1984). *Heterocycles* **22**, 549. c. Moore, H. W., and Hughes, G. M. (1982). *Tetrahedron Lett.* **23**, 4003; Moore, H. W., Hughes, G., Srinivasachar, K., Kernandez, M., Nguyen, N. V., Schoon, D., and Tranne, A. (1985). *J. Org. Chem.* **50**, 4231. d. Duran, F., and Ghosez, L. (1970). *Tetrahedron Lett.*, 245; see also Mohan, S., Jumar, B., and Sandhu, J. S. (1971). *Chem. Ind.*, 671; Zamboni, R., and Just, G. (1979). *Can. J. Chem.* **57**, 1945; Doyle, T. W., Belleau, B., Luh, B.-Y., Ferrari, C. F., and Cunningham, M. P. (1977). *Can. J. Chem.* **55**, 468; Kato, T., and Chiba, T. (1969). *J. Pharm. Soc. Jpn.* **89**, 1464; Sakamoto, M., and Tomimatsu, Y. (1970). *J. Pharm. Soc. Jpn.* **90**, 1386.
26. Fitton, A. O., Frost, J. R., Houghton, P. G., and Suschitzky, H. (1977). *J. Chem. Soc., Perkin Trans. 1*, 1450.
27. Tsuge, O., and Iwanami, S. (1971). *Bull. Chem. Soc. Jpn.* **44**, 2750; Tsuge, O., and Iwanami, S. (1970). *Bull. Chem. Soc. Jpn.* **43**, 3543.
28. a. Garashchenko, Z. M., Skvortsova, G. G., and Shestova, L. A. USSR Patent 370 208

1973. (1973). *Chem. Abstr.* **79**, 31900. b. Alberola, A., Gonzalez, A. M., Gonzalez, B., Laguna, M. A., and Pulido, F. I. (1986). *Tetrahedron Lett.* **27**, 2027.

29. a. Komatsu, M., Ohgishi, H., Oshiro, Y., and Agawa, T. (1976). *Tetrahedron Lett.*, 4589. b. Komatsu, M., Ohgishi, H., Takamatsu, S., Ohshiro, Y., and Agawa, T. (1982). *Angew. Chem. Int. Ed. Engl.* **21**, 213. c. Komatsu, M., Takamatsu, S., Uesaka, M., Yamamoto, S., Ohshiro, Y., and Agawa, T. (1984). *J. Org. Chem.* **49**, 2691. d. See also Fuks, R., Buijle, R., and Viehe, H. G. (1966). *Angew. Chem. Int. Ed. Engl.* **5**, 585.

30. Monson, R. S., and Baraze, A. (1975). *Tetrahedron* **31**, 1145.

31. a. Poncin, B. S.-, Frisque, A.-M. H.-, and Ghosez, L. (1982). *Tetrahedron Lett.* **23**, 3261. b. Ohno, M., and Sasaki, T., unpublished observations, see Ref. 5b. c. Tamura, Y., Tsugoshi, T., Nakajima, Y., and Kita, Y. (1984). *Synthesis*, 930. See also d. Potts, K. T., Bhattacharjee, D., and Walsh, E. B. (1984). *J. Chem. Soc., Chem. Commun.*, 114. e. Ota, T., Masuda, S., and Tanaka, H. (1981). *Chem. Lett.*, 411.

32. a. Cheng, Y.-S., Fowler, F. W., and Lupo, A. T., Jr., (1981). *J. Am. Chem. Soc.* **103**, 2090. b. Cheng, Y.-S., Lupo, A. T., Jr., and Fowler, F. W. (1983). *J. Am. Chem. Soc.* **105**, 7696. c. Hwang, Y. C., and Fowler, F. W. (1985). *J. Org. Chem.* **50**, 2719.

33. Whitesell, M. A., and Kyba, E. P. (1984). *Tetrahedron Lett.* **25**, 2119.

34. Bruhn, J. Zsindely, J., and Schmid, H. (1978). *Helv. Chim. Acta* **61**, 2542.

35. Nitta, M., Sekiguchi, A., and Koba, H. (1981). *Chem. Lett.*, 933.

36. Ciganek, E. (1980). *J. Org. Chem.* **45**, 1497.

37. a. Ihara, M., Kirihara, T., Kawaguchi, A., Fukumoto, K., and Kametani, T. (1984). *Tetrahedron Lett.* **25**, 4541. b. Ihara, M., Tsuruta, M., Fukumoto, K., and Kametani, T. (1985). *J. Chem. Socl, Chem. Commun.*, 1159. c. Ihara, M., Kirihara, T., Fukumoto, K., and Kametani, T. (1985). *Heterocycles* **23**, 1097.

38. a. Ito, Y., Miyata, S., Nakatsuka, M., and Saegusa, T. (1981). *J. Am. Chem. Soc.* **103**, 5250. b. Ito, Y., Nakajo, E., Nakatsuka, M., and Saegusa, T. (1983). *Tetrahedron Lett.* **24**, 2881. c. Ito, Y., Nakajo, E., and Saegusa, T. (1986). *Synth. Commun.* **16**, 1073.

39. a. Fishwick, C. W. G., Storr, R. C., and Manley, P. W. (1984). *J. Chem. Soc., Chem. Commun.*, 1304. b. Hodgetts, I., Noyce, S. J., and Storr, R. C. (1984). *Tetrahedron Lett.* **25**, 5435; Bowen, R. D., Davies, D. E., Fishwick, C. W. G., Glasbey, T. O., Noyce, S. J., and Storr, R. C. (1982). *Tetrahedron Lett.* **23**, 4501. c. Early studies: Bandaranayake, W. M., Begley, M. J., Brown, B. O., Clarke, D. G., Crombie, L., and Whiting, D. A. (1974). *J. Chem. Soc., Perkin Trans. 1*, 998; Narasimhan, N. S., and Kelkar, S. L. (1976). *Indian J. Chem. Sect. B.* **14**, 430.

40. a. Burgess, E. M., and McCullagh, L. (1966). *J. Am. Chem. Soc.* **88**, 1580. b. Lancaster, M., and Smith, D. J. H. (1980). *J. Chem. Soc., Chem. Commun.*, 471. c. Fisher, M., and Wagner, F. (1969). *Chem. Ber.* **102**, 3486; see also Ref. 41c,d.

41. a. Kobylecki, R. J., and McKillop, A. (1976). *Adv. Heterocycl. Chem.* **19**, 215. b. Herlinger, H. (1974). *Angew. Chem. Int. Ed. Engl.* **3**, 378; Crabtree, H. E., Smalley, R. K., and Suschitzky, H. (1968). *J. Chem. Soc. C*, 2730. c. Mao, Y.-L., and Boekelheide, V. (1980). *J. Org. Chem.* **45**, 1547. d. Dopp, D., and Weiler, H. (1979). *Chem Ber.* **112**, 3950. e. Kametani, T., Higa, T., Fukumoto, K., and Koizumi, M. (1976). *Heterocycles* **4**, 23.

42. a. Heine, H. W., Barchiesi, B. J., and Williams, E. A. (1984). *J. Org. Chem.* **49**, 2560. b. McKillop, A., and Sayer, T. S. B. (1976). *J. Org. Chem.* **41**, 1079. c. 1,4-Diazadienes: this chapter, Section 7.

43. *o*-Quinodimethanes (*o*-xylylenes): a. Funk, R. L., and Vollhardt, K. P. C. (1980). *Chem. Soc. Rev.* **9**, 41. b. Oppolzer, W. (1978). *Synthesis*, 793.

44. For recent work, see Chapter 7.

45. Gompper, R., and Paul, K. P., unpublished observations; see Ref. 8a, p. 324.

46. Khattak, I., and Ketcham, R. (1983). *J. Chem. Soc., Chem. Commun.,* 260.
47. Aue, D. H., and Thomas, D. (1975). *J. Org. Chem.* **40,** 1349.
48. Demoulin, A., Gorissen, H., Hesbain-Frisque, A.-M., and Ghosez, L. (1975). *J. Am. Chem. Soc.* **97,** 4409.
49. a. Gompper, R., and Heinemann, U. (1980). *Angew. Chem.* **92,** 207, 208. b. Gompper, R., private communication.
50. Wipf, P., and Heimgartner, H. (1984). *Chimia* **38,** 357.
51. Sainte, F., Serckx-Poncin, B., Hesbain-Frisque, A.-M., and Ghosez, L. (1982). *J. Am. Chem. Soc.* **104,** 1428.
52. Nomura, Y., Takeuchi, Y., Tomoda, S., and Ito, M. M. (1970). *Chem. Lett.,* 187; Nomura, Y., Takeuchi, Y., Tomoda, S., and Ito, M. M. (1981). *Bull. Chem. Soc. Jpn.* **54,** 2779.
53. Gompper, R., and Heinemann, U. (1981). *Angew. Chem. Int. Ed. Engl.* **20,** 296.
54. a. Ripoll, J. L., Lebrun, H., and Thuillier, A. (1980). *Tetrahedron* **36,** 2497; Malecot, Y. M., Ripoll, J. L., and Thuillier, A. (1983). *J. Chem. Res.,* 959; Ranganathan, L. D., and Bamezai, S. (1983). *Tetrahedron Lett.* **24,** 1067. b. Hunter, D. H., and Steiner, R. P. (1975). *Can. J. Chem.* **53,** 355. c. Worley, S. D., Taylor, K. G., Venugopalam, B., and Clark, M. S., Jr. (1978). *Tetrahedron* **34,** 833. Hickmott, P. W. (1984). *Tetrahedron* **40,** 2989.
55. a. Review: Steglich, W., Jescke, R., and Buschmann, E. (1986). *Gazz. Chim. Ital.* **116,** 361. b. Steglich, W., Buschmann, E., and Hollitzer, O. (1974). *Angew. Chem. Int. Ed. Engl.* **13,** 533. c. Steglich, W., and Buschmann, E. (1974). *Angew. Chem. Int. Ed. Engl.* **13,** 484. d. Hofle, G., Hollitzer, O., and Steglich, W. (1972). *Angew. Chem. Int. Ed. Engl.* **11,** 720. e. Steglich, W., and Hollitzer, O. (1973). *Angew. Chem. Int. Ed. Engl.* **12,** 495.
56. a. Kranz, A., and Hoppe, B. (1975). *J. Am. Chem. Soc.* **97,** 6590. b. Kricheldorf, H. R. (1972). *Angew Chem. Int. Ed. Engl.* **11,** 128. c. Stajer, G., Szabo, A. E., Fulop, F., and Bernath, G. (1984). *Synthesis,* 345; Bernath, G., Stajer, G., Szabo, A. E., and Fulop, F. (1984). *Heterocycles* **21,** 575; Stajer, G., Mod, L., Szabo, A. E., Fulop, F., and Bernath, G. (1984). *Tetrahedron* **40,** 2385. d. Beccalli, E. M., La Rosa, C., and Marchesini, A. (1984). *J. Org. Chem.* **49,** 4287; Beccalli, E. M., Marchesini, A., and Molinari, H. (1986). *Tetrahedron Lett.* **27,** 627. e. Maier, G., and Schafer, U. (1980). *Liebigs Ann. Chem.,* 798. f. Eiden, F., and Nagor, B. S. (1963). *Naturwissenschaften* **50,** 403. g. de Mayo, P., Weedon, A. C., and Zabel, R. W. (1981). *Can. J. Chem.* **59,** 2328. h. Gotze, S., and Steglich, W. (1976). *Chem. Ber.* **109,** 2327; Lepschy, J., Hofle, G., Wilshowitz, L., and Steglich, W. (1974). *Liebigs Ann. Chem.,* 1753. i. Risitano, F., Grassi, G., Foti, F., Caruso, B., and Lo Vecchio, G. (1979). *J. Chem. Soc., Perkin Trans. 1,* 1522. j. Lapkin, I. I., Semenov, V. I., and Belonovich, M. I. (1977). *Zh. Org. Khim.* **13,** 1328. For additional preparations, see Ref. 55a and references cited therein.
57. a. Jung, M. E., and Shapiro, J. J. (1980). *J. Am. Chem. Soc.* **102,** 7862. b. Daniels, P. H., Wong, J. L., Atwood, J. G., Canada, L. B., and Rogers, R. D. (1980). *J. Org. Chem.* **45,** 435. c. Eicher, T., Abdesakan, F., Franke, G., and Weber, J. L. (1975). *Tetrahedron Lett.,* 3915.
58. Eddaif, A., Laurent, A., Mison, P., and Pellissier, N. (1984). *Tetrahedron Lett.* **25,** 2779.
59. Hajicek, J., and Trojanek, J. (1981). *Tetrahedron Lett.* **22,** 1823.
60. Hasan, I., and Fowler, F. W. (1978). *J. Am. Chem. Soc.* **100,** 6696.
61. Nitta, M., Sekiguchi, A., and Koba, H. (1981). *Chem. Lett.,* 933.
62. Brooke, G. M., Matthews, R. S., and Robson, N. S. (1980). *J. Chem. Soc., Perkin Trans. 1,* 102.

63. a. Povarov, L. S. (1963). *Izv. Akad. SSSR* **955**, 2039 [*Chem. Abstr.* **59**, 7489 (1963)];
    Povarov, L. S. (1964). *Izv. Akad. SSSR* 179, 1693, 2221 [*Chem. Abstr.* **60**, 5451, 9256,
    (1964)]; Povarov, L. S. (1965. *Izv. Akad. SSSR* 365. [*Chem. Abstr.* **62**, 7723, 14624
    (1965)]; Povarov, L. S. (1966). *Izv. Akad. SSSR* 337 [*Chem. Abstr.* **64**, 17539 (1966)];
    Griggs, V. I., Povarov, L. S., and Mikhailov, B. M. (1965). *Izv. Akad. Nauk SSSR,
    Ser. Khim.*, 2163 [*Chem. Abstr.* **64**, 9680 (1966)] For related studies, see Kametani, T.,
    Takada, H., Suzuki, Y., and Honda, T. (1985). *Synth. Commun.* **15**, 449; Kozlov, N.
    S., and Pinegina, L. Yu. (1963). *Zh. Obsheh. Khim.* **33**, 1079 [*Chem. Abstr.* **59**, 9976
    (1963)]; Miyajima, S., Ito, K., Kashiwagura, I., Kitamura, C. (1979). *Nippon Kagaku
    Zasshi,* 1514 [*Chem. Abstr.* **92**, 180963 (1980)]; Schmidt, R. R. (1964). *Angew. Chem.*
    76, 991; Nakayama, J., Midorikawa, H., and Yoshida, M. (1975). *Bull. Chem. Soc.
    Jpn.* **48**, 1063; Elslager, E. F., and Worth, D. F. (1969). *J. Heterocycl. Chem.* **6**, 597;
    Perricone, S. C., Worth, D. F., and Elslager, E. F. (1970). *J. Heterocycl. Chem.* **7**, 538,
    1353; Trifonov, L. S., and Orahovats. A. S. (1984). *Heterocycles* **22**, 355; Hagihara,
    N., and Jo, S. (1972). Jpn. Patent 720067 [*Chem. Abstr.* **76**, 85715 (1972)]; Joh, T., and
    Hagihara, N. (1967). *Tetrahedron Lett.*, 4199. For related work, see Destro, F., Luc-
    chini, V., and Prato, M. (1984). *Tetrahedron Lett.* **25**, 5573. For observations which
    may be interpreted as the acid-catalyzed participation of a 2-azadiene in a [4 + 2]
    cycloaddition, see Hartman, G. D., Halczenko, W., and Phillips, B. T. (1985). *J. Org.
    Chem.* **50**, 2427. b. Cheng, Y. S., Ho, E., Mariano, P. S., and Ammon, H. L. (1985). *J.
    Org. Chem.* **50**, 5678. c. Barluenga, J., Joglar, J., Fustero, S., Gotor, V., Kruger, C.,
    and Romao, M. J. (1985). *Chem. Ber.* **118**, 3652.
64. Nomura, Y., Kimura, M., Takeuchi, Y., and Tomoda, S. (1978). *Chem. Lett.*, 267;
    Shono, T., Matsumura, Y., Inoue, K., Ohmizu, H., and Kashimura, S. (1982). *J. Am.
    Chem. Soc.* **104**, 5753.
65. a. Ghosez, L., and de Perez, C. (1971). *Angew. Chem. Int. Ed. Engl.* **10**, 184. b.
    Sonveau, E., and Ghosez, L. (1973). *J. Am. Chem. Soc.* **95**, 5417.
66. Dondoni, A., Battaglia, A., and Giorgianni, P. (1982). *J. Org. Chem.* **47**, 3998; Don-
    doni, A., Battaglia, A., and Giorgianni, P. (1980). *J. Org. Chem.* **45**, 3766; Dondoni,
    A., Battaglia, A., Giorgianni, P., Gilli, G., and Sacerdoti, M. (1977). *J. Chem. Soc.,
    Chem. Commun.*, 43; Bernardi, F., Bottoni, A., Battaglia, A., Distefano, G., and
    Dondoni, A. (1980). *Z. Naturforsch., A: Phys., Phys. Chem., Kosmophys.* **35**, 521;
    Dondoni, A. (1980). *Heterocycles* **14**, 1547.
67. a. Kuehne, M. E., and Sheeran, P. J. (1968). *J. Org. Chem.* **33**, 4406. b. Sheehan, J. C.,
    and Daves, G. D. (1965). *J. Org. Chem.* **30**, 3247. c. Nieuwenhuis, J., and Arens, J. F.
    (1957). *Recl. Trav. Chim. Pays-Bas* **76**, 999. d. Gais, H.-J., Hafner, K., and Neuensch-
    wander, M. (1969). *Helv. Chim. Acta* **52**, 2641. e. Rigby, J. H., and Balasubramanian,
    N. (1984). *J. Org. Chem.* **49**, 4569.
68. Fuks, R. (1970). *Tetrahedron* **26**, 2161; Dondoni, A., Kniezo, L., and Medici, A.
    (1982). *J. Org. Chem.* **47**, 3994.
69. Goerdeler, J., Laqua, A., and Lindner, C. (1980). *Chem. Ber.* **113**, 2509; Schaumann,
    E., Bauch, H.-G., Sieveking, S., and Adiwidjaja, G. (1982). *Chem. Ber.* **115**, 3340;
    Kuzuya, M., Ito, S., Miyaka, F., and Okuda, T. (1982). *Chem. Pharm. Bull. Tokyo* **30**,
    1980.
70. a. Hafner, K., Haring, J., and Jakel, W. (1970). *Angew. Chem. Int. Ed. Engl.* **9**, 159. b.
    Hasnaoui, A., El Messaoudi, M., and Lavergne, J. P. (1985). *J. Heterocycl. Chem.* **22**,
    25.
71. Desimoni, G., and Tacconi, G. (1975). *Chem. Rev.* **75**, 651; Zaugg, H. E. (1970).
    Synthesis, 49; Schmidt, R. R. (1972). *Synthesis,* 333; Kato, T., Katagiri, N., and
    Yamamoto, Y. (1980). *Heterocycles* **14**, 1333. For additional work, see Safronova, Z.

V., Sumonyan, L. A., Zeifman, Yu. V., and Gambaryan, N. P. (1979). *Izv. Akad. Nauk SSSR, Ser. Khim.,* 1826; Burger, K., and Simmerl, R. (1983). *Synthesis,* 127; Hall, H. K., and Miniotti, D. L. (1984). *Tetrahedron Lett.* **25,** 943.

72. a. Gokou, C. T., Pradere, J.-P., and Quiniou, H. (1985). *J. Org. Chem.* **50,** 1545; Pradere, J.-P., Roze, J. C., Duguay, G., Guevel, A., Tea Gokou, C., and Quiniou, H. (1983). *Sulf. Lett.* **1,** 115; Meslin, J.-C., and Quiniou, H. (1979). *Bull. Soc. Chim. Fr. 2,* 347; Meslin, J.-C., and Quiniou, H. (1975). *Tetrahedron,* **31,** 3055; Meslin, J.-C., and Quiniou, H. (1974). *Synthesis,* 298. Meslin, J.-C., Reliquet, A., Reliquet, F., and Quiniou, H. (1980). *Synthesis,* 453; Roze, J.-C., Pradere, J.-P., Duguay, G., Guevel, A., Quiniou, H., and Poignant, S. (1983). *Can. J. Chem.* **61,** 1169; Giordano, C., Belli, A., and Abis, L. (1979). *Tetrahedron Lett.,* 1537; Giordano, C., Belli, A., and Bellotti, V. (1978). *Synthesis,* 443; Giordano, C., Belli, A., Erbea, R., and Panossian, S. (1979). *Synthesis,* 801; Giordano, C. (1975). *Gazz. Chim. Ital.* **105,** 1265; Burger, K., Huber, E., Schontag, W., and Ottlinger, R. (1983). *J. Chem. Soc., Chem. Commun.,* 945; Burger, K., and Goth, H. (1980). *Angew. Chem. Int. Ed. Engl.* **19,** 810; Burger, K., and Ottlinger, R. (1978). *J. Fluorine Chem.* **11,** 29; Burger, K., Ottlinger, R., and Albanbauer, J. (1977). *Chem. Ber.* **110,** 2114; Burger, K., Gott, H., Schoentag, W., and Firl, J. (1982). *Tetrahedron* **38,** 287; Burger, V. K., Partscht, H., Huber, E., Gieren, A., Hubner, T., and Kaerlein, C.-P. (1984). *Chem. Zeit.* **108,** 209; Burger, K., Albanbauer, J., and Foag, W. (1975). *Angew. Chem. Int. Ed. Engl.* **14,** 767. b. O-oka, M., Kitamura, A., Okazaki, R., and Inamoto, N. (1978). *Bull. Soc. Chem. Jpn.* **51,** 301; Okazaki, R., O-oka, M., and Inamoto, N. (1976). *J. Chem. Soc., Chem. Commun.,* 562; Okazaki, R., Ishii, F., and Inamoto, N. (1978). *Bull. Chem. Soc. Jpn.* **51,** 309; Okazaki, R., Ishii, F., Ozawa, Kazumi, Ozawa, Kenji, and Inamoto, N. (1975). *J. Chem. Soc., Perkin Trans. 1,* 270.

73. Tschaen, D. M., Turos, E., and Weinreb, S. M. (1984). *J. Org. Chem.* **49,** 5058; Weinreb, S. M., and Scola, P. M. (1986). *J. Org. Chem.* **51,** 3250.

74. Kober, R., Papadopoulos, K., Miltz, W., Enders, D., and Steglich, W. (1985). *Tetrahedron* **41,** 1693.

75. Gilchrist, T. L. (1983). *Chem. Soc. Rev.* **12,** 53.

76. a. Faragher, R., and Gilchrist, T. L. (1979). *J. Chem. Soc., Perkin Trans. 1,* 249. b. Faragher, R., and Gilchrist, T. L. (1976). *J. Chem. Soc., Chem. Commun.,* 581. c. Gilchrist, T. L., Roberts, T. G., and Lingham, D. A. (1979). *J. Chem. Soc., Chem. Commun.,* 1089. d. Gilchrist, T. L., and Roberts, T. G. (1979). *J. Chem. Soc., Chem. Commun.,* 1090. e. Gilchrist, T. L., and Roberts, T. G. (1978). *J. Chem. Soc., Chem. Commun.,* 847. f. Gilchrist, T. L., Iskander, G. M., and Yagoub, A. K. (1981). *J. Chem. Soc., Chem. Commun.,* 696. g. Gilchrist, T. L., and Roberts, T. G. (1983). *J. Chem. Soc., Perkin Trans. 1,* 1283; Davies, D. E., Gilchrist, T. L., and Roberts, T. G. (1983). *J. Chem. Soc., Perkin Trans. 1,* 1275. h. Ottenheijm, H. C. J., Plate, R., Noordik, J. H., and Herscheid, J. D. M. (1982). *J. Org. Chem.* **47,** 2147; Plate, R., Hermkens, P. H. H., Smits, J. M. M., Ottenheijm, H. C. J. (1986). *J. Org. Chem.* **51,** 309; Plate, R., and Ottenheijm, H. C. J. (1986). *Tetrahedron Lett.* **27,** 3755. i. Iskanderl, G. M., and Gulta, V. S. (1982). *J. Chem. Soc., Perkin Trans. 1,* 1891. j. Mackay, D., and Watson, K. N. (1982). *J. Chem. Soc., Chem. Commun.,* 775, 777. k. Nakanishi, S., Higuchi, M., and Flood, T. C. (1986). *J. Chem. Soc., Chem. Commun.,* 30.

77. a. Nakanishi, S., Shirai, Y., Takahashi, K., and Otsuji, Y. (1981). *Chem. Lett.,* 869. b. Brava, P., Gaudiano, G., Ponti, P. P., and Umani-Ronchi, A. (1970). *Tetrahedron* **26,** 1315; Oppolzer, W., Battig, K., and Hudlicky, T. (1981). *Tetrahedron* **36,** 4359.

78. Viehe, H.-G., Merenyi, R., Francotte, E., Van Meerssche, M., Germain, G., Declercq, J. P., and Bodart-Gilmont, J. (1977). *J. Am. Chem. Soc.* **99,** 2340; Francotte,

E., Merenyi, R., Vlandenbulcke-Coyette, B., and Viehe, H.-G. (1981). *Helv. Chim. Acta* **64**, 1208.

79. Denmark, S. E., Dappen, M. S., and Sternberg, J. A. (1984). *J. Org. Chem.* **49**, 4741; Denmark, S. E., and Dappen, M. S. (1984). *J. Org. Chem.* **49**, 798.
80. Bonini, B. F., Foresti, E., Maccagnani, G., Mazzanti, G., Sabatino, P., and Zani, P. (1985). *Tetrahedron Lett.* **26**, 2131.
81. Keck, G. E., and Nickell, D. G. (1980). *J. Am. Chem. Soc.* **102**, 3632, and references cited therein.
82. Mackay, D., Watson, K. N., and Dao, L. H. (1977). *J. Chem. Soc., Chem. Commun.*, 702; Dobbin, C. J. B., Mackay, D., Penney, M. R., and Dao, L. H. (1977). *J. Chem. Soc., Chem. Commun.*, 703; Mackay, D., Dao, L., and Dust, J. M. (1980). *J. Chem. Soc., Perkin Trans. 1*, 2408; Mackay, D., Neeland, E. G., and Taylor, N. J. (1986). *J. Org. Chem.* **51**, 2351.
83. Weiss, R. (1967). *Chem. Ber.* **100**, 685.
84. a. Goerdeler, J., and Schenk, H. (1965). *Chem. Ber.* **98**, 3831; Goerdeler, J., and Schulze, A. (1982). *Chem. Ber.* **115**, 1259; Goerdeler, J., Tiedt, M.-L., and Nandi, K. (1981). *Chem. Ber.* **114**, 2713. b. Goerdeler, J., and Schenk, H. (1965). *Chem. Ber.* **98**, 2954. c. Goerdeler, J., and Nandi, K. (1975). *Chem. Ber.* **108**, 3066. d. Goerdeler, J., and Nandi, K. (1981). *Chem. Ber.* **114**, 549.
85. a. Goerdeler, J., and Schenk, H. (1963). *Angew. Chem. Int. Ed. Engl.* **2**, 552; Goerdeler, J., and Schulze, A. (1982). *Chem. Ber.* **115**, 1252. b. Goerdeler, J., and Jonas, K. (1966). *Chem. Ber.* **99**, 3572. c. Goerdeler, J., and Schenk, H. (1965). *Chem. Ber.* **98**, 2954. d. Tsuge, O., Tashiro, M., Mizuguchi, R., and Kanemasa, S. (1966). *Chem. Pharm. Bull. Tokyo* **14**, 1055.
86. a. Tsuge, O., and Iwanami, S. (1971). *Bull. Chem. Soc. Jpn.* **44**, 2750. b. Tsuge, O., and Kanemasa, S. (1972). *Tetrahedron* **28**, 4737. c. Wedekind, E., and Schenk, D. (1911). *Chem. Ber.* **44**, 198. d. Tsuge, O., and Kanemasa, S. (1972). *Bull. Chem. Soc. Jpn.* **45**, 2877. e. Niess, R., and Robins, R. K. (1970). *J. Heterocycl. Chem.* **7**, 243.
87. Tsuge, O., and Sakai, K. (1972). *Bull. Chem. Soc. Jpn.* **45**, 1534.
88. Nair, V., and Kim, K. H. (1974). *Tetrahedron Lett.*, 1487; Nair, V., and Kim, K. H. (1974). *J. Org. Chem.* **39**, 3763.
89. Tsuge, O., and Inaba, A. (1976). *Bull. Chem. Soc. Jpn.* **49**, 2828.
90. Tsuge, O., and Kanemasa, S. (1972). *Bull. Chem. Soc. Jpn.* **45**, 3591.
91. Tsuge, O., and Kanemasa, S. (1974). *Bull. Chem. Soc. Jpn.* **47**, 2676.
92. Goerdeler, J., and Sappelt, R. (1967). *Chem. Ber.* **100**, 2064.
93. Goerdeler, J., and Weiss, R. (1967). *Chem. Ber.* **100**, 1627.
94. Schulze, A., and Goerdeler, J. (1974). *Tetrahedron Lett.*, 221; Schulze, A., and Goerdeler, J. (1982). *Chem. Ber.* **115**, 3063; Goerdeler, J., Schimpf, R., and Tiedt, M.-L. (1972). *Chem. Ber.* **105**, 3322.
95. Milzner, K., and Seckinger, K. (1974). *Helv. Chim. Acta* **57**, 1614.
96. Goerdeler, J., and Ludke, H. (1970). *Chem. Ber.* **103**, 3393.
97. Ratton, S., Moyne, J., and Longeray, R. (1979). *Bull. Soc. Chim. Fr. 2*, 499.
98. a. Hunig, S., and Hubner, K. (1962). *Chem. Ber.* **95**, 937. b. Carney, R. W. J., Wojtkunski, J., and deStevens, G. (1964). *J. Org. Chem.* **29**, 2887. c. deStevens, G., Smolinsky, B., and Dorfman, L. (1964). *J. Org. Chem.* **29**, 1115.
99. Huisgen, R., Morikawa, M., Breslow, D. S., and Grashey, R. (1967). *Chem. Ber.* **100**, 1602.
100. L'abbe, G., and Dekerk, J.-P. (1979). *Tetrahedron Lett.*, 3213.
101. Goerdeler, J., and Bischoff, M. (1972). *Chem. Ber.* **105**, 3566.
102. Goerdeler, J., and Ludke, H. (1968). *Tetrahedron Lett.*, 2455; Goerdeler, J., and Ludke, H. (1970). *Chem. Ber.* **103**, 3393.

103. a. Review: Kresze, G., and Wucherpfennig, W. (1967). *Angew. Chem. Int. Ed. Engl.* **6**, 149. b. Kataev, E. G., and Plemenkov, V. V. (1968). *Z. Org. Chim.* **4**, 1094. c. Collins, G. R. (1964). *J. Org. Chem.* **29**, 1688. d. Macaluso, A., and Hamer, J. (1967). *J. Org. Chem.* **32**, 506. e. Kresze, G., Maschke, A., Albrecht, R., Bederke, K., Patzchke, H. P., Smalla, H., Trede, A., and Munchen, T. H. (1960). *Angew. Chem.* **74**, 135; Hanson, P., and Stone, T. W. (1984). *J. Chem. Soc., Perkin Trans. 1*, 2429.

104. a. Beecken, H. (1967). *Chem. Ber.* **100**, 2159, 2167. b. Horhold, H.-H., and Eibisch, H. (1968). *Chem. Ber.* **101**, 3567; Atkins, G. M., Jr., and Burgess, E. M. (1972). *J. Am. Chem. Soc.* **94**, 6135; Burgess, E. M., and Williams, W. M. (1973). *J. Org. Chem.* **38**, 1249; Kloek, J. A., and Leschinsky, K. L. (1980). *J. Org. Chem.* **45**, 721; Kobelt, D., Paulus, E. F., and Kampe, K.-D. (1970). *Tetrahedron Lett.*, 123; Kobelt, D., Paulus, E. F., and Kampe, K.-D. (1971). *Tetrahedron Lett.*, 1211.

105. a. Sommer, S. (1977). *Tetrahedron Lett.*, 117; Sommer, S. (1977). *Chem. Lett.*, 583; Schantl, J. (1977). *Z. Naturforsch, B: Anorg. Chem., Org. Chem.* **32**, 72. b. The reaction of electron-deficient azoalkenes with enamines has been shown to provide products derived from [3 + 2] cycloaddition and not the reported [4 + 2] cycloaddition products: Sommer, S. (1979). *Angew. Chem. Int. Ed. Engl.* **18**, 695; Burger, K., and Rottegger, S. (1984). *Tetrahedron Lett.* **25**, 4091; Schultz, A. G., Hagmann, W. K., and Shen, M. (1979). *Tetrahedron Lett.*, 2965.

106. a. Clarke, S. J., Davies, D., and Gilchrist, T. L. (1983). *J. Chem. Soc., Perkin Trans. 1*, 1803; Gilchrist, T. L., Parton, B., and Stevens, J. A. (1981). *Tetrahedron Lett.*, 1059; Gilchrist, T. L., and Stevens, J. A. (1985). *J. Chem. Soc., Perkin Trans. 1*, 1741. b. Gilchrist, T., and Richards, P. (1983). *Synthesis*, 153; see also Heckendorn, R. (1986). *Bull. Soc. Chim. Belg.* **95**, 921.

107. Sprio, V., and Plescia, S. (1971). *Ann. Chim. (Rome)* **61**, 391, 655; Sprio, V., and Plescia, S. (1972). *Ann. Chim. (Rome)* **62**, 345; Plescia, S., Daidone, G., Sprio, V., and Marino, M. L. (1976). *J. Heterocycl. Chem.* **13**, 629; Plescia, S., Daidone, G., and Sprio, V. (1979). *J. Heterocycl. Chem.* **16**, 805.

108. Caglioti, L., Rosini, G., Tundo, P., and Vigevani, A. (1970). *Tetrahedron Lett.*, 2349.

109. Sommer, S. (1979). *Angew. Chem. Int. Ed. Engl.* **18**, 696.

110. Sommer, S. (1977). *Synthesis*, 305.

111. Bonini, B. F., Maccagnani, G., Mazzanti, G., Rosini, G., and Foresti, E. (1981). *J. Chem. Soc., Perkin Trans. 1*, 2322.

112. Sommer, S. (1977). *Angew. Chem. Int. Ed. Engl.* **16**, 58. Sommer, S. (1976). *Angew. Chem. Int. Ed. Engl.* **15**, 432.

113. Moody, C. J. (1982). *Adv. Heterocycl. Chem.* **30**, 1.

114. von Gustorf, E. K., and Kim, B. (1964). *Angew. Chem.* **76**, 592; von Gustorf, E. K., White, D. V., Kim, B., Hess, D., and Leitich, J. (1970). *J. Org. Chem.* **35**, 1155; von Gustorf, E. K., White, D. V., and Leitich, J. (1969). *Tetrahedron Lett.*, 3109; von Gustorf, E. K., White, D. V., Leitich, J., and Henneberg, D. (1969). *Tetrahedron Lett.*, 3113.

115. Huebner, C. F., Strachan, P. L., Donoghue, E. M., Cahoon, N., Dorfman, L., Margerison, R., and Wenkert, E. (1967). *J. Org. Chem.* **32**, 1126; Huebner, C. F., Donoghue, E. M., Novak, C. J., Dorfman, L., and Wenkert, E. (1970). *J. Org. Chem.* **35**, 1149.

116. Firl, J., and Sommer, S. (1970). *Tetrahedron Lett.*, 1925, 1929; Firl, J., and Sommer, S. (1971). *Tetrahedron Lett.*, 4193; Firl, J., and Sommer, S. (1969). *Tetrahedron Lett.*, 1133, 1137; Firl, J., and Sommer, S. (1972). *Tetrahedron Lett.*, 4713.

117. Tufariello, J. J., Mich, T. F., and Miller, P. S. (1966). *Tetrahedron Lett.*, 2293; Moriarty, R. M. (1963). *J. Org. Chem.* **28**, 2385; Cristol, S. J., Allred, E. L., and Wetzel, D. L. (1962). *J. Org. Chem.* **27**, 4058.

118. Landis, M. E., and Mitchell, J. C. (1979). *J. Org. Chem.* **44**, 2288.
119. Fahr, E., and Lind, H. (1966). *Angew. Chem. Int. Ed. Engl.* **5**, 372; Fahr, E., Keil, K. H., Scheckenbach, F., and Jung, A. (1964). *Angew. Chem. Int. Ed. Engl.* **3**, 646.
120. Hall, J. H., and Wojciechowska, M. (1979). *J. Org. Chem.* **44**, 38; Hall, J. H., and Wojciechowska, M. (1978). *J. Org. Chem.* **43**, 3348, 4869.
121. Marchetti, L., Serantoni, E. F., Mongiorgi, R., and di Sanseverino, L. R. (1973). *Gazz. Chim. Ital.* **103**, 615; Marchetti, L. (1977). *J. Chem. Soc., Perkin Trans. 2,* 1977; Marchetti, L. (1978). *J. Chem. Soc., Perkin Trans. 2,* 382; Marchetti, L., and Colonna, M. (1969). *Gazz. Chim. Ital.* **99**, 14; Marchetti, L., and Tosi, G. (1971). *Tetrahedron Lett.,* 3071.
122. Bigotta, A., Forchiassin, M., Risaliti, A., and Russo, C. (1979). *Tetrahedron Lett.,* 4761; Forchiassin, M., Risaliti, A., and Russo, C. (1981). *Tetrahedron* **37**, 2921; Forchiassin, M., Pitacco, G., Russo, C., and Valentin, E. (1982). *Gazz. Chim. Ital.* **112**, 335; Forchiassin, M., Pitacco, G., Risaliti, A., Russo, C., and Valentin, E. (1983). *J. Heterocycl. Chem.* **20**, 305; Forchiassin, M., Russo, C., and Nardin, G. (1983). *Tetrahedron Lett.* **24**, 2909.
123. a. Campbell, J. A., Harris, I., Mackay, D., and Sauer, T. D. (1975). *Can. J. Chem.* **53**, 535. b. Mackay, D., Campbell, J. A., and Jennison, C. P. R. (1970). *Can. J. Chem.* **48**, 81. c. Campbell, J. A., Mackay, D., and Sauer, T. D. (1972). *Can. J. Chem.* **50**, 371. d. Mackay, D., and Pilger, C. W. (1974). *Can. J. Chem.* **52**, 1114. e. Jennison, C. P. R., and Mackay, D. (1973). *Can. J. Chem.* **51**, 3726. f. Chung, C. Y.-J., Mackay, D., and Sauer, T. D. (1972). *Can. J. Chem.* **50**, 1568. g. Mackay, D., Pilger, C. W., and Wong, L. L. (1973). *J. Org. Chem.* **38**, 2043.
124. Matsuda, I., Yamamoto, S., and Ishii, Y. (1976). *J. Chem. Soc., Perkin Trans. 1,* 1523, 1528.
125. Weidinger, H., and Kranz, J. (1963). *Chem. Ber.* **96**, 2070.
126. Weidinger, H., and Sturm, H. J. (1968). *Liebigs Ann. Chem.* **716**, 143.
127. a. Morel, G., Marchand, E., and Foucaud, A. (1985). *J. Org. Chem.* **50**, 771. b. Kapran, N. A., Lukmanov, V. G., Yagupolskil, L. M., and Cherkosov, V. M. (1977). *Khim. Geterotsikl. Soedin,* 122 [*Chem. Abstr.* **86**, 171389 (1977)].
128. Matsuda, I., Itoh, K., and Ishii, Y. (1972). *J. Chem. Soc., Perkin Trans. 1,* 1678.
129. Burger, K. V., Partscht, H., Huber, E., Gieren, A., Hubner, T., and Koerlein, C.-P. (1984). *Chem. Zeit.* **108**, 209; Burger, K. V., Wassmuth, U., Partscht, H., Gieren, A., Hubner, T., and Kaerlein, C.-P. (1984). *Chem. Zeit.* **108**, 205; Burger, K., and Penninger, S. (1978). *Synthesis,* 524.
130. Blatter, H. M., and Lukaszewski, H. (1966). *J. Org. Chem.* **31**, 722; Blatter, H. M., and Lukaszewski, H. (1964). *Tetrahedron Lett.,* 855, 1087; Fairfull, A. E. S., and Peak, D. A. (1955). *J. Chem. Soc.,* 796; Howard, J. C., and Michels, J. G. (1960). *J. Org. Chem.* **25**, 829; Goerdeler, J., and Weber, D. (1964). *Tetrahedron Lett.,* 799; Ugi, J., and Rosendahl, F. K. (1963). *Liebigs Ann. Chem.* **670**, 80; Goerdeler, J., and Richter, R. (1978). *Synthesis,* 760.
131. Sakamoto, M., Miyazawa, K., and Tomimatsu, Y. (1976). *Chem. Pharm. Bull. Tokyo* **24**, 2532.
132. Richter, R., and Ulrich, H. (1970). *Chem. Ber.* **103**, 3525.
133. Kato, T., and Masuda, S. (1975). *Chem. Pharm. Bull. Tokyo* **23**, 2251.
134. Katagiri, N., Niwa, R., and Kato, T. (1983). *Chem. Pharm. Bull. Tokyo* **31**, 2899.
135. Kato, T., and Masuda, S. (1974). *Chem. Pharm. Bull. Tokyo* **22**, 2251.
136. Sakamoto, M., Shibano, M., Miyazawa, K., Suzuki, M., and Tomimatsu, Y. (1976). *Chem. Pharm. Bull. Tokyo* **24**, 2889.
137. Pummerer, R., and Fiesselmann, H. (1940). *Liebigs Ann. Chem.* **544**, 206; Pummerer,

R., and Stieglitz, E. (1942). *Chem. Ber.* **75,** 1072; Pummerer, R., and Reuss, F. (1947). *Chem. Ber.* **80,** 242.

138. Sayed, G. H., Elhalim, M. S. A., El-Kady, M. Y., and Elwahab, L. M. A. (1982). *Indian J. Chem. B* **21,** 589.

139. a. Lora-Tamayo, M., Ossorio, R. P., and Burata, M. S. (1954). *An. Soc. Espan. B.* **50,** 765; Lora-Tamayo, M. (1958). *Tetrahedron* **4,** 17; Adams, R., and Way, J. W. (1954). *J. Am. Chem. Soc.* **54,** 2763; Adams, R., and Reifschneider, W. (1958). *Bull. Soc. Chim. Fr.,* 23. b. Review: Friedrichsen, W., and Bottcher, A. (1981). *Heterocycles* **16,** 1009. c. Friedrichsen, W., and Oeser, H.-G. (1978). *Liebigs Ann. Chem.,* 1146; Friedrichsen, W., and Oeser, H.-G. (1975). *Chem. Ber.* **108,** 31. d. Fujita, S. (1985). *J. Syn. Org. Chem. Jpn.* **43,** 153.

140. Begland, R. W., Cairncross, A., Donald, D. S., Hartter, D. R., Sheppard, W. A., and Webster, O. W. (1971). *J. Am. Chem. Soc.* **93,** 4953; Begland, R. W., and Hartter, D. H. (1972). *J. Org. Chem.* **37,** 4136.

141. a. Pfleger, R., and Jager, A. (1957). *Chem. Ber.* **90,** 2460; Burpitt, R. D., Brannock, K. C., Nations, R. G., and Martin, J. C. (1971). *J. Org. Chem.* **36,** 2222; Sakamoto, M., Miyazawa, K., Ishihara, Y., and Tomimatsu, Y. (1974). *Chem. Pharm. Bull. Tokyo* **22,** 1419; Sakamoto, M., and Tomimatsu, Y. (1970). *Yakugaku Zasshi* **90,** 1386. b. von Pechmann, H., and Bauer, W. (1900). *Chem. Ber.* **33,** 644; von Pechmann, H., and Bauer, W. (1909). *Chem. Ber.* **42,** 659; Petersen, S., and Heitzer, H. (1970). *Angew. Chem. Int. Ed. Engl.* **9,** 67.

142. a. Trost, B. M., and Whitman, P. J. (1974). *J. Am. Chem. Soc.* **96,** 7421; Trost, B. M., and Whitman, P. J. (1972). *J. Am. Chem. Soc.* **94,** 8634; Trost, B. M., and Whitman, P. J. (1969). *J. Am. Chem. Soc.* **91,** 7534. b. Fagan, P. J., Neidert, E. E., Nye, M. J., O'Hare, M. J., and Tang, W.-P. (1979). *Can. J. Chem.* **57,** 904. c. Sasaki, T., Kanematsu, K., and Kataoka, T. (1975). *J. Org. Chem.* **40,** 1201. d. Harano, K., Yasuda, M., Ban, T., and Kanematsu, K. (1980). *J. Org. Chem.* **45,** 4455. e. Harano, K., Ban, T., and Kanematsu, K. (1979). *Heterocycles* **12,** 453. f. Ban, T., and Kanematsu, K. (1981). *Heterocycles* **15,** 273. g. Nye, M. J., and Fagan, P. J. (1971). *J. Chem. Soc., Chem. Commun.,* 537.

143. Evnin, A. B., and Arnold, D. R. (1968). *J. Am. Chem. Soc.* **90,** 5330.

144. Review: Christl, M. (1986). *Gazz. Chim. Ital.* **116,** 7. a. Steglich, W., Buschmann, E., Gansen, G., and Wilshowitz, L. (1977). *Synthesis,* 252. b. Christl, M., Lanzendorfer, U., and Freund, S. (1981). *Angew. Chem. Int. Ed. Engl.* **20,** 674; Christl, M., Lanzendorfer, U., Peters, K., Peters, E.-M., and von Schnering, H. G. (1983). *Tetrahedron Lett.* **24,** 353; Christl, M., Lanzendorfer, U, Hegmann, J., Peters, K., Peters, E. M., and von Schnering, H. G. (1985). *Chem. Ber.* **118,** 2940. c. Padwa, A., and Eisenbarth, P. (1984). *Tetrahedron Lett.* **25,** 5489; Padwa, A., and Eisenbarth, P. (1985). *J. Heterocycl. Chem.* **22,** 61; Padwa, A., and Eisenbarth, P. (1985). *Tetrahedron* **41,** 283.

145. Berning, W., Hunig, S., and Prokschy, F. (1984). *Chem. Ber.* **117,** 1455.

146. a. Ege, G., Gilbert, K., and Franz, H. (1977). *Synthesis,* 556; Ege, G., and Gilbert, K. (1979). *Tetrahedron Lett.,* 4253; Ege, G., and Gilbert, K. (1981). *J. Heterocycl. Chem.* **18,** 675; Durr, H., and Schmitz, H. (1978). *Chem. Ber.* **111,** 2258; Norison, T. (1976). *J. Med. Chem.* **19,** 517; Elnagdi, M. H., Elmoghayar, M. R. H., Kandeel, E. M., and Ibrahim, M. K. A. (1977). *J. Heterocycl. Chem.* **14,** 227. b. Cherkasov, V. M., Nasyr, I. A., and Tsyba, V. T. (1970). *Khim. Geterotsikl. Soedin,* 1704.

147. Garanti, L., and Zecchi, G. (1980). *Tetrahedron Lett.,* 559; Bruche, L., Garanti, L., and Zecchi, G. (1982). *J. Chem. Soc., Perkin Trans. 1,* 755.

148. a. For the first report of an isolated diene system containing a quaternary salt, acridizinium bromide, participating in a [4+ + 2] cycloaddition see Bradsher, C. K., and

Solomons, T. W. G. (1958). *J. Am. Chem. Soc.* **80**, 933. b. Bradsher, C. K., and harvan, D. J. (1971). *J. Org. Chem.* **36**, 3778; Bradsher, C. K., and Stone, J. A. (1968). *J. Org. Chem.* **33**, 519; Burnham, W. S., and Bradsher, C. K. (1972). *J. Org. Chem.* **37**, 355; Parham, M. E., Bradsher, C. K., and Frazer, M. G. (1972). *J. Org. Chem.* **37**, 358; Bradsher, C. K., and Westerman, I. J. (1971). *J. Org. Chem.* **36**, 969; Bradsher, C. K., Wallis, T. G., Westerman, I. J., and Porter, N. A. (1977). *J. Am. Chem. Soc.* **99**, 2588; Bradsher, C. K., and Stone, J. A. (1969). *J. Org. Chem.* **34**, 1700; Bradsher, C. K., Carlson, G. L. B., Porter, N. A., Westerman, I. J., and Wallis, T. G. (1978). *J. Org. Chem.* **43**, 822; Bradsher, C. K., Miles, C. R., Porter, N. A., and Westerman, I. J. (1972). *Tetrahedron Lett.*, 4969. c. Bradsher, C. K., and Day, F. H. (1971). *Tetrahedron Lett.*, 409; Bradsher, C. K., Day, F. H., McPhail, A. T., and Wong, P.-S. (1971). *Tetrahedron Lett.*, 4205; Bradsher, C. K., Day, F. H., McPhail, A. T., and Wong, P.-S. (1973). *J. Chem. Soc., Chem. Commun.*, 156; Bradsher, C. K., and Day, F. H. (1974). *J. Heterocycl. Chem.* **11**, 23. d. Day, F. H., Bradsher, C. K., and Chen, T.-K. (1975). *J. Org. Chem.* **40**, 1195; Wimmer, T. L., Day, F. H., and Bradsher, C. K. (1975). *J. Org. Chem.* **40**, 1198; Bradsher, C. K., and Chen, T. K. (1979). *J. Org. Chem.* **44**, 4680. e. Westerman, I. J., and Bradsher, C. K. (1979). *J. Org. Chem.* **44**, 727; Westerman, I. J., and Bradsher, C. K. (1978). *J. Org. Chem.* **43**, 3002.
149. Fields, D. L., Regan, T. H., and Dignan, J. C. (1968). *J. Org. Chem.* **33**, 390; Fields, D. L., and Regan, T. H. (1970). *J. Org. Chem.* **35**, 1870; Fields, D. L., and Regan, T. H. (1971). *J. Org. Chem.* **36**, 2986, 2991; Fields, D. L., and Miller, J. B. (1970). *J. Heterocycl. Chem.* **7**, 91; Fields, D. L., Regan, T. H., and Graves, R. E. (1971). *J. Org. Chem.* **36**, 2995; Fields, D. L. (1971). *J. Org. Chem.* **36**, 3002.
150. Bradsher, C. K. (1974). *Adv. Heterocycl. Chem.* **16**, 289; Saraf, S. D. (1980). *Heterocycles* **14**, 2047; Bradsher, C. K. (1981). *Heterocyclic Comp.* **38**, Part 1, 381.
151. Alder, K., and Stein, G. (1937). *Angew. Chem.* **50**, 510.
152. Falck, J. R., Manna, S., and Mioskowski, C. (1983). *J. Am. Chem. Soc.* **105**, 631; Falck, J. R., Manna, S., and Mioskowski, C. (1982). *J. Org. Chem.* **47**, 5021.
153. Franck, R. W., and Gupta, R. B. (1985). *Tetrahedron Lett.* **26**, 293; Franck, R. W., and Gupta, R. B. (1984). *J. Chem. Soc., Chem. Commun.*, 761.
154. Stevens, R. V., and Pruitt, J. R. (1983). *J. Chem. Soc., Chem. Commun.*, 1425.
155. Gisby, G. P., Sammes, P. G., and Watt, R. A. (1982). *J. Chem. Soc., Perkin Trans. 1*, 249; Sammes, P. G., and Watt, R. A. (1976). *J. Chem. Soc., Chem. Commun.*, 367.
156. Fuks, R., King, G. S. D., and Viehe, H. G. (1969). *Angew. Chem. Int. Ed. Engl.* **8**, 675.
157. Adachi, I. (1969). *Chem. Pharm. Bull. Tokyo* **17**, 2209.
158. Marchetti, L., and Tosi, G. (1967). *Ann. Chem. (Rome)* **57**, 1414; Haddadin, M. J., and Issidorides, C. H. (1965). *Tetrahedron Lett.*, 3253; McFarland, J. W. (1971). *J. Org. Chem.* **36**, 1842.
159. Ley, K., Seng, F., Eholzer, U., Nast, R., and Schubart, R. (1969). *Angew. Chem. Int. Ed. Engl.* **8**, 596.
160. Swan, G. A. (1969). *J. Chem. Soc., Chem. Commun.*, 20.
161. Hesse, K.-D. (1970). *Liebigs Ann. Chem.* **741**, 117.
162. Shono, T., Matsumura, Y., Inoue, K., Ohmizu, H., and Kashimura, S. (1982). *J. Am. Chem. Soc.* **104**, 5753.
163. Fischer, O., Muller, A., and Vilsmeier, A. (1925). *J. Prakt. Chem.* **109**, 69.
164. Meth-Cohn, O. (1985). *Tetrahedron Lett.* **26**, 1901.
165. Hartman, G. D., Halczenko, W., and Phillips, B. T. (1985). *J. Org. Chem.* **50**, 2427.
166. Franke, H., Grabhoff, H., and Scherowsky, G. (1979). *Chem. Ber.* **112**, 3623.
167. Schmidt, R. R. (1965). *Angew. Chem. Int. Ed. Engl.* **3**, 804; Grundmann, C., Weisse, G., and Seide, S. (1952). *Liebigs Ann. Chem.* **577**, 77; Meerwein, H., Laasch, P.,

Mersch, R., and Nentwig, J. (1956). *Chem. Ber.* **89**, 224; Schmidt, R. R. (1968). *Tetrahedron Lett.*, 3443; Lora-Tamayo, M., Madronero, R., and Garcia-Munoz, G. (1961). *Chem. Ber.* **94**, 208.

168. For additional related studies, see Boyd, G. V. (1966). *Tetrahedron Lett.*, 3369; Boyd, G. V., and Dando, S. R. (1970). *J. Chem. Soc. C*, 1397; Boyd, G. V., and Dando, S. R. (1971). *J. Chem. Soc. C*, 2314; Boyd, G. V., and Summers, A. J. H. (1971). *J. Chem. Soc. C*, 410, 2311.

169. Smith, M. B., and Shroff, H. N. (1985). *Heterocycles* **23**, 2229.

170. a. Schmidt, R. R. (1965). *Angew. Chem. Int. Ed. Engl.* **4**, 241. b. Schmidt, R. R. (1970). *Chem. Ber.* **103**, 3242.

171. Seeliger, W., and Diepers, W. (1966). *Liebigs Ann. Chem.* **697**, 171; Schmidt, R. R. (1965). *Chem. Ber.* **98**, 334; Schmidt, R. R. (1969). *Angew. Chem. Int. Ed. Engl.* **8**, 602; Schmidt, R. R., and Machat, R. (1970). *Angew. Chem. Int. Ed. Engl.* **9**, 311.

172. Ben-Ishai, D., and Hirsch, S. (1983). *Tetrahedron Lett.* **24**, 955.

173. Kempe, U. M., Das Gupta, T. K., Blatt, K., Gygax, P., Felix, D., and Eschenmoser, A. (1972). *Helv. Chim. Acta* **55**, 2187; Das Gupta, T. K., Felix, D., Kempe, U. M., and Eschenmoser, A. (1972). *Helv. Chim. Acta* **55**, 2198; Gygax, P., Das Gupta, T. K., and Eschenmoser, A. (1972). *Helv. Chim. Acta* **55**, 2205; Shatzmiller, S., Gygax, P., Hall, D., and Eschenmoser, A. (1973). *Helv. Chim. Acta* **56**, 2961; Petrzilka, M., Felix, D., and Eschenmoser, A. (1973). *Helv. Chim. Acta* **56**, 2950; Shatzmiller, S., and Eschenmoser, A. (1973). *Helv. Chim. Acta* **56**, 2975.

174. Riediker, M., and Graf, W. (1979). *Helv. Chim. Acta* **62**, 205.

175. a. Shatzmiller, S., and Neidlein, R. (1976). *Tetrahedron Lett.*, 4151. Hardegger, B., and Shatzmiller, S. (1976). *Helv. Chim. Acta* **59**, 2765. b. Shalom, E., Zenou, J.-L., and Shatzmiller, S. (1977). *J. Org. Chem.* **42**, 4213. c. Shatzmiller, S. (1982). *Liebigs Ann. Chem.*, 1933. d. Marciano, D., and Shatzmiller, S. (1982). *Liebigs Ann. Chem.*, 1495. e. Neidlein, R., Shatzmiller, S., and Walter, E. (1980). *Liebigs Ann. Chem.*, 686. f. Hepp, L. R., Bordner, J., and Bryson, R. A. (1985). *Tetrahedron Lett.* **26**, 595.

176. Denmark, S. E., Dappen, M. S., and Cramer, C. J. (1986). *J. Am. Chem. Soc.* **108**, 1306; Denmark, S. E., Cramer, C. J., and Sternberg, J. A. (1986). *Tetrahedron Lett.* **27**, 3693; Denmark, S. E., and Cramer, C. J., unpublished observations.

Chapter **10**

Heteroaromatic Azadienes*

INTRODUCTION

Since the early demonstrations of the participation of substituted 1,2,4,5-tetrazines[1] and oxazoles[2] in Diels–Alder reactions with olefinic and acetylenic dienophiles, the potential utilization of the [4 + 2] cycloaddition reactions of heteroaromatic systems possessing a suitable azadiene arrangement has been the focus of extensive investigations.[3–8] In general, the heteroaromatic systems which possess an electron-deficient azadiene

---

* Adapted with permission from Boger, D. L. (1986). *Chem. Rev.* **86**, 781.

are ideally suited for participation in inverse electron demand ($LUMO_{diene}$ controlled) Diels–Alder reactions. The recognition of the electron-deficient nature of heteroaromatic azadienes led to the proposed potential and subsequently demonstrated reversal of the diene/dienophile electronic properties in the Diels–Alder reaction and led to the full investigations of the inverse electron demand ($LUMO_{diene}$ controlled) Diels–Alder reaction.[1] Since the initial investigations, several general approaches have been investigated and shown to promote or accelerate the participation of a wide range of electron-deficient heterocyclic azadienes in Diels–Alder reactions.

The additional substitution of the heterocyclic azadiene system with electron-withdrawing groups accents the electron-deficient nature of the heterodiene and permits the use of electron-rich, strained, or even simple olefins as dienophiles.[3,4b,6] Substitution of the heterocyclic azadiene with strongly electron-donating substituents in many instances is sufficient to overcome the electron-deficient nature of the azadiene and permits the use of conventional electron-deficient dienophiles in normal ($HOMO_{diene}$ controlled) Diels–Alder reactions.[4,6] The entropic assistance provided by the intramolecular Diels–Alder reaction is sufficient in most instances to override the reluctant azadiene participation in Diels–Alder reactions.[7] The incorporation of the heterocyclic azadiene, or the dienophile, into a reactive system, e.g., heterocumulene, allows a number of specialized [4 + 2] cycloaddition processes which are best characterized as stepwise addition–cyclization [4 + 2] cycloadditions.[8]

1. OXAZOLES

Since the initial report that alkyloxazoles participate in Diels–Alder reactions with maleic anhydride,[2] extensive efforts have defined the scope and synthetic utility of the [4 + 2] cycloadditions of oxazole derivatives. This work has been the subject of several reviews.[3,9]

The course of and facility with which the Diels–Alder reaction of oxazoles proceed are dependent on the dienophile structure, the oxazole/dienophile substitution, as well as the reaction conditions. Olefinic dienophiles provide pyridine products derived from the fragmentation of the initial [4 + 2] cycloadducts 2 to provide 3 which subsequently aromatize to provide the substituted pyridines [Eq. (1)].

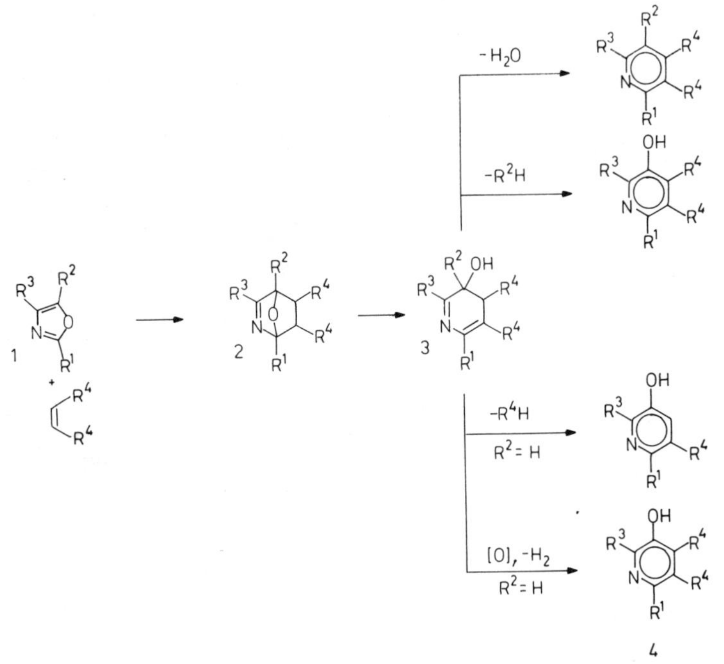

$$(1)$$

Simple dehydration of **3** provides pyridines ($R^2 = R, H$), and 3-hydroxy-pyridines are formed from **3** by the elimination of R^2H (e.g., EtOH, $R^2 = OEt$) or R^4H (e.g., HCN, $R^4 = CN$ and $R^2 = H$) or by simple dehydrogenation ($R^2 = H, -H_2$). In many instances, more than one pathway is followed and a mixture of pyridine products is obtained. Appropriate selection of the substituted oxazole (e.g., $R^2 = OEt, OSiMe_3, CN$), complementary selection of an olefinic dienophile (e.g., $R^4 = $ alkyl versus CN), and conduction of the reaction under defined reaction conditions (HOAc versus C_6H_6) can determine the observed course of the reaction. The addition of electron-donating substituents to the oxazole nucleus (OR > alkyl > 4-phenyl > $COCH_3$ > CO_2R >> 2-phenyl, 5-phenyl) increases its rate of participation in normal ($HOMO_{diene}$ controlled) Diels–Alder reactions with typical electron-deficient and simple olefinic dienophiles. Although the number of studies of the regioselectivity of the Diels–Alder reaction of oxazoles with unsymmetrical olefinic dienophiles are limited, the generalization has been made that the strongest electron-withdrawing olefinic substituent is found at position C-4 of the pyridyl products. A number of exceptions to this generalization have been observed.[3,9]

Since the annual commercial requirements for pyridoxol (**5**), one of the three equally active forms of vitamin B_6, are in excess of 200,000 pounds

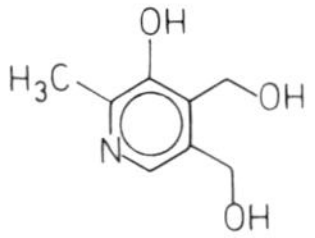

5 pyridoxol

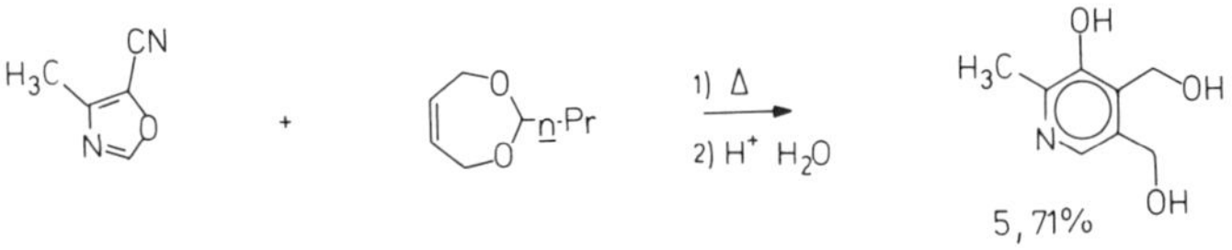

per year, the potential for the development of a commercially viable process for the preparation of pyridoxol based on the Diels–Alder reactions of substituted oxazoles played a major role in the initial investigation and development of the scope of the oxazole [4 + 2] cycloadditions. Much of this effort is summarized in Table 10-I.[10–37]

Despite the early recognition that heterocyclic azadiene systems are typically electron-deficient,[1] little effort has been devoted to the exploration of the potential participation of electron-deficient oxazoles in inverse electron demand ($LUMO_{diene}$ controlled) Diels–Alder reactions with electron-rich or simple olefinic dienophiles. One study has demonstrated the potential of such investigations (Table 10-I, entry 26).[34] Breslow and co-workers have adapted the oxazole olefin Diels–Alder reaction for the preparation of a tetrahydroquinoline-based analog of pyridoxamine with the stereochemically defined placement of a catalytic group [Eq. (2)].[37b]

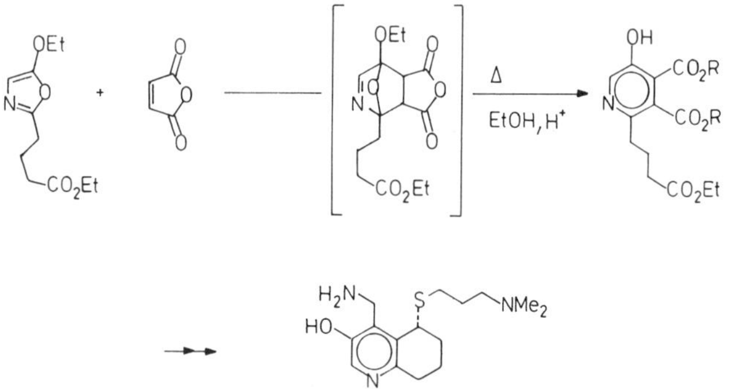

$$(2)$$

The pyridoxamine analog was shown to be capable of converting keto acids to amino acids with good enantioselectivity.

A key step employed in the Kozikowski–Hasan approach to the antitumor agent ellipticine (**6**) was the regioselective [4 + 2] cycloaddition of acrylonitrile with the oxazole **7** [Eq. (3)].

TABLE 10-I

Oxazole Olefin Diels–Alder Reactions Employed in the Total Synthesis of Pyridoxol

		Oxazole		Dienophile			Diels–Alder product					
Entry	R¹	R²	R³	R	X	Y	R²	R³	X	Y	R¹	Ref.
1	H	$COCH_3$	CH_3	H	$C(O)NHC(O)$		$COCH_3$	CH_3	$C(O)NHC(O)$		H	10, 11
2	H	H	CH_3	CN	CH_2OCH_3	CH_2OCH_3	OH	CH_3	CH_2OCH_3	CH_2OCH_3	H	12
3	H	H	CH_3	CN	H	CN	OH	CH_3	H	CN	H	13
4	H	H	CH_3	H	CO_2R	CO_2R	OH	CH_3	CO_2R	CO_2R	H	17
5	H	H	CH_3	H	CN	CN	OH	CH_3	CN	CN	H	14
6	H	H	CH_3	H	$C(O)NHC(O)$		OH	CH_3	$C(O)NHC(O)$		H	15
7	H	H	CH_3	H	CN	CH_2OH	OH	CH_3	CN	CH_2OH	H	16
8	H	H	CH_3	H	CHO	CH_2OAc	OH	CH_3	CHO	CH_2OAc	H	17a
9	H	H	CH_3	RSO_2	CH_2OCH_2	CH_2OCH_2	OH	CH_3	CH_2OCH_2	CH_2OCH_2	H	18
10	H	OEt (OR)	CH_3	H	CH_2OCH_2	CH_2OCH_2	OH	CH_3	CH_2OCH_2	CH_2OCH_2	H	17
11	H	OEt (OR)	CH_3	H	CO_2Et	CO_2Et	OH	CH_3	CO_2Et	CO_2Et	H	17, 19, 20

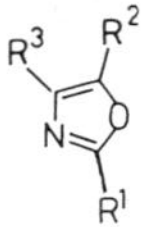

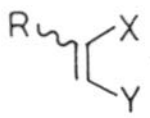

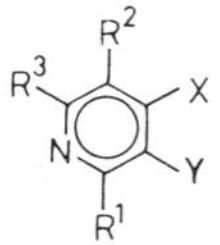

No.										Ref
12	H	OEt (OR)	CH₃	H	C(O)OC(O)	CH₃	OH	C(O)OC(O)	H	17
13	H	OEt (OR)	CH₃	H	CN / CN	CH₃	OH	CN / CN	H	21
14	H	OEt (OR)	CH₃	H	C(O)NHC(O)	CH₃	OH	C(O)NHC(O)	H	10, 21
15	H	OEt (OR)	CH₃	H	CH₂OCH(R)OCH₂	CH₃	OH	CH₂OCH(R)OCH₂	H	22
16	H	OEt (OR)	CH₃	H	CH(OMe)OCH(OMe)	CH₃	OH	CH(OMe)OCH(OMe)	H	23
17	H	OEt (OR)	CH₃	H	CH₂OR / CH₂OR	CH₃	OH	CH₂OR / CH₂OR	H	17, 24–26
18	H	OCO₂Et	CH₃	H	CN / CN	CH₃	OH	CN / CN	H	27
19	H	OCO₂Et	CH₃	H	CO₂Et / CO₂Et	CH₃	OH	CO₂Et / CO₂Et	H	27
20	H	OSiMe₃	CH₃	H	C(O)NHC(O)	CH₃	OH	C(O)NHC(O)	H	28
21	H	OSiMe₃	CH₃	H	CO₂Et / CO₂Et	CH₃	OH	CO₂Et / CO₂Et	H	28
22	H	OCH₂CH₂OR	CH₃	H	CH₂OCH(R)OCH₂	CH₃	OH	CH₂OCH(R)OCH₂	H	29
23	H	OPr	CH₃	H	CH₂OCH(R)OCH₂	CH₃	OH	CH₂OCH(R)OCH₂	H	30
24	CO₂R	OEt	CH₃	H	CH₂OCH₂	CH₃	OH	CH₂OCH₂	H	31
25	CO₂R	OEt	CH₃	H	CH₂OCH(R)OCH₂	CH₃	OH	CH₂OCH(R)OCH₂	H	31
26	H	CN	CH₃	H	CH₂OCH(R)OCH₂	CH₃	OH	CH₂OCH(R)OCH₂	H	34
27	H	OEt	CH₂CO₂Et	H	CO₂Et / CO₂Et	CH₂CO₂Et	OH	CO₂Et / CO₂Et	H	32, 33
28	H	OEt	CH₂CO₂Et	H	CN / CN	CH₂CO₂Et	OH	CN / CN	H	34
29	H	OEt	CH₂CO₂Et	H	CN / CN	CH₃	OH	CN / CN	H	35
30	H	OEt	CH₂CO₂Et	H	CH₂OCH(R)OCH₂	CH₂CO₂Et	OH	CH₂OCH(R)OCH₂	H	36 a
31	H	OEt	CH₂CO₂Et	H	CH₂OH / CH₂OH	CH₃	OH	CH₂OH / CH₂OH	H	35
32	H	OEt	CH₂CO₂H	H	CH₂OH / CN	CH₃	OH	CH₂OH / CN	H	26
33	H	OEt	CH₃	H	CH(OEt)OCH₂	CH₃	OH	CH(OEt)OCH₂	H	36b

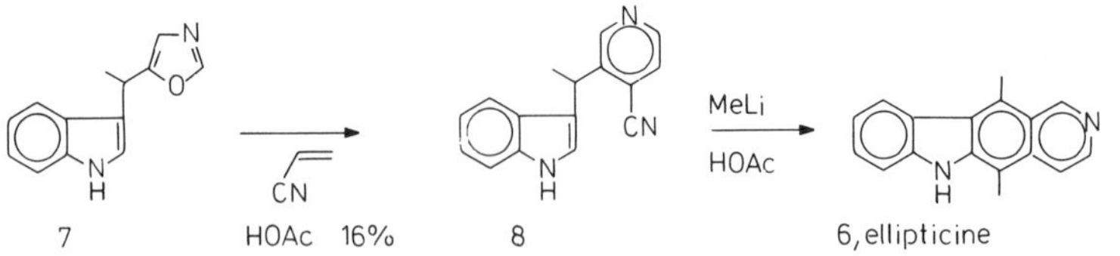

The cycloaddition provided the expected 4-cyanopyridine **8** albeit in modest yields.[38]

Weinreb and Levin have detailed a total synthesis of eupolauramine (**10**), an azaphenanthrene alkaloid, which is based on an intramolecular[39,40] alkene oxazole Diels–Alder reaction and illustrates the condition-dependent fragmentation of the initially formed olefin oxazole Diels–Alder products [Eq. (1)]. Thermal cycloaddition of **11**, even with the rigorous exclusion of oxygen from the reaction mixture, provided a mixture of **13a** and **13b** resulting from the previously observed but rare dehydrogenation of the dihydropyridine intermediate [Eq. (1), $-H_2$]. The desired and expected Kondrat'eva fragmentation involving the elimination of water [Eq. (1), $-H_2O$] could be observed if the reaction was conducted with a nonnucleophilic base (DBU) present in the thermolysis reaction mixture. Furthermore, the thermal cycloaddition of **11** represented a rare example of a 2-phenyloxazole effectively participating in an olefin oxazole Diels–Alder reaction, and it is the entropic assistance provided by the intramolecular reaction which accounts for the observed reaction [Eq. (4)].

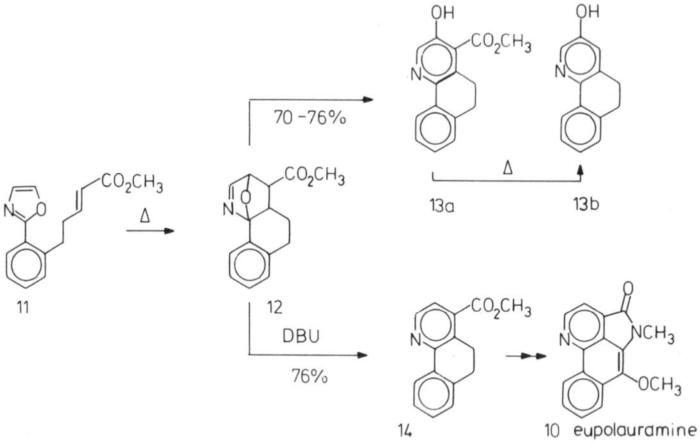

In contrast, acetylenic dienophiles including benzynes[51] participate in a well-defined [4 + 2] cycloaddition with oxazoles to provide substituted furans **15** which arise from the retro-Diels–Alder reaction with loss of

R^3CN from the initial alkyne oxazole Diels–Alder adduct **16** [Eq. (5)].[3,9,41–58]

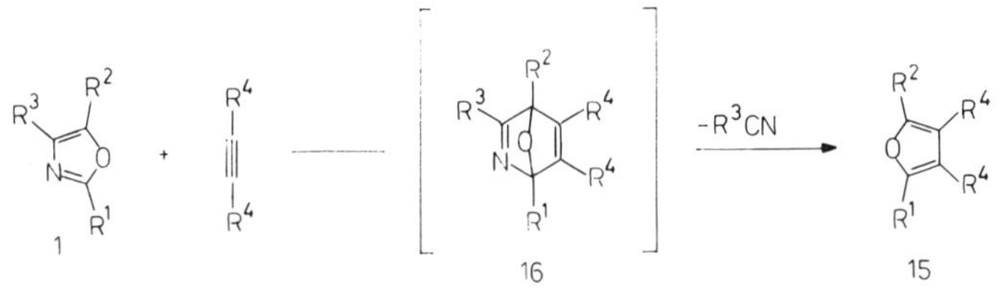

$$(5)$$

Efforts to isolate the initial alkyne oxazole [4 + 2] cycloadducts have been unsuccessful, and only the furan cycloaddition products have been detected in the reaction mixtures. The Diels–Alder reaction of alkyl- or aryl-substituted oxazoles with neutral, conjugated, or electron-deficient alkynes displays little regioselectivity whereas polarized, electron-rich oxazoles (e.g., **1**, R^2 = OEt) do participate in regioselective, intermolecular alkyne oxazole Diels–Alder reactions. In such instances, mixtures of isomeric products are still often observed, and the alkyne carbon bearing the strongest electron-withdrawing substituent preferentially attaches to oxazole C-5.[3,9,41] The facility with which the substituted oxazoles participate in [4 + 2] cycloadditions with electron-deficient alkynes follows the identical order of oxazole reactivity observed with electron-deficient olefins (OR > alkyl > 4-phenyl > $COCH_3$ > CO_2CH_3 >> 2-, 5-phenyl). Table 10-II summarizes representative work.[41–58] Jacobi and co-workers have systematically investigated the scope of the intramolecular alkyne oxazole Diels–Alder reaction and have applied their findings in the total syntheses of (±)-ligularone (**17**) and (±)-petasalbine (**18**) [Eq. (6)],

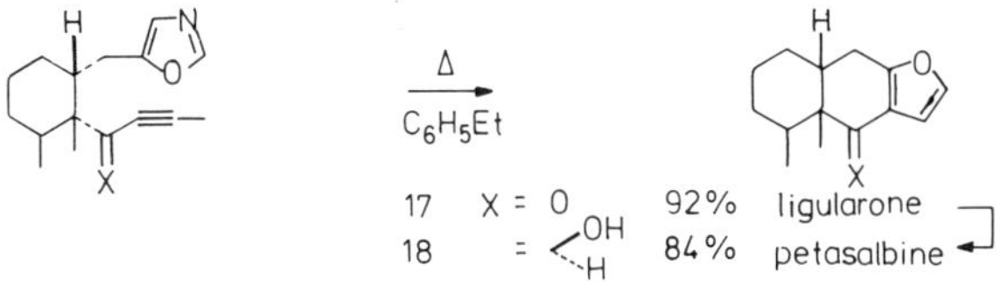

$$(6)$$

(±)-paniculide A (**19**) [Eq. (7)],

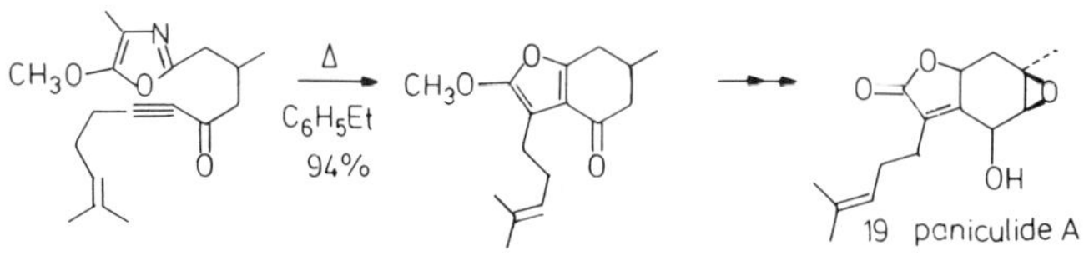

$$(7)$$

TABLE 10-II

Intermolecular Alkyne Oxazole Diels–Alder Reactions

Oxazole: R^3,R^4–(ring)–R^2/R^1 (with N, O). Alkyne: R^4—≡—R^5. Diels–Alder product (furan): R^2/R^1, R^5, R^4.

Oxazole R^1	Oxazole R^2	Oxazole R^3	Alkyne R^4	Alkyne R^5	Product R^1	Product R^2	Product R^4	Product R^5	Ref.
H	OPr/OBu	CH_3	CHO	H	H	OPr/OBu	CHO	H	42
H	OPr/OBu	CH_3	$COCH_3$	H	H	OPr/OBu	$COCH_3$	H	42
H	OPr/OBu	CH_3	$COCH_3$	$COCH_3$	H	OPr/OBu	$COCH_3$	$COCH_3$	42
H	OPr/OBu	CH_3	$CH(OEt)_2$	$CH(OEt)_2$	H	OPr/OBu	$CH(OEt)_2$	$CH(OEt)_2$	42
H	Ph	CH_3	H	$COCH_3$	H	Ph	H	$COCH_3$	42
H	Ph	CH_3	$CH(OEt)_2$	$CH(OEt)_2$	H	Ph	$CH(OEt)_2$	$CH(OEt)_2$	42
H	Ph	CH_3	H	CO_2Et	H	Ph	H	CO_2Et	44
H	$OEt/OPr/OBu$	CH_3	CO_2Et	CO_2Et	H	$OEt/OPr/OBu$	CO_2Et	CO_2Et	43, 50
H	CH_3	CH_3	CO_2Et	CO_2Et	H	CH_3	CO_2Et	CO_2Et	43
H	CH_3	CH_3	CO_2Et	H	H	CH_3	CO_2Et	H	44
H	H	tBu	CO_2Et	CO_2Et	H	H	CO_2Et	CO_2Et	43
H	H	tBu	CO_2Et	H	H	H	CO_2Et	H	44
H	CO_2Et	CH_3	CO_2Et	CO_2Et	H	CO_2Et	CO_2Et	CO_2Et	43
H	CO_2Et	CH_3	CO_2Et	H	H	CO_2Et	CO_2Et	H	44
H	OPr/OBu	CH_3	CO_2Et	CH_3	H	OPr/OBu	CO_2Et	CH_3	44
H	OPr/OBU	CH_3	CO_2Et	H	H	OPr	CO_2Et	H	44
H	OPr/OBu	CH_3	CO_2Et	Et	H	OPr/OBu	CO_2Et	Et	44
H	OEt	$H/CH_3/Ph$	CO_2CH_3	CO_2CH_3	H	OEt	CO_2CH_3	CO_2CH_3	45, 46
H	OEt	$H/CH_3/Ph$	Ph	CO_2Et	H	OEt	Ph/CO_2Et	Ph/CO_2Et	47
H	OEt	CH_3	CHO	$CH(OEt)_2$	H	OEt	CHO	$CH(OEt)_2$	48
H	OPr	CH_3	$P(O)(OEt)_2$	R	H	OEt	$P(O)(OEt)_2$	R	49
H	$(CH_2)_4$		CO_2CH_3	CO_2CH_3	H	$(CH_2)_4CN$	CO_2CH_3	CO_2CH_3	43
CH_3	H	CH_3	CH_2OCH_3	CH_2OCH_3	CH_3	H	CH_2OCH_3	CH_2OCH_3	50

Table continuation (substituent/reference listing). Columns read left-to-right as R1 | R2 | R3 | R4 | R5 | Ref.

R1	R2	R3	R4	R5	Ref
CH₃	H	OCH₃	CH₂OAc	CH₂OAc	50
CH₃	H	CO₂Et	CH₂OAc	CH₂OAc	50, 52
CH₃	H	CO₂Et	CO₂R	CO₂R	52
CH₃	H	CO₂Et	CH(OCH₃)₂	CH(OCH₃)₂/H	50
CH₃	H	H	H	H	50
CH₃	H	H	CH(OCH₃)₂	CH(OCH₃)₂	50
CH₃	H	H	CH₂OH	CH₂OH	50
CH₃	H	H	CH₂OAc	CH₂OAc	50
CH₃	H	H	CO₂Et	CO₂Et	50, 52
CH₃	H	Ph	Ph	Ph	52
CH₃	H	Ph	Ph	Ph	50
CH₃	H	CN	CH₂OAc	CH₂OAc/H	50, 52
CH₃	H	CN	CO₂Et	CO₂Et	50
CH₂Ph	CH₂Ph	H	CH₂OAc	CH₂OAc/H	50
CH₃/Ph	CH₂Ph	H	CH(OCH₃)₂	CH(OCH₃)₂/H	51
Benzynes					
Ph	H	H	Ph	H	54
Ph	H	H	C≡CTMS	H	54
Ph	H	H	TMS	CH₂OTMS	54
Ph	H	H	(CH₂)₂OAc	H	54
Ph	H	H	CH₂OH	H	54
Ph	H	H	CH₂OAc	H	54
Ph	H	H	CO₂H	H	55a
Ph	H	H	COCH₃	H	55a
Ph	H	H	CH₂OAc	CH₂OAc	56
Ph	H	H	CHO	(CH₂)₆CO₂CH₃	57
CH₃	H	CH₂Ph	CO₂CH₃	CO₂CH₃	53
CH₃	H	CH₂Ph	CO₂R	CO₂R/H	53
H	H	CH₂Ph	CO₂R	CO₂R/H	53
Ph	H	CH₂PH	CO₂R	CO₂R/H	53
CH₂Ph	H	Ph	CO₂R	H	53
OH	H	Ph	CO₂CH₃	CO₂CH₃	58
OH	H	Ph	COPh	CO₂CH₃	58
CH₃	H	H	CF₃	H	55b

309

and (±)-gnididione (**20**) and (±)-isognididione (**21**) [Eq. (8)].[59]

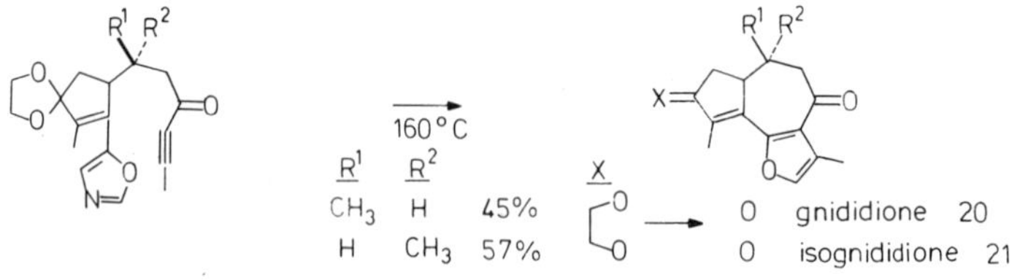

R^1	R^2		X			
CH$_3$	H	45%		0	gnididione	20
H	CH$_3$	57%		0	isognididione	21

$$(8)$$

2. THIAZOLES

Thiazoles have been shown to be much more reluctant than oxazoles to participate in Diels–Alder reactions, and only two successful intermolecular Diels–Alder reactions have been described. Like oxazoles, olefinic dienophiles provide pyridine products via a Kondrat'eva dehydration [Eq. (9)].[60]

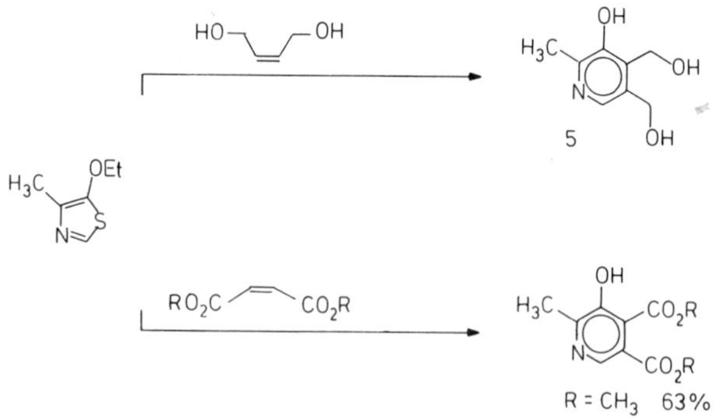

R = CH$_3$ 63%

$$(9)$$

A selected intramolecular Diels–Alder reaction of an acetylenic thiazole provided the expected thiophene product [Eq. (10)].[59g]

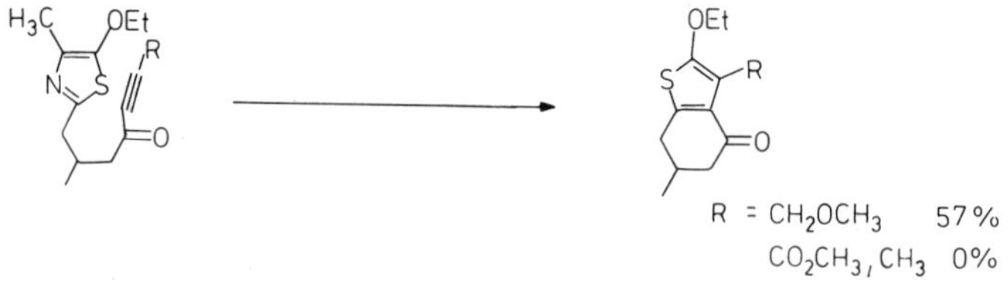

R = CH$_2$OCH$_3$ 57%
CO$_2$CH$_3$, CH$_3$ 0%

$$(10)$$

Unlike the case of oxazoles, the attempted intramolecular Diels–Alder reactions of thiazoles bearing electron-deficient alkynes have proved unsuccessful.[59g]

3. ISOXAZOLES

There are no reports of simple isoxazoles participating in [4 + 2] cycloadditions in which the isoxazole nucleus functions as a heteroaromatic 1-azadiene system. However, one of the early reports of the successful participation of a 1-azadiene system in [4 + 2] cycloadditions does include the Diels–Alder reactions of benzisoxazoles (Chapter 9, Section 1, Table 9-I, entries 1–7).[61] An additional report has detailed the participation of substituted 4,5-dihydro-3-vinylisoxazoles in [4 + 2] cycloadditions (Chapter 9, Section 1, Table 9-I, entry 8).[62a] Isoxazolium salts react with enamines to provide pyridinium salts via a [4+ + 2] reaction (Chapter 9, Section 10).[62b]

4. PYRROLES

There are a number of examples of substituted pyrroles participating as 4π components of all-carbon Diels–Alder reactions.[63] However, their reported participation as a 1- or 2-azadiene system in [4 + 2] cycloadditions are limited to selected pyrroles.[64,65] Pentachloro-2H-pyrrole has been shown to participate in inverse electron demand Diels–Alder reactions with electron-rich dienophiles exclusively in the form of the 2-azadiene system, pentachloro-3H-pyrrole,[64] and one example of an intramolecular Diels–Alder reaction of a substituted, *in situ* generated 3H-pyrrole has been described (Chapter 9, Section 2).[65]

5. PYRAZOLES

There are no examples of pyrazoles participating as 4π components of a Diels–Alder reaction with cycloaddition occurring across N-2/C-5 of the pyrazole nucleus. The reports of their participation in [4 + 2] cycloadditions have been shown to have been interpreted incorrectly.[66]

6. IMIDAZOLES

Two imidazoles bearing selectively disposed functionality have been shown to participate in [4 + 2] cycloadditions. Dimethyl imidazole-4,5-dicarboxylate has been reported to behave as a well-defined, electron-deficient 1,4-diazadiene in an inverse electron demand Diels–Alder reaction [Eq. (11)].[67]

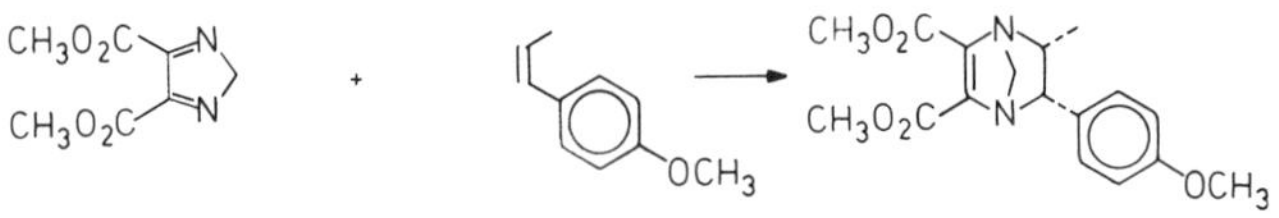

$$\tag{11}$$

and a series of fused imidazoles have been shown to participate in Diels–Alder reactions with dimethyl acetylenedicarboxylate [Eq. (12)].[68]

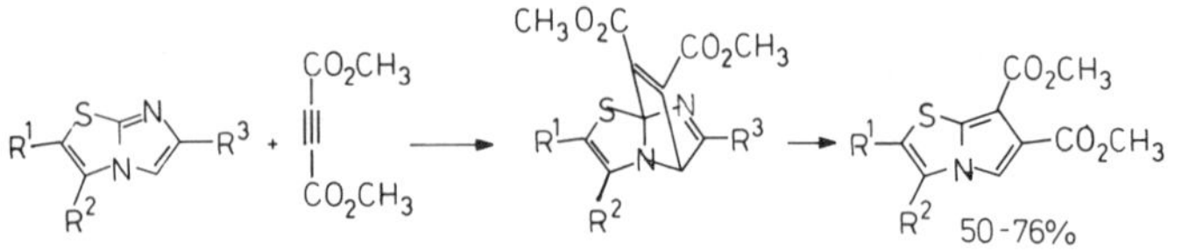

$$\tag{12}$$

7. PYRIDINES

The independent observations of Neunhoeffer and Lehmann[69] and Gompper and Heinemann[70] that dimethyl acetylenedicarboxylate is sufficiently reactive to participate in an apparent Diels–Alder reaction with dimethyl 2,6-bis(dimethylamino)pyridine-3,4-dicarboxylate represents the first evidence that pyridine systems appropriately substituted with strong electron-donating groups (pyridyl C-2/C-6) may function as a 2-azadiene system in [4 + 2] cycloadditions with reactive, electron-deficient dienophiles [Eq. (13)].

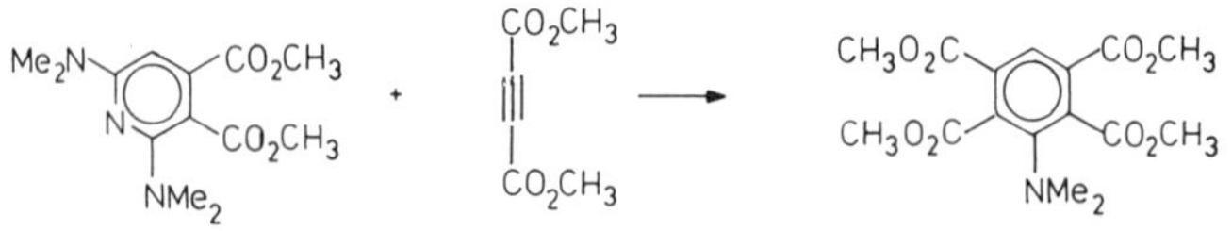

$$\tag{13}$$

The generality of this process as well as confirmation that the reaction proceeds by a [4 + 2] cycloaddition remain to be determined.

The studies of the Diels–Alder reactions of 2-pyridones (**22**) and 3-

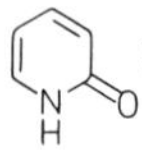

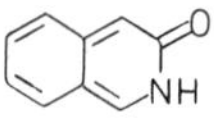

hydroxyisoquinolines (**23**) have been reviewed.[71] While constituting members of the heteroaromatic systems capable of participating in [4 + 2] cycloadditions, reactions, the systems participate as all-carbon dienes in the Diels–Alder reactions.

The participation of aromatic quaternary salts, including the acridizinium and isoquinolinium salts, in inverse electron demand [4⁺ + 2] cycloadditions with electron-rich, nucleophilic dienophiles as well as typical, electron-deficient dienophiles represent well-defined examples of activated heterocyclic azadiene systems containing a pyridyl nucleus which are capable of useful 4π participation in Diels–Alder reactions. Details of this work may be found in Chapter 9, Section 10.

8. PYRIDAZINES (1,2-DIAZINES)

Pyridazines, 1,2-diazines, substituted with additional electron-withdrawing substituents participate in well-defined inverse electron demand [4 + 2] cycloadditions with electron-rich dienophiles.[72,73] In nearly all reports, cycloaddition occurs across C-3/C-6 of the 1,2-diazine nucleus, and the regioselectivity of the reaction can be predicted based on the 1,2-diazine substitution pattern [Eq. (14)].[72]

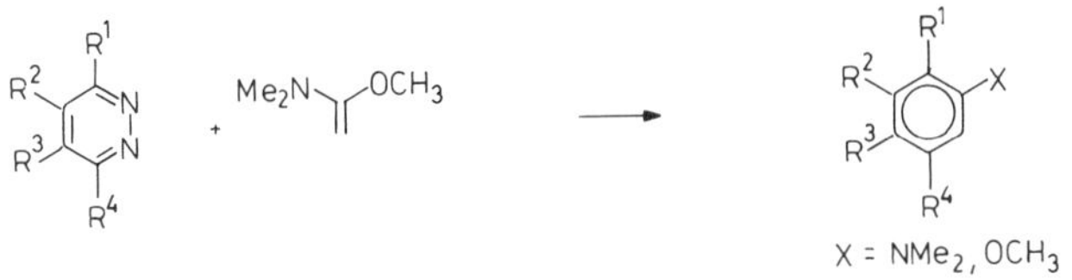

$$ (14) $$

$$ X = NMe_2, OCH_3 $$

Exceptions to this generalization are restricted to the reactions of ynamines with electron-deficient 1,2-diazines, where both C-3/C-6 and C-4/N-1 1,2-diazine cycloaddition have been observed [Eq. (15)].[73]

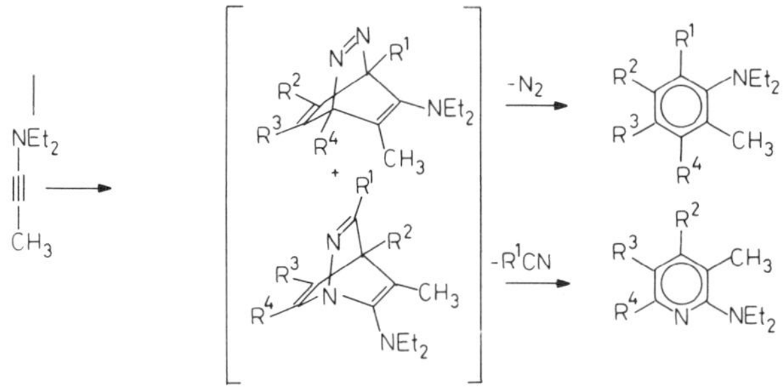

$$ (15) $$

The number and position of electron-withdrawing substituents on the 1,2-diazine nucleus control the mode and regioselectivity of the ynamine 1,2-diazine cycloaddition, and in each case studied, only one reaction product was detected.[73] Much of this work is summarized in Table 10-III.

Intramolecular Diels–Alder reactions of unactivated and highly substituted alkenyl[74] and alkynyl[75] 1,2-diazines have been explored [Eq. (16)].

$$ (16) $$

TABLE 10-III

Intermolecular Diels–Alder Reactions of 1,2-Diazines

1,2-Diazine				Dienophile	Product(s)		

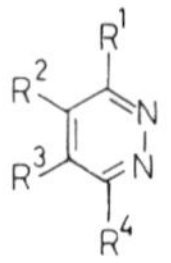

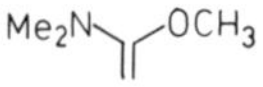

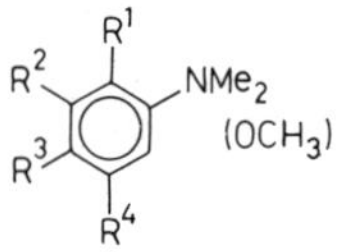

R^1	R^2	R^3	R^4		Yield	$NMe_2 : OCH_3$	Ref.
CO_2CH_3	H	H	H		60%	15:1	72
H	H	CO_2CH_3	H		57%	12:1	72
CO_2CH_3	H	H	CO_2CH_3		90%	8:1	72
H	CO_2CH_3	CO_2CH_3	H		80%	3:1	72
CO_2CH_3	H	CO_2CH_3	CO_2CH_3		93%	2:1	72
CO_2CH_3	CO_2CH_3	CO_2CH_3	CO_2CH_3		88%	10:1	72

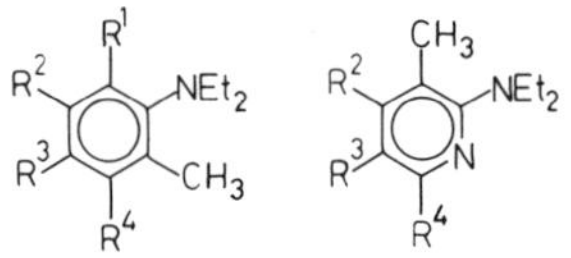

R^1	R^2	R^3	R^4		Yield		Ref.
CO_2CH_3	H	H	H		23%	—	73
H	H	CO_2CH_3	H		—	30%	73
CO_2CH_3	H	H	CO_2CH_3		—	75%	73
H	CO_2CH_3	CO_2CH_3	H		74%	—	73
CO_2CH_3	H	CO_2CH_3	CO_2CH_3		—	85%	73
CO_2CH_3	CO_2CH_3	CO_2CH_3	CO_2CH_3		73%	—	73

The reaction is sensitive to the diene/dienophile spacer as well as subtle features apparently important for substrate/product stability under the reaction conditions (200–230°C). Table 10-IV summarizes the intramolecular Diels–Alder reactions of 1,2-diazines. These observations have found application in the total synthesis of the cAMP phosphodiesterase inhibitors PDE-I and PDE-II (**24**), constituting the central and right-hand segments of the potent antitumor antibiotic CC-1065 [Eq. (17)].[76]

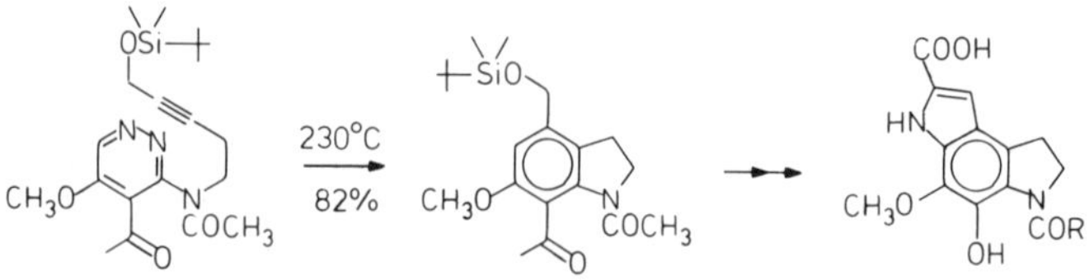

(17)

24   PDE-I   R = NH₂
     PDE-II   = CH₃

TABLE 10-IV

Intramolecular Diels–Alder Reactions of 1,2-Diazines

	1,2-Diazine			Product(s)	
R^1	R^2	X	n	Yield	Ref.
H	H	NCO_2CH_3	1	85%	75
H	CH_3	NCO_2CH_3	1	77%	75
H	$CH_2OTBDMS$	NCO_2CH_3	1	92%	75
Cl	H	NCO_2CH_3	1	91%	75
Cl	H	NCO_2CH_3	2, 3	0%	75
Cl	H	O	1–3	0%	75
Cl	CH_3	NCO_2CH_3	1	85%	75
Cl	$CH_2OTBDMS$	NCO_2CH_3	1	72%	75
CH_3	H	$S/SO/SO_2$	1	0%	99d

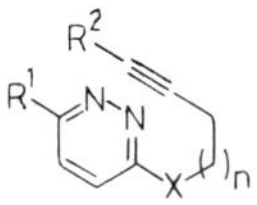

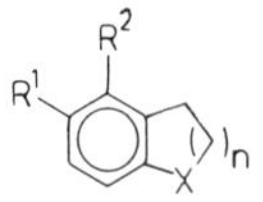

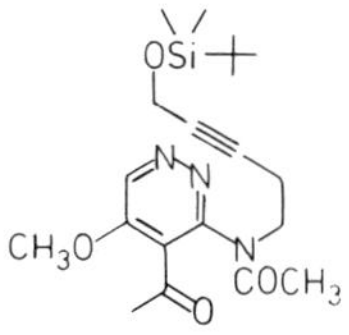	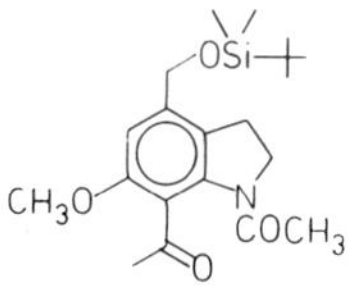 82%	76

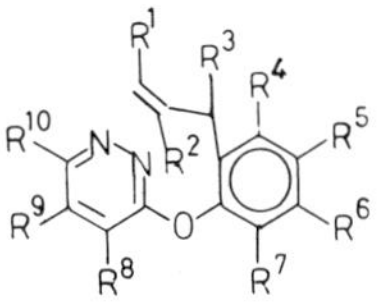

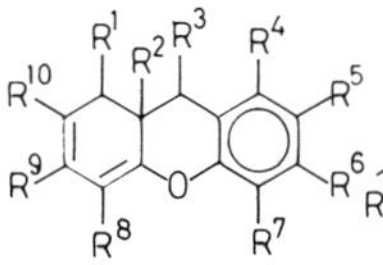

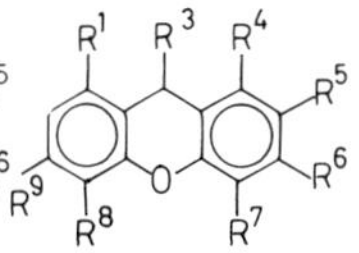

$R^1-R^9 = H$	R^{10}	Yields	Ref.
H	Cl	—	80%	74
$R^5 = Cl$	Cl	—	58%	74
$R^7 = Cl$	Cl	—	75%	74
$R^7 = OH$	Cl	—	30%	74
$R^5 = CH_3$	H	—	28%	74
H	CH_3	—	35%	74
$R^1 = CH_3$	Cl	—	20%	74
$R^2 = CH_3$	Cl	99%	—	74
$R^3 = CH_3$	Cl	—	98%	74
$R^5 = CH_3$	Cl	—	79%	74
$R^7 = CH_3$	Cl	—	64%	74
$R^8 = CH_3$	Cl	—	89%	74
$R^9 = CH_3$	Cl	—	65%	74
$R^3 = CH_3,\ R^5 = Cl$	Cl	—	60%	74

(continued)

TABLE 10-IV (*Continued*)

1,2-Diazine		Product(s)			
R^1–R^9 = H	R^{10}		Yields		Ref.
R^5 = OCH$_3$	Cl	—		87%	74
R^7 = OCH$_3$	Cl	—		78%	74
R^8 = OCH$_3$	Cl	—		—	74
R^9 = OCH$_3$	Cl	—		18%	74
R^1 = CH$_3$, R^7 = OH	Cl	—		—	74
R^3 = CH$_3$, R^7 = OH	Cl	—		14%	74
R^2 = CH$_3$	CH$_3$	60%		—	74
R^2 = R^7 = CH$_3$	Cl	94%		—	74
R^2 = R^5 = CH$_3$	Cl	79%		—	74
R^3 = R^5 = CH$_3$	Cl	—		66%	74
R^3 = R^7 = CH$_3$	Cl	—		61%	74
R^4 = R^6 = CH$_3$	Cl	—		76%	74
R^5 = R^7 = CH$_3$	Cl	—		72%	74
R^2 = CH$_3$, R^7 = OCH$_3$	Cl	44%		—	74
R^3 = CH$_3$, R^7 = OCH$_3$	Cl	—		100%	74
R^7 = CO$_2$CH$_3$	Cl	13%		—	74
R^7 = CH(CH$_3$)$_2$	Cl	—		45%	74
R^7 = CH=CHCH$_2$	Cl	—		59%	74
R^5 = CO$_2$Et	Cl	13%		—	74
R^5 = tBu	Cl	—		70%	74
R^8 + R^9 = (CH=CH)$_2$	Cl	—		64%	74
R^8 + R^9 = (CH$_2$)$_4$	Cl	—		90%	74
R^2 = CH$_3$, R^5 = CO$_2$Et	Cl	11%		—	74
R^2 = CH$_3$, R^4 + R^5 = (CH=CH)$_2$	Cl	72%		—	74
R^5 = CH$_3$, R^8 + R^9 = (CH=CH)$_2$	Cl	—		88%	74
H	Ph	83%		—	74
R^5 = Ph	Cl	—		45%	74
R^7 = Ph	Cl	—		91%	74
R^9 = Ph	Cl	—		98%	74
R^5 = Cl	Ph	68%		—	74
R^3 = Ph, R^7 = OH	Cl	—		26%	74
R^5 = NO$_2$	Ph	46%		—	74
R^2 = CH$_3$	Ph	88%		—	74
R^3 = CH$_3$	Ph	73%		—	74
R^5 = CH$_3$	Ph	100%		—	74
R^2 = CH$_3$, R^7 = Ph	Cl	79%		—	74
R^3 = Ph, R^5 = CH$_3$	Cl	—		65%	74
R^3 = Ph, R^7 = OCH$_3$	Cl	—		87%	74
R^5 = SCH$_3$	Ph	73%		—	74
R^4 = R^6 = CH$_3$	Ph	78%		—	74
R^5 = C(CH$_3$)$_2$Ph	Cl	—		69%	74
R^5 = CO$_2$Et	Ph	100%		—	74
R^8 + R^9 = (CH=CH)$_2$	Ph	97%		—	74
R^2 = CH$_3$, R^8 + R^9 = (CH=CH)$_2$	Ph	60%		—	74

TABLE 10-V
Intermolecular Diels–Alder Reactions of Pyrimidines (1,3-Diazines)

1,3-Diazine				Dienophile	Product			

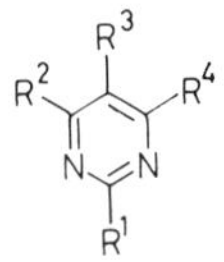

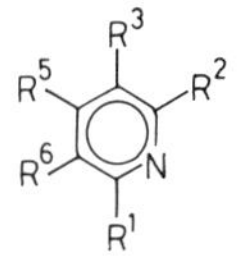

R^1	R^2	R^3	R^4		R^5	R^6	Yield	Ref.
H	CO_2Et	H	H		NEt_2	CH_3	10%	77a
H	H	CO_2Et	H		NEt_2	CH_3	90%	77a
H	CO_2Et	CO_2Et	H		NEt_2	CH_3	80%	77a
CO_2Et	CO_2Et	H	H		CH_3	NEt_2	81%	77a
H	CO_2Et	H	CO_2Et		NEt_2/CH_3, 2.5:1		89%	77a
CN	CN	H	H		CH_3	NEt_2	97%	77b
CN	CN	CH_3	H		CH_3	NEt_2	36%	77b

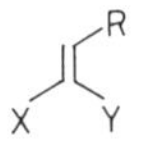

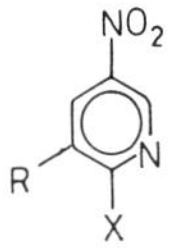

		Yield	Ref.
X = morpholine, R = H		49–57%	78
X = piperidine, R = H		52%	78
X = OCH_3, R = CH_3		17%	78

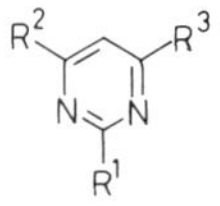

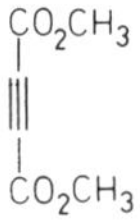

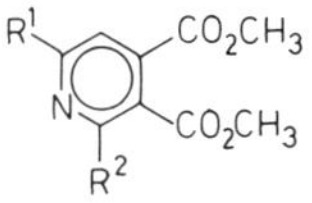

R^1	R^2	R^3		Yield	Ref.
NMe_2	NMe_2	NMe_2		53%	79
NMe_2	NMe_2	H		10%	79
NMe_2	NMe_2	OCH_3		70%	79
OCH_3	NMe_2	NMe_2		30%	79
OCH_3	NMe_2	OCH_3		55%	79
OCH_3	OCH_3	OCH_3		12%	79
NMe_2	OCH_3	OCH_3		40%	79
NMe_2	OCH_3	H		16%	79
CH_3	OH	OH		62%	80

9. PYRIMIDINES (1,3-DIAZINES)

The addition of strong electron-withdrawing substituents to the pyrimidine nucleus increases the facility and rate with which the system participates in inverse electron demand [4 + 2] cycloadditions with electron-rich dienophiles. The mode of cycloaddition (C-2/C-5[77] versus C-4/N-1[78]) and the observed regioselectivity of the C-2/C-5 cycloaddition are dependent on the dienophile employed as well as the position, type, and number of electron-withdrawing substituents present on the pyrimidine nucleus [Eq. (18) and Table 10-V].

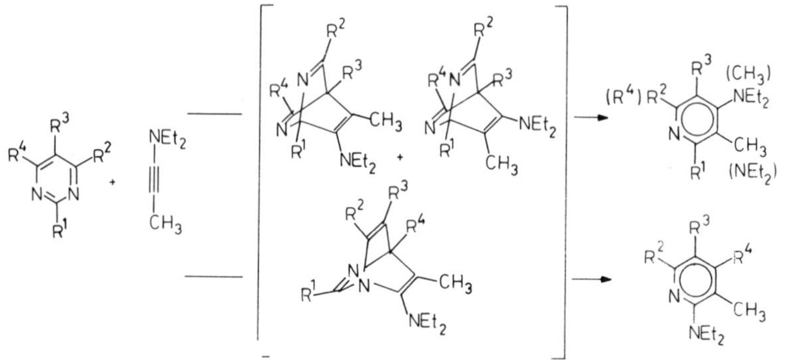

(18)

Complementary substitution of the pyrimidine nucleus with two (or three) strong electron-donating groups at C-2 and C-4 (and C-6) is sufficient to permit the 4π participation of the pyrimidine in apparent normal (HOMO-diene controlled) Diels–Alder reactions with dimethyl acetylenedicarboxylate [Eq. (19)[79,80] and Table 10-V].

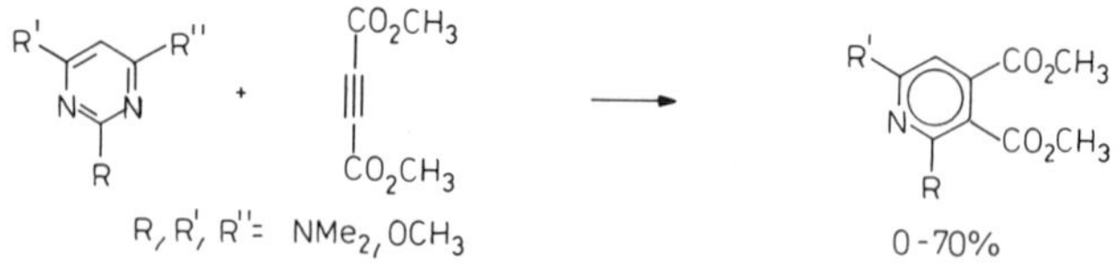

(19)

The intramolecular Diels–Alder reactions of simple, 4-hydroxy- or 4-alkyl-6-oxopyrimidines[80,81] bearing olefinic, acetylenic, or nitrile dienophiles have been investigated (Table 10-VI). In many instances, the initial intramolecular pyrimidine Diels–Alder products are thermally stable due to the formation of a bicyclic amide. The elimination of cyanic acid (HNCO) apparently requires thermal tautomerization of the amide to the hydroxyimine and subsequent retro-Diels–Alder reaction. The application of these observations in the total synthesis of ($\pm$)-acetinidine has been detailed [Eqs. (20) and (21).[82]

(20)

TABLE 10-VI

Intramolecular Diels–Alder Reactions of Pyrimidines (1,3-Diazines)

1,3-Diazine			Product(s)		
R¹	R²	n	Yields		Ref.
OH	CH₃	1	98%	0%	80
OH	Ph	1	59%	0%	80
CH₃	CH₃	1	0%	65%	80
CH₃	CH₃	2	0%	56%	80
Ph	CH₃	1	0%	51%	80
OH	CH₃	2	44%	0%	80

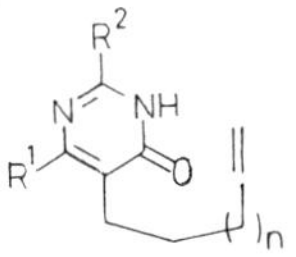

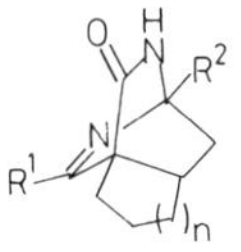

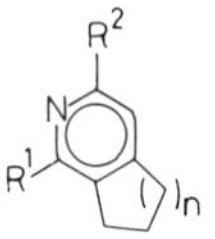

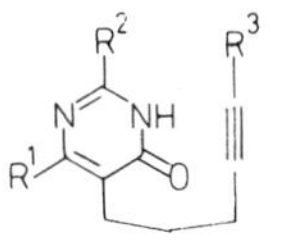

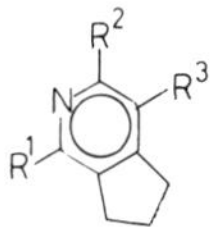

R¹	R²	R³	Yield	Ref.
OH	CH₃	H	46%	80
CH₃	CH₃	H	60%	80
OH	Ph	H	40%	80

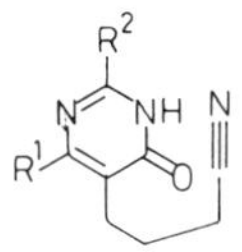

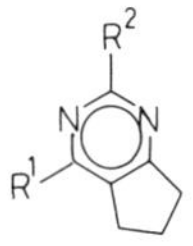

R¹ = OH, R² = CH₃   Yield not reported   80

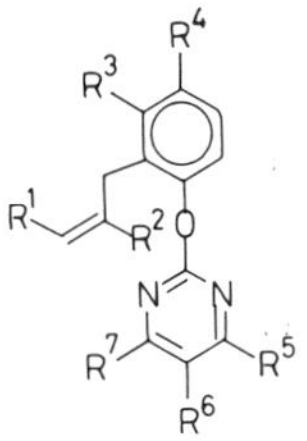

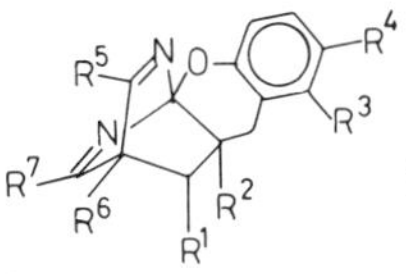

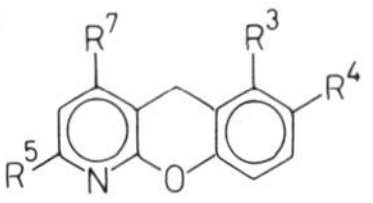

R¹–R⁷ = H	Yields		Ref.
R⁶ = Cl	24%	—	81
R⁶ = Br	—	27%	81
R⁵ = OCH₃, R⁶ = Br, R⁷ = CH₃	25%	—	81
R⁴ = NO₂, R⁵ = CH₃, R⁶ = Br, R⁷ = CH₃	24%	—	81
R³ + R⁴ = (CH=CH)₂, R⁵ = CH₃, R⁶ = Br, R⁷ = CH₃	43%	—	81

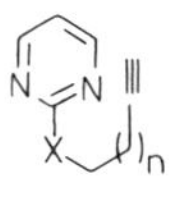

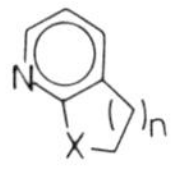

X = S, SO, SO₂; n = 1   71%, 0%, 53% yield   99d

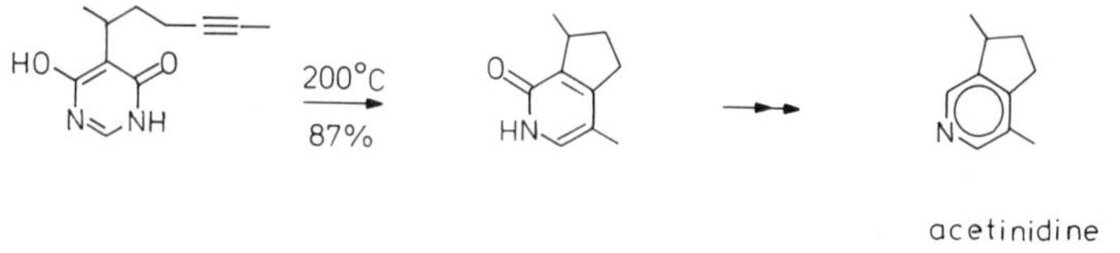

$$\text{(21)}$$

acetinidine

10. PYRAZINES (1,4-DIAZINES)

The addition of electron-withdrawing substituents to the 1,4-diazine nucleus increases its participation in inverse electron demand [4 + 2] cycloadditions with electron-rich dienophiles.[83] Both the rate and regioselectivity of the reaction are dependent on the number and position of the electron-withdrawing substituents present on the 1,4-diazine nucleus [Eq. (22) and Table 10-VII].

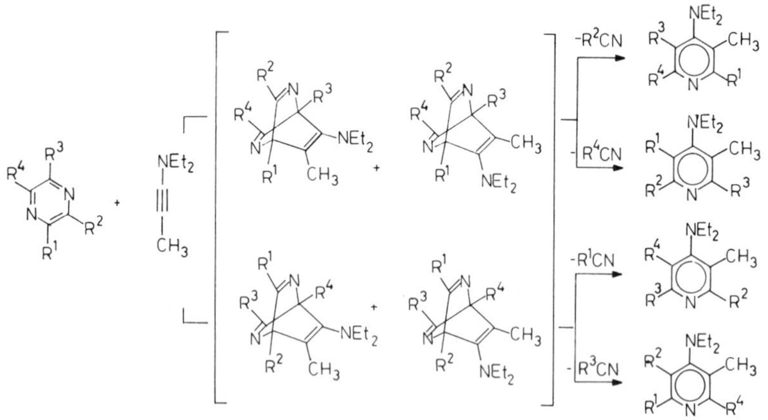

$$\text{(22)}$$

The addition of strong electron-donating substituents to the 1,4-diazine nucleus permits the observation of [4 + 2] cycloadditions with electron-deficient or strained, reactive dienophiles[84] [Eq. (23) and Table 10-VII].

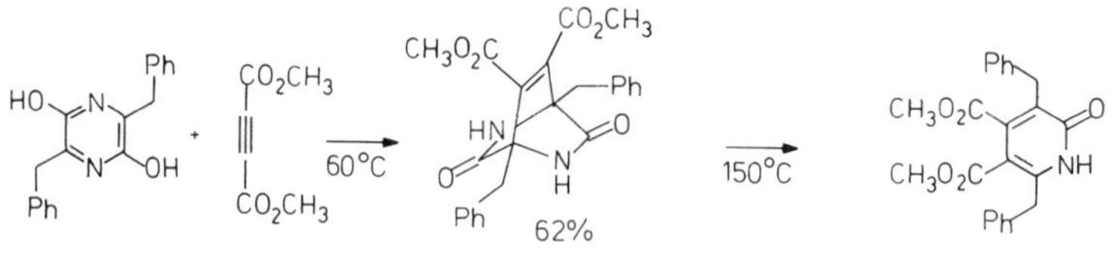

$$\text{(23)}$$

11. 1,2,3-TRIAZINES

The first successful preparation of 1,2,3-triazine and substituted derivatives has been realized, and preliminary studies have confirmed their

TABLE 10-VII

Diels–Alder Reactions of 1,4-Diazines

1,4-Diazine	Dienophile	Product(s)	Yield	Ref.
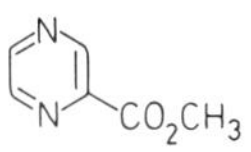		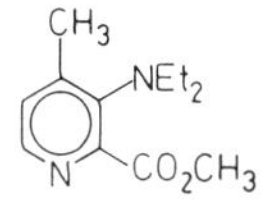	72%	83
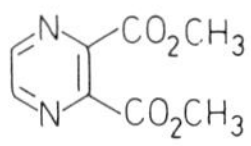		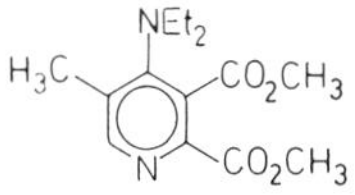	75%	83
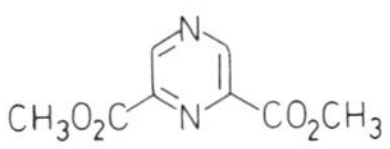		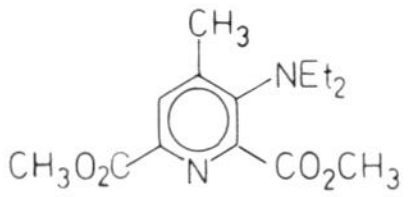	—	83
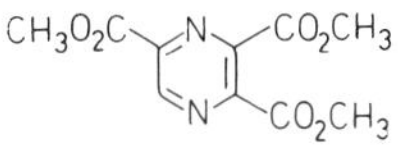			78%	83
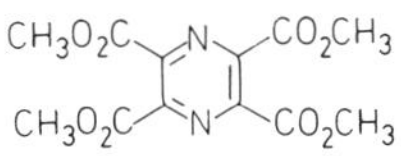		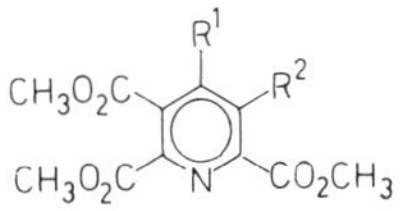	—	83
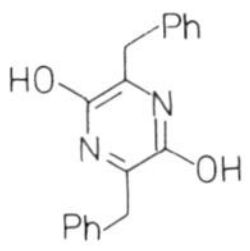		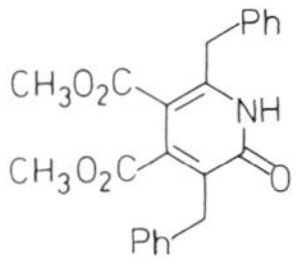	—	84a
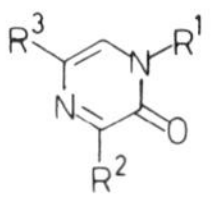	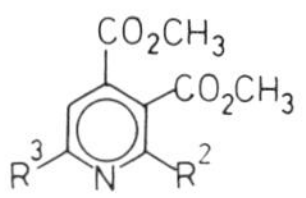	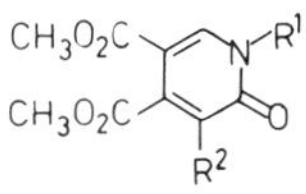		84b

$R^1 = Ph, R^2 = CN, R^3 = Cl$　　　　　　30 : 1　　70%
$R^1 = Ph, R^2 = Cl, R^3 = Cl$　　　　　　1 : 20　　63%
$R^1 = Ph, R^2 = OCH_3, R^3 = Cl$　　　15 : 1　　83%
$R^1 = CH_3, R^2 = CN, R^3 = Cl$　　　5.5 : 1　　92%
$R^1 = Ph, R^2 = CN, R^3 = H$　　　　　4 : 5　　59%

potential for participation in inverse electron demand [4 + 2] cycloadditions [Eq. (24)].[85]

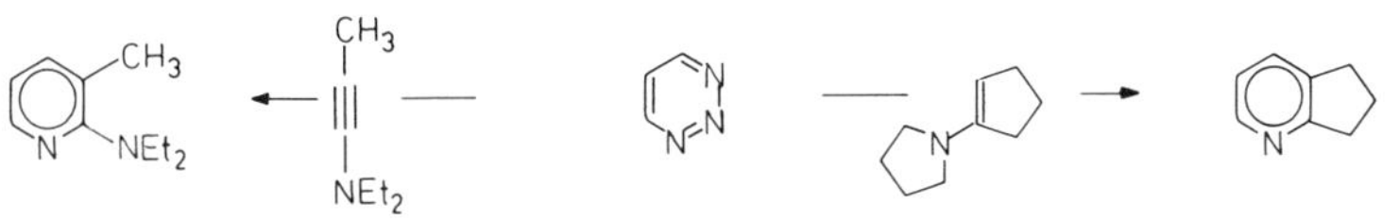

$$\qquad\qquad\qquad\qquad\qquad\qquad\qquad\qquad\qquad\qquad\qquad (24)$$

A total synthesis of fabianine, a simple terpenoid alkaloid, based on the implementation of an enamine 4-methyl-1,2,3-triazine Diels–Alder reaction has been detailed.[85c]

12. 1,3,5-TRIAZINES

The 1,3,5-triazine nucleus is sufficiently electron-deficient and susceptible to nucleophilic attack that it is well-suited for participation in [4 + 2] cycloaddition processes with electron-rich dienophiles,[86,87] and addition of electron-withdrawing substituents to the 1,3,5-triazine nucleus accelerates the rate of 1,3,5-triazine participation in inverse electron demand cycloaddition reactions [Eq. (25) and Table 10-VIII].[86]

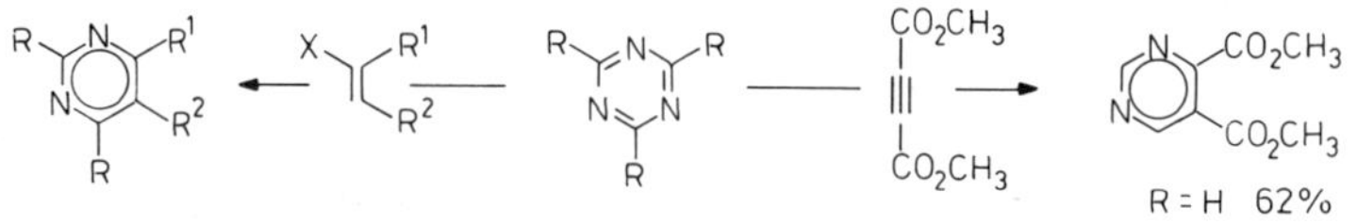

$$R = H \quad 62\% \tag{25}$$

TABLE 10-VIII
[4 + 2] Cycloadditions of 1,3,5-Triazines

1,3,5-Triazine	Dienophile	Product(s)	Yield	Ref.
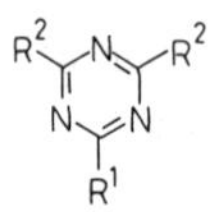		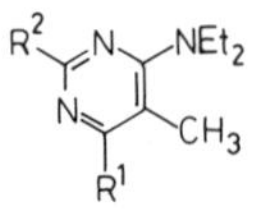		
R^1 = H, R^2 = H			97%	86
R^1 = H, R^2 = CH$_3$			80%	86
R^1 = CO$_2$CH$_3$, R^2 = CO$_2$CH$_3$			88%	86
	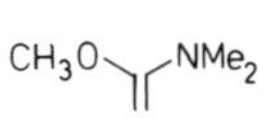	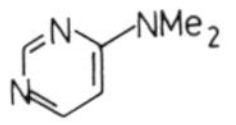		
R^1 = H, R^2 = H			90%	86
	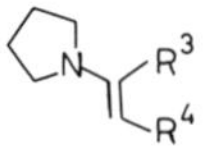	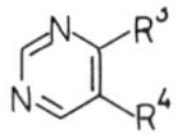		
R^1 = H, R^2 = H		R^3 = Et, R^4 = CH$_3$	80%	87
R^1 = H, R^2 = H		R^3 + R^4 = (CH$_2$)$_3$	75–93%	87
R^1 = H, R^2 = H		R^3 + R^4 = (CH$_2$)$_4$	47%	87
R^1 = H, R^2 = H		R^3 + R^4 = (CH$_2$)$_5$	76%	87
R^1 = H, R^2 = H		R^3 = cC_6H_{11}, R^4 = H	66%	87
R^1 = H, R^2 = H		R^3 = Ph, R^4 = H	80%	87

There are sufficient examples of the interception of dipolar intermediates in the observed or attempted inverse electron demand [4 + 2] cycloadditions of selected 1,3,5-triazines to infer that the reaction proceeds with the generation of discrete dipolar intermediates.[86,88] The intrinsic reactivity of the 1,3,5-triazine nucleus does not preclude its participation in Diels–Alder reactions with typical electron-deficient dienophiles[86] [Eq. (25)], and the full scope of this process remains to be investigated.

13. 1,2,4-TRIAZINES

In addition to the [4 + 2] cycloadditions of oxazoles (Section 1) and substituted 1,2,4,5-tetrazines (Section 14), the Diels–Alder cycloadditions of substituted 1,2,4-triazines constitute one of the most thoroughly investigated heteroaromatic azadiene systems capable of 4π diene participation.[3,89] In contrast to the oxazole or *sym*-tetrazine series, two potential and observed modes of cycloaddition are open to 1,2,4-triazines: cycloaddition across C-3/C-6 or C-5/N-2 of the 1,2,4-triazine nucleus, and the former is subject to 1,2,4-triazine substituent control of the observed regioselectivity.[90] The complementary addition of electron-withdrawing substituents to the 1,2,4-triazine nucleus generally increases its rate of participation in inverse electron demand Diels–Alder reactions, influences the mode of [4 + 2] cycloaddition (C-3/C-6 versus C-5/N-2 cycloaddition), and controls the observed regioselectivity. In addition, the reactivity of the electron-rich dienophile as well as the reaction conditions, polar versus nonpolar solvent, have a pronounced effect on the observed course of the [4 + 2] cycloadditions.[89]

All electron-rich dienophiles including *O,O*-ketene acetals, *O,S*-ketene acetals, *S,S*-ketene thioacetals, *O,N*-ketene acetals, *N,S*-ketene acetals, *N,N*-ketene aminals, enol ethers, enamines, and reactive or strained olefins cycloadd exclusively across C-3/C-6 of the 1,2,4-triazine nucleus [Eq. (26)].[90–92, 94]

$$(26)$$

The exception to this generalization is the cycloadditions of ynamines with 1,2,4-traizines where the C-5/N-2 cycloaddition process is generally observed [Eq. (27)].[93]

Since C-5 of the simple 1,2,4-triazine nucleus is the site of attack by conventional nucleophiles, it is likely that the observed C-5/N-2 ynamine 1,2,4-triazine [4 + 2] cycloadditions proceed in two steps with the generation of discrete, dipolar intermediates.

The regioselectivity of the C-3/C-6 cycloaddition process is subject to control by the electronic and steric properties of the 1,2,4-triazine substituents, the electronic and steric properties of the electron-rich dienophile, as well as the reaction conditions. There is a strong preference for the nucleophilic carbon of the electron-rich dienophile to attach to C-3 of the 1,2,4-triazine nucleus. The complementary positioning of additional electron-withdrawing substituents on the 1,2,4-triazine nucleus (e.g., either C-6 or C-3/C-5/and C-6) increases the rate of 1,2,4-triazine participation in the [4 + 2] cycloaddition and can enhance (e.g., C-6) the observed regioselectivity. In addition, the correct positioning of strongly electron-withdrawing substituents on the 1,2,4-triazine nucleus (e.g., C-3 or C-3/C-5) is sufficient to reverse this normal regioselectivity and illustrates the [4 + 2] regiocontrol available through proper selection and positioning of the 1,2,4-triazine substituents.[91,92] Moreover, the electron-rich dienophile can control or alter the expected course of the [4 + 2] cycloaddition reaction.

The more reactive, electron-rich olefins participate in the [4 + 2] cycloadditions under mild conditions generally with increased regioselectivity although a number of unanticipated observations have been detailed including the loss or reversal of the expected regioselectivity which accompanied the addition of alkyl substitution to the dienophile,[95,96] the use of morpholino enamines,[95,96] and the use of ketene aminals.[91b] Many of these observations have been attributed to steric effects of the dienophile which may preclude an endo transition state required for the expected [4

+ 2] cycloaddition. Figure 10-I summarizes many of these generalizations. An extensive study by Sauer has illustrated the expected and unanticipated, subtle features of dienophile reactivity which effect the 1,2,4-triazine C-3/C-6 cycloaddition regioselectivity [Eq. (28) and Table 10-IX].[91b]

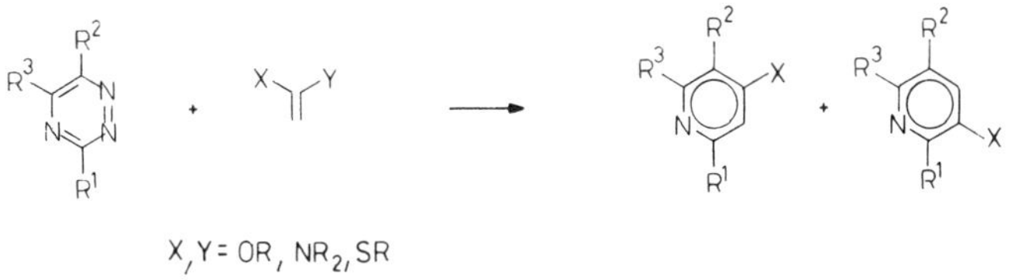

$$X, Y = OR, NR_2, SR \qquad (28)$$

These observations have found application in the formal total synthesis of streptonigrin (25) [Eq. (29)],[95]

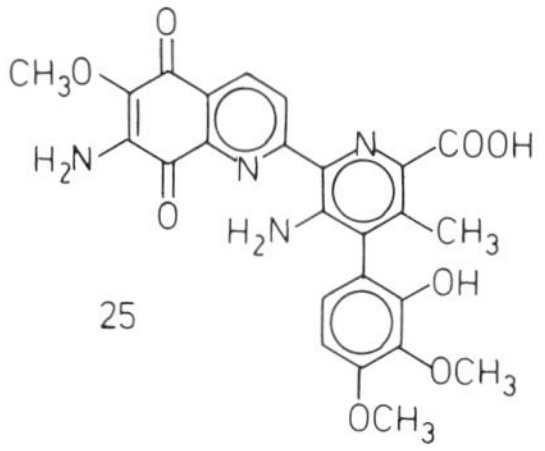

	Conditions				
X = morpholine	CHCl$_3$, 45 - 80°C				trace
	CHCl$_3$, 120°C	1	:	1	68%
	CH$_3$CN, 80°C	4	:	1	20%
	CH$_2$Cl$_2$, 25°C 6.2 kbar	1.4:		1	58%
= pyrrolidine	CHCl$_3$, 45 - 120°C				0%
	CH$_3$CN, 45 - 80°C				0%
	CH$_2$Cl$_2$, 25°C 6.2 kbar	3	:	1	65%
= OSiMe$_3$	-				0%

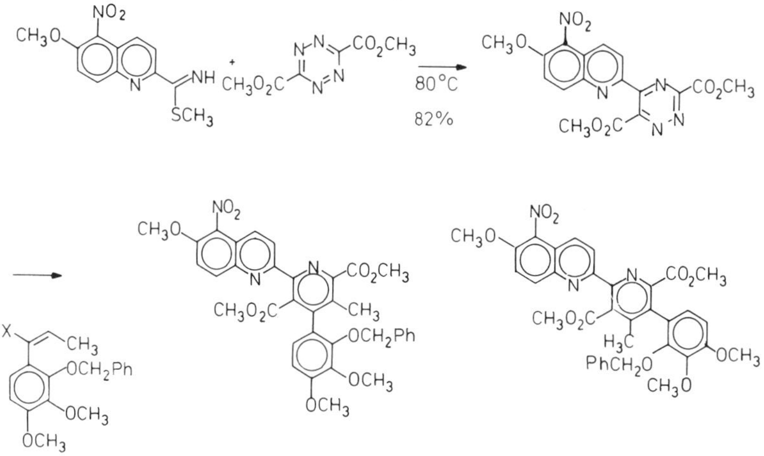

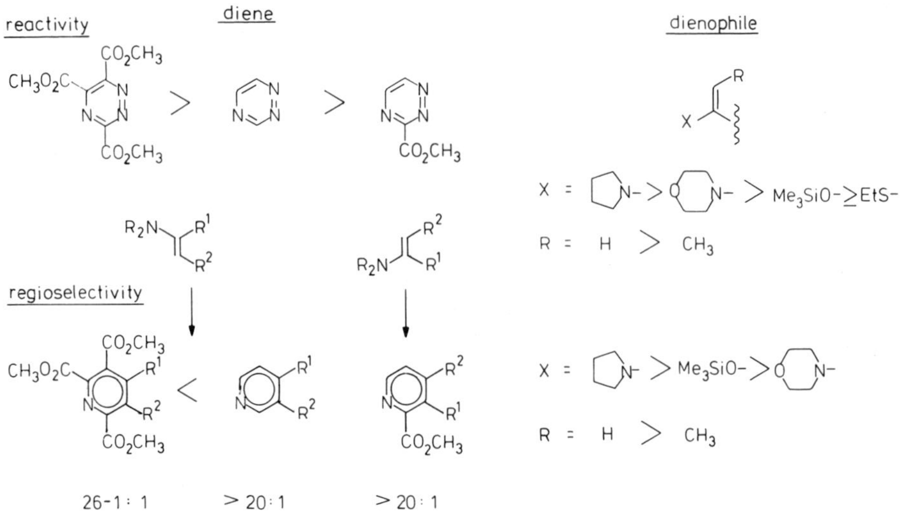

Fig. 10-I

in the total synthesis of lavendamycin methyl ester (**26**) [Eq. (30)],[96]

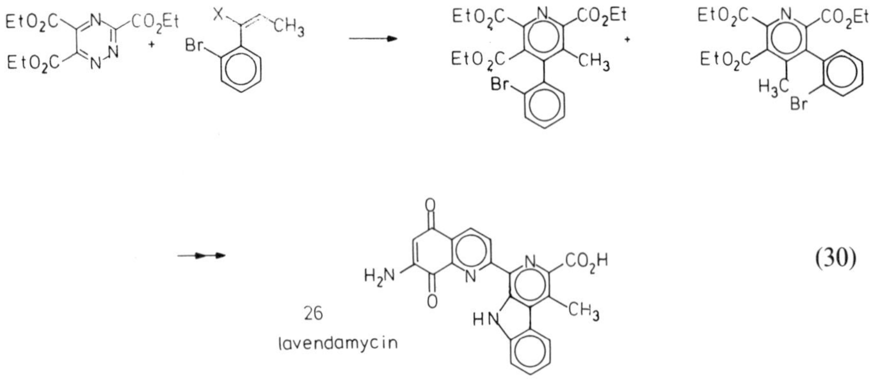

(30)

26
lavendamycin

X = pyrrolidine	CHCl$_3$, 60°C	3 : 1	51%
	CH$_2$Cl$_2$, 40°C	6.5: 1	54%
	CH$_2$Cl$_2$, 25°C	7.5: 1	50%
X = morpholine	CHCl$_3$, 60°C	1 : 1	57%
	CHCl$_3$, 45°C	–	trace

and in the preparation of pyridine-substituted pyrimidine nucleosides.[153]

The additional substitution of the 1,2,4-triazine nucleus with strong electron-donating substituents (OCH$_3$, NMe$_2$) increases its rate of participation in normal (HOMO$_{diene}$ controlled) Diels–Alker reactions [Eq. (31)].[97,98]

(*Text continues on p. 334*)

TABLE 10-IX
Representative Intermolecular 1,2,4-Triazine Diels–Alder Reactions

1,2,4-Triazine			Dienophile			Product(s)		
R^1	R^2	R^3	X	Y	Conditions	ratio	Yield	Ref.
CO_2CH_3	CO_2CH_3	CO_2CH_3	OEt	OEt	C_6H_6, 20°C	100 : 0	19–48%	91b
CO_2CH_3	CO_2CH_3	CO_2CH_3	NMe_2	OEt	Dioxane, 20°C	100 : 0	21% X = NMe_2; 70% X = OEt	91b
CO_2CH_3	CO_2CH_3	CO_2CH_3	NMe_2	SCH_3	Dioxane, 20°C	98 : 2	91%	91b
CO_2CH_3	CO_2CH_3	CO_2CH_3	NMe_2	NMe_2	Dioxane, 20°C	82 : 18	81%	91b
CO_2CH_3	CO_2CH_3	Ph	OEt	OEt	Dioxane, 100°C	95 : 5	100%	91b
CO_2CH_3	Ph	Ph	NMe_2	OEt	Dioxane, 100°C	97 : 3	99%	91b
CO_2CH_3	Ph	Ph	NMe_2	SCH_3	Dioxane, 100°C	92 : 8	84%	91b
CO_2CH_3	Ph	Ph	NMe_2	NMe_2	Dioxane, 100°C	22 : 78	90%	91b
CO_2CH_3	Ph	H	OEt	OEt	C_6H_6, 80°C	0 : 100	85%	91b
CO_2CH_3	Ph	H	NMe_2	OEt	C_6H_6, 40°C	8 : 92	63% X = NMe_2; 22% X = OEt	91b
CO_2CH_3	Ph	H	NMe_2	SCH_3	C_6H_6, 40°C	4 : 96	90%	91b
CO_2CH_3	Ph	H	NMe_2	NMe_2	C_6H_6, 40°C	8 : 92	91%	91b
CO_2CH_3	Ph	H	NMe_2	NMe_2	CH_3CN, 40°C	64 : 36	85%	91b
CO_2CH_3	Ph	H	NMe_2	NMe_2	CH_3OH, 25°C	0 : 100	—	91b
CO_2CH_3	Ph	H	NMe_2	NMe_2	C_6H_{12}, 25°C	1 : 99	—	91b
CO_2CH_3	Ph	H	NMe_2	NMe_2	Toluene, 25°C	6 : 94	—	91b
CO_2CH_3	Ph	H	NMe_2	NMe_2	THF, 25°C	8 : 92	—	91b
CO_2CH_3	Ph	H	NMe_2	NMe_2	CH_2Cl_2, 25°C	61 : 39	—	91b
CO_2CH_3	Ph	H	NMe_2	NMe_2	DMF, 25°C	63 : 37	—	91b
CO_2CH_3	Ph	H	NMe_2	NMe_2	CH_3CH, 25°C	63 : 37	—	91b
CO_2CH_3	Ph	Ph	NMe_2	NMe_2	C_6H_6, 80°C	45 : 55	78%	91b
CO_2CH_3	H	H	OEt	OEt	CH_3CN, 40°C	0 : 100	57%	91b
CO_2CH_3	H	H	NMe_2	OEt	CH_3CN, 40°C	0 : 100	74% X = NMe_2; 22% X = OEt	91b

(*continued*)

TABLE 10-IX (Continued)

1,2,4-Triazine / Dienophile / Product(s)

R¹	R²	R³	X	Y	Conditions	Product ratio	Yield	Ref.
CO_2CH_3	H	H	NMe_2	SCH_3	CH_3CN, 40°C	0 : 100	82%	91b
CO_2CH_3	H	H	NMe_2	NMe_2	CH_3CN, 40°C	81 : 19	92%	91b
CO_2CH_3	H	H	NMe_2	NMe_2	C_5H_5N, 40°C	65 : 35	54%	91b
CO_2CH_3	H	H	NMe_2	NMe_2	Acetone, 40°C	57 : 43	62%	91b
CO_2CH_3	H	H	NMe_2	NMe_2	C_6H_6, 40°C	22 : 78	85%	91b
H	H	H	NMe_2	OCH_3	Dioxane, 100°C	100 : 0	27%	91a
CH_3	H	H	NMe_2	OCH_3	Dioxane, 100°C	100 : 0	60%	91a
Ph	H	H	NMe_2	OCH_3	Dioxane, 100°C	0 : 100	81%	91a
H	Ph	H	NMe_2	OCH_3	Dioxane, 100°C	7 : 1	66%	91a
CH_3	H	Ph	NMe_2	OCH_3	Dioxane, 100°C	100 : 0	83%	91a
CO_2CH_3	CO_2CH_3	CO_2CH_3	NMe_2	OCH_3	Dioxane, 20°C	100 : 0	77% X = NMe_2, 23% X = OCH_3	91a

1,2,4-Triazine / Dienophile / Product(s)

R¹	R²	R³	X	R⁴	R⁵	Conditions	Product ratio	Yield	Ref.
H	H	H	Pyrrolidine	Et	CH_3	$CHCl_3$, 60°C	100 : 0	68%, 93%[a]	92
H	H	H	Pyrrolidine	$(CH_2)_3$		$CHCl_3$, 60°C	100 : 0	74%, 52%[a]	92
H	H	H	Pyrrolidine	$(CH_2)_4$		$CHCl_3$, 60°C	100 : 0	40%, 66%[a]	92
H	H	H	Pyrrolidine	$(CH_2)_5$		$CHCl_3$, 60°C	100 : 0	78%, 86%[a]	92
H	H	H	Pyrrolidine	H	cC_6H_{11}	$CHCl_3$, 60°C	100 : 0	64%, 36%[a]	92
H	H	H	Pyrrolidine	nPr	Et	$CHCl_3$, 60°C	100 : 0	71%	92
H	H	H	Pyrrolidine	Ph	H	$CHCl_3$, 60°C	100 : 0	—	92
CO_2CH_3	CO_2CH_3	CO_2CH_3	Pyrrolidine	$(CH_2)_3$		$CHCl_3$, 60°C	100 : 0	91%	90
CO_2CH_3	CO_2CH_3	CO_2CH_3	Pyrrolidine	Ph	H	$CHCl_3$, 60°C	26 : 1	79%	95

328

							Conditions	Ratio	Yield	Ref.
CO_2CH_3	CO_2CH_3	CO_2CH_3	Pyrrolidine	Ph		CH_3	$CHCl_3$, 45°C	>9 : 1	73%	95
CO_2CH_3	CO_2CH_3	CO_2CH_3	Pyrrolidine	$2\text{-}(OCH_2Ph)\text{-}3,4\text{-}(OCH_3)_3C_6H_2$		CH_3	$CHCl_3$, 45°C	9 : 1	59%	95
CO_2CH_3	CO_2CH_3	CO_2CH_3	$OSiMe_3$	Ph		H	$CHCl_3$, 60°C	7.2 : 1	72%	95
CO_2CH_3	CO_2CH_3	CO_2CH_3	$OSiMe_3$	Ph		CH_3	$CHCl_3$, 60°C	—	0%	95
CO_2CH_3	CO_2CH_3	CO_2CH_3	SEt	Ph		CH_3	$CHCl_3$, 60°C	—	0%	95
CO_2CH_3	CO_2CH_3	CO_2CH_3	OAc	H		H	$CHCl_3$, 60°C	—	78%	90
CO_2CH_3	CO_2CH_3	CO_2CH_3	OEt	H		H	$CHCl_3$, 60°C	—	66%	90
CO_2CH_3	CO_2CH_3	CO_2CH_3	H		$(CH_2)_3$		$CHCl_3$, 60°C	—	76%	90
CO_2CH_3	CO_2CH_3	CO_2CH_3	Pyrrolidine	$2\text{-}BrC_6H_4$		CH_3	$CHCl_3$, 60°C	3.1 : 1	51%	96
CO_2CH_3	CO_2CH_3	CO_2CH_3	Pyrrolidine	$2\text{-}BrC_6H_4$		CH_3	$CHCl_3$, 40°C	6.5 : 1	51%	96
CO_2CH_3	CO_2CH_3	CO_2CH_3	Pyrrolidine	$2\text{-}BrC_6H_4$		CH_3	CH_2Cl_2, 25°C	7.5 : 1	50%	96
CO_2CH_3	CO_2CH_3	CO_2CH_3	Morpholine	$2\text{-}BrC_6H_4$		CH_3	$CHCl_3$, 60°C	1 : 1	58%	96
CO_2CH_3	CO_2CH_3	CO_2CH_3	Morpholine	$2\text{-}BrC_6H_4$		H	$CHCl_3$, 45°C	—	—	96
CO_2CH_3	CO_2CH_3	CO_2CH_3	Pyrrolidinone	H		CH_3	$CHCl_3$, 60°C	100 : 0	92%	155
CO_2CH_3	CO_2CH_3	CO_2CH_3	Pyrrolidine	$2\text{-}FC_6H_4$		CH_3	$CHCl_3$, 50°C	>20 : 1	52%	95
CO_2CH_3	Ph	$5\text{-}NO_2\text{-}6\text{-}OCH_3\text{-}2\text{-}quinolyl$	Pyrrolidine		$(CH_2)_3$		See Eq. (29)	100 : 0	86%	90
CO_2CH_3	CH_3	Ph	Pyrrolidine		$(CH_2)_3$		—	100 : 0	49%	90
CO_2Et	H	H	Pyrrolidine			Ph	—	0 : 100	37%	95
CO_2Et	H	H	Pyrrolidine			H	C_6H_6, 80°C	0 : 100	49%	95
CO_2Et	H	H	Pyrrolidine			Ph	$CHCl_3$, 60°C	0 : 100	47%	95
CO_2Et	H	H	Pyrrolidine			Et	$CHCl_3$, 45°C	0 : 100	24%	95
CO_2Et	H	H	Pyrrolidine		$(CH_2)_4$		$CHCl_3$, 45°C	0 : 100	19–50%	95
H	CH_3	SCH_2Ph	Pyrrolidine		$(CH_2)_3$		$CHCl_3$, 45°C	100 : 0	95%	99d
H	CH_3	$S(O)CH_2Ph$	Pyrrolidine		$(CH_2)_3$		Dioxane, 101°C	100 : 0	88%	99d
H	CH_3	SO_2CH_2Ph	Pyrrolidine		$(CH_2)_3$		HOAc, 25°C	100 : 0	66%	99d
H	$CHMe_2$	$S(O)CH_2Ph$	Pyrrolidine		$(CH_2)_3$		HOAc, 0°C	100 : 0	68%	99d
H	$CHMe_2$	SO_2CH_2Ph	Pyrrolidine		$(CH_2)_3$		HOAc, 25°C	100 : 0	88%	99d
H	Ph	SCH_2Ph	Pyrrolidine		$(CH_2)_3$		HOAc, 25°C	100 : 0	9%	99d
H	Ph	$S(O)CH_2Ph$	Pyrrolidine		$(CH_2)_3$		Dioxane, 101°C	100 : 0	92%	99d
H	Ph	SO_2CH_2Ph	Pyrrolidine		$(CH_2)_3$		HOAc, 25°C	100 : 0	46%	99d
SO_2CH_2Ph	Ph	H	Pyrrolidine		$(CH_2)_3$		HOAc, 25°C	100 : 0	46%	99d
H	CH_3	SCH_3	Pyrrolidine		$(CH_2)_3$		HOAc, 25°C	100 : 0	48%	99d
H	CH_3	SCH_3	Pyrrolidine		CH_2OCH_2		Dioxane, 101°C	100 : 0	9%	99d
H	CH_3	$S(O)CH_3$	Pyrrolidine		CH_2OCH_2		Dioxane, 101°C	100 : 0	67%	99d
H	CH_3	SO_2CH_3	Pyrrolidine		CH_2OCH_2		HOAc, 25°C	100 : 0	61%	99d
H	CH_3	OEt	Pyrrolidine		$(CH_2)_3$		$1,2\text{-}Cl_2C_6H_4$, 132°C	100 : 0	35%	99d
H	CH_3	OEt	Pyrrolidine		CH_2OCH_2		$1,2\text{-}Cl_2C_6H_4$, 132°C	—	0%	99d

(*continued*)

TABLE 10-IX (Continued)

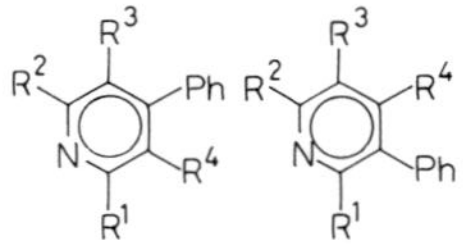

	1,2,4-Triazine		Dienophile	Product(s)		
R¹	R²	R³	R⁴		Yield	Ref.

Dienophile: Ph—≡—R⁴

Product(s):

R¹	R²	R³	R⁴	Yield	Ref.
CO_2CH_3	CO_2CH_3	CO_2CH_3	H	44 : 56 84%	90
CO_2CH_3	CO_2CH_3	CO_2CH_3	CH_3	3 : 2 62%	95
CO_2CH_3	CH_3	CH_3	H	22 : 78 28%	90

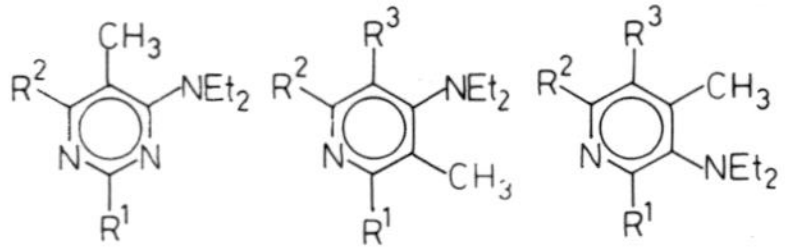

Dienophile: NR_2—≡—CH_3

Products:

R¹	R²	R³	R	Conditions	Yield	Ref.
H	H	H	Et	$CHCl_3$, 25°C	100 : 0 : 0 40%	93
CH_3	H	Ph	Et	$CHCl_3$, 25°C	100 : 0 : 0 72%	93
Ph	H	H	Et	$CHCl_3$, 25°C	100 : 0 : 0 71%	93
CO_2CH_3	H	Ph	Et	C_6H_6, 20°C	100 : 0 : 0 100%	93
CO_2CH_3	H	H	Et	C_6H_6, 20°C	100 : 0 : 0 100%	93
CO_3CH_3	Ph	H	Et	C_6H_6, 20°C	0 : 0 : 100 85%	93
CO_2CH_3	Ph	H	CH_3	C_6H_6, 80°C	0 : 0 : 100 89%	93
CO_2CH_3	H	Ph	Et	C_6H_6, 20°C	100 : 0 : 0 100%	93
CO_2CH_3	Ph	Ph	Et	C_6H_6, 80°C	45 : 35 : 20 89%	93

330

The reaction table (pyridine products, R^3, R^1, R^2 on ring):

R^1	R^2	R^3	Conditions	Ratio	Yield	Ref.
CO_2CH_3	Ph	Ph	CH_3CN, 20°C	17:75: 8	90%	93
CO_2CH_3	Ph	Ph	C_6H_6, 80°C	42:28:30	80%	93
CO_2CH_3	CH_3	CH_3	C_6H_6, 80°C	82: 0:18	67%	93
CO_2CH_3	CH_3	CH_3	CH_3CN, 20°C	87: 0:13	55%	93
CO_2CH_3	CH_3	CH_3	C_6H_6, 80°C	76: 0:24	73%	93
CO_2CH_3	CO_2CH_3	CO_2CH_3	C_6H_6, 20°C	15: 0:85	—	93
CO_2CH_3	CO_2CH_3	CO_2CH_3	CH_3CN, 20°C	22: 0:78	84%	93
CO_2CH_3	CO_2CH_3	CO_2CH_3	C_6H_6, 20°C	22: 0:78		93

Second reaction table (pyridine products with R^4, R^4, R^1, R^2):

R^1	R^2	R^3	R^4	Yield	Ref.
CO_2CH_3	CO_2CH_3	CO_2CH_3	Norbornadiene	86%	90
CO_2CH_3	Ph	Ph	Norbornadiene	70%	90
CO_2CH_3	CH_3	CH_3	Norbornadiene	94%	90
CO_2CH_3	H	H	Norbornadiene	78%	90
Cl	Cl	Cl	Norbornadiene	40–50%	94
F	F	F	Norbornadiene	40–50%	94

Third reaction table (alkene R^4, R^4):

R^1	R^2	R^3	R^4	Yield	Ref.
Cl	Cl	Cl	$(CH_2)_3$	75–80%	94
Cl	Cl	Cl	$(CH_2)_6$	75–80%	94
Cl	Cl	Cl	CH_3	75–80%	94
Cl	Cl	Cl	H	75–80%	94
			Norbornene		90, 94
			Cyclopropenes		94
			Benzocyclopropenes		94
			Cyclobutenes		100a
			Benzamidine		100b
			Formaldehyde N,N-dimethylhydrazone		

(continued)

TABLE 10-IX (Continued)

1,2,4-Triazine			Dienophile	Product(s)		
R¹	R²	R³			Yield	Ref.

First dienophile: $CH_3O_2C-C\equiv C-CO_2CH_3$

Product: pyridine with R^1, N, CO_2CH_3, CO_2CH_3, R^2

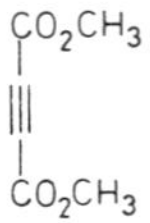

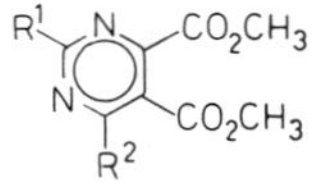

R¹	R²	R³	Yield	Ref.
NMe₂	NMe₂	H	—	97
NMe₂	NMe₂	NMe₂	57%	97
NEt₂	NEt₂	NMe₂	80%	97
NMe₂	NMe₂	OCH₃	55%	97

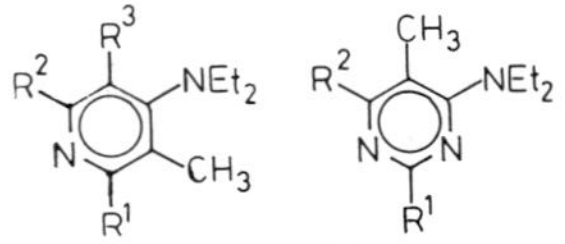

Second dienophile: $Et_2N-C\equiv C-CH_3$

Products: pyridines with CH_3/NEt_2 substituents

R¹	R²	R³	Yields		Ref.
OCH₃	H	H	—	87%	97
SCH₃	H	H	—	86%	97
OCH₃	OCH₃	—	21%	—	97
H	OCH₃	H	79%	—	97
OCH₃	OCH₃	OCH₃	—	—	97
NMe₂	H	H	—	91%	97
NMe₂	NMe₂	H	—	—	97
H	NMe₂	H	—	—	97

[a] Enamine generated in situ.

TABLE 10-X

Intramolecular Diels–Alder Reactions of 1,2,4-Triazines[99]

1,2,4-Triazine			Product
R	X	n	Yield
CH₃	S, SO, SO₂	1	69%, 69%, 75%
CH(Me)₂	S, SO, SO₂	1	nr,[a] 84%, nr
Ph	S, SO, SO₂	1	nr, 95%, nr
CH₃	S, SO, SO₂	2	57%, trace, 99%
4-ClC₆H₄	S, SO, SO₂	2	67%, 0%, 21%

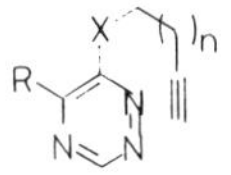

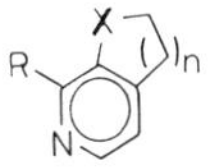

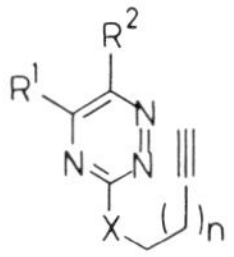

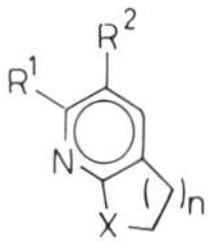

R¹	R²	X	n	Yield
Ph	H	S, SO, SO₂	1	93%, 79% ,78%
Ph	Ph	CONH	0	60%
CH₃	CH₃	CONH	0	52%
Ph	H	CONH	0	63%
Ph	Ph	O	1	97%
Ph	Ph	O	2	77%
H	H	S	1	88%
H	H	S	2	73%
H	H	O	1	96%
H	H	O	2	61%
CH₃	CH₃	O	1	100%
CH₃	CH₃	O	2	80%
Ph	Ph	O	3	8%
H	H	S	3	24%
CO₂CH₃	CO₂CH₃	NH	3	11%
H	H	NH	1	51–73%
Ph	H	NH	1	44%
4-ClC₆H₄	H	NH	1	61%
Ph	Ph	NH	1	20%
CH₃	CH₃	NH	1	—
4-ClC₆H₄	H	S, SO, SO₂	2	67%, 0%, 21%
Ph	H	CH(CO₂CH₃)₂	1	41%
4-ClC₆H₄	H	CH(CO₂CH₃)₂	1	49%
4-ClC₆H₄	H	CH(CN)₂	1	80%
4-ClC₆H₄	H	CH(CN)₂	2	84%
4-ClC₆H₄	H	CH(CN)CO₂Et	2	74%

(continued)

TABLE 10-X (*Continued*)

	1,2,4-Triazine			Product

R^1	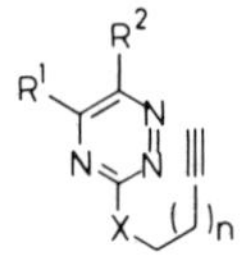 R^2	X	n	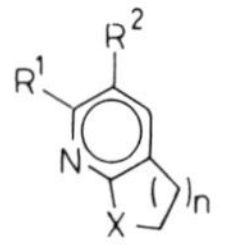 Yield
Ph	H	O	1	95%
Ph	H	O	2	68%
4-ClC$_6$H$_4$	H	O	1	90%
4-ClC$_6$H$_4$	H	O	2	20%

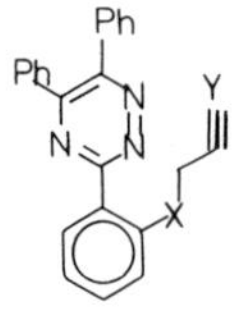 X	Y			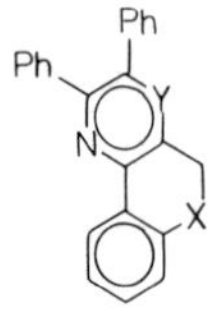 Yield
O	CH			64%
NCOCF$_3$	CH			74%
O	N			46%
NCOCF$_3$	N			36%
				(+47%, 5,6-C=N)

^{a} nr, Not reported.

$$\tag{31}$$

55 – 80%

This additional substitution of the 1,2,4-triazine nucleus with electron-donating substituents does not preclude the ability of the 1,2,4-triazine to participate in [4 + 2] cycloaddition processes with electron-rich dienophiles including ynamines.[97] The [4 + 2] cycloadditions of electron-rich 1,2,4-triazines are summarized in Table 10-IX.

Recent studies have explored and detailed the utility of the intramolecular Diels–Alder reactions of alkyne 1,2,4-triazines [Eq. (32) and Table 10-X].[99]

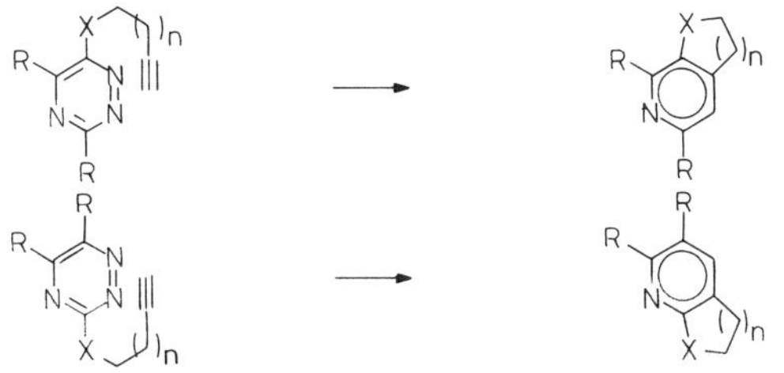

$$(32)$$

Two reports of the use of heterodienophiles, amidines[100a] and aldehyde N,N-dimethylhydrazones,[100b] in [4 + 2] cycloadditions with 1,2,4-triazines have been detailed (Table 10-IX).

14. 1,2,4,5-Tetrazines

Since the initial report of symmetrical 3,6-perfluoroalkyl-1,2,4,5-tetrazines participating in [4 + 2] cycloadditions with representative olefins in a study which constituted the first demonstration of the viability of the inverse electron demand Diels–Alder reaction,[1] extensive investigations have defined the scope and potential of 1,2,4,5-tetrazine participation in [4 + 2] cycloadditions.[101–152] For most purposes, only a limited number of symmetrical 1,2,4,5-tetrazines have been investigated, and to a large extent this reflects the current difficulty in the preparation[101a–b] and stability[104] of the 1,2,4,5-tetrazine system. A wide range of dienophiles (Table 10-XI) and heterodienophiles (Table 10-XII) are capable of participation in Diels–Alder reactions with the electron-deficient 1,2,4,5-tetrazines and include electron-rich, neutral, and electron-deficient olefins, acetylenes, allenes, dienes, enol ethers and acetates, enamines, ynamines, ketene acetals, enolates, benzynes, selected aromatics, imidates, amidines, thioimidates, aldehyde N,N-dimethylhydrazones, imines, azirines, and cyanamides. Electron-rich dienophiles usually participate in the 1,2,4,5-tetrazine [4 + 2] cycloadditions at room temperature, and the simple, neutral olefinic or typical electron-deficient dienophiles require higher reaction temperatures (50–200°C) [Eq. (33)].

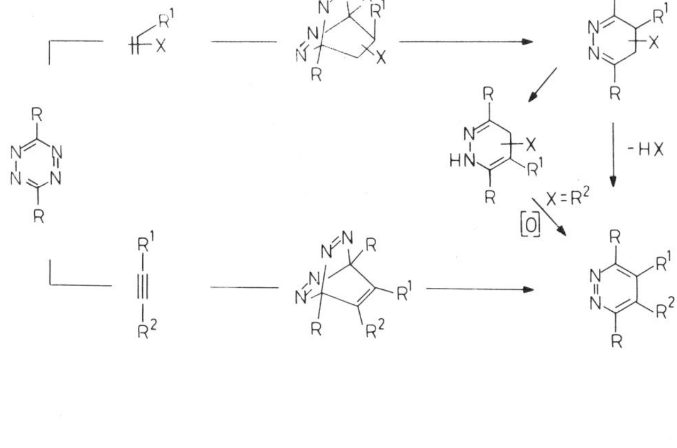

$$(33)$$

In the few cases studied,[103] unsymmetrical and electron-deficient 1,2,4,5-tetrazines participate in predictably regioselective Diels–Alder reactions with electron-rich dienophiles [Eq. (34)].

$$(34)$$

90%

Recent reports have described the first examples of intramolecular alkyne 1,2,4,5-tetrazine Diels–Alder reactions.[99b,c]

Extensive reviews have summarized[101a–b] and compiled[101c] the results of many of the studies to date, and Table 10-XI summarizes representative inter- and intramolecular Diels–Alder reactions of 1,2,4,5-tetrazines with carbon dienophiles. Extensive investigations of the participation of electron-deficient 1,2,4,5-tetrazines in inverse electron demand Diels–Alder reactions with electron-rich heterodienophiles have been described, and much of this work is summarized in Table 10-XII.

The use of the Diels–Alder reactions of dimethyl 1,2,4,5-tetrazine-3,6-dicarboxylate in the formal total synthesis of streptonigrin [Eq. (35)],[95,144]

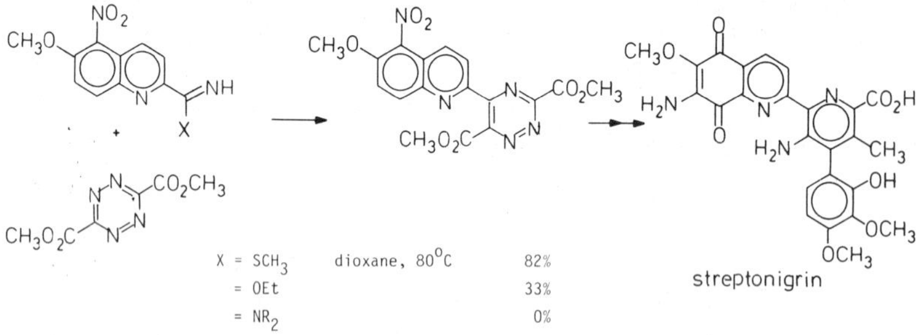

X = SCH$_3$	dioxane, 80°C	82%
	= OEt	33%
	= NR$_2$	0%

$$(35)$$

streptonigrin

in the total synthesis of PDE-I, PDE-II, and CC-1065 [Eq. (36)],[76,116]

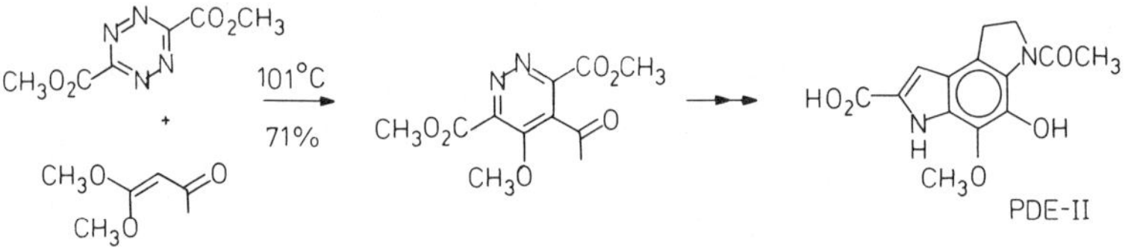

101°C

71%

$$(36)$$

PDE-II

octamethylporphin (OMP) [Eq. (37)],[102]

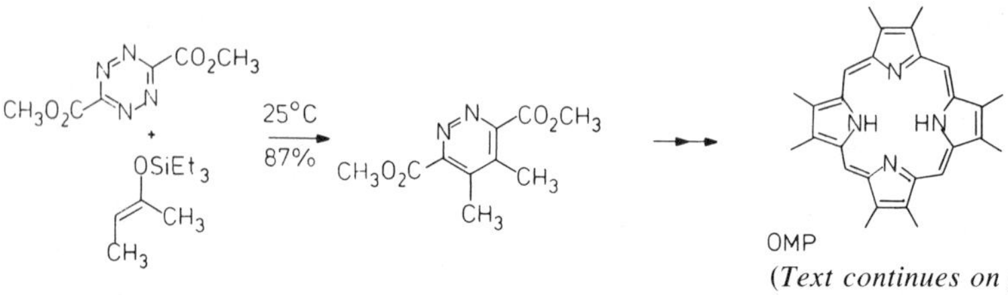

25°C

87%

$$(37)$$

OMP

(Text continues on p. 348)

TABLE 10-XI
Diels–Alder Reactions of 1,2,4,5-Tetrazines with Dienophiles

1,2,4,5-Tetrazine	Dienophile	Product(s)	Ref.
(tetrazine structure, R)	(aryl dienophile structure, X, R')	(product structure, R, R¹, Ar)	
R = CHFCF₃	Rate of reaction		1
	X = pOCH₃, R¹ = H > X = R = H >>		
	X = pOCH₃, R¹ = CH₃ > X = oOCH₃, R¹ = CH₃ >		
	X = H, R¹ = CH₃ >>> X = pOCH₃, R¹ = CO₂Et;		
	X = H, R¹ = Ph		

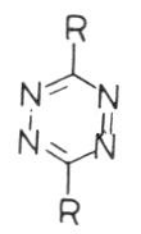

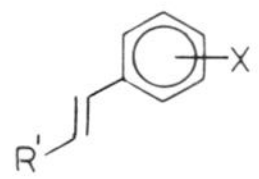

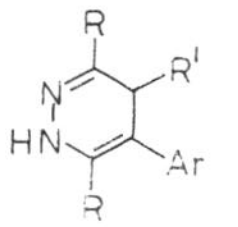

R	X	R¹	Conditions	Yield	Ref.
CHFCF₃	H	H	Et₂O, 25°C	60%	1
CF₃	H	H	Neat, 0°C	100%	105
CO₂CH₃	H	H	Dioxane, 25°C	80%	107
CO₂CH₃	pOCH₃	H	Dioxane, 25°C	98%	107
CO₂CH₃	pNO₂	H	Dioxane, 25°C	62%	107

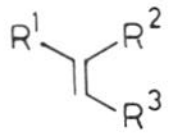

(dienophile structure: R¹, R², R³ on alkene)

R	R¹	R²	R³	Conditions	Yield	Ref.
CHFCF₃	CH₃	C(CH₃)=CH₂	H	Neat, 25°C	62%	1
CHFCF₃		=CH₂	H	—	—	1
CF₃	H	(CH₂)₄		Neat, 25°C	88%	105
Ph	H	H	CN	Toluene, 100°C	64%	1
CO₂CH₃	H	H	H	CH₂Cl₂, 25°C	98%	106

(continued)

TABLE 10-XI (*Continued*)

1,2,4,5-Tetrazine

Dienophile

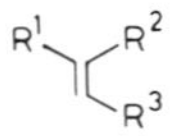

R	R¹	R²	R³	Conditions	Product(s) Yield	Ref.
CO_2CH_3	CH_3	CH_3	H	CH_2Cl_2, 25°C	81%	106
CO_2CH_3	Ph	CH_3	H	CH_2Cl_2, 25°C	91%, 88%	106, 107
CO_2CH_3	H		$(CH_2)_2$	CH_2Cl_2, 25°C	83%	106
CO_2CH_3	H		$(CH_2)_4$	CH_2Cl_2, 25°C	71%	106
CO_2CH_3	H		$(CH_2)_5$	CH_2Cl_2, 25°C	50%	106
CO_2CH_3	H		$(CH_2)_6$	CH_2Cl_2, 25°C	57%	106
CO_2CH_3	H	H	CO_2CH_3	Dioxane, 101°C	87%	107
CO_2CH_3	$(CH_2)_3CH_3$	H	H	CCl_4, reflux	99%	107
CO_2CH_3	Ph	Ph	H	Dioxane, 25°C	88%	107
CO_2CH_3	(CH_2SCH_2)		H	CH_2Cl_2, 25°C	55–95%	113
CO_2CH_3	$(CH_2)_3$		H	CH_2Cl_2, 25°C	55–95%	113
CO_2CH_3	$(CH_2)_4$		H	CH_2Cl_2, 25°C	55–95%	113
CO_2CH_3	$(CH_2)_5$		H	CH_2Cl_2, 25°C	55–95%	113
CO_2CH_3	$CH_2CH_2SCH_2CH_2$		H	CH_2Cl_2, 25°C	55–95%	113
CO_2CH_3	$(CH_2)_3OH$	H	H	Dioxane, 25°C	92%	115
2-Py	H	NH(CO)O	H	EtOAc, 25°C	95%	140

R	 R¹	R²	Conditions	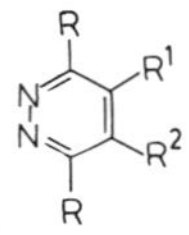 Yield	Ref.
Ph	Ph	Ph	Toluene, 110°C	86%	1
Ph	CO_2CH_3	H	Xylene, reflux	83%	72, 73

CF$_3$	H	Neat, 25°C	79%	105
CF$_3$	CH$_3$	Neat, 25°C	91%	105
CF$_3$	SnMe$_3$	CH$_2$Cl$_2$, 0°C	78%	105
CO$_2$CH$_3$	Ph	Dioxane, 101°C	44–52%	107
CO$_2$CH$_3$	SiMe$_3$	C$_6$H$_6$, 75°C	70%	111
CO$_2$CH$_3$	SiMe$_3$	C$_6$H$_5$NO$_2$, 110°C	85%	111
CO$_2$CH$_3$	SiMe$_3$	BrCH$_2$CH$_2$Br, 75–110°C	69%	111
CO$_2$CH$_3$	Et	BrCH$_2$CH$_2$Br, 75–110°C	36%	111
CO$_2$CH$_3$	(CH$_2$)$_2$CH$_3$	BrCH$_2$CH$_2$Br, 75–100°C	40%	111
CO$_2$CH$_3$	(CH$_2$)$_3$CH$_3$	BrCH$_2$CH$_2$Br, 75–100°C	34%	111
CO$_2$CH$_3$	(CH$_2$)$_4$CH$_3$	BrCH$_2$CH$_2$Br, 75–100°C	31%	111
CO$_2$CH$_3$	(CH$_2$)$_5$CH$_3$	BrCH$_2$CH$_2$Br, 75–100°C	25%	111
CO$_2$CH$_3$	CO$_2$CH$_3$	Xylene, reflux	53%	72, 73
CO$_2$CH$_3$	OEt	Dioxane, 45–101°C	63%	108
CO$_2$CH$_3$	OCH$_2$Ph	Dioxane, 25°C	82%	102, 116
CO$_2$CH$_3$	HC(OH)(CH$_2$)$_3$CH$_3$	Dioxane, 101°C	84% (lactone)	115
CO$_2$CH$_3$	C(O)(CH$_2$)$_3$CH$_3$	Dioxane, 101°C	70%	115
CO$_2$CH$_3$	CH$_2$CH$_2$OH	Dioxane, 80°C	58% (lactone)	115
CO$_2$CH$_3$	CH$_2$CH$_2$OTBDMS	Dioxane, 80°C	79%	115
CO$_2$CH$_3$	(CH$_2$)$_3$OH	Dioxane, 50°C	83%	115
CO$_2$CH$_3$	CH$_2$OTBDMS	Mesitylene, 140°C	59%	115
Ph	NEt$_2$	Toluene, 110°C	65%	108
CO$_2$CH$_3$	NEt$_2$	Dioxane, 25°C	73%	108
CH$_3$	NEt$_2$	Dioxane, 25°C	89%	109
CH$_3$	Benzyne	CH$_2$Cl$_2$, 25°C	73%	107, 110
Ph	Benzyne	CH$_2$Cl$_2$, 25°C	67%	107, 110
CO$_2$CH$_3$	2-deoxy-pyrimidine	Dioxane, 60°C	71%	153
SCH$_3$	NEt$_2$	Dioxane, 25°C	99%	154
SCH$_3$	Ph	Mesitylene, 160°C	61%	154
SCH$_3$	C$_6$H$_{13}$	Mesitylene, 160°C	66%	154

(continued)

339

TABLE 10-XI (Continued)

1,2,4,5-Tetrazine	Dienophile				Product(s)		
R	X	R¹	R²	Conditions	Yield	![pyridazine product structure: R, R¹, R², R on ring with N–N]	Ref.
CO₂CH₃	OEt	H	H	Dioxane, 25°C	100%		107
CO₂CH₃	OAc	H	H	Dioxane, 25°C	95%		107
CO₂CH₃	OEt	CH₃	H	Dioxane, 25°C	100%		107
CO₂CH₃	OEt	OEt	H	Dioxane, 25°C	100%		107
CO₂CH₃	OCH₃	Ph	H	Dioxane, 25°C	—		107
CO₂CH₃	Morpholine	Ph	H	Dioxane, 25°C	91%		107
CO₂CH₃	OEt	(CH₂)₃		Dioxane, 25°C	64%		107
CO₂CH₃	Pyrrolidine	(CH₂)₃		Dioxane, 25°C	59%		102, 107
CO₂CH₃	OCH₃	OCH₃	H	Dioxane, 25°C	95%		102, 115, 116
CO₂CH₃	OCH₃	OCH₃	COCH₃	Dioxane, 101°C	71%		76, 102, 116
CO₂CH₃	OSiMe₃	CH₃	CH₃	Dioxane, 25°C	87%		102
CO₂CH₃	Morpholine	Et	CH₃	Dioxane, 25°C	70%		102
CO₂CH₃	Pyrrolidine	Et	CH₃	Dioxane, 25°C	Trace		102
CO₂CH₃	Pyrrolidine	(CH₂)₄		Dioxane, 25°C	85%		102
CO₂CH₃	Pyrrolidine	Ph	H	Dioxane, 25°C	Trace		102
CO₂CH₃	Morpholine	Ph	H	Dioxane, 25°C	87%		102
CO₂CH₃	OSiMe₃	Ph	H	Dioxane, 25°C	92%		102
CO₂CH₃	OSiMe₃	OCH₂Ph	H	Dioxane, 25°C	33%		102
	Additional enamine and enol ether dienophiles						142, 143
	Enol lactones						108, 146
CH₃	H	CH₃	H	CH₂Cl₂, 25°C (with in situ oxidation)	35%		109

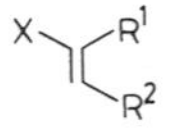

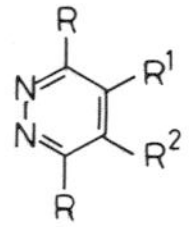

340

CH₃	H	(CH₂)₃CH₃	H	CH₂Cl₂, 25°C (with *in situ* oxidation)	48%	109
CH₃	H	C(O)OC(O)	H	Dioxane, 101°C (with *in situ* oxidation)	36%	109
Ph	OEt	H	H	Toluene, 110°C	94–97%	107
Ph	OAc	H	H	Toluene, 110°C	92–96%	107
Ph	OEt	H	OEt	Toluene, 110°C	100%	107
Ph	OCH₃	H	Ph	Toluene, 110°C	99%	107
Ph	Morpholine	H	Ph	Toluene, 110°C	91%	107
Ph	Morpholine	(CH₂)₃		C₆H₅NO₂, 30°C (with *in situ* oxidation)	80%	107
Ph	H	(CH₂)₃		Toluene, 110°C	—	107
Ph	SOPh	H	H	Toluene, 110°C	97%	114
Ph	O⁻K⁺	H	H	THF, 25°C	88%	112
Ph	O⁻K⁺	CH₃	H	THF, 25°C	31%	112
Ph	O⁻K⁺	Et	H	THF, 25°C	41%	112
Ph	O⁻K⁺	H	CH(OMe)₂	THF, 25°C	56%	112
Ph	O⁻K⁺	Ph	H	THF, 25°C	74%	112
Ph	O⁻K⁺	H	Ph	THF, 25°C	58%	112
Ph	O⁻K⁺	PhCH₂	H	THF, 25°C	72%	112
Ph	O⁻K⁺	CH₃	CH₃	THF, 25°C	51%	112
Ph	O⁻K⁺	CH₃	Et	THF, 25°C	52%	112
Ph	O⁻K⁺	Et	CH₃	THF, 25°C	30%	112
Ph	O⁻K⁺	(CH₂)₃		THF, 25°C	67%	112
Ph	O⁻K⁺	(CH₂)₄		THF, 25°C	58%	112
Ph	O⁻K⁺	(CH₂)₅		THF, 25°C	61%	112
Ph	O⁻K⁺	(CH₂)₆		THF, 25°C	11%	112
2-Py	H	N(Ac)CO₂	H	THF, 25°C	56% R¹ = NHAc, R² = H	140
2-Py	H	OC(CH₃)₂O	H	THF, 25°C	— R¹ = OH, R² = H	141
Ph, Ar, 2-Py, Ar/H	NR₂	CN,NO₂	H	THF, 25°C	34–60%	143b

(continued)

TABLE 10-XI (*Continued*)

R	X	R¹	R²	Conditions	Yield	Ref.
SCH₃	OAc	H	H	Xylene, 140°C	87%	154
SCH₃	OEt	H	H	Xylene, 130°C	87%	154
SCH₃	—NCOCH₂CH₂—	H	H	Dioxane, 100°C	80%	154
SCH₃	OSiMe₃	Ph	H	Xylene, 140°C	75%	154
SCH₃	OSiMe₃	Ph	H	CH₂Cl₂, 13 kbar, 25°C	88%	154
SCH₃	Morpholine		(CH₂)₃	CH₂Cl₂, 25°C	97%	154
SCH₃	OCH₃	OCH₃	H	Dioxane, 80°C	85%	154
SCH₃	Morpholine	Ph	H	CHCl₃, 65°C	65%	154
SCH₃	Pyrrolidine		(CH₂)₃	CHCl₃, 60°C	81%	154
SCH₃	OSiMe₃		(CH₂)₃	Mesitylene, 150°C	61%	154
SCH₃	Morpholine	Et	CH₃	CHCl₃, 65°C	67%	154

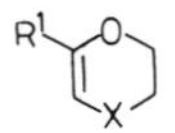

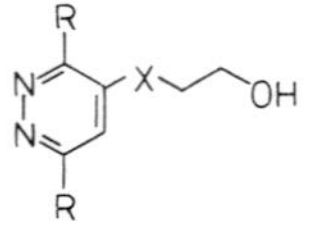

R	X	R¹	Conditions	Yield	Ref.
Ph	O	H	Toluene, 110°C	78%	108
Ph	CH₂	H	Toluene, 110°C	71%	108
CH₃	O	H	Dioxane, 101°C	90%	109
CO₂CH₃	O	H	Dioxane, 60°C	80%	108
CO₂CH₃	CH₂	H	Dioxane, 25°C	80%	108
CO₂CH₃	—	CH₃	Dioxane, 85°C	77%	108

R		Conditions	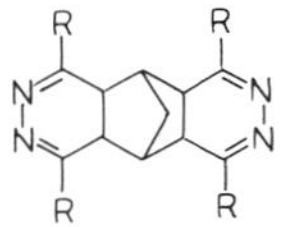	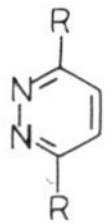 Yield	Ref.
CFHCF₃	Norbornadiene	Neat, 25°C	100 : 0	—	1
CF₃	Norbornadiene	Neat, 25°C	0 : 100	59%	105
CO₂CH₃	Norbornadiene	CH₂Cl₂, 25°C	0 : 100	87%	110a
Ph	Norbornadiene	CH₂Cl₂, 25°C	0 : 100	95%	110

Additional, reactive and strained olefins[118-136]

R		Yield	Ref.
Ph	Acenaphthylene	80%	118
2-Py	Acenaphthylene	60%	118
CO₂CH₃	Cyclopentadiene	43%	110a
CO₂CH₃	Norbornene	43%	110a
Ph	Norbornene	94%	110a
2-Py	Cyclobutene	—	117, 121
Ph	Cyclobutene	—	120, 121
CO₂CH₃	Cyclobutene	—	124
Ph	Benzocyclobutene	85%	119
CH₃	Benzocyclobutene	63%	119
2-Py	Benzocyclobutene	—	122, 123 j–k
Ph	Cyclobutadiene	76%	119
CH₃	Cyclobutadiene	21%	119
CO₂CH₃	(Cyclo)heptatrienes	—	125–127
CO₂CH₃	Cyclopropenes	—	128, 129
2-Py, Ph, CO₂CH₃	Other strained olefins	—	130–133
2-Py, Ph, CO₂CH₃	Fulvenes	—	134–136
2-Py, Ph, CO₂CH₃	Benzofulvenes, isoindoles, isobenzofurans (reactions and generation)	—	123

(continued)

343

TABLE 10-XI (*Continued*)

1,2,4,5-Tetrazine	Dienophile	Product(s)	Ref.
R	Heterocycles		
CO_2CH_3	Thiophene		137a
CO_2CH_3	2-Methylimidazole		137a,c
CO_2CH_3	2,5-Dimethylfuran		127a
CO_2CH_3	N-Methylpyrrole		137a
CO_2CH_3	Benzothiophene		137b
CO_2CH_3	Benzofuran		137b, 139
CO_2CH_3	Indole		137, 138

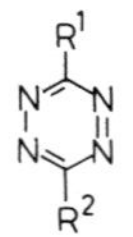

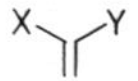

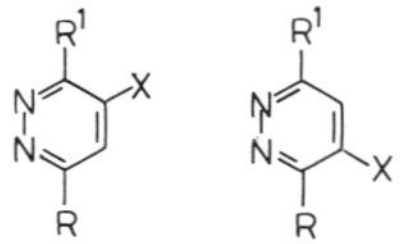

Tetrazine: R^1, R^2 substituted

Dienophile: $X-Y$ (alkene)

Products:

R^1 / X substituted pyridazine and R^1 / X regioisomer

R^1	R^2	X	Y		Conditions	Yield	Ref.
Ph	H	OEt	OEt	100 : 0	Dioxane, 50°C	90%	103a
Ph	H	NMe_2	OEt	100 : 0	Neat, 20°C	96%	103a
Ph	H	NMe_2	SCH_3	100 : 0	Dioxane, 20°C	85%	103a
Ph	H	NMe_2	NMe_2	0 : 100	CH_2Cl_2, 20°C	91%	103a
Ph	H	NMe_2	NMe_2	0 : 100	C_6H_6, 20–80°C	86%	103a
Ph	H	NMe_2	NMe_2	0 : 100	CH_3CN, 20°C	91%	103a

Dienophile: $X-\equiv$ (alkyne)

R^1	R^2	X		Conditions	Yield	Ref.
Ph	H	Ph	19 : 1	Toluene, 25°C	76%	103b
Ph	H	CH_3	0 : 100	Toluene, 25°C	20%	103b

344

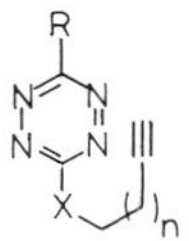

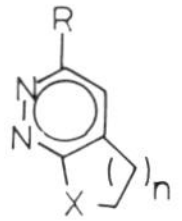

R	X		Conditions	Yield	Ref.
SCH_3	O	1	C_6H_6, 80°C	43%	99b
SCH_3	O	2	C_6H_6, 80°C	57%	99b
Ph	O	1	C_6H_6, 80°C	37%	99b
Ph	O	2	C_6H_6, 80°C	48%	99b
SCH_3	NH	3	Xylene, reflux	68%	99c
$S(O)CH_3$	NH	3	Xylene, reflux	72%	99c
SO_2CH_3	NH	3	Xylene, reflux	85%	99c

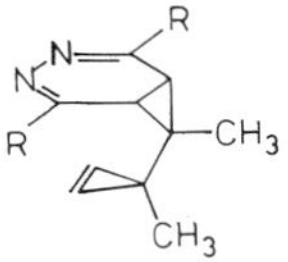

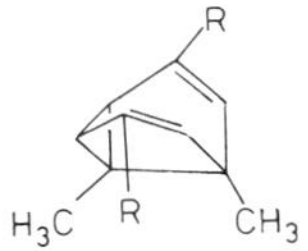

R	Conditions	Yield	Ref.
CH_3	—	38%	129
CO_2CH_3	—	70%	129
Ph	$CHCl_3$, 50–60°C	74%	129
$mClC_6H_4$	$CHCl_3$, 50–60°C	32%	129
$pCH_3C_6H_4$	$CHCl_3$, 50–60°C	84%	129
$mCH_3OC_6H_4$	$CHCl_3$, 50–60°C	33%	129
$mCF_3C_6H_4$	$CHCl_3$, 50–60°C	77%	129

TABLE 10-XII
Diels–Alder Reactions of 1,2,4,5-Tetrazines with Heterodienophiles

1,2,4,5-Tetrazine	Heterodienophile			Product(s)	
R	X	R′	Conditions	Yield	Ref.
Ph	OEt	Ph	Dioxane, 60°C	0%	108
Ph	NH_2	Ph	Dioxane, 60°C	34%	144b
CO_2CH_3	OEt	CH_3	Dioxane, 60°C	13%	108
CO_2CH_3	OEt	Ph	Dioxane, 60°C	27%	95, 108, 144a
CO_2CH_3	NH_2, NR_2	2-Py	Dioxane, 25°C	0%	144a
CO_2CH_3	OEt	2-Py	Dioxane, 80°C	43%	95, 144a
CO_2CH_3	SCH_3	2-Py	Dioxane, 80°C	68%	95, 144a
CO_2CH_3	OEt	2-Quinolyl	Dioxane, 80°C	33%	95, 144a
CO_2CH_3	SCH_3	2-Quinolyl	Dioxane, 80°C	70%	95, 144a
2-Py	NH_2	Ph	Toluene, 110°C	33%	144b
R	X	R′	Conditions	Yield	Ref.
Ph	Piperidine	$CH(Me)_2$	Toluene, 110°C	89%	147b
Ph	Piperidine	*t*Bu	Toluene, 110°C	22%	147b
2-Py	Piperidine	$CH(Me)_2$	Toluene, 110°C	78%	147b
2-Py	Piperidine	*t*Bu	Toluene, 110°C	76%	147b
CO_2CH_3	OCH_3	CH_3	Toluene, 110°C	19%	147a
CO_2CH_3	OCH_3	Ph	Toluene, 110°C	21%	147a
R = CO_2CH_3	n = 1–3			8–47% yield	Ref. 147a
R	R^1	R^2	Conditions	Yield	Ref.
CO_2CH_3	NMe_2	H	$CHCl_3$, 25°C	90%	145c
CO_2CH_3	NMe_2	CH_3	$CHCl_3$, 25°C	81%	145a
CO_3CH_3	NMe_2	Et	$CHCl_3$, 25°C	79%	145a

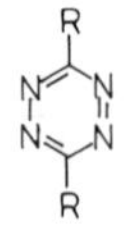

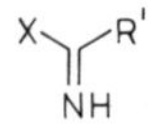

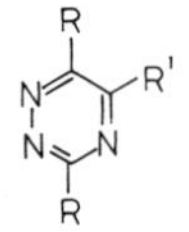

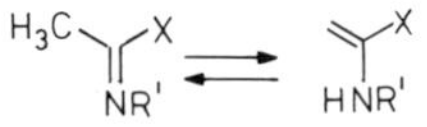

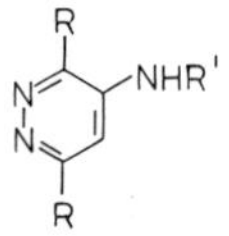

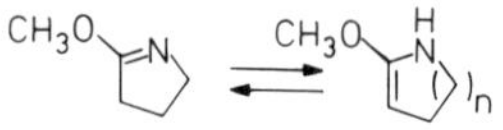

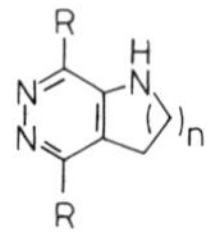

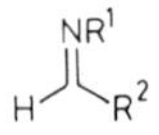

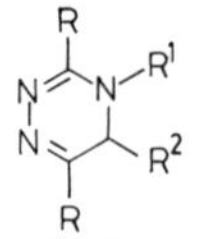

TABLE 10-XII (Continued)

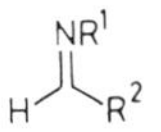

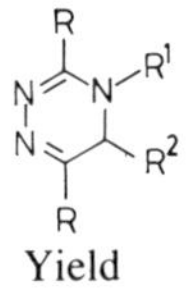

R	R^1	R^2	Conditions	Yield	Ref.
CO_2CH_3	NMe_2	iPr	$CHCl_3$, 25°C	76%	145a
CO_2CH_3	NMe_2	nPr	$CHCl_3$, 25°C	56%	145a
CO_2CH_3	NMe_2	Ph	$CHCl_3$, 25°C	70%	145a
CO_2CH_3	Piperidine	H	$CHCl_3$, 25°C	99%	145d
CO_2CH_3	Morpholine	H	$CHCl_3$, 25°C	99%	145d
CO_2CH_3	NMe_2	CH=CHPh	$CHCl_3$, 25°C	78%	145b
CO_2CH_3	Ph	Ph	$CHCl_3$, 25°C	90%	145e
CO_2CH_3	pCH$_3$OC$_6$H$_4$	Ph	$CHCl_3$, 25°C	85%	145c
CO_2CH_3	pCH$_3$OC$_6$H$_4$	pCH$_3$OC$_6$H$_4$	$CHCl_3$, 25°C	90%	145c
	Heterocyclic imines				145c

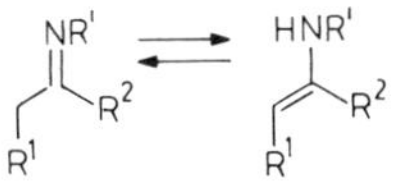

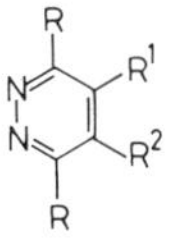

R	R′	R^1	R^2	Yield	Ref.
CO_2CH_3	OH	H	CH_3	43%	146
CO_2CH_3	NMe_2	H	CH_3	—	145a
CO_2CH_3	NMe_2	$(CH_2)_3$		—	145a
CO_2CH_3	NMe_2	$(CH_2)_4$		—	145a

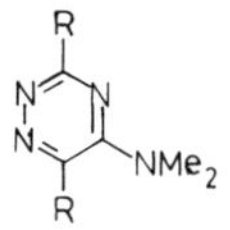

Conditions: C_6H_5Cl, reflux — 78% yield — Ref. 148

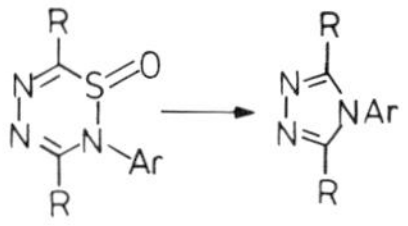

Conditions: C_6H_6, 80°C — 10–30% yield — Ref. 149

Thiocarbonyl compounds — Ref. 148, 150

Azirines — Ref. 151

Azetines — Ref. 152

and prodigiosin [Eq. (38)],[115]

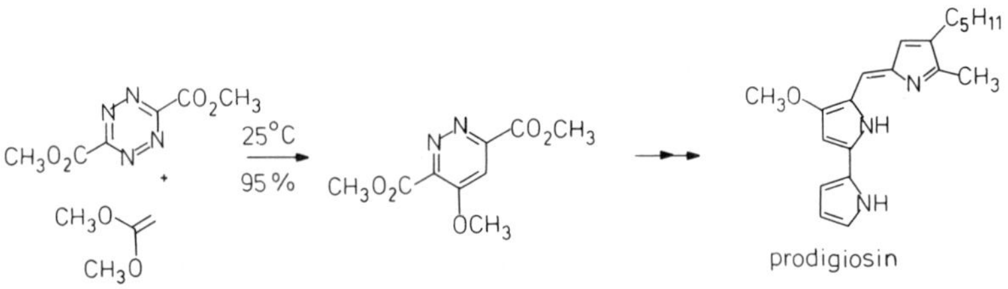

(38)

and in the preparation of anthraquinones [Eq. (39)][123]

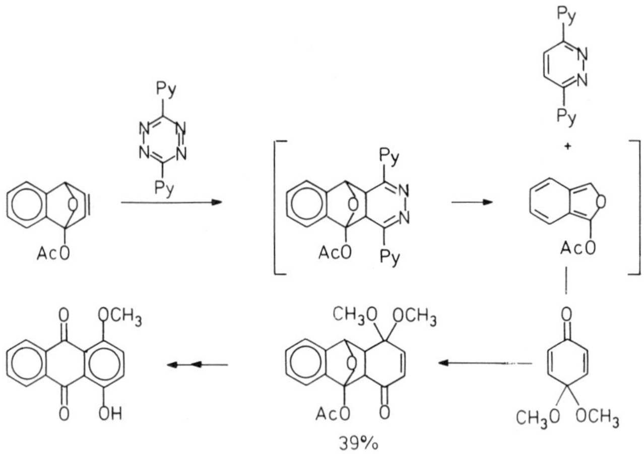

(39)

and pyridazine-substituted pyrimidine nucleosides [Eq. (40)][153]

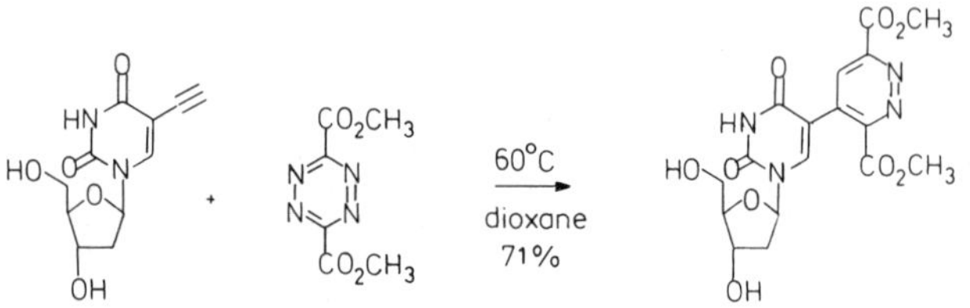

(40)

have been detailed.

15. CATIONIC HETEROAROMATIC AZADIENES

A select set of cationic heteroaromatic azadienes including acridizinium, isoquinolinium, quinolinium, and isoxazolium salts have been shown to participate as useful 4π components of $[4^+ + 2]$ cycloadditions. These systems have been discussed in Chapter 9, Section 10.

REFERENCES

1. The potential of reversing the diene/dienophile polarity of the normal Diels–Alder reaction was first discussed in the course of the early work on the [4 + 2] cycloaddition reaction: Bachmann, W. E., and Deno, N. C. (1949). *J. Am. Chem. Soc.* **71**, 3062. The first experimental demonstration of the inverse electron demand Diels–Alder reaction employed electron-deficient perfluoroalkyl-1,2,4,5-tetrazines: Carboni, R. A., and Lindsey, R. V., Jr. (1959). *J. Am. Chem. Soc.* **81**, 4342. A subsequent study confirmed the [4 + 2] cycloaddition rate acceleration accompanying the complementary inverse electron demand diene/dienophile substituent effects: Sauer, J., and Wiest, H. (1962). *Angew. Chem. Int. Ed. Engl.* **1**, 269.
2. Kondrat'eva, G. Ya. (1957). *Khim. Nauka Prom.* **2**, 666 [*Chem. Abstr.* **52**, 6345 (1958)]; Kondrat'eva, G. Ya. (1959). *Izv. Akad. Nauk SSSR, Ser. Khim.*, 484.
3. Boger, D. L. (1983). *Tetrahedron* **39**, 2869; Boger, D. L. (1986). *Chem. Rev.* **86**, 781.
4. a. Wollweber, H. (1971). In "Methoden der Organische Chemie (Houben-Weyl)" (E. Muller, ed.), Tiel 3, Vol. V/lc, p. 981. Thieme, Stuttgart; Jager, V., and Viehe, H. G. (1977). In "Methoden der Organische Chemie (Houben-Weyl)" (E. Muller, ed.), Tiel 4, Vol. V/2a, p. 807. Thieme, Stuttgart. b. Woodward, R. B., and Hoffmann, R. (1970). "The Conservation of Orbital Symmetry." Academic Press, New York; Oppolzer, W. (1984). *Angew. Chem. Int. Ed. Engl.* **23**, 876; Wurziger, H. (1984). *Kantakte* **2**, 3. Paquette, L. A. (1984). In "Asymmetric Synthesis" (J. D. Morrison, ed.), Vol. 3, p. 455. Academic Press, Orlando; Martin, J. G., and Hill, R. K. (1961). *Chem. Rev.* **61**, 537; Huisgen, R. (1963). *Angew. Chem. Int. Ed. Engl.* **2**, 565; Huisgen, R. (1968). *Angew. Chem. Int. Ed. Engl.* **7**, 321; Sauer, J. (1966). *Angew. Chem. Int. Ed. Engl.* **5**, 211; Sauer, J. (1967). *Angew. Chem. Int. Ed. Engl.* **6**, 16.
5. Weinreb, S. M., and Staib, R. R. (1982). *Tetrahedron* **38**, 3087.
6. Sauer, J., and Sustmann, R. (1980). *Angew. Chem. Int. Ed. Engl.* **19**, 779; Sauer, J. (1984). *Naturwissenschaften* **71**, 37; Houk, K. N. (1973). *J. Am. Chem. Soc.* **95**, 4092; Houk, K. N. (1975). *Acc. Chem. Res.* **8**, 361; Fukui, K. (1970). *Fortschr. Chem. Forsch.* **15**, 1; Fukui, K. (1971). *Acc. Chem. Res.* **4**, 57; Burnier, J. S., and Jorgensen, W. L. (1983). *J. Org. Chem.* **48**, 3923.
7. Ciganek, E. (1984). *Org. React.* **32**, 1; Fallis, A. G. (1984). *Can. J. Chem.* **62**, 183; Brieger, G., and Bennett, J. N. (1980). *Chem. Rev.* **80**, 63; Oppolzer, W. (1969). *Angew. Chem. Int. Ed. Engl.* **16**, 10; Carlson, R. G. (1974). *Ann. Rep. Med. Chem.* **9**, 270.
8. Ulrich, H. (ed.) (1967). "Cycloaddition Reactions of Heterocumulenes." Academic Press, New York; Gompper, R. (1969). *Angew. Chem. Int. Ed. Engl.* **8**, 312; Schmidt, R. R. (1973). *Angew. Chem. Int. Ed. Engl.* **12**, 212.
9. a. Karpeiskii, M. Ya., and Florent'ev, V. L. (1969). *Russ. Chem. Rev. (Engl. Ed.)* **38**, 540. b. Lakhan, R., and Ternai, B. (1967). *Adv. Heterocycl. Chem.* **17**, 99. c. Turchi, I. J., and Dewar, M. J. S. (1975). *Chem. Rev.* **75**, 389. d. Turchi, I. J. (1981). *Ind. Eng. Chem. Prod. Res. Dev.* **20**, 32. e. Boyd, G. V. (1984). In "Comprehensive Heterocyclic Chemistry" (K. T. Potts, ed.), Vol. 6, pp. 195–197. Pergamon, Oxford.
10. Kondrat'eva, G. Ya., and Huang, C.-H. (1961). *Dokl. Akad. Nauk SSSR* **141**, 628.
11. Chih-heng, H., and Kondrat'eva, G. Ya. (1962). *Izv. Akad. Nauk SSSR, Ser. Khim.*, 525; Naito, T., and Yoshikawa, T. (1966). *Chem. Pharm. Bull. Tokyo* **14**, 918.
12. Yoshikawa, T., Ishikawa, F., and Naito, T. (1965). *Chem. Pharm. Bull. Tokyo* **13**, 878.
13. Yoshikawa, R., Ishikawa, F., Omura, Y., and Naito, T. (1965). *Chem. Pharm. Bull. Tokyo* **13**, 873; Velde, V. W. V., Mackenzie, N. E., and Scott, A. I. (1985). *J. Label Compd. Radiopharm.* **22**, 595; Naito, T., Yoshikawa, T., Ishikawa, F., and Omura, H. (1966). Jpn. Patent 65 22,740 [*Chem. Abstr.* **64**, 3496 (1966)]; Naito, T., Yoshikawa, T.,

Ishikawa, F., and Omura, H. (1966). Jpn. Patent 65 23,908 [*Chem. Abstr.* **64**, 3495 (1966)].

14. Puchnova, V. A., and Luk'yanets, E. A. (1970). *Khim. Geterotsikl. Soedin* **2**, 327.
15. Naito, T., and Yoshikawa, T. (1966). *Chem. Pharm. Bull. Tokyo* **14**, 918.
16. Takeda Chem. Ind. Ltd. (1965). Belgian Patent 648,226 [*Chem. Abstr.* **63**, 18036 (1965)].
17. a. Pfister, K., Harris, E. E., and Firestone, R. A. (1966). U.S. Patent 3,227,722 [*Chem. Abstr.* **65**, 16949 (1966)]; Pfister, K., Harris, E. E., and Firestone, R. A. (1966). U.S. Patent 3,227,721 [*Chem. Abstr.* **64**, 9689 (1966)]. b. Firestone, R. A., Harris, E. E., and Reuter, W. (1967). *Tetrahedron* **23**, 943; Harris, E. E., Firestone, R. A., Pfister, K., Boettcher, R. R., Cross, F. J., Currie, R. B., Monaco, M., Peterson, E. R., and Rueter, W. (1962). *J. Org. Chem.* **27**, 2705; Pfister, K., Harris, E. E., and Firestone, R. A. (1963). Belgian Patent 617,500 [*Chem. Abstr.* **59**, 581 (1963)]; Pfister, K., Harris, E. E., and Firestone, R. A. (1966). U.S. Patent 3,227,724 [*Chem. Abstr.* **64**, 8149 (1966)]; Merck and Co., Inc. (1968). Netherland Patent 6,614,801 [*Chem. Abstr.* **68**, 87190n (1968)]; Harris, E. E., Zabriskie, J. L., Chamberlin, E. M., Crane, J. P., Peterson, E. R., and Reuter, W. (1969). *J. Org. Chem.* **34**, 1993; Chamberlin, E. M., Harris, E. E., and Zabriskie, J. L., Jr. (1972). U.S. Patent 3,658,846 [*Chem. Abstr.* **77**, 48534u (1972)].
18. Boell, W., and Koenig, H. (1979). *Liebigs Chem. Ann.*, 1657; Boell, W., and Koenig, H. (1973). Ger. Patent 2,143,989 [*Chem. Abstr.* **78**, 147815p (1973)].
19. Kondrat'eva, G. Ya., Kazanskii, B. A., Proshchina, N. G., and Oshueva, N. A. (1968). U.S.S.R. Patent 213,879 [*Chem. Abstr.* **69**, 52013a (1968)]; Kondrat'eva, G. Ya., Medvedskaya, L. B., and Ivanova, Z. N. (1976). U.S.S.R. Patent 487,072 [*Chem. Abstr.* **84**, 43869s (1976)].
20. Balyakina, M. V., Zhukova, Z. N., and Zhdanovich, E. S. (1968). *Zh. Prikl. Khim.* **41**, 2324.
21. Kondrat'eva, G. Ya., and Huang, C.-H. (1961). *Dokl. Akad. Nauk SSSR* **141**, 861; Itov, Z. I., L'vova, S. D., and Gunar, V. J. (1976). *Khim.-Farm. Zh.* **10**, 100 [*Chem. Abstr.* **85**, 62921q (1976)].
22. Kimel, W., and Leimgruber, W. (1965). French Patent 1,384,099 [*Chem. Abstr.* **63**, 4263 (1965)]; Kimel, W., and Leimgruber, W. (1966). Netherland Patent 6,506,703 [*Chem. Abstr.* **64**, 15851 (1966)]; Kimel, W., and Leimgruber, W. (1965). Netherland Patent 6,404,750 [*Chem. Abstr.* **62**, 11818 (1965)]; Szlompek-Nesteruk, D., Rudnicki, A., and Sikorska, T. (1975). *Przem. Chem.* **54**, 238 [*Chem. Abstr.* **83**, 79037n (1975)]; Szlompek-Nesteruk, D., Rudnicki, A., Kazimiera, W., Spychala, S., Suwalska, K., and Adamus, M. (1978). Poland Patent 93,375 [*Chem. Abstr.* **89**, 109108e (1978)].
23. Naito, T., Ueno, K., Sano, M., Omura, Y., Itoh, I., and Ishikawa, F. (1968). *Tetrahedron Lett.*, 5767.
24. Miki, H., and Saikawa, H. (1973). Ger. Patent 2,218,739 [*Chem. Abstr.* **78**, 43288 (1973)].
25. Morita, Y., Onishi, S., and Yamagami, T. (1973). Jpn. Patent 73 15,949 [*Chem. Abstr.* **79**, 53287m (1973)]; Morita, Y., Onishi, S., and Fujiwara, T. (1974). Jpn. Patent 73 30,636 [*Chem. Abstr.* **80**, 82700s (1974)].
26. Miki, T., and Matsua, T. (1967). *Yakugaku Zasshi* **87**, 323.
27. Murakamo, M., Iwanami, M., and Ozawa, I. (1970). Jpn. Patent 69 32,574 [*Chem. Abstr.* **72**, 66839y (1970)].
28. Takagaki, H., Yasuda, N., Asaoka, M., and Takei, H. (1979). *Chem. Lett.*, 183.
29. Naito, T., Ueno, K., and Miki, T. (1970). Jpn. Patent 70 07,747 [*Chem. Abstr.* **73**, 25318e (1970)]; Naito, T., Ueno, K., Morita, Y., Shimada, S., and Omura, H. (1970). Jpn. Patent 70 11,906 [*Chem. Abstr.* **73**, 25317d (1970)].

30. L'vova, S. D., Itov, Z. I., and Gunar, V. I. (1978). *Khim.-Farm. Zh.* **12**, 106 [*Chem. Abstr.* **90**, 121560z (1979)].

31. Ajinomoto, Co., Inc. (1969). Fr. Patent 1,533,817 [*Chem. Abstr.* **71**, 101839b (1969)]; Takehara, M., Togo, K., Maeda, Y., and Yoshida, Y. (1972). Jpn. Patent 72 07,553 [*Chem. Abstr.* **77**, 114447w (1972)].

32. Tanabe Seiyaku Co., Ltd. (1966). Belgian Patent 671,385 [*Chem. Abstr.* **65**, 15348 (1966)]; Matsuo, T., and Miki, T. (1971). *Chem. Pharm. Bull. Tokyo* **19**, 858; Matsuo, T., and Miki, T. (1972). *Chem. Pharm. Bull. Tokyo* **20**, 669,806.

33. Kawazu, M. (1968). Jpn. Patent 67 18,627 [*Chem. Abstr.* **69**, 10366n (1968)].

34. Daiichi Seiyaku Co., Ltd. (1967). Netherlands Patent 6,607,005 [*Chem. Abstr.* **67**, 32593v (1967)].

35. Miki, T., and Matsuo, T. (1969). U.S. Patent 3,413,297 [*Chem. Abstr.* **70**, 68172h (1969)].

36. a. Miki, T., and Matsuo, T. (1973). Jpn. Patent 72 39,115 [*Chem. Abstr.* **78**, 4293f (1973)]. b. Bringmann, G., and Schneider, S. (1986). *Tetrahedron Lett.* **27**, 175.

37. a. Additional studies including the preparation of pyridoxine analogs utilizing the Diels–Alder reaction of substituted oxazoles: Florent'ev, V. L., Drobinskaja, N. A., Ionova, L. V., and Karpeiskii, M. Ya. (1967). *Tetrahedron Lett.*, 1747; Drobinskaya, N. A., Ionova, L. V., Karpeisky, M. Ya., Turchin, K. F., and Florent'ev, V. L. (1967). *Dokl. Akad. Nauk SSSR* **177**, 617; Muhlradt, P. F., Morino, Y., and Snell, E. E. (1967). *J. Med. Chem.* **10**, 341; Doktorova, N. D., Ionova, L. V., Karpeisky, M. Ya., Padyukova, N. Sh., Turchin, K. F., and Florent'ev, V. L. (1969). *Tetrahedron* **25**, 3527; Kozikowski, A. P., and Isobe, K. (1978). *Heterocycles* **9**, 1271; Morisawa, Y., Kataoka, M., and Watanabe, T. (1976). Jpn. Patent 75 58,075 [*Chem. Abstr.* **84**, 178832z (1976)]; Morisawa, Y., Kataoka, M., and Watanabe, T. (1976). Jpn. Patent 75 105,669 [*Chem. Abstr.* **84**, 150509w (1976)]; Morisawa, Y., Kataoka, M., and Watanabe, T. (1976). Jpn. Patent 75 69,081 [*Chem. Abstr.* **84**, 43853g (1976)]; Morisawa, Y., Kataoka, M., and Watanabe, T. (1976). *Chem. Pharm. Bull. Tokyo* **24**, 1089; Usui, Y., Hara, Y., Shimamoto, N., Yurugi, S., and Masuda, T. (1975). *Heterocycles* **3**, 155; Johnsen, B. A., and Undheim, K. (1983). *Acta Chem. Scand. Ser. B* **37**, 907; Stokker, G. E., Smith, R. L., Cragoe, E. J., Jr., Ludden, C. T., Russo, H. F., Sweet, C. S., and Watson, L. S. (1981). *J. Med. Chem.* **24**, 115. b. Breslow, R., Czarnik, A. W., Lauer, M., Leppkes, R., Winkler, J., and Zimmerman, S. (1986). *J. Am. Chem. Soc.* **108**, 1969.

38. Kozikowski, A. P., and Hasan, N. M. (1977). *J. Org. Chem.* **42**, 2039.

39. Shimada, S., and Tojo, T. (1983). *Chem. Pharm. Bull. Tokyo* **31**, 4236, 4247.

40. Weinreb, S. M., and Levin, J. I. (1984). *J. Org. Chem.* **49**, 4325; Weinreb, S. M., and Levin, J. I. (1983). *J. Am. Chem. Soc.* **105**, 1397.

41. Medvedskaya, L. B., Makarov, M. G., and Kondrat'eva, G. Ya. (1973). *Izv. Akad. Nauk SSSR, Ser. Khim.*, 1311.

42. Kondrat'eva, G. Ya., Medvedskaya, L. B., Ivanova, Z. N., and Shmelev, L. V. (1971). *Izv. Akad. Nauk SSSR, Ser. Khim.*, 1363.

43. Kondrat'eva, G. Ya., Medvedskaya, L. B., and Ivanova, Z. M. (1971). *Izv. Akad. Nauk SSSR, Ser. Khim.*, 2276.

44. Kondrat'eva, G. Ya., Medvedskaya, L. B., Ivanova, Z. N., and Shmelev, L. V. (1971). *Dokl. Akad. Nauk SSSR* **200**, 1358.

45. Grigg, R., Hayes, R., and Jackson, J. L. (1969). *J. Chem. Soc., Chem. Commun.*, 1167.

46. Grigg, R., and Jackson, J. L. (1970). *J. Chem. Soc. C.*, 552.

47. Gotthardt, H., Huisgen, R., and Bayer, H. O. (1970). *J. Am. Chem. Soc.* **92**, 4340.

48. Gorgues, A., and Le Coq, A. (1979). *Tetrahedron Lett.* **20**, 4829.

49. Stepanova, S. V., L'vova, S. D., Filippova, T. M., and Gunar, V. I. (1976). *J. Org. Chem. USSR* **12**, 1544.
50. Koenig, H., Graf, F., and Weberndorfer, V. (1981). *Liebigs Ann. Chem.,* 668; Koenig, H., and Graf, F. (1971). Ger. Patent 1,935,009 [*Chem. Abstr.* **74**, 64201 (1971)].
51. Reddy, G. S., and Bhatt, M. V. (1980). *Tetrahedron Lett.* **21**, 3627.
52. Jaworski, T., and Mizerski, T. (1976). *Rocz. Chem.* **50**, 359 [*Chem. Abstr.* **85**, 32735q (1976)].
53. Novak, J. J. K. (1975). *Coll. Czech. Chem. Commun.* **40**, 2855.
54. Liotta, D., Saindane, M., and Ott, W. (1983). *Tetrahedron Lett.* **24**, 2473.
55. a. Turner, S., and Ohlsen, S. R. (1971). *J. Chem. Soc. C.,* 1632. b. Kawada, K., Kitagawa, O., and Kobayashi, Y. (1985). *Chem. Pharm. Bull. Tokyo* **33**, 3670.
56. Hutton, J., Potts, B., and Southern, P. F. (1979). *Synth. Commun.* **9**, 789.
57. Ansell, M. F., Caton, M. P. L., and North, P. C. (1981). *Tetrahedron Lett.* **22**, 1727.
58. Potts, K. T., and Marshall, J. (1972). *J. Chem. Soc., Chem. Commun.,* 1000.
59. a. Jacobi, P. A., and Walker, D. G. (1981). *J. Am. Chem. Soc.* **103**, 4611. b. Jacobi, P. A., Craig, T. A., Walker, D. G., Arrick, B. A., and Frechette, R. F. (1984). *J. Am. Chem. Soc.* **106**, 5585. c. Jacobi, P. A., and Craig, T. A. (1978). *J. Am. Chem. Soc.* **100**, 7748. d. Jacobi, P. A., Walker, D. G., and Odeh, I. M. A. (1981). *J. Org. Chem.* **46**, 2065. e. Jacobi, P. A., and Selnick, H. G. (1984). *J. Am. Chem. Soc.* **106**, 3041. f. Jacobi, P. A., Kaczmarek, C. S. R., and Udodong, U. E. (1984). *Tetrahedron Lett.* **25**, 4859. g. Jacobi, P. A., Weiss, K. T., and Egbertson, M. (1984). *Heterocycles* **22**, 281.
60. Takeda Chem. Ind., Ltd. (1965). Fr. Patent 1,400,843 [*Chem. Abstr.* **63**, 9922 (1965)].
61. Nanitzescu, C. D., Cioranescu, E., and Birladeanu, L. (1958). *Commun. Acad. Rep. Populare Romine* **8**, 775 [*Chem. Abstr.* **53**, 18003 (1958)]; Taylor, E. C., Eckroth, D. R., and Bartulin, J. (1967). *J. Org. Chem.* **32**, 899; Wilk, M., Schwab, H., and Rochlitz, J. (1966). *Liebigs Ann. Chem.* **698**, 149; Campbell, C. D., and Rees, C. W. (1969). *J. Chem. Soc. C.,* 748.
62. a. Wade, P. A., Amin, N. V., Yen, H.-K., Price, D. T., and Huhn, G. F. (1984). *J. Org. Chem.* **49**, 4595. b. Adachi, I. (1969). *Chem. Pharm. Bull. Tokyo* **17**, 2209.
63. Bird, C. W., and Cheeseman, G. W. H. (1984). "Comprehensive Heterocyclic Chemistry," Vol. 4, p. 261. Pergamon, London.
64. Jung, M. E., and Shapiro, J. J. (1980). *J. Am. Chem. Soc.* **102**, 7862. Daniels, P. H., Wong, J. L., Atwood, J. G., Canada, L. B., and Rogers, R. D. (1980). *J. Org. Chem.* **45**, 435. Eicher, T., Abdesakan, F., Franke, G., and Weber, J. L. (1975). *Tetrahedron Lett.,* 3915.
65. Eddaif, A., Laurent, A., Mison, P., and Pellissier, N. (1984). *Tetrahedron Lett.* **25**, 2779.
66. Elguero, J. (1984). In "Comprehensive Heterocyclic Chemistry" (K. T. Potts, ed.), Vol. 5, p. 247. Pergamon, London.
67. Tamayo, M. L., Madronero, R., and Munoz, G. G. (1960). *Chem. Ber.* **93**, 289.
68. Abe, N., Nishiwaki, T., and Komoto, N. (1980). *Bull. Chem. Soc. Jpn.* **53**, 3308; Abe, N., Nishiwaki, T., and Komoto, N. (1980). *Chem. Lett.,* 223.
69. Neunhoeffer, H., and Lehmann, B. (1975). *Liebigs Ann. Chem.,* 1113.
70. Gompper, R., and Heinemann, U. (1980). *Angew. Chem. Int. Ed. Engl.* **19**, 216, 217.
71. a. Shusherina, N. P. (1974). *Russ. Chem. Rev.* **43**, 851; Shusherina, N. P., and Pilipenko, V. S. (1984). *Khim. Geterotsikl. Soedin,* 3. b. Matsumoto, K., Ikemi, Y., Nakamura, S., Uchida, T., and Acheson, R. M. (1982). *Heterocycles* **19**, 499; Gompper, R., and Schmidt, A. (1980). *Angew. Chem. Int. Ed. Engl.* **19**, 463; Mruk, N. J., and Tieckelmann, H. (1970). *Tetrahedron Lett.,* 1209; Tomisawa, H., Fujita, R., Naguchi, K., and Hongo, H. (1970). *Chem. Pharm. Bull. Tokyo* **18**, 941; Tomisawa, H., and

Hongo, H. (1970). *Chem. Pharm. Bull. Tokyo* **18**, 925; Tomisawa, H., and Hongo, H. (1969). *Tetrahedron Lett.,* 2465; Bauer, L., Bell, C. L., and Wright, G. E. (1966). *J. Heterocycl. Chem.* **3**, 393; Sheinin, E. B., Wright, G. E., Bell, C. L., and Bauer, L. (1968); *J. Heterocycl. Chem.* **5**, 859. Jones, D. W. (1969). *J. Chem. Soc. C.,* 1729.

72. Neunhoeffer, H., and Werner, G. (1972). *Tetrahedron Lett.,* 1517; Neunhoeffer, H., and Werner, G. (1973). *Liebigs Ann. Chem.,* 1955.

73. Neunhoeffer, H., and Werner, G. (1973). *Liebigs Ann. Chem.,* 437.

74. Jojima, T., Takeshiba, H., and Konotsune, T. (1972). *Chem. Pharm. Bull. Tokyo* **20**, 2191; Jojima, T., Takeshiba, H., and Konotsune, T. (1976). *Chem. Pharm. Bull. Tokyo* **24**, 1581, 1588; Jojima, T., Takeshiba, H., and Konotsune, T. (1980). *Chem. Pharm. Bull. Tokyo* **28**, 198.

75. Boger, D. L., and Coleman, R. S. (1984). *J. Org. Chem.* **49**, 2240. See also: Boger, D. L., and Sakya, S. M. (1987). *J. Org. Chem.* **52**, in press.

76. Boger, D. L., and Coleman, R. S. (1986). *J. Org. Chem.* **51**, 3250. See also: Boger, D. L., and Coleman, R. S. (1987). *J. Am. Chem. Soc.* **109**, 2717.

77. a. Neunhoeffer, H., and Werner, G. (1974). *Liebigs Ann. Chem.,* 1190. b. Martin, J. C. (1980). *J. Heterocycl. Chem.* **17**, 1111.

78. van der Plas, H. C., and Charushin, V. N. (1982). *Tetrahedron Lett.* **23**, 3965; Charushin, V. N., and van der Plas, H. C. (1983). *J. Org. Chem.* **48**, 2667; Marcelis, A. T. M., and van der Plas, H. C. (1986). *J. Org. Chem.* **51**, 67; de Bie, D. A., Guertsen, B., and van der Plas, H. C. (1986). *J. Org. Chem.* **51**, 71.

79. Neunhoeffer, H., and Lehmann, B. (1975). *Liebigs Ann. Chem.,* 1113.

80. Machin, P. J., Porter, A. E. A., and Sammes, P. G. (1973). *J. Chem. Soc., Perkin Trans. 1,* 404; Porter, A. E. A., and Sammes, P. G. (1970). *J. Chem. Soc., Chem. Commun.,* 1103; Kappe, T., and Lube, W. (1971). *Angew. Chem. Int. Ed. Engl.* **10**, 925; Potts, K. T., and Sorm, M. (1971). *J. Org. Chem.* **36**, 8; Sammes, P. G., and Watt, R. A. (1975). *J. Chem. Soc., Chem. Commun.,* 502; Davies, L. B., Sammes, P. G., and Watt, R. A. (1977). *J. Chem. Soc., Chem. Commun.,* 663; Davies, L. B., Leci, O. A., Sammes, P. G., and Watt, R. A. (1978). *J. Chem. Soc., Perkin Trans. 1,* 1293.

81. Jojima, T., Takeshiba, H., and Kinoto, T. (1979). *Heterocycles* **12**, 665.

82. Davies, L. B., Greenberg, S. G., and Sammes, P. G. (1981). *J. Chem. Soc., Perkin Trans. 1,* 1909.

83. Neunhoeffer, H., and Werner, G. (1972). *Liebigs Ann. Chem.* **761**, 39.

84. a. Machin, P. J., Porter, A. E. A., and Sammes, P. G. (1973). *J. Chem. Soc., Perkin Trans. 1,* 404. b. Tutanda, M., Vanderzande, D., Vekemans, J., Toppet, S., and Hoornaert, G. (1986). *Tetrahedron Lett.* **27**, 2509.

85. a. Neunhoeffer, H., Clausen, H.-D., Votter, H.-D., Ohl, H., Kruger, C., and Angermund, K. (1985). *Liebigs Ann. Chem.,* 1732; Ohsawa, A., Arai, H., Ohnishi, H., and Igeta, H. (1981). *J. Chem. Soc., Chem. Commun.,* 1174; Ohsawa, A., Arai, H., Ohnishi, H., Itoh, T., Kaihoh, T., Okada, M., and Igeta, H. (1985). *J. Org. Chem.* **50**, 5520. b. See also Hearn, M. J., and Levy, F. (1984). *Org. Prep. Proc. Int.* **16**, 199. c. Sugita, T., Koyama, J., Tagahara, K., and Suzata, Y. (1986). *Heterocycles* **24**, 29; Sugita, T., Koyama, J., Tagahara, K., and Suzuta, Y. (1985). *Heterocycles* **23**, 2789.

86. Neunhoeffer, H., and Bachmann, M. (1975). *Chem. Ber.* **108**, 3877.

87. Boger, D. L., Schumacher, J., Mullican, M. D., Patel, M., and Panek, J. S. (1982). *J. Org. Chem.* **47**, 2673; Boger, D. L., Patel, M., and Mullican, M. D. (1982). *Tetrahedron Lett.,* 4559; Taylor, E. C., Fletcher, S. R., and Fitzjohn, S. (1985). *J. Org. Chem.* **50**, 1010.

88. Gall, M., private communication.

89. Extensive reviews of the [4 + 2] cycloaddition reactions of 1,2,4-triazines: a.

Neunhoeffer, H., and Wiley, P. F. (1978). "Chemistry of Heterocyclic Compounds" (A. Weissberger and E. C. Taylor, eds.), Vol. 33, pp. 226–228. Wiley, New York. b. Neunhoeffer, H. (1984). *In* "Comprehensive Heterocyclic Chemistry" (A. J. Boulton and A. McKillop, eds.), Vol. 3, pp. 421–429. Pergamon, London. c. Boger, D. L. (1983). *Tetrahedron* **39**, 2912–2922; Boger, D. L. (1986). *Chem. Rev.* **86**, 781. d. See also Ref. 85b.

90. For preliminary studies on the scope of the [4 + 2] cycloaddition reactions of 1,2,4-triazines with electron-rich dienophiles including enol ethers, enamines, ketene acetals, ynamines and strained or reactive olefins, see Dittmar, W., Sauer, J., and Steigel, A. (1969). *Tetrahedron Lett.*, 5171; Reim, H., Steigel, A., and Sauer, J. (1975). *Tetrahedron Lett.*, 2901.

91. [4 + 2] Cycloaddition reactions of 1,2,4-triazines with ketene acetals: a. Neunhoeffer, H., and Fruhauf, H.-W. (1972). *Liebigs Ann. Chem.* **758**, 120; Neunhoeffer, H., and Frey, G. (1973). *Liebigs Ann. Chem.*, 1963; Oeser, O. (1973). *Liebigs Ann. Chem.*, 1970; Neunhoeffer, H., and Werner, G. (1973). *Liebigs Ann. Chem.*, 1955. b. Burg, B., Dittmar, W., Reim, H., Steigel, A., and Sauer, J. (1975). *Tetrahedron Lett.*, 2897; Muller, K., and Sauer, J. (1984). *Tetrahedron Lett.* **25**, 2541.

92. [4 + 2] Cycloaddition reactions of 1,2,4-triazines with enamines: Boger, D. L., and Panek, J. S. (1981). *J. Org. Chem.* **46**, 2179; Boger, D. L., Panek, J. S., and Meier, M. M. (1982). *J. Org. Chem.* **47**, 895; Metz, H.-J., and Neunhoeffer, H. (1982). *Chem. Ber.* **114**, 2807; see also Refs. 95, 96.

93. [4 + 2] Cycloaddition reactions of ynamines with 1,2,4-triazines: Neunhoeffer, H., and Fruhauf, H.-W. (1972). *Liebigs Ann. Chem.* **758**, 125; Neunhoeffer, H., and Fruhauf, H.-W. (1969). *Tetrahedron Lett.*, 3151; Neunhoeffer, H., and Fruhauf, H.-W. (1970). *Tetrahedron Lett.*, 3355. Alder, J., Bohnisch, V., and Neunhoeffer, H. (1978). *Chem. Ber.* **111**, 240; Steigel, A., and Sauer, J. (1970). *Tetrahedron Lett.*, 3357; Reim, H., Steigel, A., and Sauer, J. (1975). *Tetrahedron Lett.*, 2901.

94. [4 + 2] Cycloaddition reactions of 1,2,4-triazines with olefins: (cyclopropenes) Steigel, A., Sauer, J., Kleier, D. A., and Binsch, G. (1972). *J. Am. Chem. Soc.* **94**, 2770; Gockel, U., Hartmannsgruber, U., Steigel, A., and Sauer, J. (1980). *Tetrahedron Lett.*, 595, 599; Oeser, O., Neunhoeffer, H., and Frufhauf, H.-F. (1975). *Liebigs Ann. Chem.*, 1445; (benzocyclopropenes) Maddox, M. L., Martin, J. C., and Muchowski, J. M. (1980). *Tetrahedron Lett.*, 7; Martin, J. C., and Muchowski, J. M. (1984). *J. Org. Chem.* **49**, 1040; (cyclobutenes) Elix, J. A., Wilson, W. S., and Warrener, R. N. (1970). *Tetrahedron Lett.*, 1837; Elix, J. A., Wilson, W. S., Warrener, R. N., and Calder, I. C. (1972). *Aust. J. Chem.* **25**, 865; (norbornene, norbornadiene, and simple olefins) Barlow, M. G., Haszeldine, R. N., and Simpkin, D. J. (1979). *J. Chem. Soc., Chem. Commun.*, 658; Seitz, G., Dhar, R., and Kampchen, T. (1982). *Arch. Pharm.* **315**, 697; Balear, J., Chrisam, G., Huber, F. X., and Sauer, J. (1983). *Tetrahedron Lett.* **24**, 1481; see also Ref. 90.

95. a. Boger, D. L., and Panek, J. S. (1982). *J. Org. Chem.* **47**, 3763. b. Boger, D. L., and Panek, J. S. (1983). *J. Org. Chem.* **48**, 681. c. Boger, D. L., and Panek, J. S. (1985). *J. Am. Chem. Soc.* **107**, 5745. d. Martin, J. C. (1982). *J. Org. Chem.* **47**, 3761.

96. a. Boger, D. L., and Panek, J. S. (1984). *Tetrahedron Lett.* **25**, 3175. b. Boger, D. L., Duff, S. R., Panek, J. S., and Yasuda, M. (1985). *J. Org. Chem.* **50**, 5790. c. Boger, D. L., Duff, S. R., Panek, J. S., and Yasuda, M. (1985). *J. Org. Chem.* **50**, 5782.

97. Neunhoeffer, H., and Lehmann, B. (1977). *Liebigs Ann. Chem.*, 1413; Ewald, H., Neunhoeffer, H., and Lehmann, B. (1977). *Liebigs Ann. Chem.*, 1718.

98. Neunhoeffer, H., Lehmann, B., and Ewald, H. (1977). *Liebigs Ann. Chem.*, 1421.

99. a. Taylor, E. C., and Macor, J. E. (1985). *Tetrahedron Lett.* **26**, 2415, 2419; Seitz, G.,

and Dietrich, S. (1984). *Arch. Pharm.* **317,** 379; Taylor, E. C., and Macor, J. E. (1986). *Tetrahedron Lett.* **27,** 431, 2107; Taylor, E. C., and French, L. G. (1986). *Tetrahedron Lett.* **27,** 1967. b. Seitz, G., Gorge, L., and Dietrich, S. (1985). *Tetrahedron Lett.* **26,** 4355. c. Seitz, G., Dietrich, S., Gorge, L., and Richter, J. (1986). *Tetrahedron Lett.* **27,** 2747. d. Taylor, E. C., and Macor, J. E., unpublished observations; see also Macor, J. E. (1986). Ph.D. thesis, Princeton University.

100. a. Figeys, H. P., and Mathy, A. (1981). *Tetrahedron Lett.* **22,** 1393. b. Seitz, G., Char, R., and Huhnermann, W. (1982). *Chem. Zeit.* **106,** 100.

101. Comprehensive reviews of the [4 + 2] cycloadditions of 1,2,4,5-tetrazines: a. Neunhoeffer, H. (1984). *In* "Comprehensive Heterocyclic Chemistry" (A. J. Boulton and A. McKillop, eds.), Vol. 3, pp. 550–555. Pergamon, London. b. Neunhoeffer, H., and Wiley, P. F. (1978). "Chemistry of Heterocyclic Compounds" (A. Weissberger and E. C. Taylor, eds.), Vol. 33, pp. 1095–1097. Wiley, New York.

102. Boger, D. L., Coleman, R. S., Panek, J. S., and Yohannes, D. (1984). *J. Org. Chem.* **49,** 4405.

103. a. Burg, B., Dittmar, W., Reim, H., Steigel, A., and Sauer, J. (1975). *Tetrahedron Lett.,* 2897; Muller, K., and Sauer, J. (1984). *Tetrahedron Lett.* **25,** 2541; Balcar, J., Chrisam, G., Huber, F. X., and Sauer, J. (1983). *Tetrahedron Lett.* **24,** 1481. b. Meresz, O., and F.-Verner, P. A. (1972). *J. Chem. Soc., Chem. Commun.,* 950.

104. Boger, D. L., Coleman, R. S., Panek, J. S., Huber, F. X., and Sauer, J. (1985). *J. Org. Chem.* **50,** 5377.

105. Barlow, M. G., Haszeldine, R. N., and Pickett, J. A. (1978). *J. Chem. Soc., Perkin Trans. 1,* 378.

106. Avram, M., Dinulescu, J. G., Marica, E., and Nenitzescu, C. D. (1962). *Chem. Ber.* **95,** 2248.

107. Sauer, J., Mielert, A., Lang, D., and Peter, D. (1965). *Chem. Ber.* **98,** 1435.

108. Roffey, P., and Verge, J. P. (1969). *J. Heterocycl. Chem.* **6,** 497.

109. Skorianetz, W., and Kovats, E. (1971). *Helv. Chim. Acta* **54,** 1922.

110. a. Sauer, J., and Heinrichs, G. (1966). *Tetrahedron Lett.,* 4980. b. Martin, J. C., and Bloch, D. R. (1971). *J. Am. Chem. Soc.* **93,** 451.

111. a. Birkofer, L., and Stilke, R. (1974). *J. Organomet Chem.* **74,** C1–C3. b. Birkofer, L., and Hansel, E. (1981). *Chem. Ber.* **114,** 3154.

112. Haddadin, M. J., Firsan, S. J., and Nader, B. S. (1979). *J. Org. Chem.* **44,** 629; Haddadin, M. J., Agha, B. J., and Salka, M. S. (1984). *Tetrahedron Lett.* **25,** 2577.

113. Seitz, G., Kampchen, T., Overheu, W., and Martin, U. (1975). *Arch. Pharm.* **308,** 237.

114. Paquette, L. A., Moerck, R. E., Harirchain, B., and Magnus, P. D. (1978). *J. Am. Chem. Soc.* **100,** 1597.

115. Boger, D. L., and Patel, M. (1987). *Tetrahedron Lett.* **28,** in press. Boger, D. L., and Patel, M. (1987). *J. Org. Chem.* **52,** in press.

116. Boger, D. L., and Coleman, R. S., unpublished observations.

117. Warrener, R. N., and Kretschmer, G. (1977). *J. Chem. Soc., Chem. Commun.,* 806.

118. Sasaki, T., Kanematsu, K., and Hiramatsu, T. (1974). *J. Chem. Soc., Perkin Trans. 1,* 1213.

119. a. Paquette, L. A., Short, M. R., and Kelly, J. F. (1971). *J. Am. Chem. Soc.* **93,** 7179. b. Paquette, L. A., and Kelly, J. F. (1969). *Tetrahedron Lett.,* 4509. c. Martin, H.-D., and Hekman, M. (1978). *Tetrahedron Lett.,* 1183.

120. Anastassiou, A. G., and Reichmanis, E. (1976). *J. Chem. Soc., Chem. Commun.,* 313.

121. Warrener, R. N., Wilson, W. S., and Elix, J. A. (1973). *Aust. J. Chem.* **26,** 389; Warrener, R. N., Wilson, W. S., and Elix, J. A. (1972). *Synth. Commun.* **2,** 73; McCay, I. W., Paddon-Row, M. N., and Warrener, R. N. (1972). *Tetrahedron Lett.,* 1401;

Paddon-Row, M. N., and Warrener, R. N. (1972). *Tetrahedron Lett.*, 1405; Paddon-Row, M. N. (1972). *Tetrahedron Lett.*, 1409; Wilson, W. S., and Warrener, R. N. (1970). *Tetrahedron Lett.*, 4787.

122. McCay, I. W., and Warrener, R. N. (1970). *Tetrahedron Lett.*, 4779.
123. a. Warrener, R. N., Paddon-Row, M. N., Russell, R. A., and Watson, P. L. (1981). *Aust. J. Chem.* **34**, 397; Watson, P. L., and Warrener, R. N. (1973). *Aust. J. Chem.* **26**, 1725. b. Harrison, P. A., Russell, R. A., and Warrener, R. N. (1971). *Tetrahedron Lett.*, 3921. c. Paddon-Row, M. N., and Warrener, R. N. (1974). *Tetrahedron Lett.*, 3797. d. Sasaki, T., Manabe, T., and Hayakawa, K. (1981). *Tetrahedron Lett.*, 2579. e. Tanida, H., Irie, T., and Tori, K. (1972). *Bull. Chem. Soc. Jpn.* **45**, 1999. f. Warrener, R. N., Russell, R. A., and Collin, G. J. (1978). *Tetrahedron Lett.*, 4447. g. Warrener, R. N. (1971). *J. Am. Chem. Soc.* **93**, 2346. h. Priestly, G. M., and Warrener, R. N. (1972). *Tetrahedron Lett.*, 4925. i. Sasaki, T., Manabe, T., and Nishida, S. (1980). *J. Org. Chem.* **45**, 476. j. Wilson, W. S., and Warrener, R. N. (1972). *J. Chem. Soc., Chem. Commun.*, 211. k. Sasaki, T., Hayakawa, K., Manabe, T., Nishida, S., and Wakabayashi, E. (1981). *J. Org. Chem.* **46**, 2021. l. Russell, R. A., Evans, D. A. C., and Warrener, R. N. (1984). *Aust. J. Chem.* **37**, 1699; for related work, see Haddadin, M. J., Agha, B. J., and Tabri, R. F. (1979). *J. Org. Chem.* **44**, 494; Moursounidis, J., and Wege, D. (1986). *Tetrahedron Lett.* **27**, 3045. m. Katada, T., Eguchi, S., and Sasaki, T. (1986). *J. Org. Chem.* **51**, 314.
124. Neuberg, R., Schroder, G., and Oth, J. F. M. (1978). *Liebigs Ann. Chem.*, 1368.
125. a. Seitz, G., Kampchen, T., and Overheu, W. (1978). *Arch. Pharm.* **311**, 786. b. Seitz, G., and Kampchen, T. (1975). *Chem. Zeit.* **99**, 503.
126. Ban, T., Nagai, K., Miyamoto, Y., Harano, K., Yasuda, M., and Kanematsu, K. (1982). *J. Org. Chem.* **47**, 110.
127. Seitz, G., Kampchen, T., Overheu, W., and Martin, U. (1981). *Arch. Pharm.* **314**, 892.
128. Reaction of 1,2,4,5-tetrazines with cyclopropenes: a. Huber, F. X., Sauer, J., McDonald, W. S., and Noth, H. (1982). *Chem. Ber.* **115**, 444. b. Fuhlhuber, H. D., Gousetis, C., Troll, T., and Sauer, J. (1978). *Tetrahedron Lett.*, 3903. c. Steigel, A., Sauer, J., Kleier, D. A., and Binsch, G. (1972). *J. Am. Chem. Soc.* **94**, 2770. d. Steigel, A., Sauer, J., Kleier, D. A., and Binsch, G. (1970). *J. Am. Chem. Soc.* **92**, 3787. e. Dittmar, W., Heinrichs, G., Steigel, A., Troll, T., and Sauer, J. (1970). *Tetrahedron Lett.*, 1623. f. Heinrichs, G., Krapf, H., Schroder, B., Steigel, A., Troll, T., and Sauer, J. (1970). *Tetrahedron Lett.*, 1617. g. Battiste, M. A., and Barton, T. J. (1967). *Tetrahedron Lett.*, 1227. h. Sauer, J., and Heinrichs, G. (1966). *Tetrahedron Lett.*, 4979. i. Beynon, G., Figeys, H. P., Lloyd, D., and Mackie, R. K. (1979). *Bull. Soc. Chim. Belg.* **88**, 905. j. Moerck, R. E., and Battiste, M. A. (1972). *J. Chem. Soc., Chem. Commun.*, 1171. k. Steigel, A., and Sauer, J. (1973). *Tetrahedron Lett.*, 1213. l. Fuhlhuber, H. D., and Sauer, J. (1977). *Tetrahedron Lett.*, 4393. m. Paquette, L. A., and Epstein, M. J. (1971). *J. Am. Chem. Soc.* **93**, 5936. n. Kolbinger, H., Reissenweber, G., and Sauer, J. (1976). *Tetrahedron Lett.*, 4321; Tsuge, O., Kamata, K., and Yogi, S. (1977). *Bull. Chem. Soc. Jpn.* **50**, 2153.
129. Paske, D., Ringshandl, R., Sellner, I., Sichert, H., and Sauer, J. (1980). *Angew. Chem. Int. Ed. Engl.* **19**, 456; Schuster, H., Sichert, H., and Sauer, J. (1983). *Tetrahedron Lett.*, 1485; Schuster, H., and Sauer, J. (1983). *Tetrahedron Lett.* **24**, 4087; Sellner, I., Schuster, H., Sichert, H., Sauer, J., and Noth, H. (1983). *Chem. Ber.* **116**, 3751.
130. Ashe III, A. J., Chan, W. T., and Smith, T. W. (1978). *Tetrahedron Lett.*, 2537; Ashe III, A. J., Bellville, D. J., and Friedman, H. S. (1979). *J. Chem. Soc., Chem. Commun.*, 880.

131. Paddon-Row, M. N., Patney, H. K., and Warrener, R. N. (1978). *J. Chem. Soc., Chem. Commun.*, 296.

132. Christl, M., Luddeke, H.-J., N.-Neppel, A., and Freitag, G. (1977). *Chem. Ber.* **110**, 3745; Christl, M. (1973). *Angew. Chem. Int. Ed. Engl.* **12**, 660.

133. Semmelhack, M. F., Weller, H. N., and Clardy, J. (1978). *J. Org. Chem.* **43**, 3791; Inoue, H., Origuchi, T., and Umano, K. (1984). *Bull. Chem. Soc. Jpn.* **57**, 806.

134. Friedrichsen, W., and von Wallis, H. (1978). *Tetrahedron* **34**, 2509.

135. Sasaki, T., Kanematsu, K., and Kataoka, T. (1975). *J. Org. Chem.* **40**, 1201.

136. Bachmann, M., and Neunhoeffer, H. (1979). *Liebigs Ann. Chem.*, 675.

137. a. Seitz, G., and Kampchen, T. (1975). *Chem. Zeit.* **99**, 292. b. Seitz, G., and Kampchen, T. (1976). *Arch. Pharm.* **309**, 679. c. Seitz, G., and Kampchen, T. (1978). *Arch. Pharm.* **311**, 728.

138. Takahashi, M., Ishida, H., and Kohmoto, M. (1976). *Bull. Chem. Soc. Jpn.* **49**, 1725.

139. Reinhoudt, D. N., and Kouvenhoven, C. G. (1974). *Recl. Trav. Chim. Pays-Bas* **93**, 321.

140. Deyrup, J. A., and Gingrich, H. L. (1977). *Tetrahedron Lett.*, 3115.

141. Scharf, H. D., and Mattay, J. (1976). *Tetrahedron Lett.*, 3509.

142. a. Neunhoeffer, H., and Bachmann, M. (1975). *Chem. Ber.* **108**, 3877. b. Neunhoeffer, H., and Frey, G. (1973). *Liebigs Ann. Chem.*, 1963. c. Oeser, E. (1973). *Liebigs Ann. Chem.*, 1970; Bachmann, M., and Neunhoeffer, H. (1979). *Liebigs Ann. Chem.*, 675.

143. a. Bach, N. J., Kornfeld, E. C., Jones, N. D., Chaney, M. O., Dorman, D. E., Paschal, J. W., Clemens, J. A., and Smalstig, E. B. (1980). *J. Med. Chem.* **23**, 481. b. Marcelis, A. T. M., and van der Plas, H. C. (1985). *Heterocycles* **23**, 683.

144. a. Boger, D. L., and Panek, J. S. (1983). *Tetrahedron Lett.* **24**, 4511. b. Figeys, H. P., and Mathy, A. (1981). *Tetrahedron Lett.*, 1393.

145. a. Seitz, G., and Overheu, W. (1979). *Arch. Pharm.* **312**, 452. b. Seitz, G., and Overheu, W. (1981). *Arch. Pharm.* **314**, 376. c. Seitz, G., Dhar, R., and Huhnermann, W. (1982). *Chem. Zeit.* **106**, 100. d. Seitz, G., Dhar, R., and Dietrich, S. (1983). *Arch. Pharm.* **316**, 472. e. Seitz, G., Dhar, R., Mohr, R., and Overheu, W. (1984). *Arch. Pharm.* **317**, 237.

146. Seitz, G., Dhar, R., and Mohr, R. (1983). *Chem. Zeit.* **107**, 172.

147. a. Seitz, G., and Overheu, W. (1977). *Arch. Pharm.* **310**, 936. b. Figeys, H. P., Mathy, A., and Dralants, A. (1981). *Synth. Commun.* **11**, 655.

148. Seitz, G., and Overheu, W. (1979). *Chem. Zeit.* **103**, 230.

149. Seitz, G., and Kampchen, T. (1977). *Arch. Pharm.* **310**, 269.

150. Seitz, G., Mohr, R., Overheu, W., Allmann, R., and Nagel, M. (1984). *Angew. Chem. Int. Ed. Engl.* **23**, 890.

151. Anderson, D. J., and Hassner, A. (1974). *J. Chem. Soc., Chem. Commun.*, 45; Anderson, D. J., and Hassner, A. (1975). *Synthesis*, 483; Takahashi, M., Suzuki, N., and Igari, Y. (1975). *Bull. Chem. Soc. Jpn.* **48**, 2605; Nair, V. (1975). *J. Heterocycl. Chem.* **12**, 183; Moerck, R. E., and Battiste, M. A. (1974). *J. Chem. Soc., Chem. Commun.*, 782; Johnson, G. D., and Levin, R. H. (1974). *Tetrahedron Lett.*, 2303.

152. Adger, B. M., Rees, C. W., and Storr, R. C. (1975). *J. Chem. Soc., Perkin Trans. 1*, 45.

153. Maggiora, L., and Mertes, M. P. (1986). *J. Org. Chem.* **51**, 950.

154. Boger, D. L., and Sakya, S. M. (1987). *J. Org. Chem.* **52**, in press.

155. Boger, D. L., Panek, J. S., and Yasuda, M. (1987). *Org. Syn.* **67**.

Index